Stem Cell Biology and Regenerative Medicine

Volume 74

Series Editor

Kursad Turksen, Ottawa Hospital Research Institute, Ottawa, ON, Canada

Our understanding of stem cells has grown rapidly over the last decade. While the apparently tremendous therapeutic potential of stem cells has not yet been realized, their routine use in regeneration and restoration of tissue and organ function is greatly anticipated. To this end, many investigators continue to push the boundaries in areas such as the reprogramming, the stem cell niche, nanotechnology, biomimetics and 3D bioprinting, to name just a few. The objective of the volumes in the Stem Cell Biology and Regenerative Medicine series is to capture and consolidate these developments in a timely way. Each volume is thought-provoking in identifying problems, offering solutions, and providing ideas to excite further innovation in the stem cell and regenerative medicine fields.

Oguzhan Gunduz · Christophe Egles ·
Román A. Pérez · Denisa Ficai ·
Cem Bulent Ustundag
Editors

Biomaterials and Tissue Engineering

Springer

Editors
Oguzhan Gunduz
Center for Nanotechnology
and Biomaterials Application and Research
(NBUAM)
Marmara University
Istanbul, Turkey

Department of Metallurgical and Materials
Engineering, Faculty of Technology
Marmara University
Istanbul, Turkey

Health Biotechnology Joint Research
and Application Center of Excellence
Istanbul, Turkey

Román A. Pérez
Bioengineering Institute of Technology
International University of Catalonia
Barcelona, Spain

Cem Bulent Ustundag
Department of Bioengineering, Faculty
of Chemical and Metallurgical Engineering
Yildiz Technical University
Istanbul, Turkey

Health Biotechnology Joint Research
and Application Center of Excellence
Istanbul, Turkey

Christophe Egles
Univ Rouen Normandie, INSA Rouen
Normandie, CNRS, Normandie Univ
Polymères Biopolymères Surfaces (PBS)
UMR
Évreux, France

Denisa Ficai
Polytechnic University of Bucharest
Bucharest, Romania

ISSN 2196-8985 ISSN 2196-8993 (electronic)
Stem Cell Biology and Regenerative Medicine
ISBN 978-3-031-35831-9 ISBN 978-3-031-35832-6 (eBook)
https://doi.org/10.1007/978-3-031-35832-6

This Springer imprint is published by the registered company Springer Nature Switzerland AG
The registered company address is: Gewerbestrasse 11, 6330 Cham, Switzerland

To Mrs. Sema Gündüz, whose exceptional leadership as the project manager and unwavering guidance throughout the entire process have been instrumental in bringing this book to fruition

Preface

The exponential growth of publications in biomaterials for tissue engineering serves as a testament to the field's continuous expansion. This growth is closely tied to the progress made in biomaterial selection, production methods, processing, and characterization techniques. The use of biomaterials in tissue engineering has also played a vital role in advancing micro- and nanotechnology. Biomaterials have emerged as a cornerstone in developing various scientific and engineering disciplines, including medicine, dentistry, and related fields. The production of biomaterials has experienced significant growth, and new types of biostructures have emerged, leading to advancements in 3D printing technology and ongoing theoretical achievements. Using biomaterials for tissue engineering has proven to be an innovative and promising approach toward regenerative medicine, enabling the repairing or replacing damaged tissues or organs. As research in this area expands, new challenges arise, and further biomaterials and tissue engineering advancements are anticipated.

This book aims to provide a comprehensive overview of the latest developments in biomaterials and tissue engineering, focusing on the production, characterization, and application of various biomaterials. With 13 in-depth chapters, it presents a range of reviews and experimental findings, offering a broad and detailed view of the subject. Due to its comprehensive nature, this book is an excellent resource for graduate students and a valuable reference for undergraduates studying any discipline related to biomaterials for tissue engineering.

The initial two chapters of this publication are dedicated to exploring the use of biomaterials and their fundamental attributes in creating scaffolds that can effectively emulate the intricacies of the tissue microenvironment. Chapter 1 "Introduction to Biomaterials and Tissue Engineering" reported that the history of materials for medical applications dates back to ancient times, with natural materials being the first used. Nowadays, synthetic materials such as metals, polymers, ceramics, and composites loaded with biological components are being developed. The challenge is to design improved biomaterials with biomimetic structures that emulate the micro- and nanostructure of natural materials to enhance their properties. Chapter 2 "Tissue Regeneration Processing and Mimicking" provides a comprehensive analysis of the strategies used to replicate the complex tissue microenvironment, focusing on the critical attributes required of a scaffold to promote optimal tissue regeneration.

The following three chapters focus on tissue engineering, which seeks to develop advanced platforms for regenerative medicine. Specifically, they focus on critical aspects of this field, including the use of decellularized matrices and scaffolds, the selection of cells for tissue engineering applications and the cutting-edge technologies and approaches for tissue engineering biomaterials. Chapter 3 "Cell Sources for Tissue Engineering" investigated the careful selection of cells and growth factors to regenerate fully functional tissue constructs. The choice of cells and factors should be tailored to the specific tissue targeted for regeneration. Chapter 4 "Biomaterials" investigated the biomaterials that are designed to emulate natural biological structures and exhibit precise replication of biological materials' behaviors, thus resulting in biomaterials that are tailored for specific applications. Chapter 5 "Micro and Nanotechnology" covered five different sections of the history of micro- and nanotechnology, where we looked at the definitions of both worlds and X-rayed their closely knitted differences.

The following three chapters provide a concise and accessible overview of the history of bioceramics from the past to the present, the development of drug delivery systems, and the fundamental concepts of cell interaction with extracellular matrix (ECM) and other cells, respectively. Chapter 6 "Bioceramics" delved into the classification of bioceramics, ranging from bioinert to bioactive and bioabsorbable materials. Chapter 7 "Drug Delivery Systems for Tissue Engineering" explored the evolution of drug delivery devices over time and discussed recent advancements in the smart drug delivery systems field. Chapter 8 "Cell-Materials Interaction" provided an in-depth exploration of the fundamental concepts underlying cellular interactions with ECM and other cells. Additionally, we examine how these same principles apply to cell-surface interactions with biomaterials, highlighting the similarities and differences between them.

The following four chapters of this book are dedicated to advancing the field of bioreactors design, the different scaffold fabrication processes for tissue engineering, biological characterization for new biomaterials, and additive manufacturing techniques for biomaterials. Chapter 9 "Bioreactors for Tissue Engineering" explored the bioreactors design to more accurately mimic the physiological pathways of cells and tissues and their interactions with their microenvironment. Chapter 10 "Scaffolds Fabrication Processes: From Classical to Advanced Techniques" detailed the various fabrication techniques utilized in tissue engineering, focusing on the different design parameters that can be modified and controlled to achieve successful scaffold fabrication. Chapter 11 "Characterization of the Biological Response to Scaffolds" examined the critical in vitro and in vivo characterization techniques that are essential to advancing our understanding of biomaterials and scaffold properties. Chapter 12 "Additive Manufacturing of Biomaterials" provides the advantages and drawbacks of additive manufacturing techniques for biomaterials and explores why they may be preferred over traditional manufacturing methods.

Finally, Chapter 13 "Bioprinting" highlighted the importance of selecting the most appropriate technique and bioink for different application situations, including assessing various bioprinting processes and bionics, such as inkjet, extrusion, and laser-assisted bioprinting. It then explored the advancements made in 3D bioprinting using bioinks for producing specific tissues, such as the cornea, skin, bone, and cartilage.

Istanbul, Türkiye
May 2023

Oguzhan Gunduz

Acknowledgments We would like to acknowledge the support of the Erasmus+ 2020 Call Strategic Partnerships in Response of the COVID-19 Situation: Partnerships for Digital Education Readiness in the field of Higher Education (KA226) program, under grant agreement N° 2020-1-TR01-KA226-HE-098239. The creating a virtual laboratory for online teaching project (CVL4OT), which is the basis for this book, is a ERASMUS+ Strategic Partnership Project made up of 5 partners from Romania, Spain, France, and Turkey. We are grateful for the contribution of all project partners in making this book possible. We also extend our heartfelt thanks to the Turkish National Agency for their generous support for this book.

Contents

Chapter 1
Introduction to Biomaterials and Tissue Engineering

Ludmila Motelica, Ovidiu Oprea, Denisa Ficai, and Anton Ficai

Abstract Biomaterials represent one of the most dynamic domains of the medical research. Humans have employed biomaterials, without naming them so, for millennia. Only in the last century the domain become organized and defined as biomaterials and medical devices. The latest advances permit doctors and scientist involved in the design and production of biomaterials to be one step closer to the deity concept. This chapter is especially devoted to general aspects related to biomaterials, historical evolution, principles, challenges and benefits. Being a complementary chapter to the Chap. 4. Biomaterials and Tissue Engineering, the most of the examples related to biomaterials are related to metals. Biomaterials are used to manufacture various devices needed to aid, substitute or replace a part or the entire function from a body organ, therefore augmenting, recovering, improving quality and prolonging life expectation of the patient. Nowadays, 3D printing can process complex body parts (like hearth valves or skin graft) and help patients to recover a normal life after surgery. But, biomaterials are also used in implants, spinal rods or bionic limbs. In addition, there is a whole world of biomaterials at the bottom of metric scale. Drug delivery systems based on nanocarriers have revolutionized the medicine with innovative therapies in which the drugs are delivered at specific

---

L. Motelica · O. Oprea (✉) · D. Ficai · A. Ficai
National Centre for Food Safety, University Politehnica of Bucharest, 313 Splaiul Independenței, 060042 Bucharest, Romania
e-mail: ovidiu.oprea@upb.ro

National Centre for Micro and Nanomaterials and National Centre for Food Safety, University Politehnica of Bucharest, 313 Splaiul Independenței, 060042 Bucharest, Romania

O. Oprea · D. Ficai
Department of Inorganic Chemistry, Physical Chemistry and Electrochemistry, Faculty of Chemical Engineering and Biotechnologies, University Politehnica of Bucharest, 1-7 Gh. Polizu Street, 011061 Bucharest, Romania

O. Oprea · A. Ficai
Academy of Romanian Scientists, 3 Ilfov Street, 050045 Bucharest, Romania

A. Ficai
Department of Science and Engineering of Oxide Materials and Nanomaterials, Faculty of Chemical Engineering and Biotechnologies, University Politehnica of Bucharest, 1-7 Gh. Polizu Street, 011061 Bucharest, Romania

O. Gunduz et al. (eds.), *Biomaterials and Tissue Engineering*, Stem Cell Biology and Regenerative Medicine 74, https://doi.org/10.1007/978-3-031-35832-6_1

targeted tissues, released under controlled external or internal stimuli, with more potent results and fewer side effects. The benefits of biomatcrials research are so great that are influencing other domains as well. Here we can briefly mention the surface designs originating from geckos, shark skin or lotus effect and structural coloring, all with impact in automotive, clothing, painting, or military industry.

Keywords Composites · Natural biomaterials · Synthetic biomaterials · 3D printing

Introduction

Considering the history of the materials as presented by Ashby [1], on ages, the stone age lasted the longest because of practically no technological advances and tools available, the later ages decreases as duration while de developed materials were diverse and superior. It can see that cooper, bronze, iron and steel ages took almost 12.000 years but most of this period is associated with the cooper and bronze age while steel age lasted only about ~ 150 years (Fig. 1.1).

A radical change regarding the use of metallic materials was associated with the World War II when, the need of metal increased very much and thus, in almost all the other fields substitutes were identified and this was the starting of the diversification of the plastics, ceramics and even composite materials. An important feature is that the properties of the materials are intrinsically related to their nature and thus they have specific applications.

As seen, metals were the first generation of the materials widely used in repair internal parts of the body but they were soon found to induce some problems, especially because of the corrosion. Even if the corrosion can induce inflammation and even rejection, even if these materials have to be replaced after a while, they are still widely used, especially in orthopedic applications.

The interest in the domain of biomaterials and tissue engineering is dated back in the antiquity. Wood was used as "incipient biomaterials"—their use in "biomedical applications" being dated back to the Neanderthal age. Later, 7th—fourth century, the Greeks and the Romans started to use metals and natural materials in tissue engineering. Since the sixteenth century, in Europe, silver and gold were used for dental repair and later, metals (iron based materials) were used for orthopedic applications. A more scientific approach in designing, fabrication and use of materials in tissue engineering can be observed after 1850 [2].

As a consequence of the intensive research in the field, one can observe the development of materials with improved properties including physico-chemical, mechanical but also biological. Based on the diagram of Ashby, the density-strength diagram of the materials is represented in Fig. 1.2. It can see that polymers have lowest strength and density and Metals and Ceramics have the highest density and strength but, the specific strength (strength/density) of the composites are usually higher comparing

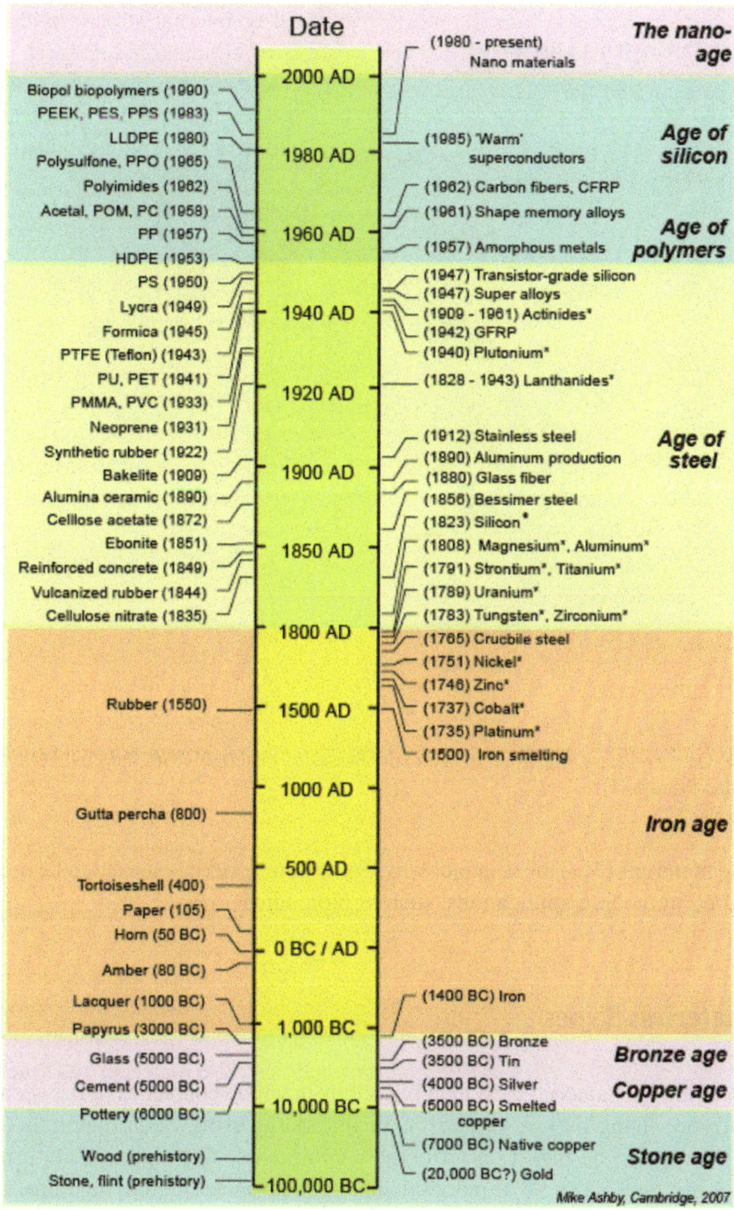

Fig. 1.1 The evolution of the materials from stone age to current days (with the kind permission of Taylor and Francis [1])

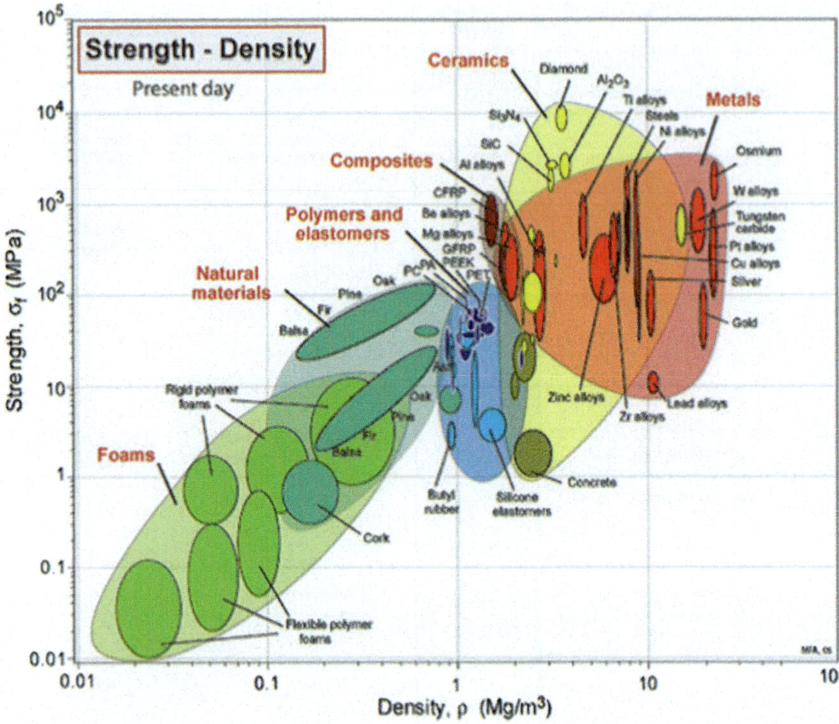

Fig. 1.2 Density-strength diagram of the main classes of materials (with the kind permission of Taylor and Francis [1])

to these materials [3]. This is in fact why, composite materials started to be more and more used in modern applications, such as biomedical ones.

Biomaterials Types

By definition is a material used in applications in which contact exist between it and one or more living tissues, organisms or microorganisms. Such materials (named biomaterials) can be based on an organic compound, like alginate, chitosan or a polymer mixture, it can be of inorganic nature, be it metallic (e.g. titanium, magnesium, steel etc.) or a metallic compound (oxide or salt) like magnetite, hydroxyapatite, titanium dioxide etc. In addition, a biomaterial can be a composite resembling the natural tissue like collagen scaffold with deposited hydroxyapatite, or a titanium piece covered with titanium dioxide.

Natural biopolymers and other small molecules are a logical choice as building blocks for biomaterials and here one can mention the cellulose, chitosan, alginate, hyaluronic acid, silk, collagen, biomimetic peptides and others. In addition, inorganic nanoparticles that are used for drug delivery systems (e.g. SiO_2, Fe_3O_4, Au etc.) are falling in the definition area.

A search of Web of Science database for "biomaterial" keyword returns ~ 33.000 hits, which is rather low when compared with "tissue engineering" with 110.000 hits or "drug delivery" with 272.000 hits. This indicates that not everyone working in the domain of biomaterials is assigning the obtained results to this broader field, but rather to the narrower sub-domains of 3D printing, drug delivery systems or tissue engineering (Table 1.1). Moreover, it can also conclude that the researchers are no more emphasize on materials but on "materials design" or on improving these materials for certain applications such as tissue engineering and certainly this can be also obtained by loading specific biological active agents and thus to induce or enhance the desired properties.

Nevertheless, the field of biomaterials is enjoying a solid increase of ~ 10% each year, specific to the top research domains. The 3D printing section of biomaterials, being still in its infancy is exhibiting the faster growing rate being expected to represent a milestone in the future.

As a scientific domain, the field of biomaterials dates back to mid twentieth century, but the first use of biomaterials can be traced back to the Egyptians in antiquity when they used animal sinew to stitch open wounds. The Chinese, Aztecs and Romans used gold as a malleable, inert metal in dentistry. On the other hand, glass as a hard ceramic or other materials were used to replace a missing eye in a purely cosmetic fashion. A special case of medical use of a biomaterial is the one of the Amerindians who used ant pincers to suture wounds (Fig. 1.3) [4]. Nowadays three-dimensional (3D) printing leading technology brings major advantages in designing and manufacturing biomedical devices or in case of tissue engineering [5].

Therefore, a nonviable material that is introduced in a medical procedure or device can be considered a biomaterial, as long as there will be an interaction with the biological systems. Biomaterials are used to manufacture various devices needed to aid, substitute or replace a part or the entire function from a body organ. This process must take place in a reliable, safe, physiologically acceptable manner and as affordably possible. A synthetic or natural substance (except drugs) that can be used to augment, replace or treat body functions, organs or tissues can be considered a biomaterial.

Table 1.1 Scientific papers in web of science in domain of biomaterials and related sub-domains

Year	2017	2018	2019	2020	2021	2022[a]	Total
Count/keywords	2018	2230	2552	2791	3117	2772	33.000
Tissue engineering	7140	7441	7877	8286	9033	8183	110.000
3D printing + medical	250	412	467	627	749	645	3.551
Drug delivery	19,560	21,067	21,157	23,342	24,233	22,311	272,000

[a] Consulted in 20.01.2023, with papers appearing towards 2022 end not being indexed yet

Fig. 1.3 Ant's mandible (Atta laevigata) used as suture (with the kind permission of SciELO—CC-BY-NC type license, [4])

The need for biomaterials stems from the inability of the body to treat many diseases, large injuries and conditions with other therapies or procedures:

- by manufacturing replacement of body parts that has lost function (e.g. total hip—Fig. 1.4, artificial organs—Fig. 1.6);
- by correcting abnormalities with spinal rods, contact lenses, etc.;
- by improving functions of some body parts like in case of pacemaker, stents (Fig. 1.5) [7], or artificial valves,

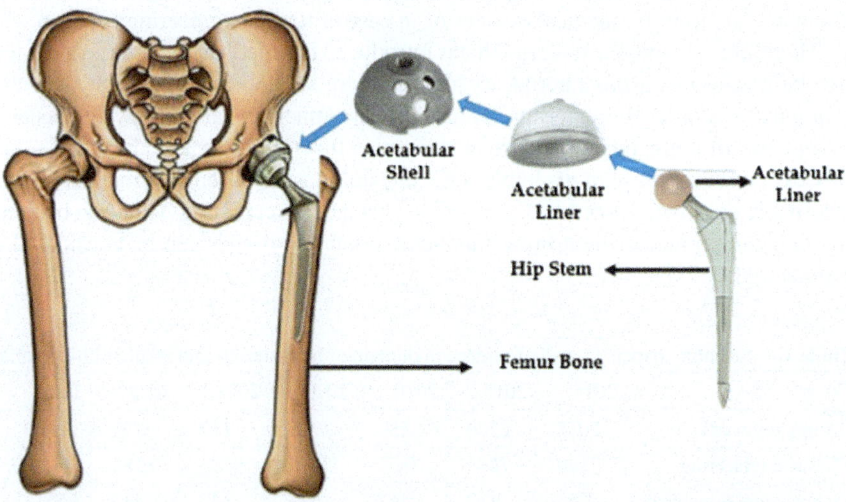

Fig. 1.4 The use of biomaterials in a prosthesis or implant (hip replacement system) (reproduced from [6], with kind permission of MDPI)

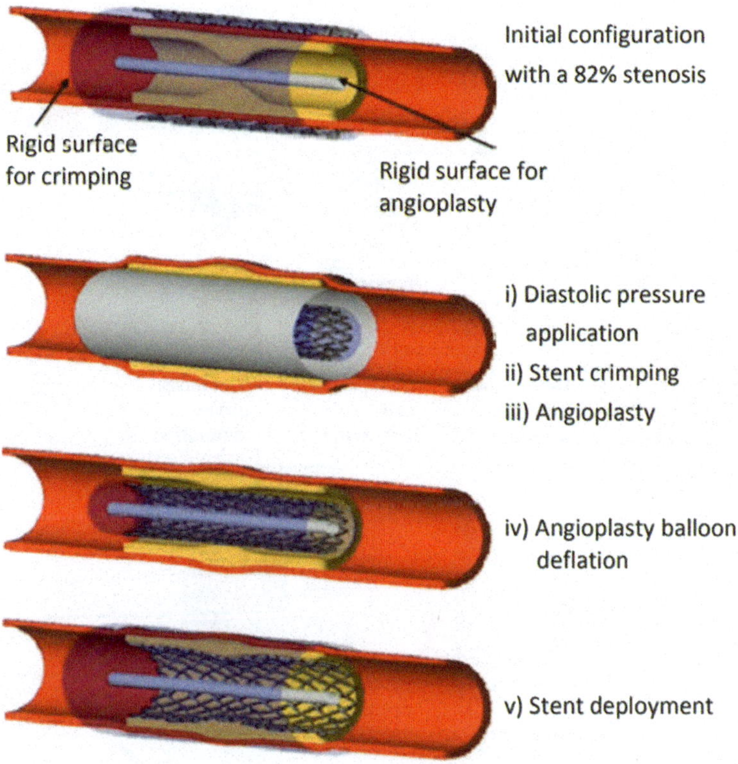

Fig. 1.5 Stenting procedure using metal stent (reprinted with the kind permission of World Scientific, [7])

- by using the biomaterials to assist in healing by structural or pharmaceutical effects, like in case of sutures, wound dressing, bone grafting—including drug delivery capacity);
- to augment or replace tissues for aesthetic reason or as a consequence of diseases (dental restorations, surgical amputation of tumoral tissue, etc.).

Specific medical devices, temporary-used can also be made from iron-based alloys, such as cardiovascular stents (Fig. 1.5). They can be used to quickly restore the functionality to a clogged blood vessel, which will stop the narrowing and eventual, total failure [9]. The possibility of using iron-based alloys in cardiovascular devices, like stents, derive from their high ductility and strength. The high value of ductility is an insurance that the stent will not break because of plastic deformations that occurs during implantation procedures. The blood vessel will be maintained fully open by the strength of the material [10]. In addition, such iron-base stents are fully biodegradable, a trait that can help alleviate the long-term adverse, side-effects. Other important properties for cardiovascular stents can be found in case of iron-based alloys, which are hemocompatible and biocompatible. Polymeric stents are also used in clinical

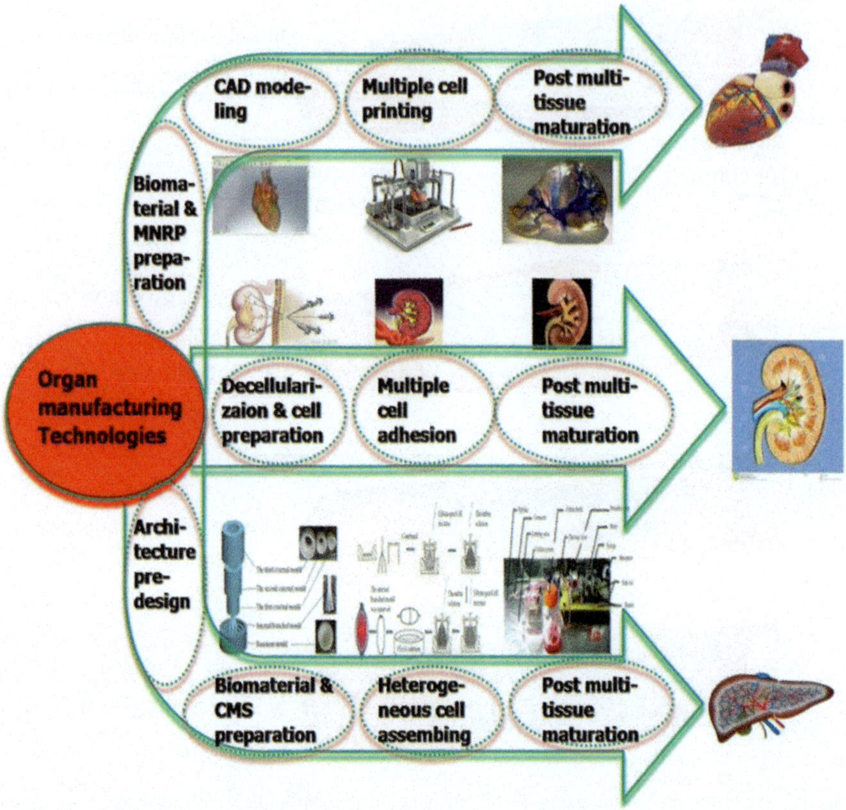

Fig. 1.6 Organ manufacturing technologies based on 3D printing (with kind permission of Sage, [8])

applications, some of these being able to resorb in a predetermined period of time and simultaneously assure the release of biological active agents which reduce the risk of restenosis. For instance, PLA, PLG, PLGA are extensively used as coating but also as bulk material to develop stents, including drug eluting stents with resorbable features [11]. Coating can be also realized using ceramic materials, especially oxides, nitride, carbide but also oxynitride [12]. These coatings are essential in tailoring the surface properties: stability (corrosion, ion release), surface adherence (bacterial adherence, protein adsorption, encrustation, selective cell adherence of endothelial cells but not fibroblasts), biocompatibility, etc.

Usually the biomaterials are solid (e.g. rods, dental screws, valves etc.), but this is not mandatory, as for example artificial blood is a liquid regardless of composition. This artificial blood can be made from expired human or bovine hemoglobin, which is stabilized in a polymeric form by cross-linking with glutaraldehyde, for example as in case of Hemopure or Oxyglobin (Fig. 1.7). Furthermore, the hemoglobin polymer

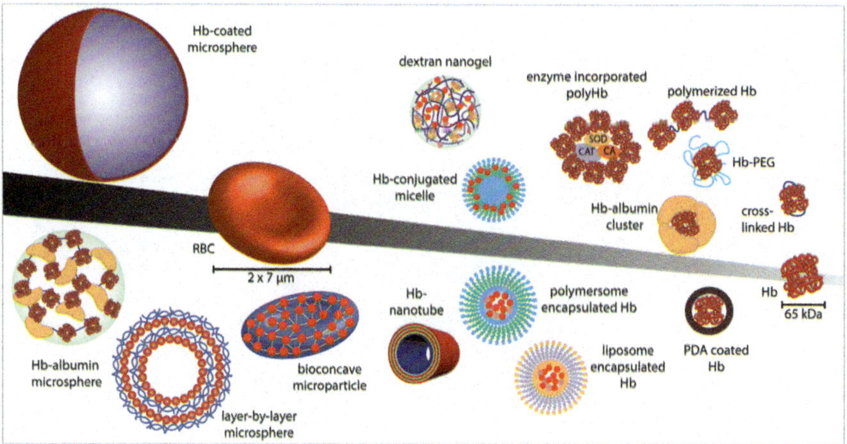

Fig. 1.7 Hemoglobin derivatives proposed to be used as oxygen carriers (with kind permission of Elsevier [14])

can be functionalized with polyethylene glycol and encapsulated in a liposome as in the case of Polyheme [13].

Other strategies developed to obtain artificial blood are based on the alteration of the genetic code of a virus or bacteria, with the specific hemoglobin production gene. Such microorganisms can be further used to infect the host cells. The microorganisms will spread and will create artificial red cells-like structures [15].

Perfluorocarbons (PFCs) are at the other end of artificial blood technologies. They constitute a class of liquid compounds that can dissolve high quantities of oxygen, which can be made available to a living organism. The potential application of PFCs in medicine, as an oxygen carrier medium, was first discovered back in 1966 by L. Clark when his team demonstrated that a mice can spontaneously breath the liquid when immersed in perfluorocarbons, surviving for up to 3 h in the baker [16–18].

If the artificial blood can be considered a system to deliver oxygen, for various drugs the delivery systems based on nanoparticles are also falling in the biomaterials domain. Such delivery systems can be considered targeted when they will unload the drug in a specific organ, tissue, or cell type, but the carrier can respond to external stimuli, like magnetic field, temperature or light to release the encapsulated drug, in such case being called intelligent drug delivery systems.

The carrier particles can be inorganic or organic in nature. The inorganic nanoparticles can be metallic (like Au, Ag, Cu), oxide (like SiO_2, Fe_3O_4, ZnO), can have magnetic properties like iron oxides or can be mesoporous like MCM-41, SBA-15, montmorillonite, etc. The organic particles can be simple polymeric ones like calcium alginate, chitosan or cellulose, or can be more complex lipid-based systems generating solid lipid nanoparticles (SNL) or nanostructured lipid carriers (NLC) particles (Fig. 1.8).

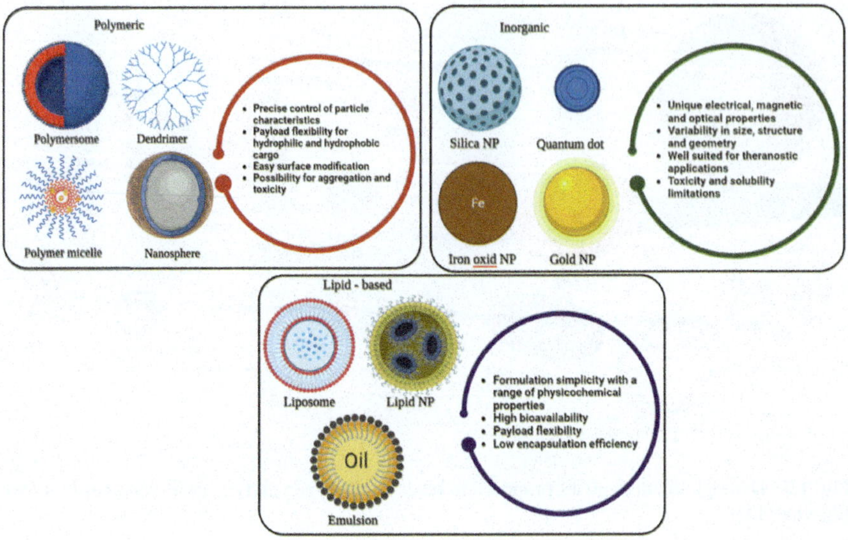

Fig. 1.8 Nanoparticles (NP) types function of loading and delivery features (adapted from [19])

The main area in which nanoparticles-based drug delivery systems are researched is related to cancer treatment for which very potent drugs are available, but with many unwanted systemic side effects that limits their usability. By encapsulating them in a drug delivery system capable of releasing the antitumor agent exactly on target such side effects can be minimized and the therapy improved.

The fields of application for biomaterials are very diverse and go from orthopedic and prosthetics like hip replacement, fixing rods and plates, prosthetics limbs and ends with drug delivery systems for innovative therapies and each of this domains have specific requirements for biomaterials used. In cardiovascular domain, the stents and artificial valves are the main applications. In ophthalmology, the contact lens or artificial crystalline are well known examples. In dentistry the filling, dental cap or dental braces are examples of biomaterials. In addition, any sutures or grafts that are used in wound healing falls in the domain of biomaterials.

Such diversity of application fields leads to a high diversity of materials. From the above examples, a spinal rod can be made from titanium, as well as the screw used for a tooth implant. The actual tooth can be ceramic, while a suture or a catheter is usually made from polymers. At the same time composites can be found in bone repair or dentistry materials. Therefore, the biomaterials have been classified into four different types (Table 1.2): metals, polymers, ceramics and composites.

In the last decades, a special attention was paid to micro and nanomaterials. This tendency can be easily explained considering the new or improved properties which can be obtained when the size is getting down below 10 nm. This is, for instance, the case of the metal-based nanoparticles. Au, Ag and Cu nanoparticles exhibit strong antimicrobial and also antitumoral activity when the size is in the nanometric range

Table 1.2 Biomaterials types and related applications

Materials	Characteristic advantages	Disadvantages	Applications
Metals (titanium, steel, silver, gold etc.)	Ductile, strong, resilient	Difficult to obtain in the correct/desired shape; some may corrode; high density leads to excessive weight	Bone plates, rods and screws; dental implants; stents; joint replacements
Polymers (silicon, Teflon, polyesters, rubber, nylon, chitosan, alginate etc.)	Easy to manufacture in various shapes, can be 3D printed easily, resilient	Low strength, can be deformed in time, some are biodegradable by the body fluids	Sutures, 3D printed prosthetics (ear, nose etc.), soft tissue engineering, drug delivery systems
Ceramics (calcium phosphates like hydroxyapatite, aluminum oxide, carbonates, silica based materials etc.)	Highly biocompatible, hardness, usually chemical inert, porous	Brittle, with low resilience, requires special methods to manufacture	Orthopedic implants, dentistry implants, head of hip, drug delivery systems
Composites (carbon fibers reinforced bone cement, scaffolds, reinforced polymers etc.)	Strong, superior mechanical properties	Complex manufacture process	Heart valves, joint implants

(especially a few nanometers). Similar results were observed also for metal oxides such as ZnO, TiO_2 when the antimicrobial activity is enhanced with the decrease of the size. Considering these enhancement, these nanoparticles are extensively used in many formulations, in bulk or as a coating on various surfaces [20–27].

Silica is known in many forms, from sand—an inert material to mesoporous silica. In the case of mesoporous materials, the highly porous nature is leading to a very high surface area (over 1000–1500 m^2/g) which increase very much the interaction of these materials with the biological active agents and thus, the inert silica (if comparing to sand) became very reactive, mesoporous silica being a remarkable drug delivery system as well a bioresorbable material when implanted into bone defects, being easily transformed into wollastonite and finally in apatite [28].

Among characteristics of biomaterials, we can mention the biocompatibilitywhich is a fundamental concept. One must remember that a biomaterial can be biocompatible in some applications but not in others. Biomaterials of course must be nontoxic and non-carcinogenic. They must exhibit the required physical mechanical properties, but which are different from a hip replacement to the artificial skin, and must be considered in relation with the application domain. Polymers are versatile materials which can be practically used in many fields, alone or in association with other components, from soft to hard tissue engineering, from inert to bioactive and bioresorbable surfaces, etc. (Fig. 1.9). As a consequence of the large variety of the

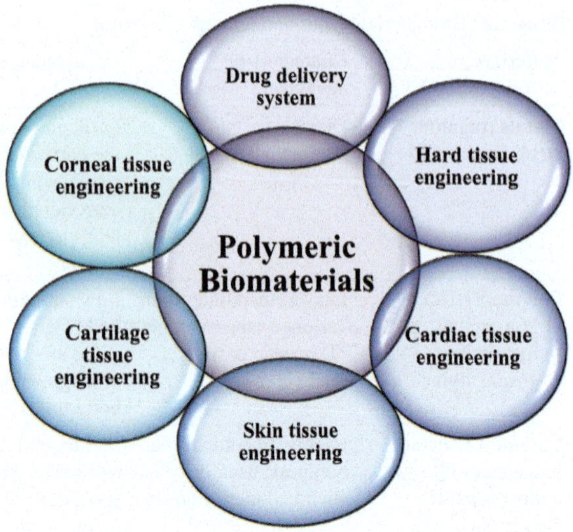

Fig. 1.9 Polymeric-based biomaterials and their application field (adapted from [29])

existent natural or synthetic polymers there are hydrophilic or hydrophobic, adherent or nonadherent, plastic or elastic, antimicrobial polymers.

From economic point of view, the biomaterials should be affordable, which means low cost and readily available. As different applications and different patients have specific needs, the biomaterials should have the capacity to be molded into different shapes (Fig. 1.9). In addition, a biomaterial must be in general resistant to degradation; but this is not always a requirement, as for example the absorbable stitches must dissolve in time. Acceptable strength and resistance to wear can be understand in relation with dentistry materials. For example, a tooth implant is not supposed to be much tougher than the natural tooth, as this will wear the opposite one.

Implants

Metallic implants (Fig. 1.10) represent an important class of biomaterials. Before using a material in health related applications, some factors need to be accounted for. In the case of biomaterials of metallic nature, some mechanical properties need to be obtained (e.g. low friction, high wear resistance, adequate strength), but also corrosion resistance is important because this can be further associated with inflammation and even rejection because of the ion release [30]. Custom implants or other 3D personalized devices can be obtained from metallic biomaterials, the interest and the potential increase of the domain being reflected in the growing number of US hospitals that opened 3D printing facility between 2010 and 2019, (more than 100) [31].

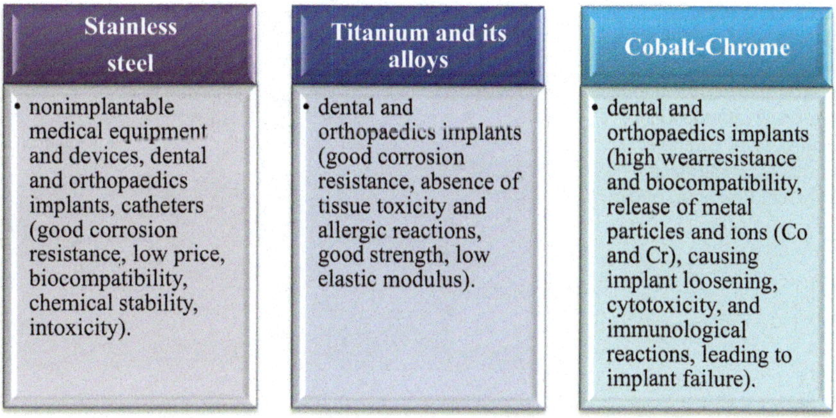

Fig. 1.10 Uses of some metallic biomaterials (adapted from [32])

Long term implants can have a negative impact on the body (especially inflammation) and this is why the biocompatibility (which is many time correlated with the chemical stability in the body) is very important [10]. Among other metallic materials, iron occupy a special place as it is biodegradable and nontoxic. Nevertheless, the rate of degradation for pure iron is low. By alloying iron with other elements, a faster degradation rate, and a uniform etching can be obtained. Porous structures can be created from iron by 3D printing, increasing the available contact surface, which will promote cell adhesion and ingrowth but also the degradation rate [10]. Iron can withstand high mechanical loads which makes it suitable for tissue engineering in orthopedic applications. Pure iron has a higher mechanical strength when comparing with the bone tissue, but alloying it and manufacturing porous structures by 3D printing can alleviate the interface problems. The porous nature of the iron-based alloys used in orthopedic implants can help the osseointegration, which will lead to a healthier bone growth.

Stainless steel is a biomaterial found in many devices like artificial heart valves, otolaryngology ear scope nozzles, bone fixations or orthopedic implants (Fig. 1.11), catheters, needles and syringes, sensor probes and other medical equipment. By 3D printing new devices can be fabricated to match exactly the patient's anatomy (e.g. bit splits, bridges and crowns, all used for customizable orthopedic implants). Special care must be taken for the metal fatigue and corrosion that might lead to implant fail.

Another biodegradable/resorbable metal, which is successfully used for implants, is magnesium. The magnesium mechanical strength is comparable with the natural human bone and it is biocompatible. The main problem is that implants fabricated from magnesium present a high corrosion rate inside human body, because the existing contact with the body fluids. To solve this drawback the magnesium is usually used as alloy and thus higher degradation resistance is achieved or, the magnesium implant is simply coated with a protective layer [34]. Nevertheless, surgical staples can be made from magnesium, as it has a good biocompatibility and is biodegradable.

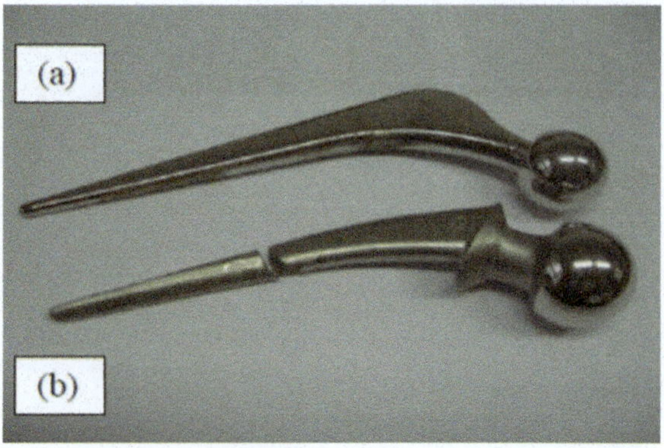

Fig. 1.11 Failure of stainless steel implants in human body: **a** corrosion on surface; **b** mechanical fatigue (reprinted with the kind permission of Biomedical Journal of Scientific & Technical Research, [33])

On the other hand, titanium and many of its alloys are considered for many implant types due to their biocompatibility, chemical stability and mechanical performances (e.g. dental or orthopedic implants). Its successful story stems from its high stability and capacity to integrate with bone [32]. By 3D printing the titanium implant can be engineered perfectly to the patients' specific needs. In the preoperative phase of the medical intervention the medical team can plan and envisage the required intervention stages, therefore improving the preparedness and the successful outcome of the operation (Fig. 1.12).

Conventional alloys present some downsides like incompatibility with bone tissue, bacterial infection, ease of corrosion, prone to fracture etc. Aiming to overcome such drawbacks a new class of materials, namely high-entropy alloys (HEAs) have been investigated as potential biomaterials (Fig. 1.13).

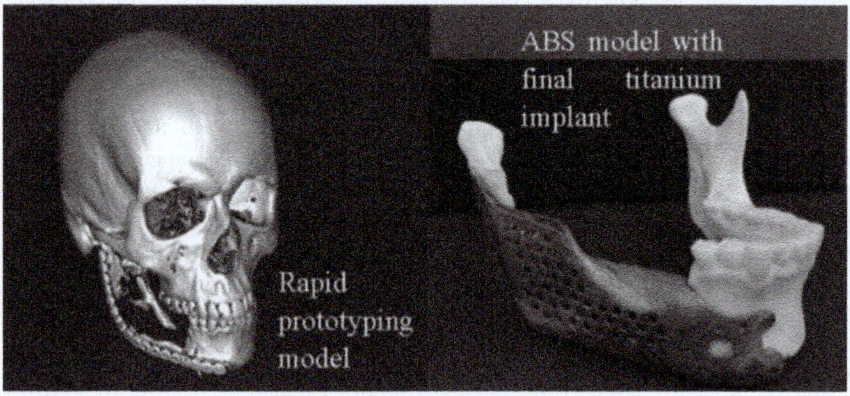

Fig. 1.12 Planning of a mandibular implant (with the kind permission of Korean Society of Medical and Biological Engineering and Springer-Verlag GmbH Germany, part of Springer Nature 2018, [35])

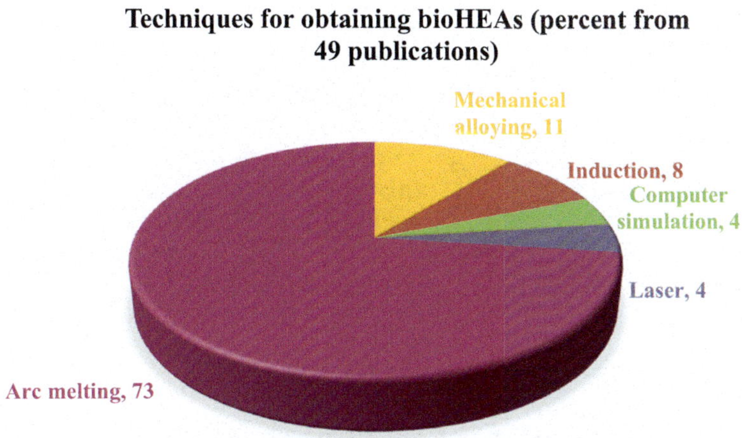

Fig. 1.13 Percentage of biocompatible high-entropy alloys (bioHEAs) development techniques evaluating 49 selected publications (adapted from [36])

Tissue Engineering and 3D Printing

An ideal material used to manufacture an implant should possess a series of attributes. It must be easily moldable but mechanically durable, should be inert but biocompatible. The 3D printing is a revolutionary technology for the pharmaceutical and medical domains, as it has the ability to manufacture implants tailored for specific needs of the patient. Such implants can also be incorporated with proteins, cells or bioactive drugs (Fig. 1.14) [37].

Daily there are an important number of people that die still waiting for an organ transplant, with sufficient tissue compatibility. For them tissue engineering can bring hope by scaffold formation, that can overcome the issue of the compatibility. The literature reports each year new or improved methods for scaffold manufacturing, with the aim to fulfill the specific requirements of each application. Among the popular fabrication methods of scaffolds can be included: additive manufacturing, solution casting, self-assembly, electrospinning, phase separation, foaming or extrusion [38]. Such scaffolds represent a substratum for cells migration that leads to new tissue formation.

The 3D printing has the capability to produce one-of-a-kind parts or limited series of elements, on-demand, based on specific needs of the patient (Fig. 1.15). These abilities come with negligible additional costs at design change or adjustments, as the pieces usually are different among patients [39]. Nevertheless, the major advantage is the flexibility in choosing the starting materials. The surface roughness represents another advantage of the 3D printed materials. The microscopic finish of the surface generated by the layer-by-layer fabrication process generates this roughness. This stair-stepped surface offers anchoring points for cells, leading to proliferation and integration, therefore being an useful trait [39].

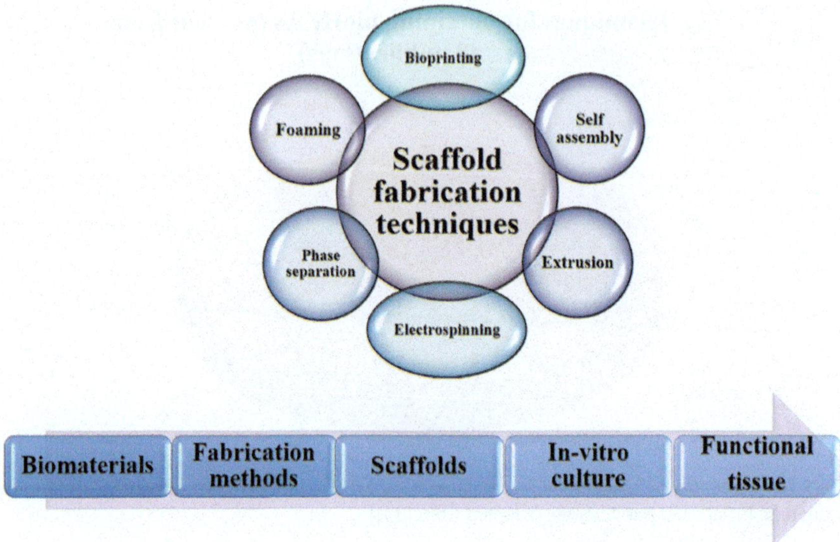

Fig. 1.14 Scaffolds' fabrication techniques

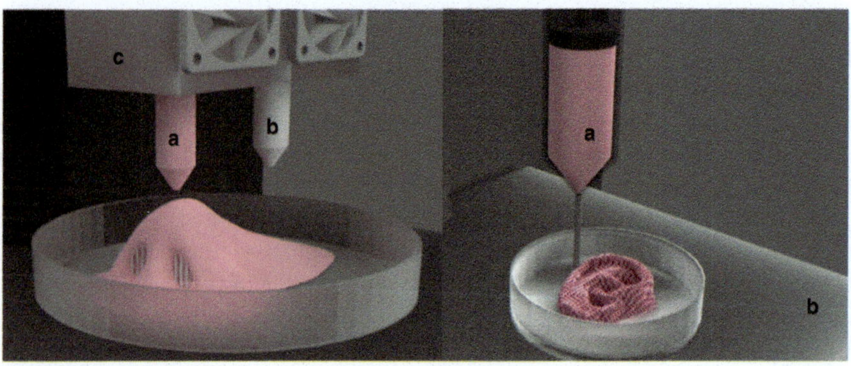

Fig. 1.15 **a** 3D printer with dual head; **b** Bioprinting process (with kind permission of MDPI, CC BY 4.0, [37])

One or the most important applications of 3D printing are related to the manufacturing of prostheses and implants tailored for the specific patients, generating scaffolds used in tissue regeneration or growth of biosynthetic organs, fabrication of drug delivery systems for specific diseases or making anatomical models in preparation of the real surgical procedure (Fig. 1.16). All these are constitutive parts of personalization medicine and leads to a better ratio cost/efficiency and increased productivity [41].

The types of biodegradable polymers used in 3D printing are still limited, as most of them are used as biomaterials for space-filling applications or as scaffold for drug

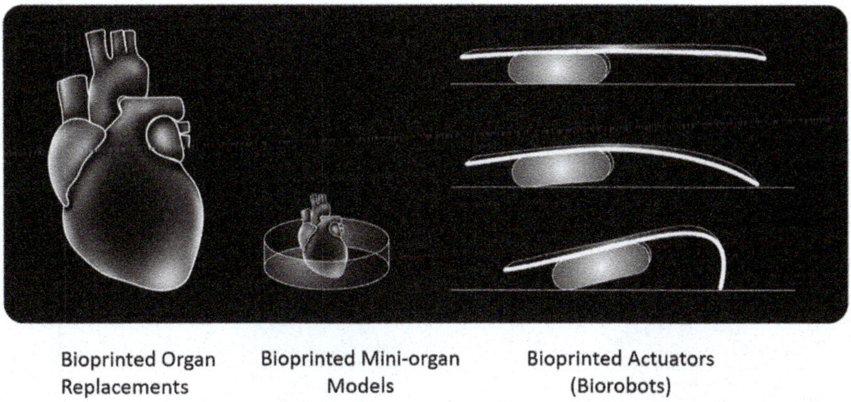

Fig. 1.16 Potential uses of bioprinting (with the kind permission of Elsevier, [40])

delivery systems (Fig. 1.17). These limitations led to the new research effort in the domain of biopolymers that can exhibit adjustable bio-properties, which can help recover the organs' functionality after implementation. Example of inexpensive and low cost biopolymers that are suitable for 3D printing applications and technologies include polycaprolactone and polylactic acid. Such polymers exhibit necessary mechanical properties for the implants and are also biodegradable. By mixing these polymers with other well-known biomaterials, like tricalcium phosphate or hyaluronic acid, one can obtain composite biomaterials with superior mechanical stability, good tissues biocompatibility and integration and excellent printability that can be used in orthopedic devices (Fig. 1.18) [37].

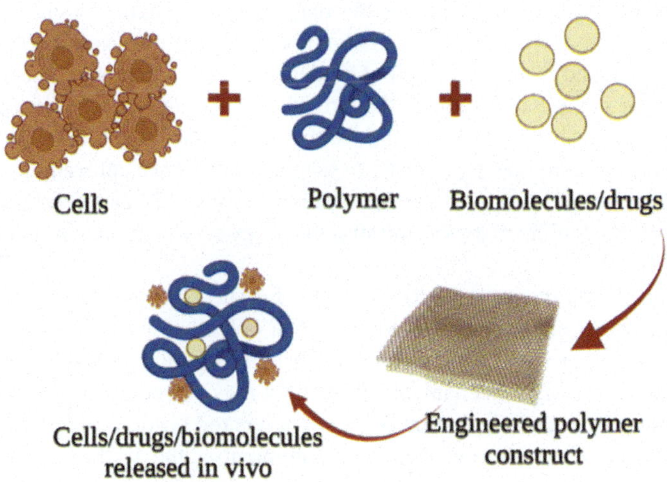

Fig. 1.17 Scaffold with biomolecules, drugs or cells

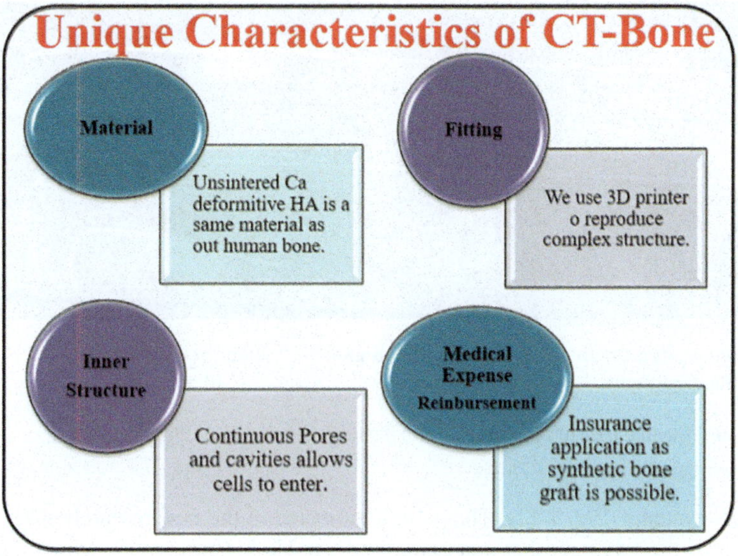

Fig. 1.18 Characteristics of bones obtained by 3D printing [43] (adapted from)

If the first generation of biomaterials had as a mandatory trait the passive nature, in the second-generation of biomaterials, the bioactive nature prevailed, the passive/inertness trait being erased from the traits. Such bioactive materials can be integrating with the biological environment and can actively interact with the body tissue.

The biomaterials from the third-generation were created at the start of this century. The researchers combined the bioactive with resorbable materials aiming to create new biomaterials that can stimulate the regeneration of living tissue at a molecular level by activating genes [42]. This third generation of biomaterials, despite being bioactive and resorbable, are not biodegradable in a homogeneous pattern. In addition, their biological responses that are not optimal and far from ideal.

The newest generation of biomaterials, the fourth, must exhibit the correct interactions between components, in order to help and encourage cells and body tissues to adopt the required behavior to regenerate tissue with high successful rate, thus, mimicking the signaling, reaction and structure of extracellular matrix characteristics component [42].

As a conclusion the most important factor for successfully use of a biomaterial is the match with the application domain. The polymers are easy to manufacture and are resilient but can deform in time and usually are biodegradable. At the same time metals have a high mechanical strength but may corrode, while the ceramic is normally inert from chemical point of view but it is too brittle. In order to retain the desired traits and even improve them, while eliminating the weak spots, composites biomaterials can be engineered. They are usually stronger and better than any of the component materials, and quite often they exhibit some unique properties. As such, composite biomaterials are engineered specially to minimize the individual disadvantages but they are difficult to manufacture.

Biomaterials engineering is an extremely dynamic domain, with a market value of more than 100 billion and a fast growth pace (of more than 10% compound annual growth rate).

Biomimicry

A special topic in materials science is related to mimicking the solutions available in the nature. This is why many natural materials are implemented as biomaterials in tissue engineering and, most probably, collagen, chitosan, cellulose, alginate, hydroxyapatite, calcium carbonate are some of the most studied. At the same time, many biomimetic surfaces are used in the quest to obtain high performance materials. Here can be mentioned the lotus effect (which means the surface will repel dirt and water), sharkskin surface for lowering water friction or gecko effect for the superior adhesive capacity. Such biomimetic surfaces are studied as selective support for various tissues regeneration, like cartilage, bone, blood vessel or nerve.

Ultra-hydrophobicity of the lotus flower leaves and their self-cleaning properties are known as the lotus effect (Fig. 1.19). Besides the obvious applications like paints, windows or cloths, patterned ultra-hydrophobic surfaces are also promising for "lab-on-a-chip" microfluidic devices, which are promising new development in surface-based bioanalysis [44].

Biological surfaces are the result of hundred of million years of evolution, be it from plant or animal regnum. Such surfaces present a functional role and the nature has modeled them to the best form for the assigned tasks. We need only to understand the natural design and try to replicate it as well as we can, to obtain the same result that was achived by the living organisms during evolution time. Hydrophobic, antiadhesive, oxygen permeable surfaces can be found also in the insect world (Fig. 1.20). These structural surface are water-repellent, self-cleaning and antimicrobial, all being feature sought for biofunctional materials [46].

The sharkskin effect represents the diminishing of the fluid drag during fast speed moving through a fluid and the associated protection of the material surface against biofouling. The effect is caused by the presence of microstructures at surface level (on the sharkskin originally). The applications are multiple, from naval industry to the high performance swimsuits. Sharkskin effect also inspired designs that improve aerodynamic performance for wings and planes (Fig. 1.21) [47]. It is also important that such surfaces are suitable for medical applications because some of them have antiadherent, antimicrobial and/or antibiofilm activity which is important in medical applications.

The gecko effect (coming from the lizard that can climb on any surface, even glass), is translated in real world in integrating the processes and mechanisms of attachment with features of the configuration of the adhesive systems (Fig. 1.22) [48].

Fig. 1.19 A Lotus plant (Nelumbo nucifera) (**a**). Surface of a Lotus leaf with dust (**b**) Cleaning of a dirt from Lotus leaf by water (**c**). The SEM images (**d–f**) of a Lotus leaf. Water droplet placed on Lotus leaf surface (**g**). Sudan-red (fat-soluble) is removed by water from a Lotus leaf (**h**). The SEM image of a water droplet on Lotus leaf surface (**i**) (reprinted with the kind permission of Elsevier, [45])

By 4D printing approach multimaterial shape memory polymer can be obtained [49]. Such constituents have enabled controlled shape memory that results in the desired thermomechanical behavior, for example mimicking the blooming process of a flower (Fig. 1.23).

Structural coloration is another surface effect that can be encounter at some animals or plants. The structural color represents the generation of a color by microscopically structures presented on the surfaces, that are so small that they can directly interfere certain wavelengths from visible light. Sometimes the structural coloration

Fig. 1.20 Cuticle patterns of different life forms. A–A″ Entomobryomorpha species. B–B″ Isotomidae (Entomobryomorpha). C–C″ Poduromorpha. D–D″ Symphypleona. Scale bars A′–D′ = 2000 nm, A″–D″ = 500 nm (with the kind permission of Springer, [46])

is also combined with the presence of pigments. A classic example is that of feathers from peacock tail, which despite their brown pigment, are structurally colored, often iridescent, with green, blue and turquoise light reflections (Fig. 1.24).

The *Morpho rhetenor* butterfly has the brightest structural coloration in animal kingdom. The color comes from the special structure of its scales, like a fir-tree (Fig. 1.25). The scales are so spaced that they can reflect only the blue light. As a consequence, the butterfly is bright blue only on the dorsal side, while on the ventral side is colored brown [51].

In a similar fashion the *Papilio palinurus* or emerald peacock, exhibit a green color only on the dorsal side [53]. The iridescent green bands of this butterfly are not produced by pigments, instead they are a generated by the microstructures presented on the wing scales' surface, being therefore a structural coloration. The microstructures bowl shaped and they can refract the light in two ways. The bottom of the bowl reflects the yellow part of the visible spectrum, while the sides of each bowl reflects

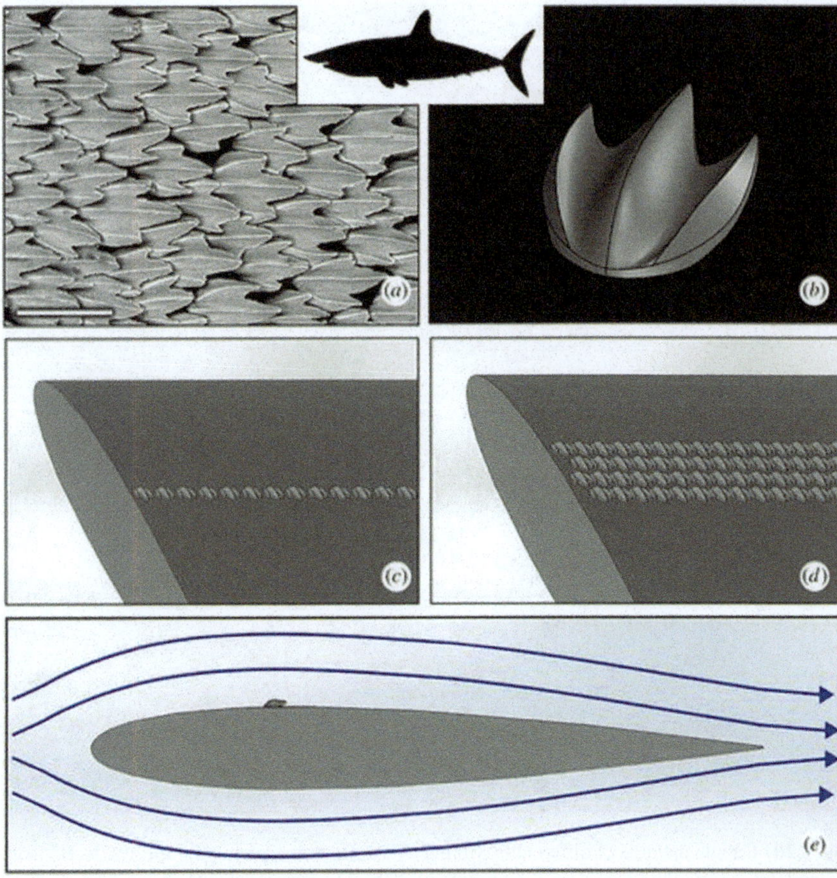

Fig. 1.21 Shark skin structure (**a**); shark denticle model (**b**); denticles arranged in different patterns (**c**, **d**); **e** aerofoils with denticles evaluated for lift and drag (with the kind permission of the Royal Society, [47])

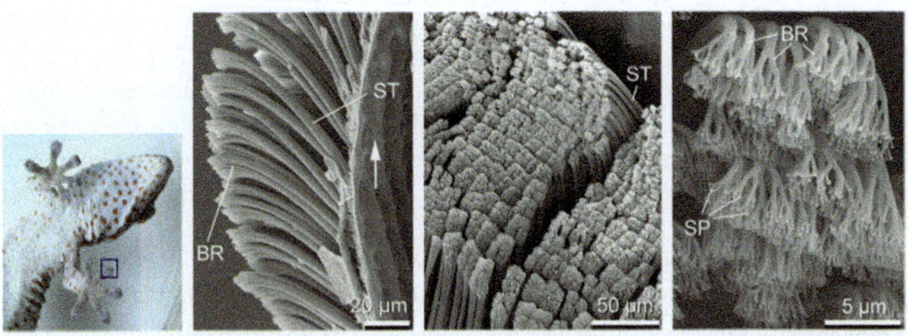

Fig. 1.22 A Gekko gecko. SEM images of setae (**a**, **b**); SEM image of spatulae (**c**); ST: seta; SP: spatula; BR: branch (reprinted with kind permission of Elsevier, [48])

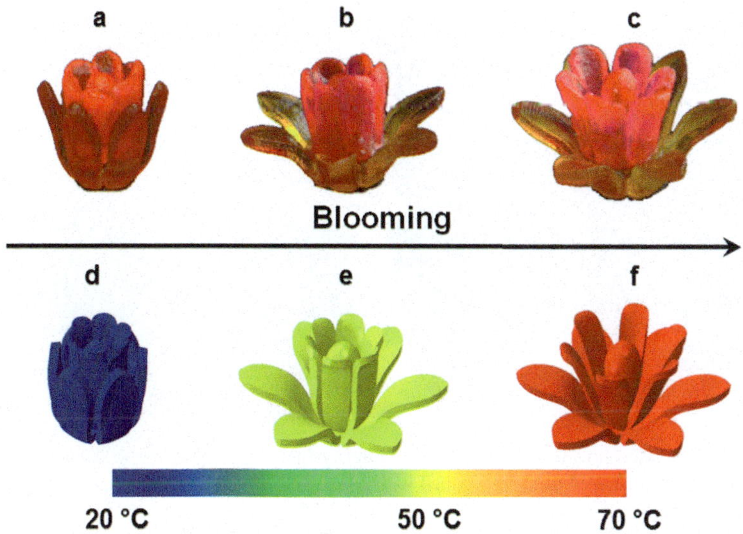

Fig. 1.23 Multimaterial flower at different temperature: bud stage at 20 °C (**a**), opening of outer petals at 50 °C (**b**) and fully bloomed at 70 °C (**c**). FE simulations of the corresponding flower blooming process (**d–f**) (reproduced with the kind permission of Springer Nature, [49])

Fig. 1.24 Peacock tail feathers. Eye region of peacock feather (**a–c–e**). TEM images of Panels **b–d–f** show TEM images of transverse cross section of a brown/white barbule in the eyespot (reprinted with permission from [50] © The Optical Society)

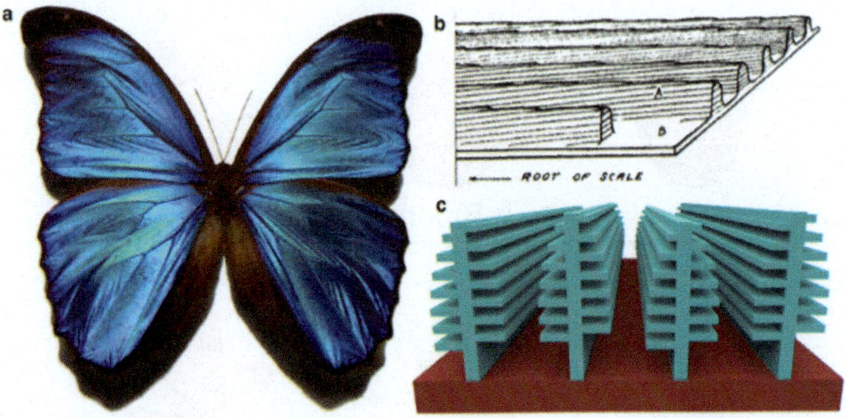

Fig. 1.25 Blue Morpho (*Morpho menelaus*) butterfly (**a**). Diagram of the scale surfaces' structures (fir—tree like shape) (**b, c**). Reprinted with permission from [52] Copyright 1926, American Chemical Society

the blue light. The green structural coloration we can perceive is in fact the additive mixing of those two reflected colors, yellow and blue (Fig. 1.26).

In case of the berries from the *Pollia condensata*, the blue color of the fruit is created by structural coloration, and it is considered the most intense of any living organism (Fig. 1.27) [55]. The cells wall contain cellulose micro-fibrils that are spirally stacked, which generates the structural coloration as a Bragg reflection. This kind of helicoidally arrangements of cellulose generates not only the light refection, but forces the wavelengths to be modified, to converge within a narrow interval prior to reflection. This mechanism acts like a light amplifier for that specific wavelength. The result of this constructive interference process is the most intense coloration of any known biological material. Mimicking the nature in this case has several net advantages. As there are no pigments involved in the obtaining of the color, there

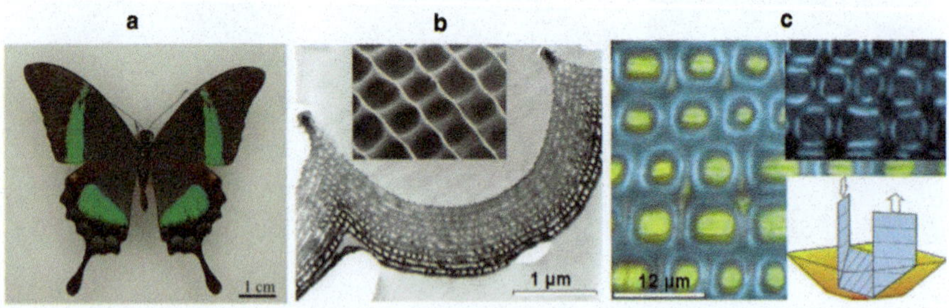

Fig. 1.26 Papilio palinurus butterfly (**a**). TEM micrograph of a concavity cross section (**b**). Color image presenting the dual-color from the surface of the *P. palinurus* butterfly iridescent scale (**c**) [53, 54] [ask permission from RSC Advances figure (**a**) https://doi.org/10.1039/c3ra41096j and Nature Springer for (**b**) and (**c**)]

is no associated toxicity—the color is obtained from the arrangements of the chitin or cellulose structures, and therefore as long the material exist, it retains the color. Such colors do not fade in time, under sun light action, samples from eighteenth century still retaining the original color. This offers new opportunities in cosmetics, car industry or for obtaining security elements on banknotes.

Bioinspired surfaces were also used to develop new surfaces with antiadherent and antimicrobial properties. Thus, nanopatterned surfaces of different materials were proved to show antimicrobial activity even if the bulky/flat material is not antimicrobial. For instance, silicon nanopillars with specific blunt/sharp ends and their relative disposal (distance and arrangements) were found to mimick the natural defense solutions already known to work well for the plants' surfaces, sharks, lizards or insect wings [56]. Such a representative surface is presented in Fig. 1.28.

Fig. 1.27 Pollia condensata single berry (**a**). Infructescence (**b**). Anatomy of *Pollia condensate* fruit (**c, d**) [55] (ask permission from PNAS-Proceedings of the National Academy of Sciences www.pnas.org)

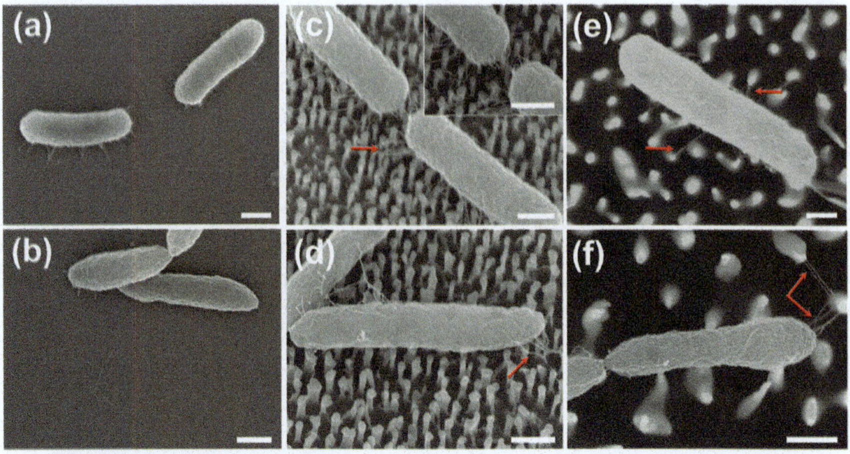

Fig. 1.28 SEM images of *E. coli* (**a**, **c**, **e**) and R. capsulatus (**b**, **d**, **f**) onto smooth (**a**, **b**) and nanopatterned black silicon surfaces (**c**, **d**, **e**, **f**). Scale bars 500 nm (reprinted with the kind permission of Royal Society of Chemistry, [56])

Conclusions

Since the ancient times, people tried to develop materials for medical applications. Certainly the first materials were some natural, available for them and thus, their performances for medical applications were limited. Later, step by step, the use of allografts can be dated back in the first century AD when firstly the use if cadaveric tooth was reported to replace missing dental element. Since the tenth and eleventh centuries, natural and artificial teeth were reported to be used in the Arabic and European countries. Nowadays, the challenge is related not only to develop new materials but also to design them such as to develop improved properties. From evolutionary point of view, the naturally available materials, such as wood, bone, teeth were replaced with synthetically produced one, from metals to polymers, ceramics or composite nature, loaded or not with biological components. A special attention in developing improved biomaterials is paid to the biomimetic structures similar to the surfaces of leaves, flies, lizards, sharks, etc. being known that both physical, chemical and biological properties are influenced by the micro and nanostructuration.

References

1. Ashby M (2008) Materials—a brief history. Philos Mag Lett 88(9–10):749–755
2. Vallet-Regi M (2022) Evolution of biomaterials. Front Mater 9
3. Engineering ToolBox (2008) Engineering materials. https://www.engineeringtoolbox.com/engineering-materials-properties-d_1225.html. Accessed Feb 2023

4. Brito TO, Elzubair A, Araujo LS, Camargo SAD, Souza JLP, Almeida LH (2017) Characterization of the mandible Atta Laevigata and the bioinspiration for the development of a biomimetic surgical clamp. Mater Res-Ibero-Am J Mater 20(6):1525–1533
5. Chia HN, Wu BM (2015) Recent advances in 3D printing of biomaterials. J Biol Eng 9
6. Soliman MM, Chowdhury MEH, Islam MT, Musharavati F, Nabil M, Hafizh M, Khandakar A, Mahmud S, Nezhad EZ, Shuzan MNI, Abir FF (2022) A review of biomaterials and associated performance metrics analysis in pre-clinical finite element model and in implementation stages for total hip implant system. Polymers 14(20):4308
7. Petrini L, Wu W, Dordoni E, Meoli A, Migliavacca F, Pennati G (2012). Fatigue behavior characterization of nitinol for peripheral stents. Funct Mater Letter 5(1)
8. Wang XH (2019) Bioartificial organ manufacturing technologies. Cell Transplant 28(1):5–17
9. Francis A, Yang Y, Virtanen S, Boccaccini AR (2015) Iron and iron-based alloys for temporary cardiovascular applications. J Mater Sci-Mater Med 26(3)
10. Chua K, Khan I, Malhotra R, Zhu D (2021) Additive manufacturing and 3D printing of metallic biomaterials. Eng Regener 2:288–299
11. Hassan S, Ali MN, Ghafoor B (2022) Evolutionary perspective of drug eluting stents: from thick polymer to polymer free approach. J Cardiothoracic Surg 17(1)
12. Beshchasna N, Saqib M, Kraskiewicz H, Wasyluk L, Kuzmin O, Duta OC, Ficai D, Ghizdavet Z, Marin A, Ficai A, Sun ZL, Pichugin VF, Opitz J, Andronescu E (2020) Recent advances in manufacturing innovative stents. Pharmaceutics 12(4)
13. Mohanto N, Park YJ, Jee JP (2022) Current perspectives of artificial oxygen carriers as red blood cell substitutes: a review of old to cutting-edge technologies using in vitro and in vivo assessments. J Pharm Investig 53:153–190
14. Jansman MMT, Hosta-Rigau L (2018) Recent and prominent examples of nano- and microarchitectures as hemoglobin-based oxygen carriers. Adv Coll Interface Sci 260:65–84. https://doi.org/10.1016/j.cis.2018.08.006
15. Zhao XR, Zhou JW, Du GC, Chen J (2021) Recent advances in the microbial synthesis of hemoglobin. Trends Biotechnol 39(3):286–297
16. Clark LC (1985) Introduction to fluorocarbons. Int Anesthesiol Clin 23(1):1–9
17. Moore RE, Clark LC (1985) Chemistry of fluorocarbons in biomedical use. Int Anesthesiol Clin 23(1):11–24
18. Persico DF, Huang HN, Lagow RJ, Clark LC (1985) A general-synthesis for symmetrical highly branched Perfluoro ethers—a new class of oxygen carriers. J Org Chem 50(25):5156–5159
19. Mitchell MJ, Billingsley MM, Haley RM, Wechsler ME, Peppas NA, Langer R (2021) Engineering precision nanoparticles for drug delivery. Nat Rev Drug Disc 20(2):101–124
20. Birkett M, Dover L, Lukose CC, Zia AW, Tambuwala MM, Serrano-Aroca A (2022) Recent advances in metal-based antimicrobial coatings for high-touch surfaces. Int J Mol Sci 23(3)
21. Ficai D, Oprea O, Ficai A, Holban AM (2014) Metal oxide nanoparticles: potential uses in biomedical applications. Curr Proteomics 11(2):139–149
22. Goral D, Goral-Kowalczyk M (2022) Application of metal nanoparticles for production of self-sterilizing coatings. Coatings 12(4)
23. Kim S, Somaratne RMDS, Grimm OC, Steeves DM, Soares JW, Pang R, Welsh EA, Whitten JE (2022) Surface segregation of zinc oxide nanoparticles in polymer films and fibers: implications for functionalized fabrics. ACS Appl Nano Mater 5(3):3992–3999
24. Marinescu L, Ficai D, Oprea O, Marin A, Ficai A, Andronescu E, Holban AM (2020) Optimized synthesis approaches of metal nanoparticles with antimicrobial applications. J Nanomater
25. Motelica L, Ficai D, Oprea O, Ficai A, Trusca RD, Andronescu E, Holban AM (2021) Biodegradable alginate films with ZnO nanoparticles and citronella essential oil—a novel antimicrobial structure. Pharmaceutics 13(7)
26. Richardson JJ, Liao WT, Li JC, Cheng BH, Wang CY, Maruyama T, Tardy BL, Guo JL, Zhao LY, Aw WP, Ejima H (2022). Rapid assembly of colorless antimicrobial and anti-odor coatings from polyphenols and silver. Sci Rep 12(1)
27. Tiplea RE, Lemnaru GM, Trusca RD, Holban A, Kaya MGA, Dragu LD, Ficai D, Ficai A, Bleotu C (2021) Antimicrobial films based on chitosan, collagen, and ZnO for skin tissue regeneration. Biointerf Res Appl Chem 11(4):11985–11995

28. Petrisor G, Motelica L, Ficai D, Trusca RD, Surdu VA, Voicu G, Oprea OC, Ficai A, Andronescu E (2022) New mesoporous silica materials loaded with polyphenols: caffeic acid, ferulic acid and p-coumaric acid as dietary supplements for oral administration. Materials 15(22):7982
29. Kalirajan C, Dukle A, Nathanael AJ, Oh TH, Manivasagam G (2021) A critical review on polymeric biomaterials for biomedical applications. Polymers 13(17):3015
30. Bazaka O, Bazaka K, Kingshott P, Crawford RJ, Ivanova EP (2021) The chemistry of inorganic biomaterials (vol. Metallic implants for biomedical applications). The Royal Society of Chemistry
31. Michas F (2021) Statista. https://www.statista.com/statistics/1211961/medical-3d-printing-facilities-hospitals-united-states/. Accessed Feb 2023
32. Resnik M, Bencina M, Levicnik E, Rawat N, Iglic A, Junkar I (2020) Strategies for improving antimicrobial properties of stainless steel. Materials 13(13):2944
33. Hryniewicz T (2018) Electropolishing processes for better implants' performance. Biomed J Sci Techn Res 11(1):8262–8266
34. Zhou HL, Li JY, Li J, Ruan QD, Peng X, Li SJ, Jin WH, Yu ZT, Chu PK, Li W (2021) A composite coating with physical interlocking and chemical bonding on WE43 magnesium alloy for corrosion protection and cytocompatibility enhancement. Surface Coat Technol 412
35. Popov VV, Muller-Kamskii G, Kovalevsky A, Dzhenzhera G, Strokin E, Kolomiets A, Ramon J (2018) Design and 3D-printing of titanium bone implants: brief review of approach and clinical cases. Biomed Eng Lett 8(4):337–344
36. de Oliveira TG, Fagundes DV, Capellato P, Sachs D, da Silva AAAP (2022) A review of biomaterials based on high-entropy alloys. Metals 12(11):1940
37. Tappa K, Jammalamadaka U (2018) Novel biomaterials used in medical 3D printing techniques. J Funct Biomater 9(1):17
38. Zaszczynska A, Moczulska-Heljak M, Gradys A, Sajkiewicz P (2021) Advances in 3D printing for tissue engineering. Materials 14(12):3149
39. Bandyopadhyay A, Bose S, Das S (2015) 3D printing of biomaterials. MRS Bull 40(2):108–112
40. Nagarajan N, Dupret-Bories A, Karabulut E, Zorlutuna P, Vrana NE (2018) Enabling personalized implant and controllable biosystem development through 3D printing. Biotechnol Adv 36(2):521–533
41. Biotechnology in Healthcare (2022) Technologies and innovations, vol 1. In: Barh D (ed). Academic Press
42. Hench LL, Thompson I (2010) Twenty-first century challenges for biomaterials. J R Soc Interface 7:S379–S391
43. Kanno Y, Nakatsuka T, Saijo H, Fujihara Y, Atsuhiko H, Chung UI, Takato T, Hoshi K (2016) Computed tomographic evaluation of novel custom-made artificial bones, "CT-bone", applied for maxillofacial reconstruction. Regener Ther 5:1–8
44. Lian ZX, Xu JK, Wang ZB, Yu HD (2020) Biomimetic superlyophobic metallic surfaces: focusing on their fabrication and applications. J Bionic Eng 17(1):1–33
45. Koch K, Bhushan B, Barthlott W (2009) Multifunctional surface structures of plants: an inspiration for biomimetics. Prog Mater Sci 54(2):137–178
46. Nickerl J, Helbig R, Schulz HJ, Werner C, Neinhuis C (2013) Diversity and potential correlations to the function of Collembola cuticle structures. Zoomorphology 132(2):183–195
47. Domel AG, Saadat M, Weaver JC, Haj-Hariri H, Bertoldi K, Lauder GV (2018) Shark skin-inspired designs that improve aerodynamic performance. J Roy Soc Interface 15(139)
48. Gao HJ, Wang X, Yao HM, Gorb S, Arzt E (2005) Mechanics of hierarchical adhesion structures of geckos. Mech Mater 37(2–3):275–285
49. Ge Q, Sakhaei AH, Lee H, Dunn CK, Fang NX, Dunn ML (2016) Multimaterial 4D printing with tailorable shape memory polymers. Sci Rep 6
50. Medina JM, Diaz JA, Vukusic P (2015) Classification of peacock feather reflectance using principal component analysis similarity factors from multispectral imaging data. Opt Express 23(8):10198–10212
51. Burg SL, Parnell AJ (2018) Self-assembling structural colour in nature. J Phys-Conden Matter 30(41)

52. Zhang SC, Chen YF (2015) Nanofabrication and coloration study of artificial Morpho butterfly wings with aligned lamellae layers. Sci Rep 5
53. Sun JY, Bhushan B, Tong J (2013) Structural coloration in nature. RSC Adv 3(35):14862–14889
54. Vukusic P, Sambles JR, Lawrence CR (2000) Structural colour—colour mixing in wing scales of a butterfly. Nature 404(6777):457–457
55. Vignolini S, Rudall PJ, Rowland AV, Reed A, Moyroud E, Faden RB, Baumberg JJ, Glover BJ, Steiner U (2012) Pointillist structural color in Pollia fruit. Proc Natl Acad Sci USA 109(39):15712–15715
56. Michalska M, Gambacorta F, Divan R, Aranson IS, Sokolov A, Noirot P, Laible PD (2018) Tuning antimicrobial properties of biomimetic nanopatterned surfaces. Nanoscale 10(14):6639–6650

Chapter 2
Tissue Regeneration Processing and Mimicking

Aysel Oktay, Busra Oktay, Elif Durasi, Hilal Calik, Ilkay Tenim, Rabia Yilmaz Ozturk, Ruveyda Aydin, Tarlan Mahouti, Hakan Yilmazer, and Rabia Cakir Koc

Abstract Tissue regeneration has been one of the comprehensive topics that underlie tissue engineering and has been researched over years. The main aim in tissue engineering is to create a tissue microenvironment produced from natural or synthetic biomaterials, to promote tissue regeneration in the injured site, thus mimicking the natural extracellular matrix (ECM) structure as much as possible, to ensure the migration of specific cells to the site, cell proliferation, and cell differentiation. In this context, it is critical to understand the difference between tissue repair and tissue regeneration, the main stages of tissue repair (hemostasis, inflammation, proliferation and remodeling), and the regeneration and repair mechanisms of the four basic tissues (connective, epithelial, muscle, and nerve tissue). Studies on tissue regeneration mainly focus on scaffolds, decellularized tissues, and their combination with cells capable of self-renewal and differentiation, such as stem cells. Herein, it is also presented in detail how to mimic the tissue microenvironment, the essential characteristics of a scaffold and why decellularized tissues are needed.

Keywords Tissue regeneration · Tissue repair · Tissue microenvironment · Mimicking · Scaffold · Decellularization

---

A. Oktay · B. Oktay · E. Durasi · H. Calik · R. Yilmaz Ozturk · R. Aydin · R. Cakir Koc
Department of Bioengineering, Faculty of Chemical and Metallurgical Engineering, Yildiz Technical University, Istanbul 34220, Turkey

A. Oktay
Department of Molecular Biotechnology, Turkish-German University, Istanbul 34820, Turkey

I. Tenim
Department of Metallurgical and Materials Engineering, Marmara University, Istanbul 34854, Turkey

T. Mahouti · H. Yilmazer (✉)
Department of Metallurgical and Materials Engineering, Faculty of Chemical and Metallurgical Engineering, Yildiz Technical University, Istanbul 34220, Turkey
e-mail: yilmazerh@gmail.com

R. Cakir Koc
Health Institutes of Turkiye (TUSEB), Istanbul, Turkey

O. Gunduz et al. (eds.), *Biomaterials and Tissue Engineering*, Stem Cell Biology and Regenerative Medicine 74, https://doi.org/10.1007/978-3-031-35832-6_2

Introduction

Tissue regeneration is a physiological process that aims to restore, maintain or improve damaged or diseased tissue and organ functions or conditions [1, 2]. The natural tissue regeneration process requires significant communication between the various cellular elements of the tissue and the extracellular matrix (ECM) [3]. In normal physiological conditions, tissue regeneration in the body includes complex cellular processes such as homeostasis, inflammation, angiogenesis, ECM synthesis, reepithelization and collagen deposition [4]. However, the body's natural tissue regeneration may not be sufficient to repair serious wounds, resulting in either a chronic wound or an excessive amount of scar tissue and therefore considerable loss of the structure and function of the original tissue [5].

The main goal of tissue regeneration studies is the development of in vitro tissues and implantation of them into the body via surgery or to stimulate natural tissue repair mechanisms using stem cells at the site of damage, a process known as in situ tissue regeneration [6]. The principle of in situ tissue regeneration is to implant tissue-specific biomaterials combined with stem cells and biomolecules in the damaged places in the body and use the in vivo microenvironment to guide cells to regenerate new tissues [7].

Maintaining the appropriate environment for tissue regeneration depends on improving the regenerative capability of native cells. To achieve this, the approach of mimicking native tissue microenvironment including natural micro-vascular networks and ECM nanostructures would be a promising approach to facilitate tissue regeneration by affecting various cell behaviors such as cell survival, attachment, migration, proliferation, and differentiation [8]. However, in order to mimic the tissue microenvironment, first the natural regeneration processes of tissues such as epithelial, muscle, and nerve tissues and their regeneration mechanisms should be comprehended. Therefore, this chapter provides a comprehensive summary of the elements of the natural tissue microenvironment, and tissue regeneration mechanisms and highlights the approaches how to mimic the tissue regeneration processes.

Tissue Microenvironment

The tissue microenvironment is a biophysical and biochemical environment that surrounds cells, creates an environment and transmits intracellular and intercellular molecular signals [9]. This highly dynamic microenvironment contributes to the scaffold and function of the tissue. As schematically shown in Fig. 2.1, the tissue microenvironment consists of many factors that directly or indirectly affect cell behaviors. The basis of microenvironments forms intracellular, intercellular and extracellular spaces and components. The microenvironmental components help maintain normal tissue function, and changes in these components can trigger abnormal cell formation, cell behavior, and the occurrence of diseases. The components required for the

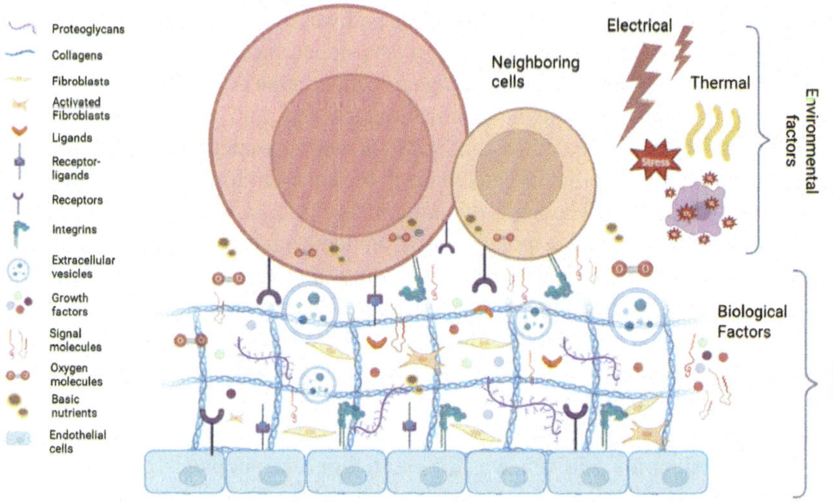

Fig. 2.1 Schematic diagram of ECM structure and tissue microenvironment components

normal physiological and biological behavior of an organ show common features in terms of their general composition and functions, although components vary in each tissue or cell environment [10]. This microenvironment helps maintain the homeostasis of the tissue in which it is located and consists of cellular and non-cellular components such as basal lamina (basement membrane), components of ECM, signal molecules, and cytokines. In general, the key components in the tissue microenvironment are classified into the ECM, neighboring cells (or cell–cell interactions), biological factors, and environmental factors [11].

Cells in tissues not only maintain the integrity of complex organs but also communicate and interact with neighboring cells to maintain their morphology and normal functions. This communication and interaction occur with the help of direct contact with neighboring cells, signal networks in the environment, environmental factors, adhesion molecules, and connection molecules [12]. Direct cell–cell interactions include close cell–cell interactions through tight junctions, anchoring junctions, gap junctions, and distant cell–cell interactions by mechanical communication with the aid of the ECM [10].

ECM, another crucial part of the microenvironment for cells, is a physiologically active, heterogeneous extracellular microenvironment surrounding tissue cells. The ECM consists of a protein network structure in which various proteins such as collagen, fibronectin, laminin, elastin, and proteoglycans are interconnected [13]. The structural component and properties of the ECM affect its cellular behavior. The structure, mechanics, components, organization, adaptation, and function of the ECM are unique to each ECM of cell type and characterize the specific functional properties of cells in that environment in their normal microenvironment. Overall, the ECM both acts as a structural scaffold support for cells and enables mechanical support for their differentiation, proliferation, circulation, and migration. In parallel, the ECM occurs through microenvironmental signaling pathways with the combination of cells and receptors for the cell to maintain its normal functions. Another significant role of the ECM is to contain soluble signaling molecules, adhesion ligands and chemical functional groups that regulate the localization, stability, and bioactivity of cells [9]. The microenvironment created by the ECM surrounding adherent cells is essential for cell survival. The inability of cells to reach the adherent form in ECM causes cell death by apoptosis [14]. Having a dynamic nature, the ECM shows physiological reactions and remodeling in response to environmental changes or cellular mutations to ensure the standard survival, growth, and other biological activities of cells [15]. The interactions of ECM components with cells are reciprocal, and cells continually produce, distinct, and remodel the components of the ECM to regulate their activity and behavior [14]. The ECM also adjusts the concentration in tissue of biological factors during its remodeling [16].

Biological factors that perform regulating cell behaviors and functions such as proliferation, migration, self-renewal, differentiation, and apoptosis, comprise one of the major factors in the tissue microenvironment and are involved in many biochemical and biophysical processes. Numerous biological components are present in the microenvironment, including cytokines, growth factors, hormones, signaling molecules, and essential nutrient molecules, and these components regulate the activation, growth, proliferation, migration and differentiation of cells. There are molecules such as oxygen, carbon dioxide, glucose, and amino acids as basic nutrient molecules in the microenvironment and these are vital components in the maintenance of cell functions. The distribution and activity of growth factors such as bone morphogenetic proteins (BMPs), epidermal growth factors (EGFs), fibroblast growth factors (FGFs), vascular endothelial growth factors (VEGFs), nerve growth factor (NGFs), and transforming growth factors (TGFs) in the microenvironment contain substantial clues for the regulation of cell behavior [17]. In addition, growth factors are essential for coordinating interactions between cells and the ECM and maintaining cell proliferation, behavior, and cellular activities [18].

Environmental factors in the microenvironment include dynamic or static mechanical forces, shear forces, pH effects, and the presence of oxygen (O_2) which are other significant parameters in cell functions [11]. Cells may experience different stress and strain fields modulated by mechanical microenvironments depending on the cell type and location [10]. In addition to stress and strain fields, cells can be affected by environmental factors such as electric, magnetic, acoustic and thermal fields. Electric fields in the cells are thought to be able to regulate cell migration, organization, proliferation, and differentiation [19, 20]. However, it is still unknown exactly how magnetic, acoustic, and thermal fields affect cells [10]. Environmental parameters such as temperature, pH, ions, and energy also make a considerable contribution to cell functions and homeostasis. Although the cell type-specific temperature response may vary, changes in the type and amount of intracellular chemicals with temperature change control cell functions [15]. In parallel with the temperature, pH has a main role in cellular functions by organizing cell cycle and growth, and functions as a significant checkpoint under normal and pathogenic conditions.

Understanding the effects of microenvironment components on cell behavior is essential both to design the cell microenvironment and to enable the development of biomimetic materials for many biomedical applications. However, it is difficult to overall comprehend the microenvironment due to the dynamic nature of the tissue microenvironment, the heterogeneity of tissues and the specific functions in organs. Therefore, it is more accurate to evaluate the physical, physiological, biological, and metabolic functions of specific tissue to understand the tissue microenvironments. Therefore, we will briefly describe the microenvironment of each of the basic types of tissues in the body. In general, the basic types of tissues found in the human body are connective tissue, epithelial tissue, muscle tissue and nerve tissue. Most human connective tissues contain migrating stem cells, fibroblasts, pericytes and tissue-associated adipocytes, tissue-specific cells, vascular system cells, lymphatic and immune system cells along with the mentioned microenvironment components [15]. Muscle tissue, another basic tissue type, is a highly organized tissue that is composed of ECM, blood vessels, nerves, connective tissue, myofibers, and soluble factors such as wnt1 inducible signaling pathway protein 1 (WISP1), bone morphogenetic protein 1 (BMP1), and follistatin [16]. The ECM structure found in muscle tissue generally consists of collagen, fibronectin and laminin proteins, fibrils, dystrophin, dystroglycan, and proteoglycan [21]. Epithelial tissue is a highly dynamic structure, generally consisting of E-cadherin, b-catenin, α-catenin and vinculin, actin filaments, and extracellular spaces [22]. Finally, in addition to the microenvironment components mentioned, the nervous tissue microenvironment consists of peripheral nerves, neurons, myelin sheaths, and connective tissues such as epineurium, perineurium, endoneurium, vessels, and immune cells [23].

Mechanisms of Regeneration and Repair Processes

Coagulation and Hemostasis Stage

After the injury, the mechanisms of coagulation and hemostasis occur in the wound to prevent exsanguination and damage to vital organs. Hemostasis is controlled by the dynamic equilibrium of endothelial cells, thrombocytes, fibrinolysis, and coagulation. Also, it determines the amount of fibrin at the lesion location, consequently impacting the course of the reparative processes. Thanks to the neuronal reflex mechanism, the vascular smooth muscle cells contract and enable the damaged vessels to rapidly constrict. The coagulation pathway is activated via extrinsic and intrinsic pathways together with hemostatic events. As bleeding starts, exposed collagen and other components of ECM contact with the blood components and platelets and this interaction induces the release of coagulation factors from the platelets. The clot additionally offers a transient matrix for cell movement during the inflammatory and hemostatic stages. Cytokines such as growth factors and transforming growth factor-β (TGF-β), platelet-derived growth factor (PDGF), insulin-like growth

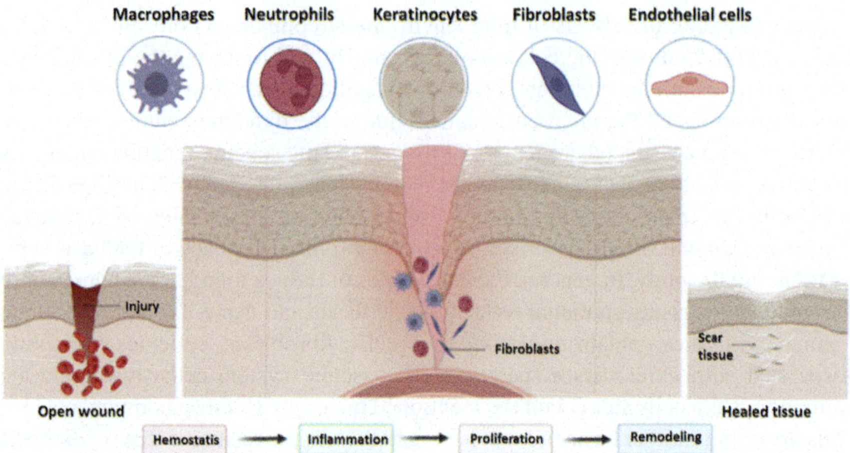

Fig. 2.2 Schematic representation of different cell types involving tissue regeneration process

factors, and EGF in granules within the cytoplasm of platelets enhance the healing of wounds by stimulating neutrophils, then macrophages, endothelial cells, and fibroblasts (Fig. 2.2) [24].

Inflammatory Stage

This stage's primary goal is to stop infection. The mechanical barrier that served as the first line of defense against encroaching microbes is broken. Neutrophils, which are moving cells and "initial responders", migrate for 48 h after injury, and the mechanism that mediates is chemotaxis. As a result, neutrophils move in a chemical gradient in the direction of the wound. Interleukin activation and TGF-β signaling work well in this situation. As a result, neutrophils move in a chemical gradient in the direction of the wound and in this process, interleukin activation and TGF-β signals are highly activated. Neutrophils have a few methods for eliminating bacteria and salvaging them. The first one is phagocytosis, which can eliminate external objects along this route. Other method includes the poisonous compounds lactoferrin, proteases, neutrophil elastase, and cathepsin that neutrophils can release which will kill germs. After finishing their job, neutrophils either apoptosis or are phagocytosed by macrophages. Macrophages are significantly larger phagocytic cells that can also endure in the more acidic wound region and macrophages attract the chemical ends and injured cells released by platelets. Growth factors found in abundance in macrophages, like TGF-β and EGF, have a crucial role in controlling the inflammatory response, promoting angiogenesis, and enhancing the production of granulation tissue. Around 72 h, lymphocytes start to show up in the wound. These factors have a crucial task in controlling the improvement of wounds by producing an ECM scaffold and modifying collagen [25].

Proliferative Stage

The proliferative stage lasts 14 days and begins four days after the injury. When the inflammation subsides, the body releases several cell types that are accountable for migration and proliferation. In order to re-establish the vascular network and produce granulation tissue, the wound region goes through several interrelated processes at this stage, including reepithelialization, angiogenesis, collagen synthesis, and ECM creation. Reepithelialization occurs by epithelial stem cells from surrounding sweat glands or hair follicles and pre-existing keratinocytes in the basal layer of the wound side migrate. Three main MAP kinase pathways are triggered via numerous stimulants, including calcium influx, TNF (tumor necrosis factor), and EGF and these pathways differentiate keratinocytes. The proliferation of keratinocytes two to three days after the injury and the development of granulation tissue, which replaces the matrix established during the homeostasis phase, cause a barrier to be formed between the wound and its surroundings (i.e., reepithelialization) [26].

In this stage, fibroblasts play a significant role and proliferate in the deeper regions of the lesion. They generate a few quantities of collagen that enables a support structure for fibroblast migration, proliferation, and ECM production. New blood vessels are created through angiogenesis to provide nutrients and oxygen to the wound. New blood vessels are constructed while endothelial cells migrate and replicate in response to growth factors such as PDGF, VEGF, basic fibroblast growth factor (bFGF), and serine protease thrombin in wounds [26].

ECM Remodeling Stage

ECM is one of the most crucial elements in the functioning of tissues and regulates the phenotypic behavior of the cells. As shown in Fig. 2.1, ECM contains neighboring cells, cytokines, and growth factors [27]. Two different ECM kinds that are basal membrane and interstitial matrix to which epithelium and endothelium are connected exist. Type IV laminins, collagen, entactin/nidogen, and heparan sulfate proteoglycans are used in its construction. The interstitial matrix consists of collagens, including type I and III, fibronectin, proteoglycans, and tenascins. Proteoglycans and tenascins give tissues the moisture necessary for cytokines and growth factors to bind to the tissue [28]. In addition to supporting tissue cells, the ECM also functions as a major regulator of tissue and cell activity through transmembrane signaling. Post-translational modifications (PTMs) also affect the mechanical properties of the ECM. The rigidity and structure of the ECM control how tissues and cell types develop, remodel, migrate, and differentiate [29].

The ECM is mainly composed of proteoglycans and collagens. Collagens is a group of proteins with three chains coiled in a triple helix or around one another. They are stabilized by intra- and intermolecular cross-links, which also contribute to their high tensile strength and tiniest extensibility. The most prevalent group of collagens, fibrillar collagens, include collagens of types I, II, III, and V [30]. Type I collagen is a heterotrimer molecule that contains $\alpha_1\alpha_2\alpha_3$(I) chains. It possesses considerable tensile strength, load-bearing capacity, stress-carrying properties, and provides tensile stiffness to the bone [31]. Type II collagen, a homotrimer, contains a lot of glycosyl and hydroxylysine residues [27]. It is found primarily in hyaline cartilage, nucleus pulposus of intervertebral discs, notochord, and corneal epithelium [30]. Type III collagen makes up a large portion of the dermis, liver, and lung interstitial tissues and increases the tissue's flexibility [32]. Type IV collagen makes up the majority of the basement membrane. It contains three primary domains: the 7S domain at the N-terminal, a triple helix, and the NC1 globular domain at the C-terminus [33]. Type V collagen is often expressed in type I collagen-containing tissues. Tyrosine sulfated residues are abundant in the N-terminal domain of type V collagen that has a significant role in increasing the stability of fibrils [34]. Finally, type VI collagen, a heterotrimeric molecule, interacts with the ECM components such as type I collagen and fibronectin [35].

Proteoglycans are made up of a protein core and negatively charged glycosaminoglycans (GAGs) [27]. Some of the proteoglycans can selectively bind to other elements of the ECM and participate in the structures of connective tissues. These proteoglycans belong to the short leucine-rich proteoglycan (SLRP) family [36, 37]. Proteoglycans can also influence tissue regeneration and tumor development, promote proliferation, adhesion and migration of cells and regulate the actions of cytokines and growth factors [27]. One of the major proteoglycans, aggrecan, is mostly present in cartilage and allows the cartilage tissue to resist deformation [38]. On the other hand, versican is a substantial chondroitin sulfate interstitial proteoglycan that is one of the essential components of the ECM of blood vessels and can interact with junction proteins and hyaluronan to create stable, high-molecular-weight aggregates. Perlecan is made of heparan sulfate widely dispersed in basement membranes and interacts with laminin and type IV collagen [39]. The most prevalent SLRP in cartilage tissue is decorin. Decorin has complicated secondary structures and develops particular interactions with ECM molecules. It also regulates collagen fibrillogenesis and upholds tissue integrity by interacting with fibronectin and thrombospondin [40]. Mimecan is a proteoglycan made of keratan sulfate, also a member of the SLRP family and is produced by the osteoglycin-encoding gene [41, 42]. Furthermore, lumican is an SLRP widely distributed in connective tissues, where it influences collagen fibrillogenesis and enhances collagen fibril stability by controlling the arrangement and diameters of collagen fibers [43].

ECM remodeling is the final stage of the tissue regeneration mechanism. The main objective of this stage is to break down, restructure, and synthesize the new ECM in order to get the maximum possible tensile strength. Granulation tissue is softly modified at this stage of healing when the development of normal tissue starts to take place. Then, scar tissue that is less cellular and vascular begins to form [44]. Because a monolayer of keratinocytes covers the surface of the lesion, epidermal migration is inhibited and a new multilayer epidermis forms, and finally the ECM content changes [45]. Type III collagen is broken down and type I collagen synthesis starts during the wound closure [44]. Together with matrix maturation, collagen diameters increase, whereas hyaluronic acid and fibronectin are degraded. The amount of collagen that forms increases the wound's tensile strength [46]. Enzymes called matrix metalloproteinases, which are produced by neutrophils, macrophages, and fibroblasts, break down collagen. As metalloproteinase activity gradually declines, the accumulation of a new matrix is encouraged. The wound edges get closer together as the underlying connective tissue gets smaller. PDGF, TGF-β, and FGF are a few of the factors that control these processes [47]. Most fibroblasts, blood vessels and inflammatory cells leave the lesion area throughout the remodeling processes due to the migration and apoptosis of cells. Therefore, scar tissue with fewer cells begins to develop. Myofibroblasts in the granulation tissue then undergo a phenotypic change and begin to momentarily express smooth muscle actin and cause the wound to contract [44, 48].

Apoptosis and Regeneration

Apoptosis is involved in the transformation of granulation tissue into scar tissue. In the remodeling process, apoptosis ensures the elimination of cells without causing tissue damage [49]. This indicates that there is a link between apoptosis and wound healing that needs to be examined.

During wound healing, an increase in the number of various cells occurs that is involved in wound maturation, tissue repair, inflammation, and collagen deposition. Once these cells have completed their task, they must move away from the wound to allow the next phase to begin. There are 3 basic mechanisms that cause the cells to move away from the damaged area. These are necrosis, apoptosis, and cell migration. Apoptosis has been found to be the main mechanism for the elimination of cells during the wound-healing process. It ensures the elimination of cells without tissue damage or any inflammatory response. Therefore, it is important to examine apoptotic signals in understanding tissue repair. Incorrect occurrence of apoptosis in the wound healing process can lead to various pathological conditions. Migration and necrosis mechanisms are not dominant in reducing cells in tissue repair. Necrosis occurs during pathological tissue repair. During normal wound healing, excessive inflammation and tissue damage are not observed. Therefore, the effective mechanism cannot be necrosis. In addition, there is insufficient evidence of cell migration from the damaged area. For this reason, it is thought that the effective mechanism in the reduction of cells is not cell migration. Apoptosis is a mechanism that can remove invading organisms and non-viable tissue at various stages of healing without causing further inflammation. Similarly, when enough collagen accumulates, fibroblast cells are also destroyed by apoptosis. In the final stage of wound healing, endothelial cells and fibroblasts remaining in the mature wound are likewise destroyed [50].

Apoptosis occurs through two mechanisms called the 'mitochondrial pathway', also known as the intrinsic apoptotic pathway and the 'extrinsic pathway' (Fig. 2.3). Bcl-2 proteins, located in the mitochondrial intermembrane space and regulating the release of various molecules, take part in the intrinsic apoptotic pathway. These proteins enter the cytosol and initiate the caspase cascade, which enables the formation of the apoptosome. The resulting multiprotein complex activates procaspase-9 to form caspase-9. Caspase-9 cleaves caspase-3 and causes cell apoptosis. The external pathway begins with the interaction between ligands and "death proteins" from the TNF family. After the interaction, the death-causing signal complex (DISC) is formed. Finally, similar to the internal pathway, caspase-8, which can cleave caspase-3, is activated and apoptosis occurs. As a result of changes in apoptotic cells, these cells are phagocytosed by the immune system and "apoptotic bodies" are formed [51].

In tissue remodeling, apoptosis regulates cell number and proliferation. It is involved at the beginning of the process to stimulate regeneration and then regulates regenerative modeling. Apoptosis acts as a signal in regenerative tissues and supports the production of cells necessary for regeneration. Apoptotic cells can exert non-autonomous effects on neighboring cells by releasing various mitogenic factors that induce cell proliferation. Therefore, it is necessary for regeneration in multiple organisms and tissues [52, 53].

The first cells to reach the wound are neutrophils, which provide defense against invading organisms. Cytokines such as tumor necrosis factor-α (TNF-α) and interleukin 1 (IL-1) are secreted from neutrophils and other inflammatory cells. These cytokines cause the "stress response". This response, caused by neutrophils, is controlled by apoptosis. After neutrophils destroy invading organisms at the injury site, they undergo apoptosis and are removed by macrophages. Apoptosis of neutrophils is thought to be regulated by TNF-α. There are also various signals that macrophages use to recognize apoptotic neutrophils. There must be a balance between apoptosis and proliferation for proper wound healing. Otherwise, hypertrophic scar or keloid formation may be seen in tissue repair. These pathological conditions are caused by an imbalance between collagen deposition and degradation [50].

Apoptosis that occurs during tissue repair includes all cell clusters. For this reason, it is called 'cluster apoptosis' or 'apoptosis-induced apoptosis (AIA)'. During cluster apoptosis, cells undergoing apoptosis release various signals belonging to the TNF family, causing their surrounding cells to undergo apoptosis. The initial apoptosis signal is thought to arise due to the activation of the c-Jun N-terminal kinase (JNK) pathway. Tissue regeneration occurs through the proliferative response of various cells to replace damaged tissue. This physiological mechanism is called compensatory proliferation. Apoptotic cells can trigger this proliferation by producing various mitogenic signals. This biological mechanism is called "apoptosis-induced compensatory proliferation" (AIP) [51]. This compensatory process was first discovered in Drosophila melanogaster. In the study, it was shown that apoptotic cells affect the mitosis of neighboring cells by secreting different mitogenic signals [54]. In addition, pro-apoptotic signals induce the generation of reactive oxygen species (ROS),

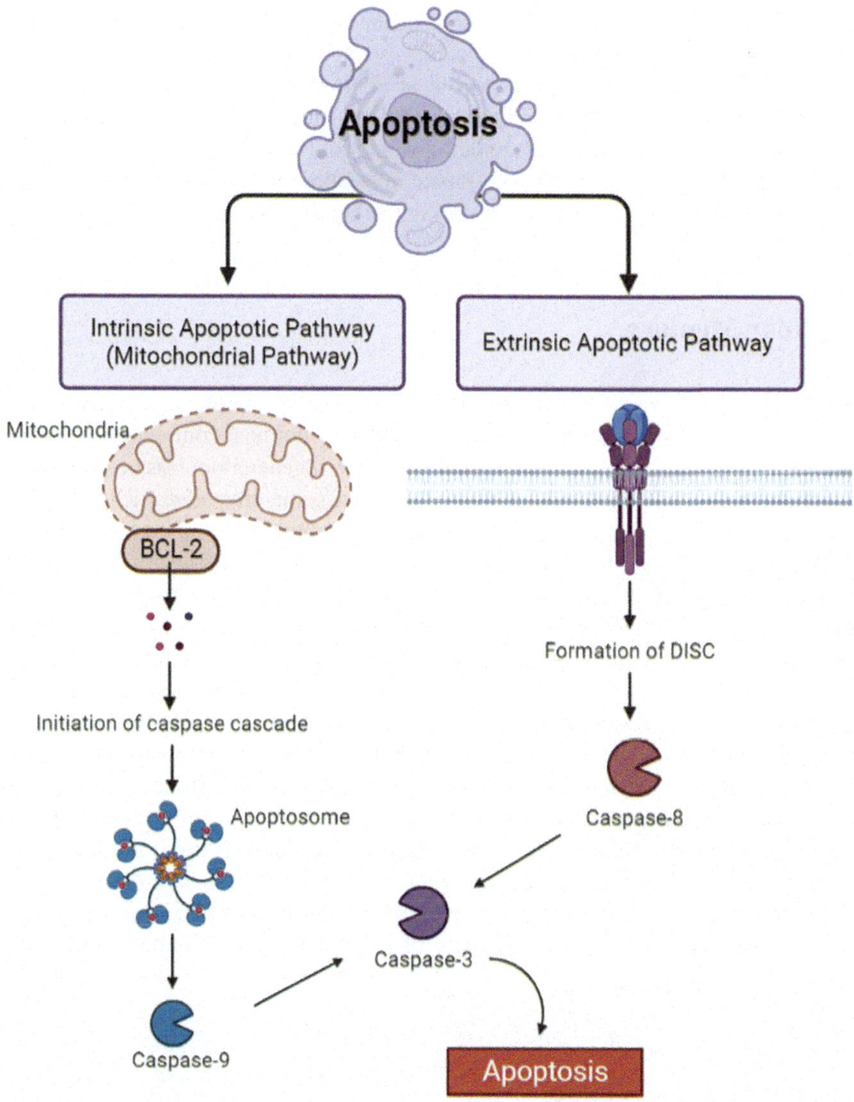

Fig. 2.3 Intrinsic and extrinsic apoptotic pathways during tissue regeneration

which is essential for effective wound healing. Produced ROS initiate tissue regeneration by activating Wnt/β-catenin and FGF signaling [51]. Hydra is also a creature that can reproduce any part of its body. In studies with Hydra, regeneration is thought to be supported by apoptosis. Moreover, apoptosis is limited in some tissues of Hyra and appears to be slower to regenerate in these tissues [55]. Another study showing that apoptosis is effective in tissue repair was carried out by Brown et al. According to the study, apoptosis of inflammatory cells occurs in the early stages of tissue repair, that is, within 12 h of injury [56].

In conclusion, apoptosis plays a role in controlling cell population in wound healing. Control of proliferation at the injury site ensures rapid closure of the wound. When this balance is disturbed, pathological tissue repair occurs. Understanding the process of apoptosis and tissue repair may allow the prevention of excessive inflammation or scarring that occurs during the healing process.

Cellular Memory

Recovery depends on the body's ability to remember the structure of damaged areas. In the case of an injury to the body, the wound may not heal completely. Although the body contains all the components necessary for regeneration, loss of structure or function can be seen in damaged areas. However, some creatures, such as lizards, have the ability to completely heal complex structures. In fact, a common mechanism plays a role in all these processes. Cells, tissues and organs have a memory that can heal or regenerate damaged tissue and know when and how to do so. In addition to repairing damaged areas, this memory is also effective in maintaining body shape for years. This memory possessed by living things directly affects the regenerative processes. While this biological function regenerates the damaged tail in some creatures, it sometimes controls wound healing. The term structural memory, which preserves and regenerates the complex structures of tissues and organs, includes all relevant biological mechanisms, various biomolecules, cellular structures, and many components, including DNA sequences. Although it is not yet understood how the structural memory of the tissues is stored, it can be thought that it is found in all body structures. Structural memory prevents exposure of cell surface antigens to the immune system and similarly, the release of growth factors from the ECM when tissue is not injured. The basic integrity of the tissues is not compromised unless there is any damage [57].

Regeneration of Tissues

The process called regeneration is used to fully or partially replace lost or damaged tissues and organs. Tissue regeneration is a physiological mechanism required to restore the functionality and shape of injured tissues or organs (Fig. 2.4). Regeneration occurs through the proliferation of cells that survive the damage and maintain the capacity to proliferate [58]. Additionally, tissue stem cells may contribute to the restoration of damaged tissues. A population of stem or progenitor cells with the ability to differentiate and reproduce is necessary for regeneration [59]. Poorly differentiated tissues retain their potential to regenerate, whereas highly differentiated tissues mostly or entirely lose this ability. Regenerative capability is inversely related

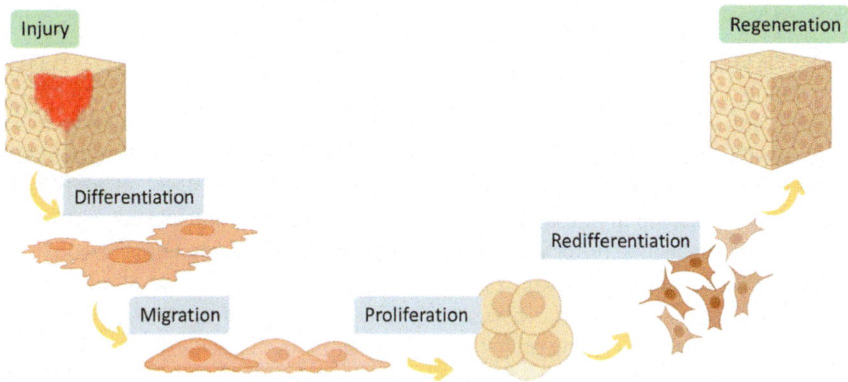

Fig. 2.4 A scheme of tissue regeneration

to the degree of differentiation (muscular and nervous tissue). Tissues with strong regeneration capacity adjust by growing in terms of cell quantity when faced with higher functional demands. The term "numerical growth" or "hyperplasia" refers to this process [60].

Humans and other mammals have four main tissue groupings that make up their bodies: epithelial, connective, muscular, and nervous. Over time, tissue damage and wear may occur in the body. As a result of the inflammatory response in the injured tissues, repair mechanisms are triggered. The repair abilities of the body are very effective in the normal functioning of the damaged structures. If the damaged tissue does not have the power to fully regenerate or if it has been severely injured, the repair occurs in the connective tissue, known as fibrous tissue ground, and heals in the form of scar tissue (scar that leaves a thick scar on the skin), which is an undesirable situation. Collagen deposition resulting from prolonged excessive inflammatory reaction or insufficient blood supply in tissues is called fibrosis. In repair, certain cells are both activated and multiplied. These are endothelial cells (in the vessel), platelets, leukocytes, fibroblasts, and stem cells. The proliferation of these cells is governed by proteins called growth factors [3, 59]. The level of proliferation capacity of the tissues is the most decisive factor in the repair of that tissue. Tissues are classified into three groups: labile (constantly proliferating and dividing), stable (low proliferating), and permanent (non-proliferating). Examples of labile cells are bone marrow, gastrointestinal epithelium, and skin. Liver, kidney, pancreas, thyroid, lung, endothelium, fibroblast, and smooth muscle cells are stable. Examples of permanent cells are heart, muscle cells, and neurons. The injuries that occur here are irreversible and result in scarring [61].

Regeneration of Connective Tissue

The connective tissue is the tissue that is composed of sparse cell communities in ECM, holding, and supporting many tissues and organs together. Connective tissue contains abundant intercellular substances. It consists of blood vessels, connective tissue cells, and fibers. The connective tissue cells include fibroblasts, macrophages, adipose cells, mesenchyme cells, mast cells, melanocytes, and plasma cells. The basic cell of connective tissue is fibroblasts. Fibroblasts form connective tissue fibers composed of collagen or elastin. These protein-structured fibers are of three types: collagen, elastic, and reticular fibers [62]. The functions of connective tissue are to play a binding, supporting and shaping role in many tissues, to act as a reserve energy store thanks to adipose tissue, and to act as a protector against microorganisms invading the body thanks to its special cells. The components of connective tissue vary from one site of the body to another. As various types can transform into each other through intermediate forms, one type of connective tissue can transform into another type of connective tissue with the change of conditions [63]. Mesenchymal connective tissue is the first connective tissue to appear in the developing embryo. This connective tissue is called embryonic connective tissue or mesenchyme. It consists of small shuttle or star-shaped mesenchymal cells that resemble each other. Cells form a cellular network in the embryo with their thin cytoplasmic extensions. Nexuses are cellular communication regions that provide this network. The cells are contained in a sticky substrate containing very thin and sparsely located collagen fibrils. Mesenchymal cells have the potential for rapid division and strong differentiation. Mucous connective tissue is similar to mesenchymal connective tissue. It is located in the umbilical cord in the embryo. It contains more ground substance than cells. The jelly-like ground substance found in the umbilical cord is called Wharton's jelly. Very thin collagen fibrils are found in the intercellular space, which occupies a large space. Its main cells are fibroblasts. It is also found in limited amounts in the pulp of the tooth in adults. Loose connective tissue consists of mesenchyme that remains after other tissues of the embryo have formed. Mesenchymal cells transform into fibroblasts. There are an unlimited number of voids covered with a small amount of ground substance in this shrinking and opening texture that looks like a sponge. These spaces can expand and contract from time to time depending on the amount of fluid in it, so this tissue is also called areolar tissue. It is not resistant to stretching and pulling. Loose connective tissue fills in between other tissues. It connects the skin to the underlying organ. It fills the spaces between the muscles. It is located under the epithelial tissue and nourishes it with the blood vessels it contains. It forms the structure of the mucous membranes in the respiratory and digestive system, which is the entry point of foreign substances into the body. It also contributes to the structure of serous membranes such as the peritoneum, pleura, and pericardium. Dense connective tissue is separated from loose connective tissue because the fibrils it contains are too much compared to cells and ground substance. Fibroblasts are most abundant. If the fibrils are distributed in such a way that they form regular and parallel bundles, it is called dense regular connective tissue (in tendons and ligaments), and if the distribution of the fibrils is scattered in different directions, it is called dense irregular connective tissue (in the dermis, in sheath around the nerves, and in many organ capsules). Specialized connective tissue is divided into four groups: elastic connective tissue, reticular connective tissue, pigment tissue, and adipose tissue. Elastic connective tissue contains abundant elastic fiber bundles. These fibrils are either arranged in parallel or form a regular interconnected network. Among the fibrils are fibroblasts and collagen fibrils. It is found in the spine between the vertebrae, in the walls of the hollow organs, and in some parts of the respiratory tract. Reticular connective tissue is found around liver sinusoids, thymus, lymph nodes, stroma of hematopoietic organs, and spleen. The reticular fibrils include reticulocytes and very fine networks of collagen fibrils. It provides support to the organ in which it is located. Pigment tissue is a type of connective tissue in which pigment cells (melanocytes) are concentrated. It is found intensely in the retina and iris of the eye, but less frequently in the skin. Pigment connective tissue protects the organ in which it is found against UV rays [64–67].

Repair occurs through two types of reactions: regeneration of damaged tissue and scar formation by deposition of connective tissue. After a slight damage, the cells are healed by regeneration. Regeneration occurs with the proliferation of the cells that remain after the damage and have not lost their capacity to divide, and the regeneration of the tissue using stem cells. However, after a more severe injury in which the connective tissue is also damaged, the repair occurs with scar formation. In case of permanent tissue damage, repair is provided by the replacement of non-regenerative parenchyma cells by connective tissue [68]. The fibrous connective tissue formed in this repair provides sufficient structural stability that the injured tissue needs to function. Stages of repair with connective tissue are hemostasis, inflammation, proliferation containing the formation of new blood vessels (angiogenesis) and the migration of fibroblasts, fibrous tissue maturation, and remodeling (Fig. 2.5). Hemostasis is the first stage of the wound healing process. Following the injury, the destroyed blood vessels in the wound site narrow immediately. When the vessel wall undergoes the damage, platelets (blood cells involved in blood coagulation and tissue repair) and erythrocytes are activated as a result of contacting the collagen in the vessel wall which is opened. Thus, adhesion of the platelets occurs both to the vessel wall and to each other. Tissue products are released from platelets to damaged tissue. With the activation of thromboxane A2, platelets form a temporary first clot plug, initially reducing bleeding. Then, a permanent clot is formed with the formation of fibrin in this plug. Erythrocytes allow clots to be contracted by platelets and form an impermeable barrier necessary for hemostasis and wound healing. In the later stage, vasoconstriction develops with serotonin and other vasoconstrictor agents released from platelets and bleeding decreases. The narrowing of the vessels is replaced by vasodilation with the activation of the prostaglandin and complement system. Enlarged vessels increase the passage of blood cells to the damaged tissue area and collect inflammatory exudate between the cells. Many chemotactic factors also have an effect on this. The release of wound stimulating substances such as PDGF, TGF-β, fibronectin, and serotonin is made by platelet alpha granules. The

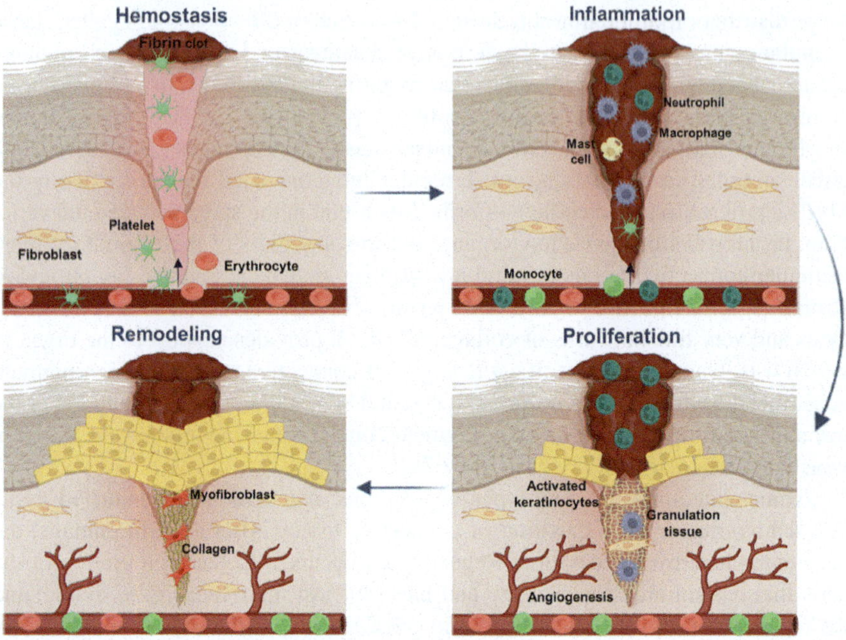

Fig. 2.5 General schematic representation of tissue repair

fibrin clot acts as a skeleton for neutrophils and monocytes to regulate hemostasis [69–72].

In the first day of tissue repair, neutrophil leukocytes are seen at the incision margins and migrate towards the fibrin clot. Basal epithelial cells at the edge of the incision begin to proliferate and shift towards the midline. Thus, a continuous but thin epithelial layer is formed. From the 2nd day onwards, the epithelial ends have merged at the level of the basal layer. Collagen fibers begin to appear on the wound edges. Epithelial proliferation is directed upwards. Endothelial cells of intact vessels around the site to be repaired proliferate towards the wound site and provide new vessel formation. This migration, replication, and new capillary tubule formation is under the influence of growth factors such as TNF-α, TGF-β, and VEGF. VEGF is mainly macrophage-sourced and its receptors are found on endothelial cells [73–75]. During this period, the tissue turns red. Since the connection areas between the endothelial cells in the newly formed vessels are not fully developed, the edema continues for a while even though the inflammation in the site decreases. After the 3rd day, macrophages start to take the place of neutrophils. Macrophages perform crucial functions such as phagocytosis and antimicrobial defense during the inflammation phase. The patch tissue begins to take shape. The patch tissue that performs the repair is called granulation tissue. Edema, inflammatory cells, new vessel formation (angiogenesis), and fibroblasts form the components of granulation tissue. Granulation tissue fully develops in 3–5 days. Collagen fibers proliferate and begin to join the incision line. Epithelial proliferation is at its maximum. Granulation tissue and angiogenesis reach their peak on the 5th day. The next step is contraction. Wound size is reduced by wound contraction. This is taken place by myofibroblasts, which can contract like smooth muscle cells. When granulation tissue is formed and epithelialization is completed, the proliferation phase ends. Fibrosis occurs in two stages. Firstly, fibroblasts migrate to the granulation site. The second stage is the deposition of ECM proteins made by fibroblasts. Fibroblasts start collagen synthesis on the 3–5th day and continue the synthesis for weeks depending on the size of the wound. Accumulation of ECM proteins increases. At the end of this stage, the granulation tissue transforms into a scar tissue composed largely of inactive spindle-shaped fibroblasts, dense collagen, elastic tissue fragments, and other ECM components [24, 76–78]. Degradation of collagen and other ECM components is provided by matrix metalloproteinases (MMPs) with zinc ion-dependent activity. Matrix metalloproteinase function at the wound site plays a role in wound debris and connective tissue remodeling. These enzymes are made by various cell types such as fibroblasts, some epithelial cells, synovial cells, macrophages, and neutrophils. They are activated by certain chemicals and proteases such as plasmin. In addition, they are rapidly inhibited by specific tissue metalloproteinase inhibitors (TIMPs) synthesized by keratinocytes, fibroblasts, smooth muscle cells, and endothelial cells. At the end of the first month, a minimal scar tissue lined with normal epithelium forms. The skin appendages disappear along the incision. The resistance of the wound increases over time, but it takes months to reach its peak. The improvement in resistance to tension is due to the fact that collagen synthesis is higher than collagen destruction in the first two months, and when the synthesis decreases later on, structural changes occur in collagen such as cross-linking and increase in fiber sizes. Wound resistance reaches an average of 70–80% compared to normal at the end of the third month [79–82].

Regeneration of Epithelial Tissue

Epithelial tissue is a specialized tissue that covers the inner and outer surfaces of the body, consisting of epithelial cells with very tight connection between them, and containing a layer called the basal membrane underneath [83]. Epithelial cells originate from ectoderm, endoderm, and mesoderm which are the three germinative layers of the developing embryo and epithelial cells can perform mitosis [77]. Epithelial tissue does not contain blood vessels so that it is nourished through diffusion in the blood vessels in the connective tissue. Functions of the epithelial tissue are to cover all body surfaces and cavities other than epithelial articular surfaces, protection from external factors by covering certain surfaces such as skin, absorption as in the epithelium lining the small intestine, producing secretion and releasing it out or into the blood, superficial, intracellular and intercellular transport of the substances, receiving senses from the environment, and contraction in myoepithelial cells [84]. Epithelial tissue is divided into three main groups: surface epithelia, glandular epithelia, and sensory epithelia. Surface epithelia are cell unions that are closed, avascular, and innervated that cover and guard the underlying connective tissue. Epithelia can be found in stratified or simple shapes. Figure 2.6 represents types of surface epithelial tissues based on shape and number of layers. Due to the highly organized, layer-by-layer structure of human skin, 3D bioprinting technologies are now widely used to create skin tissues. The interaction between the cells and the basal membrane is one of the factors used to distinguish pseudostratified from stratified epithelia. Although they may not always extend to the epithelial surface, all cells in pseudostratified epithelia are in contact with the basal membrane. Conversely, only the basal cells are in touch with the basal membrane in stratified epithelia. The names of stratified epithelia are based on the shapes of the surface-forming cells. As a result, the surface cells in stratified columnar epithelia are prismatic and those in stratified squamous epithelia are plate-like [85, 86].

Glandular epithelium or secretory epithelium is a type of secretory epithelial tissue. The cells that make up the glandular tissue take various molecules necessary for their functions from the blood and convert them into more complex products with intracellular biosynthesis mechanisms, and then release them into the blood in the same way or secrete them into the internal or external environment. The substance that is produced and secreted in the cell is called secretion. Cells and groups of cells specialized for secretory work are called glands. The chemical transformation of secretion requires energy. The glandular epithelium is divided into two according to the excretion of secretory products from the cells: endocrine and exocrine glands. Endocrines are glands that secrete secretions called hormones. These glands do not have excretory ducts. Endocrine glands have important roles in the fulfillment of almost all functions of all living things, in their control, and in the protection of homeostasis. Exocrines are glands that release their secretions through a special

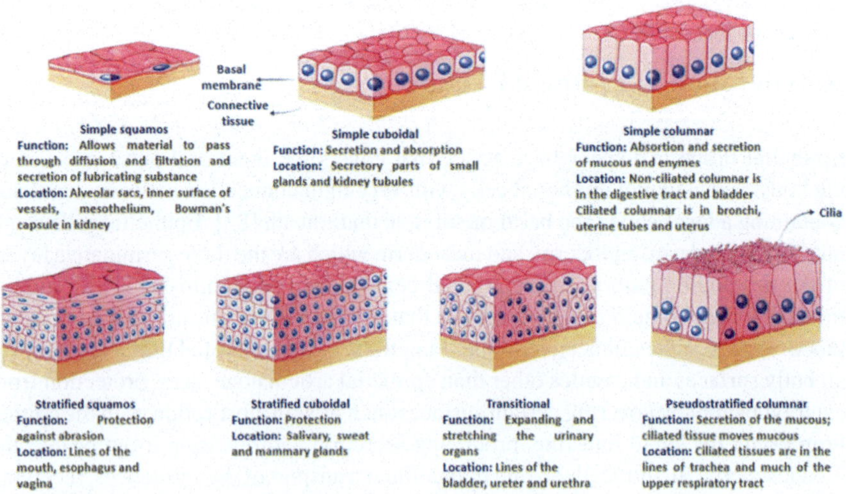

Fig. 2.6 Types of surface epithelia based on shape of cells and number of layers

duct or directly outside the body. Exocrine glands consist of two parts called stroma and parenchyma. The connective tissue part of the exocrine glands is called the stroma, while the epithelial part is called the parenchyma. Exocrine glands originate from and develop from surface epithelium during embryonic development. Salivary glands, sweat, and sebaceous glands in the skin are examples of exocrine glands [87–89]. Sensory epithelia consist of specialized epithelial cells that receive physical, chemical, and optical stimuli from the external environment. Nerve cells terminate in epithelial cells, forming the sensory epithelium. These specialized cells sense the changes in their environment and inform the nervous system. Sensory epithelial cells are of three types: taste epithelium, olfactory epithelium, and free nerve endings. Sensory epithelial cells cannot regenerate [90].

In order to maintain the functions of the surface and many glandular epithelia, it is necessary to regularly replace the cells that are lost. Mitosis, which often takes place as far away from the sites of consumption as possible, plays a significant part in replacing the destroyed cells. Sebaceous cells that break down during holocrine secretion are replaced by new sebaceous cells by the sebaceous gland's basal cells. Regeneration of epithelial tissue proceeds cyclically, similar to the monthly changes that occur in the uterine endometrium during the reproductive period of women. In the first 4–5 days of the menstrual cycle, if no blastocyst formation has occurred, most of the endometrium is desquamated. The basal layer and bases of the uterine glands remain inside the uterus. In the next 9–10 days, regeneration of the epithelial tissue and adjacent subepithelial connective tissue of the endometrium occurs with a high degree of mitotic proliferation via the epithelium of the glandular remnants. It takes approximately 14 days for the endometrium to completely regenerate after menstruation. After a skin surface wound, the lesion's germinal layer of the wound periphery starts to cover it, accompanied by intense mitotic activity that causes the epithelium to migrate. The cells of sweat gland excretory ducts and hair follicles also contribute to epidermal renewal. With larger wounds, the defect is filled with granulation tissue, which is regenerated vascular connective tissue from subepidermal layers. A skin transplant may be required to accelerate this process, as the epidermis proceeds slowly on the granulation tissue from the wound site [91–93].

Stem cells have two basic properties. They can proliferate continuously, thus creating a usable pool. When they receive the right signal, they can transform into various cell types. The most important feature that distinguishes stem cells from others is their high differentiation capability. Differentiation presents a set of processes attended by the combination of cytokines, growth and differentiation factors, ECM components, and intercellular communications. Thanks to these features, they offer a wide opportunity to renew the cells that are lost in the cellular construction and repair of the organism. They can be unipotent, capable of transforming into a single cell type; multipotent, capable of transforming into multiple cell types; pluripotent, capable of forming many tissues in the body; and totipotent, capable of forming a full embryo according to their differentiation abilities [94–96].

Stem cells from the bulb of the hair follicle and in the basal epidermal layer have the potential to be a renewable source of epidermal cells capable of differentiation, proliferation, and migration (Fig. 2.7). These cells enable angiogenesis and form new epithelium. After injury, if the epidermal basement membrane or hair bulbs are not damaged, the skin epithelium and hair follicles are regenerated from stem cells [77]. While stem cells provide their high proliferation potential thanks to their high telomerase enzyme activities, they maintain this function without differentiation and they are sustained by continued expression of transcription factors (such as Oct3, Nanog) which are responsible for self-renewal due to the activation of some signaling pathways (such as Wnt, Notch and Jak/Stat3) originating from the microenvironment (niche) in which they are located [97–100]. In the mammalian epidermis, epidermal stem cells consisting of keratinocytes with self-renewal ability are located around the pilosebaceous unit, which is the structure formed by the hair follicle and sebaceous glands. They show constant localization and have high proliferation ability. They maintain the homeostasis of the skin, promote wound healing, and prevent neoplasm formation. Epidermal stem cells differ from other cells with their surface markers. They express high levels of CD34, CD200, β integrin, and α integrin on their surface. The basal membrane of epidermal cells proliferates to the outermost layer to form the stratified squamous epithelia and begins to differentiate towards the skin surface. Epidermal stem cells move out of the niche that protects the population of stem cells against environmental injury and pigment formation and go up to the outermost layer, producing the intermediate products of the cells. There are different signaling pathways for the homeostasis, differentiation, and proliferation of epidermal stem cells. Notch, Wnt/β-catenin, c-myc, and p63 pathways form the epidermal stem cell network. The Notch signaling pathway is activated by cleavage of Notch by TNF-α converting enzyme (TACE) and β-secretase into NICD (Notch intracellular domain). NICD transforms the CBF1/CSL transcription factor from a gene repressor to a gene activator, thereby starting transcription of target genes containing the Hes/Hey gene family [97, 101–103].

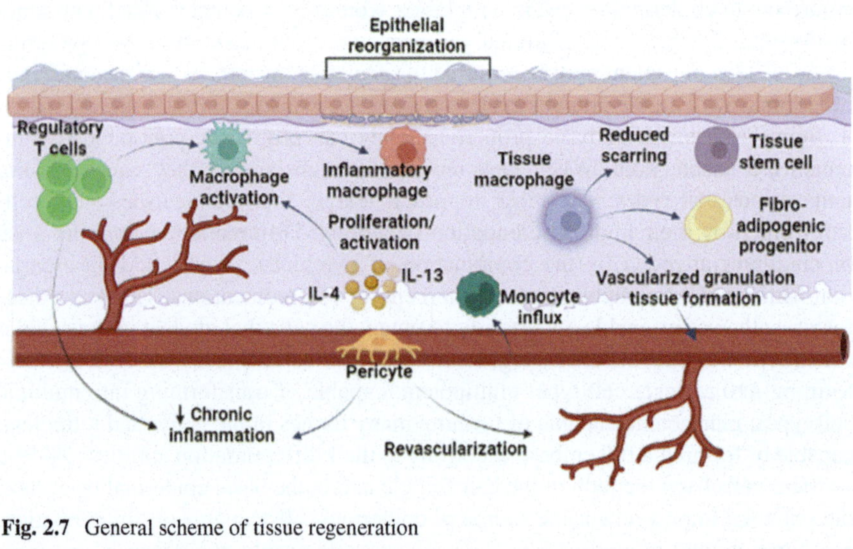

Fig. 2.7 General scheme of tissue regeneration

Stem cells go through an intermediate stage before they differentiate. Cells at this stage are called progenitor cells. Progenitor cells in fetal and adult tissue are semi-differentiated and can differentiate into mature cells by dividing. Unlike stem cells, progenitor cells are stable cells and have little or no self-renewal ability. However, they retain their ability to differentiate and proliferate. The circulating endothelial progenitor cells can differentiate into mature endothelial cells without originating from any pre-existing vessel and form a new vessel by vasculogenesis. They can regenerate hair follicles, epidermal cells, glandular cells, and epithelial crypt cells [104–108].

Regulatory T cells have a crucial role in tissue regeneration by regulating macrophage activation and decreasing the chronic inflammation. Interleukin-4 (IL-4) and interleukin-13 (IL-13) increase polarization of M2 macrophages that control excess inflammation and enhance regeneration by releasing TGF-β. Pericytes promote revascularization of damaged tissue and migration of macrophages. A number of macrophages might promote vascularized granulation, epithelialization, and reduced scarring, also supplying extra support to stem cells and fibro-adipogenic progenitors through trophic factors [69].

Regeneration of Muscle Tissue

Contraction is a feature seen in almost all cells. This feature is more developed in muscle tissue cells than in other tissue cells. Muscle tissue is a specialized tissue that converts chemical energy into mechanical work through contraction and relaxation. Muscle cells that make up muscle tissue are also called muscle fibers. Muscles are divided into three types: smooth muscle, skeletal muscle (striated muscle), and cardiac muscle according to the morphological and physiological characteristics of muscle fibers and their distribution in the body [109].

The skeletal muscle that covers the skeletal system is a controlled muscle that can move joints with strong, fast, and short contractions. They are also called voluntary muscles because they are responsible for voluntary and controllable movements. Skeletal muscles have a striated appearance due to the repetitive structure of the muscle. In this repetitive structure, there are numerous myofibrils (fibers), each formed by the repetition of muscle sarcomeres. Myofibrils contain actin and myosin filaments responsible for muscle contraction. Muscles are divided into three types: smooth muscle, skeletal muscle (striated muscle), and cardiac muscle. Skeletal muscle cells are incapable of division, but they can synthesize new protein and grow (hypertrophy). Skeletal muscles are tissues with an amazing regenerative capacity [110]. Muscle fibers are renewed intermittently or continuously. Thanks to their superior regenerative properties, a heavily damaged muscle tissue can completely return to its former structure and function after a short time. However, since the muscle fibers cannot divide, satellite cells take part in the repair process when they are damaged. Under normal conditions, satellite cells are responsible for the routine growth of skeletal muscle and for its repair and regeneration. Satellite cells are skeletal muscle-specific stem cells found in large numbers under the basal membrane. Satellite cells are mononuclear and quiescent cells when at rest. When these cells are damaged, they become active for tissue regeneration and differentiate to form myoblasts, which are mononuclear cells. Myoblasts proliferate and differentiate to form myotubes, which are multinucleated fibers (Fig. 2.8). Thus, myoblasts provide the regeneration of the damaged fiber or the growth of the fiber by combining with the existing muscle fibers. After damage, the remains of muscle fibers are phagocytosed by macrophages [111, 112]. After tissue regeneration, satellite cells that are able to withstand differentiation return to inactivity to take part in the next tissue injury of the tissue. During muscle regeneration, dynamic interactions with inflammatory cells, stromal cells, trophic signals, and ECM components drive satellite cells. Skeletal muscle can completely and spontaneously regenerate in cases of minor injury such as strain. However, after severe injury, muscle recovery cannot be completed and fibrotic tissue is formed that impairs muscle function [113]. Remodeling of connective tissue is necessary for regeneration. After the muscle is damaged, gaps form between the muscle fibers and the spaces are filled with a hematoma. If the hematoma is not removed quickly, skeletal muscle regeneration is delayed, fibrous tissue develops, and the biomechanical properties of the regenerated muscle decrease. Fibroblasts are the main collagen-producing cells in skeletal muscle. The effect of transforming growth factor on myoblasts enables myoblasts to transform into myofibroblasts. Thus, myofibroblasts produce significant levels of collagen and contribute to muscle fibrosis. In addition, revascularization is important to regenerate damaged skeletal muscle and restore blood flow. Without revascularization, muscle regeneration is incomplete and significant fibrosis occurs. Tissue hypoxia occurs in the damaged area as a result of muscle injury and rupture of blood vessels. Therefore, new capillary formation is required immediately after injury, and factors such as VEGF are secreted into this region. Thus, muscle regeneration is fully realized within 2–3 weeks. Smooth muscles are the muscles that make up the walls of the organs of the digestive tract, the walls of blood vessels, and the muscle tissue of the iris located in the pupil. They are responsible for the movements of the organs that function in mixing, advancement, absorption, and discharge of the ingested nutrients with the digestive juices. They are involuntary, long, and slow-twitch muscles. Since the contractile proteins myosin and actin are randomly arranged, they do not have a striated appearance. They are spindle-shaped cells containing an oval nucleus in the middle of the cell [64, 114, 115]. Smooth muscle cells are adhered to each other and connected by special cell junctions that are given. Smooth muscle cells have the ability to divide. Smooth muscles can also cause hypertrophy. The regeneration capacity in smooth muscles is limited and they are the only type of muscle that can be regenerated by mitosis. They differentiate from mesenchymal cells in repairing the damaged vessel wall [116].

Mesenchymal cells that will differentiate into smooth muscle cells come together and become denser. These cells that differentiate and become longer are called myoblasts. With the elongation of myoblasts, their nuclei also begin to elongate. As a result of elongation, myofilaments appear in the sarcoplasm. Myoblasts divide

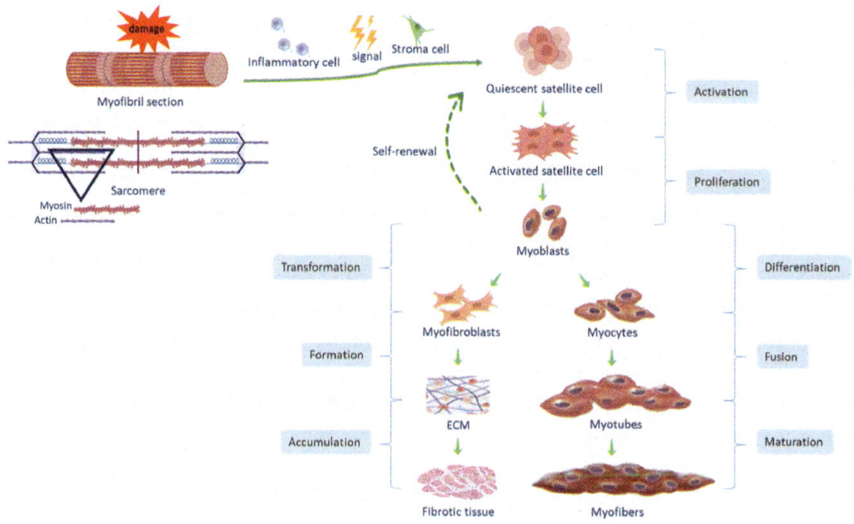

Fig. 2.8 Regeneration of muscle tissue

by mitosis and increase in number. The proliferating cells form the smooth muscle tissue by forming a tight relationship with each other [117].

Cardiac muscle forms the muscular walls of the heart (myocardium). The heart muscle, which can contract rhythmically, works involuntarily because it is not controlled. However, it has a striated appearance similar to skeletal muscle due to the repetition of muscle sarcomeres. In skeletal muscle as well as in cardiac muscle, actin and myosin are regulated in sarcomeres. Cardiac muscle cells, like smooth muscle, usually have a single central nucleus. The cells are usually branched and their ends are tightly connected to other cells by special connections called the intercalation disc. The heart muscle can also hypertrophy. However, since cells such as satellite cells, which are present in the regeneration of skeletal muscles, are not found in the heart muscle, they are not replaced when cardiac muscle cells die. Although cardiac muscle is more resistant to damage than other muscle types, it shows very little regeneration as a result of damage. Fibrosis provides connective tissue repair and scar tissue is formed. However, cardiac functions are lost in the damaged area [118, 119].

Regeneration of Nerve Tissue

Nerve tissues formed by nerve cells are responsible for controlling and regulating the functions of the body. Nervous tissue consists of two types of cells: nerve cells or neurons and glial cells. Neurons are specialized cells that respond to stimuli by generating signals through axons. Glial cells help transmit nerve impulses and provide nutrients to neurons. The brain, spinal cord, and nerves are made up of nervous tissue. They allow the signal to be transmitted quickly from one part of the body to another when a part of the body is stimulated. Nerve cells have lost their ability to divide. However, since their growth and regeneration functions are not completely lost, they can grow and regenerate (Fig. 2.9) [70]. Nerve cells that are not fully specialized continue to reproduce. Nerve cells have limited self-renewal ability. Damaged nerve cells can renew themselves with the help of glia. Glial cells circulate throughout the brain and multiply when the brain is damaged. Axons are structures that emerge from the cell in the form of long stalks. Axons communicate with other cells and transmit impulses called chemical and electrical changes that occur on a nerve fiber as a result of excitation. The cell nucleus, cytoplasm, and cell organelles make up the body of the cell. Dendrites located in the cell body are responsible for receiving information from other neurons. Dendrites provide connections between other neurons and synapses and the body of the cell. Schwann cells form myelin sheaths for the peripheral nervous system. There are two types of Schwann cells, myelinated and unmyelinated. Myelin is an insulating material in the form of a sheet. The fact that axons are surrounded by Schwann cells is called myelination. The myelin sheath in the nerve cell (neuron) wraps around the neuron in the form of a fat layer. It is responsible for the faster transmission of impulses [70, 99, 120]. As a result of damage to the nerve fibers, the ability of the neurons to transmit impulses is lost, and within a few weeks, the fibers degenerate. However, nerve fibers have the ability to regenerate under certain conditions. With the separation of an axon from the body of the nerve cell, the myelin sheath breaks down within the first 3 days and turns into plaque-like oil droplets after about 2–3 weeks. Meanwhile (simultaneously), the severed axon is fragmented. This secondary or Wallerian degeneration extends to sensory and motor nerve endings. Meanwhile, Schwann cells do not degenerate. With the separation of the axon, the nerve cell whose body is damaged swells and moves towards the nuclear cell membrane ("fish eye cells"). As long as the damage does not occur near the perikaryon, nerve cells recover quickly from the trauma. The degeneration of nerve fibers continues until the next node of Ranvier®. From the first 15 days after injury to 60 days, myelin sheath lipids are gradually broken down. Microglia in the central nervous system or macrophages in the peripheral nervous system phagocytize the degraded lipids and transport them out of the cell. Schwann cells begin to divide and converge in the first week. In this way, Bungner bands are formed, which bridge the gaps. The growing axon uses these glial filaments to locate the target organ. The axon, which travels along the glial filaments—about 1–2 mm per day—is surrounded by a myelin sheath. The renewal phase can take about 3 months. In this process, muscle fibers and effectors atrophy, as the muscle fibers are immobile. The contact with the target organ by the myelinated axon is initially thin and weak. During the maturation phase, which can last for several months, the diameter and performance of the regenerated nerve fiber increase. As a result of amputation or severe damage to the nerve fibers, the connective tissue hides between the stumps. As a result, the glial filaments cannot bridge the gap. As a result of unsuccessful regeneration after severe damage, a swollen abnormal structure called amputation neuroma is formed with dense collagen accumulation in the injury area, regenerated axon fragments, and irregular proliferation of Schwann cells. In humans, the nerve fibers of the central nervous system are not capable of regeneration [121–123].

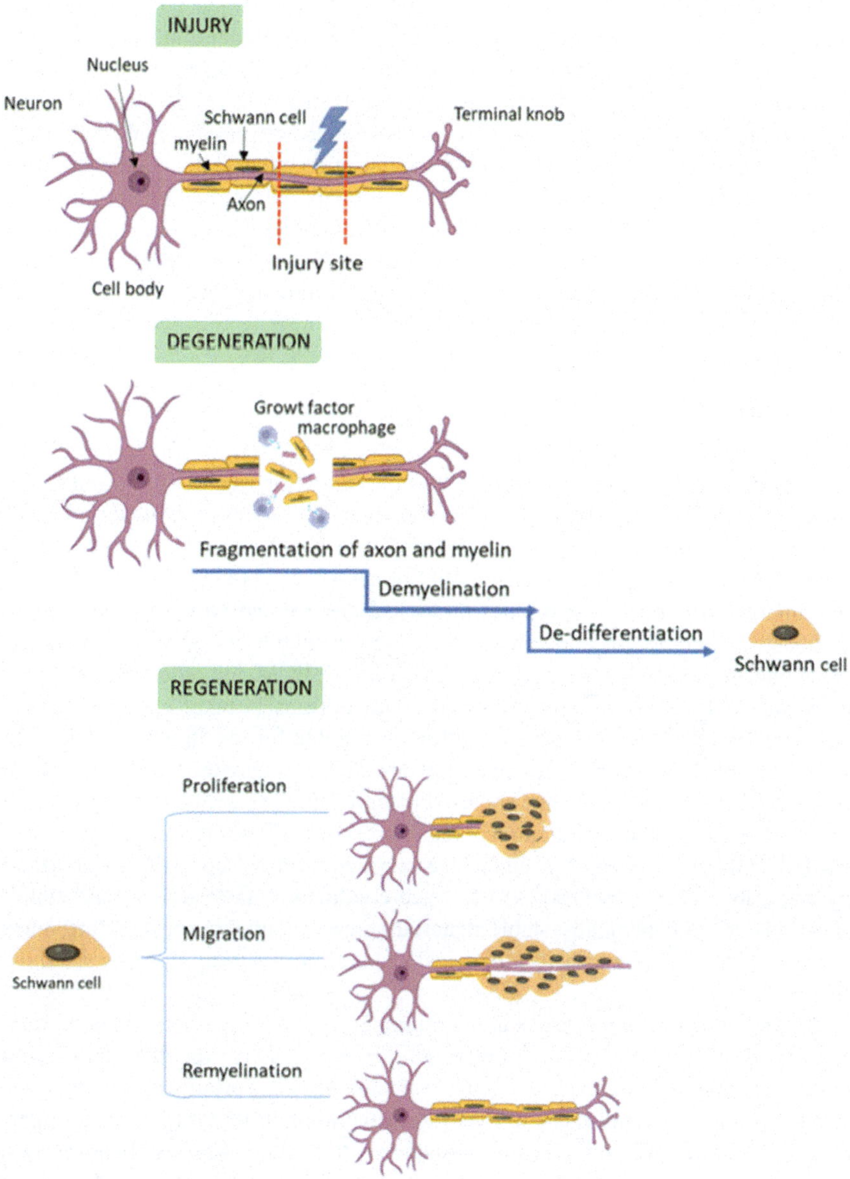

Fig. 2.9 Regeneration of nerve tissue

Mimicking Tissue Microenvironment

Three-dimensional (3D) tissue culture systems that mimic native tissue microenvironments have been studied for years, with significant advancements being made recently in terms of materials, designs, and applications. Better understanding of the roles of various tissue cells, the biocompatibility and biodegradability of scaffolding materials, the physiologically functional elements of native tissues, and the pathophysiological conditions of native tissue microenvironments have all contributed to these successes. The complex physiological microenvironments seen in human connective tissues are ideal for tissue cell growth, differentiation, proliferation, migration, and death. The modeling of human diseases, with an emphasis on different tissue types, as well as tissue healing and regeneration, depend on these constantly changing physical impacts of the tissue microenvironment [15].

Scaffolds

In order to nurture and direct cellular adhesion, migration, proliferation, and desired differentiation to particular cells in three dimensions, scaffolds provide an appropriate surface and sufficient gaps (volumes). A scaffold's design has a crucial role in the development and final functionality of neo-tissues [15]. To create an in vivo-like microenvironment that mimics biological entities and stimulates cell-specific responses to promote tissue regeneration and repair, physiologically mimetic and functionalized scaffolds, such as biologically active ECM, are required [124].

Regardless of the tissue type, a few crucial factors must be taken into account when constructing or judging whether a scaffold is suitable for use in tissue engineering [125]: (1) biocompatible [126] materials are non-immunogenic and non-toxic to living cells and tissues. (2) They are also biodegradable or capable of being modified in accordance with the regeneration or repair process [127]. (3) Porous materials offer a suitable 3D environment for cell and tissue penetration as well as the transportation of nutrients and wastes [100]. (4) Surface conductive materials enhance cellular functions. (5) Mechanically stable materials must be suitable for handling during surgery [128]. The prerequisites for a scaffold used in tissue engineering are shown in Table 2.1.

Cell adhesion, which is a crucial requirement for the long-term survival of transplanted cells, is another factor to take into account. The polypeptide families that contain arginylglycylaspartic acid (RGD) are thought to be promising candidates for treating a range of illnesses as well as for regenerating different tissues and organs. The ECM contains a cell adhesion sequence called RGD, which is believed to be a specific recognition site for integrin receptors. Fibronectin, laminin, fibrinogen, osteopontin, and vitronectin all contain significant amounts of RGD peptides [138]. For the manufacturing of adsorbents targeting various pathogenic factors depending on the class of antibody, ligand density is a programmable parameter [139]. It

Table 2.1 Requirements for the scaffold used in tissue engineering

Tissue type	Expected regeneration time for biodegradation rate	Microarchitecture (pore size)	Inductivity S/m	Biomechanics properties (rigidity)	References
Connective	10–12 weeks	Permeable interconnective pore structure Pore-size: 300–350 µm	0.51–1.14	0.01–2 GPa	[129–132
Muscle	2–4 weeks	Permeable interconnective pore structure Pore-size: 50–200 µm	0.62–0.93	1.53–12 kPa	[107, 133]
Epithelial	~ 5 days 6 months depending on the type of the epithelial tissues	Permeable interconnective pore structure Pore-size: 30–270 µm		0.1–10 kPa	[85, 104, 134–136]
Nerve	3–4 mm/day	Permeable interconnective pore structure Pore-size: 50–500 µm	0.39	0.1–8 kPa	[137, 138]

is essential for determining how many bound integrins there are. There are two components to ligand density measurement. The amount of ligand that has been adsorbed onto substrate is the first factor. The availability of the adsorbed ligand for receptor binding makes up the second factor [139]. For tissue engineering and regenerative medicine applications, creating functional scaffolds necessitates the use of material systems with exact control over cellular performance. A powerful method for producing extremely intricate, multi-component structures with clearly specified design and composition is 3D printing [140]. Various types of 3D bioprinting techniques have been developed, such as micro-extrusion, inkjet, laser-assisted, and stereolithographic (SLA) printing methods, microfluidic-based bioprinting, and extrusion-based 3D printing (EBP) [134]. These bioprinting methods have been used to fabricate various types of tissue constructs [141]. The following Table 2.2 indicates the general properties, fabricating methods, and outcomes of scaffolds in tissue engineering.

Decellularization

Due to the complexity of the cell microenvironment, the preparation and acquisition of synthetic scaffolds pose a challenge. To overcome this problem, naturally, the use of human or animal tissue-derived ECM itself has gained significant importance today. Decellularization aims to preserve the ECM's structural, biochemical, and biomechanical cues while removing native cells and genetic components such as DNA. The patient's own cells can then be infused back into the decellularized ECM to create a customized tissue [154].

The process of removing the organ from a human or animal donor and sterilizing it to the extent that only the collagen network base remains is called decellularization. The cell-free tissue remaining after the procedure is classified as a scaffold [155]. The tissue is then decellularized, revitalized, and restored to function by seeding the decellularized matrix with stem cells of the donor's corresponding tissue type [2]. In addition to serving as structural support, the cellular scaffold plays an essential role in homeostasis, regeneration, tissue growth, and organ development by interacting with molecules on the cell surface and storing growth factors. This approach is ideal for designing organs for transplantation because the ECM protects the inherent vascular architecture and enables the scaffold to withstand physiological blood pressures [156, 157].

A variety of methods can be used for this process including a mixture of chemical, physical, and enzymatic processes [105]. Exemplary substances for removing cells include detergents, salts, enzymes, and physical agents. There various methods available for different application areas [158]. Decellularization generally involves removing the cells from the tissue, which dramatically lowers its immunogenic potential. In other words, decellularization significantly reduces the immunogenic potential by removing existing cells in the tissue [159]. Figure 2.10 shows various decellularized ECMs, their applications and sources.

Table 2.2 General properties, fabricating methods and outcomes of scaffolds in tissue engineering

Materials	General properties	Fabricating method	Important outcomes	Tissue engineering applications	References
HA and HA-PCL composites, ceramics	Inorganic component, encourage bone growth, biocompatibility, bioactivity, osteoconductive, non-toxic	Fabricated by mixing with or coating on a matrix (for ex: in situ precipitation for biomimetic mineralization for composite with polymer as hydrogel)	Compressive elastic modulus: from 2.32 to 2.57 GPa Compressive stress: 195 MPa Elongitudinal: 17.4 GPa pore size: 45.90 ± 4.3^{a} (Micrometer) at porosity: %57 Degradation time: about 2 years	Connective	[142, 143]
Bioactive glasses	Ca- and possibly P-containing silica glasses, not good for load-bearing applications, better bioactivity and degradability	SLA (Stereolithography), Viper Si2TM SLA system (3D Systems)	Young's modulus of bioglass (229 MPa) Degradation time: about 100 days	Connective	[144, 145]
PLA (polylactic acid) and polymeric biomaterials (PGA, PLGA, PEG, PPF, Polyurethane)	Composite material, biodegradable, immiscibility, aging and recyclability,	Porogen leaching, gas foaming, phase separation, fiber meshing, supercritical fluid processing, microsphere sintering, and three-dimensional printing	Density: 1.252 g/cm^3 Degradation time: the order of 2 years Elastic modulus 3500 MPa Yound's modulus: 1280 MPa T_g: 55 °C T_m: 165 °C	Connective, nerve	[144, 146]

(continued)

Table 2.2 (continued)

Materials	General properties	Fabricating method	Important outcomes	Tissue engineering applications	References
Graphene, carbon-based nanomaterials	Display metallic, semi-conducting, mitigate inflammatory responses, chemical stability	Laser vaporization, arc-discharge experiment in liquid nitrogen, hydrometallurgical process	Young modulus: 1.0 TPa Intrinsic strength: 130 GP Young's modulus: 1000 GPa T_m: 3697 °C Density: 2270 kg/m^3 Degradation: a couple of days (by enzimes)	Muscle	[147, 148]
Hydrogels, films, nanofibers	Biomimetic, elastic retractivity, capable of undergoing large deformation	Cross-linking method, in situ method, blending method	Thixotropic hydrogel Yield value: 8.33 Pa Density: 1.37 g/cm^3 Hydrazone-based hydrogel Degradation time: 8–27 days at pH between 1.5 and 6	Muscle	[149, 150]
Natural materials (collagen gels, alginate, matrigel)	Stiffness can be modulated independently of architecture, time-dependent stiffening with calcium crosslinking, enables mammary epithelial cells to polarize	3D-printing, 2-D micropatterns	Density: 0.73 ml/g Degradation time (defined as RI = 0.2) was 982,263,648 s (mean[+, −]std) (strained) and 780,562,321 s (unstrained) for directional edge intensity, and 907,662,769 s (strained) and 746,761,738 s (unstrained) for total edge intensity	Epithelial	[151, 152]

(continued)

Table 2.2 (continued)

Materials	General properties	Fabricating method	Important outcomes	Tissue engineering applications	References
Synthetic materials [PDMS (polydimethylsiloxane), PLGA, polyacrylamide, synthetic–natural hybrid materials (polyethylene glycol-heparin, methacrylated hyaluronic acid (MeHA))	Flexible substrate with patterns promoting cancer cell alignment, wide stiffness range, direct conjugation of many types of adhesive ligands or degradable linkers, inert, enzymatically degradable	3D-printing, 2-D micropatterns	PLGA Density: 1.2–1.3 g/ml T_m: 262 °C T_g: 40–60 °C Degradation kinetics with the apparent rate constants of 0.0927 and 0.0483 day^{-1} for pH 2.4 and pH 7.4	Epithelial	[86, 153]

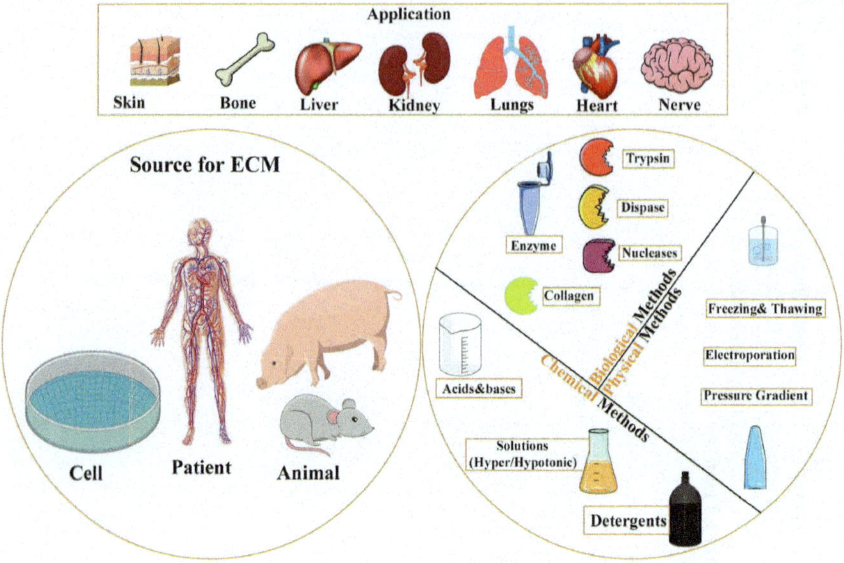

Fig. 2.10 Schematic illustration of various decellularized ECM, applications and sources

The decellularization method depends on several parameters, some of which include the cellularity, thickness, and density of the tissue. It generally relies on the breakdown of cells in the tissue while preserving the ECM. Cell lysis methods can alter the chemical and microarchitecture of the ECM. Therefore, it is necessary to improve the approaches to increase cell clearance while minimizing damage to the scaffold's physico-chemical properties [160]. Table 2.3 outlines the decellularization method and its effects commonly used for tissues.

Table 2.3 Decellularization methods commonly used in tissues

Tissue type	Method	General properties	Effect on ECM	Biomaterial type	References
Epithelial	Enzymatic, chemical, physical methods (triton X-100, SDS, trypsin, EDTA and endonuclease)	Removal of skin burns or scars, epithelial implantation	Elasticity, stiffness, printable, biocompatibility, biodegradability, stiffness	Natural materials (collagen gel), synthetic materials, PDMS (polydimethylsiloxane, synthetic–natural hybrid materials (polyethylene glycol-heparin, methacrylated hyaluronic)	[151]
Muscle	Physical, chemical methods (SDS and triton X-100)	Constructive and regenerative effects on tendons	Elasticity, strength, low viscosity, chemical stability	Carbon-based nanomaterials (mainly graphene and carbon nanotube), metal-based hydrogels, films	[147, 149]
Connective	Physical, chemical methods (SDS and triton X-100 and alveolar bone tissue allograft application)	Regenerative no rejection on tissue transplantation	Strength, immiscibility, recyclability, biocompatibility, nontoxicity, promotion of cellular proliferation and stemness, bioactivity	Polymers, wax or wax compounds, HA and HA composites, ceramics bioactive glasses, PLA (polylactic acid)	Daskalakis et al. [161]
Cartilage	Physical, chemical methods (SDS and Triton X-100 and alveolar bone tissue allograft application)	Regenerative no rejection on tissue transplantation	Chondrogenic differentiation, regenerative	Agarose, collagen, chitosan, chondroitin sulfate, fibrin glue gelatin, hyaluronic acid, silk fibroin, synthetic polymers poly (α-hydroxy esters) poly (ethylene glycol/oxide), poly (NiPAAm), poly (urethane), poly (vinyl) alcohol, self-assembling peptides	Daskalakis et al. [161]
Nerve	Chemical methods (detergent decellularization, mild wash buffers, induction of apoptosis)	Tissue-specific composition and microarchitecture on native tissue	Biocompatibility, non-immunogenicity	Polymeric biomaterials (PGA, PLA, PLGA)	[162]

Conclusion

Tissue regeneration is an emerging field that aims to correct physical tissue defects, maintain tissue function, or improve organ functionality. Regeneration occurs through the proliferation of cells that recover from damage and preserve their reproductive capacity. The natural regeneration process involves various cellular mechanisms such as homeostasis, inflammation, proliferation, ECM remodeling, and apoptosis. A population of stem cells or progenitor cells capable of differentiation and reproduction is required for regeneration. While poorly differentiated tissues retain their regenerative potential, highly differentiated tissues mostly or completely lose this ability. However, when the body's own regeneration system fails in several instances, there is a need to restore the body's regenerative capacity. Tissue regeneration studies are mainly concerned with the development of biological substitutes such as scaffolds and decellularized tissues and their combination with cells possessing differentiation and reproductive abilities as well as biological molecules, to support the regeneration of new tissue. Therefore, understanding the native tissue microenvironment, its components, and its relationships with adjacent tissues and ECM components is critical for tissue regeneration research. Scaffolds and decellularized tissues can meet the challenges of personalized medicine by providing effective treatments for tissue regeneration. It has received considerable attention for tissue scaffolds designed easily mimic the microenvironment and ECM. The tissue microenvironment is a multi-component and highly dynamic structure creating a biophysical and biochemical environment. It plays a substantial role substantial role in tissue regeneration, which is regulated by the interaction of microenvironmental components. In addition, the tissue microenvironment should be evaluated along with the tissue-specific components, their properties, and tissue-site-specific conditions for a better understanding of a particular tissue.

Decellularized matrices and scaffolds shed light on tissue regeneration and biological therapy as they shed light on in vitro ECM modeling in tissue engineering applications to accelerate the regeneration of damaged tissues. Understanding the mechanisms of tissue regeneration leads to the development of treatment modalities used in tissue engineering and regenerative medicine. Since tissue engineering applications do not require major surgical intervention, they offer significant opportunities for future tissue regeneration treatments.

References

1. Boisseau P, Loubaton B (2011) Nanomedicine, nanotechnology in medicine. Comptes Rendus Phys 12:620–636. https://doi.org/10.1016/j.crhy.2011.06.001
2. Raja TI, Mozafari M, Milan PB et al (2018) Nanoengineered biomaterials for tracheal replacement. Elsevier Inc.
3. Eming SA, Martin P, Tomic-Canic M (2014) Wound repair and regeneration: mechanisms, signaling, and translation. Sci Transl Med 6:265sr6. https://doi.org/10.1126/scitranslmed.3009337

4. Kim WJH (2000) Cellular signaling in tissue regeneration. Yonsei Med J 41:692–703
5. Vig K, Chaudhari A, Tripathi S et al (2017) Advances in skin regeneration using tissue engineering. Int J Mol Sci 18. https://doi.org/10.3390/ijms18040789
6. Albulet D, Florea DA, Boarca B et al (2017) Nanotechnology for personalized medicine: cancer research, diagnosis, and therapy. Elsevier Inc.
7. Lei D, Yang Y, Liu Z et al (2019) 3D printing of biomimetic vasculature for tissue regeneration. Mater Horizons 6:1197–1206. https://doi.org/10.1039/c9mh00174c
8. Li Q, Ma L, Gao C (2015) Biomaterials for in situ tissue regeneration: development and perspectives. J Mater Chem B 3:8921–8938. https://doi.org/10.1039/c5tb01863c
9. Hao D, Lopez JM, Chen J et al (2022) Engineering extracellular microenvironment for tissue regeneration. Bioengineering 9:1–17. https://doi.org/10.3390/bioengineering9050202
10. Huang G, Li F, Zhao X et al (2017) Functional and biomimetic materials for engineering of the three-dimensional cell microenvironment. Chem Rev 117:12764–12850. https://doi.org/10.1021/acs.chemrev.7b00094
11. Bottaro DP, Liebmann-vinson A, Heidaran MA (2002) Molecular signaling in bioengineered tissue. Ann N Y Acad Sci 961:143–153
12. Tamori Y, Suzuki E, Deng WM (2016) Epithelial tumors originate in tumor hotspots, a tissue-intrinsic microenvironment. PLoS Biol 14.https://doi.org/10.1371/journal.pbio.1002537
13. Mouw JK, Ou G, Weaver VM (2014) Extracellular matrix assembly: a multiscale deconstruction. Nat Rev Mol Cell Biol 15:771–785. https://doi.org/10.1038/nrm3902
14. Ozcelik H, Hindie M, Hasan A et al (2014) Cell microenvironment engineering and monitoring for tissue engineering and regenerative medicine: the citation accessed citable link cell microenvironment engineering and monitoring for tissue engineering and regenerative medicine: the recent advances. Biomed Res Int 2014:1–18
15. Rijal G, Li W (2018) Native-mimicking in vitro microenvironment: an elusive and seductive future for tumor modeling and tissue engineering. J Biol Eng 12:1–22. https://doi.org/10.1186/s13036-018-0114-7
16. Loreti M, Sacco A (2022) The jam session between muscle stem cells and the extracellular matrix in the tissue microenvironment. NPJ Regen Med 7:1–15. https://doi.org/10.1038/s41536-022-00204-z
17. Cezar CA, Mooney DJ (2015) Biomaterial-based delivery for skeletal muscle repair. Adv Drug Deliv Rev 84:188–197. https://doi.org/10.1016/j.addr.2014.09.008
18. Sawicka KM, Seeliger M, Musaev T et al (2015) Fibronectin interaction and enhancement of growth factors: importance for wound healing. Adv Wound Care 4:469–478. https://doi.org/10.1089/wound.2014.0616
19. Shi Z, Gao X, Ullah MW et al (2016) Electroconductive natural polymer-based hydrogels. Biomaterials 111:40–54. https://doi.org/10.1016/j.biomaterials.2016.09.020
20. Song B, Zhao M, Forrester JV, McCaig CD (2002) Electrical cues regulate the orientation and frequency of cell division and the rate of wound healing in vivo. Proc Natl Acad Sci USA 99:13577–13582. https://doi.org/10.1073/pnas.202235299
21. Zhang W, Liu Y, Zhang H (2021) Extracellular matrix: an important regulator of cell functions and skeletal muscle development. Cell Biosci 11:1–13. https://doi.org/10.1186/s13578-021-00579-4
22. Guillot C, Lecuit T (2013) Mechanics of epithelial tissue homeostasis and morphogenesis. Science 340:1185–1189. https://doi.org/10.1126/science.1235249
23. Kong L, Gao X, Qian Y et al (2022) Biomechanical microenvironment in peripheral nerve regeneration: from pathophysiological understanding to tissue engineering development. Theranostics 12:4993–5014. https://doi.org/10.7150/thno.74571
24. Velnar T, Bailey T, Smrkolj V (2009) The wound healing process: an overview of the cellular and molecular mechanisms. J Int Med Res 37:1528–1542. https://doi.org/10.1177/147323000903700531
25. Singh S, Young A, McNaught CE (2017) The physiology of wound healing. Surg (United Kingdom) 35:473–477. https://doi.org/10.1016/j.mpsur.2017.06.004

26. Abazari M, Ghaffari A, Rashidzadeh H et al (2022) A systematic review on classification, identification, and healing process of burn wound healing. Int J Low Extrem Wounds 21:18–30. https://doi.org/10.1177/1534734620924857
27. Karsdal MA, Nielsen MJ, Sand JM et al (2013) Extracellular matrix remodeling: the common denominator in connective tissue diseases possibilities for evaluation and current understanding of the matrix as more than a passive architecture, but a key player in tissue failure. Assay Drug Dev Technol 11:70–92. https://doi.org/10.1089/adt.2012.474
28. Bosman FT, Stamenkovic I (2003) Functional structure and composition of the extracellular matrix. J Pathol 200:423–428. https://doi.org/10.1002/path.1437
29. Barker HE, Erler JT (2011) The potential for LOXL2 as a target for future cancer treatment. Futur Oncol 7:707–710. https://doi.org/10.2217/fon.11.46
30. Gelse K, Pöschl E, Aigner T (2003) Collagens—structure, function, and biosynthesis. Adv Drug Deliv Rev 55:1531–1546. https://doi.org/10.1016/j.addr.2003.08.002
31. Birk DE, Fitch JM, Babiarz JP, Linsenmayer TF (1988) Collagen type I and type V are present in the same fibril in the avian corneal stroma. J Cell Biol 106:999–1008. https://doi.org/10.1083/jcb.106.3.999
32. Kadler KE, Holmes DF, Trotter JA, Chapman JA (1996) Collagen fibril formation Karl. Biochem J 11:1–11
33. Kruegel J, Miosge N (2010) Basement membrane components are key players in specialized extracellular matrices. Cell Mol Life Sci 67:2879–2895. https://doi.org/10.1007/s00018-010-0367-x
34. Fessler LI, Brosh S, Chapin S, Fessler JH (1986) Tyrosine sulfation in precursors in collagen V. J Biol Chem 261:5034–5040. https://doi.org/10.1016/s0021-9258(19)89211-6
35. Engvall E, Hessle H, Klier G (1986) Molecular assembly, secretion, and matrix deposition of type VI collagen. J Cell Biol 102:703–710. https://doi.org/10.1083/jcb.102.3.703
36. Bock HC, Michaeli P, Bode C et al (2001) The small proteoglycans decorin and biglycan in human articular cartilage of late-stage osteoarthritis. Osteoarthr Cartil 9:654–663. https://doi.org/10.1053/joca.2001.0420
37. Merline R, Schaefer RM, Schaefer L (2009) The matricellular functions of small leucine-rich proteoglycans (SLRPs). J Cell Commun Signal 3:323–335. https://doi.org/10.1007/s12079-009-0066-2
38. Wight TN, Merrilees MJ (2004) Proteoglycans in atherosclerosis and restenosis: key roles for versican. Circ Res 94:1158–1167. https://doi.org/10.1161/01.RES.0000126921.29919.51
39. Yanagishita M (1993) Function of proteoglycans in the extracellular matrix. Pathol Int 43:283–293. https://doi.org/10.1111/j.1440-1827.1993.tb02569.x
40. Monfort J, Tardif G, Reboul P et al (2006) Degradation of small leucine-rich repeat proteoglycans by matrix metalloprotease-13: identification of a new biglycan cleavage site. Arthritis Res Ther 8:1–9. https://doi.org/10.1186/ar1873
41. Funderburgh JL, Corpuz LM, Roth MR et al (1997) Mimecan, the 25-kDa corneal keratan sulfate proteoglycan, is a product of the gene producing osteoglycin. J Biol Chem 272:28089–28095.https://doi.org/10.1074/jbc.272.44.28089
42. Michelacci YM (2003) Collagens and proteoglycans of the corneal extracellular matrix. Brazilian J Med Biol Res 36:1037–1046. https://doi.org/10.1590/S0100-879X2003000800009
43. Brier J, Jayanti LD (2020) 主観的健康感を中心とした在宅高齢者における 健康関連指標に関する共分散構造分析 21:1–9
44. Gonzalez ACDO, Andrade ZDA, Costa TF, Medrado ARAP (2016) Wound healing—a literature review. An Bras Dermatol 91:614–620. https://doi.org/10.1590/abd1806-4841.20164741
45. Li J, Chen J, Kirsner R (2007) Pathophysiology of acute wound healing. Clin Dermatol 25:9–18. https://doi.org/10.1016/j.clindermatol.2006.09.007
46. Baum CL, Arpey CJ (2005) Normal cutaneous wound healing: clinical correlation with cellular and molecular events. Dermatologic Surg 31:674–686. https://doi.org/10.1097/00042728-200506000-00011

47. Pierce GF, Vande Berg J, Rudolph R et al (1991) Platelet-derived growth factor-BB and transforming growth factor beta1 selectively modulate glycosaminoglycans, collagen, and myofibroblasts in excisional wounds. Am J Pathol 138:629–646
48. Midwood KS, Williams LV, Schwarzbauer JE (2004) Tissue repair and the dynamics of the extracellular matrix. Int J Biochem Cell Biol 36:1031–1037. https://doi.org/10.1016/j.biocel.2003.12.003
49. Petreaca M, Martins-Green M (2020) The dynamics of cell-extracellular matrix interactions, with implications for tissue engineering. INC
50. Greenhalgh DG (1998) The role of apoptosis in wound healing. Int J Biochem Cell Biol 30:1019–1030. https://doi.org/10.1016/S1357-2725(98)00058-2
51. Codispoti B, Makeeva I, Sied J et al (2019) Should we reconsider the apoptosis as a strategic player in tissue regeneration? Int J Biol Sci 15:2029–2036. https://doi.org/10.7150/ijbs.36362
52. Guerin DJ, Kha CX, Tseng KAS (2021) From cell death to regeneration: rebuilding after injury. Front Cell Dev Biol 9:1–7. https://doi.org/10.3389/fcell.2021.655048
53. Tseng AS, Adams DS, Qiu D et al (2007) Apoptosis is required during early stages of tail regeneration in *Xenopus laevis*. Dev Biol 301:62–69. https://doi.org/10.1016/j.ydbio.2006.10.048
54. Fan Y, Wang S, Hernandez J et al (2014) Genetic models of apoptosis-induced proliferation decipher activation of JNK and identify a requirement of EGFR signaling for tissue regenerative responses in drosophila. PLoS Genet 10. https://doi.org/10.1371/journal.pgen.1004131
55. Chera S, Ghila L, Dobretz K et al (2009) Apoptotic cells provide an unexpected source of Wnt3 signaling to drive hydra head regeneration. Dev Cell 17:279–289. https://doi.org/10.1016/j.devcel.2009.07.014
56. Brown DL, Kao WWY, Greenhalgh DG (1997) Apoptosis down-regulates inflammation under the advancing epithelial wound edge: delayed patterns in diabetes and improvement with topical growth factors. Surgery 121:372–380. https://doi.org/10.1016/S0039-6060(97)90306-8
57. Dudas M, Wysocki A, Gelpi B, Tuan TL (2008) Memory encoded throughout our bodies: molecular and cellular basis of tissue regeneration. Pediatr Res 63:502–512. https://doi.org/10.1203/PDR.0b013e31816a7453
58. Reinke JM, Sorg H (2012) Wound repair and regeneration. Eur Surg Res 49:35–43. https://doi.org/10.1159/000339613
59. Sephel GC, Woodward S (2007) Repair, Regeneration, and fibrosis. Rubin's Pathol: 71–98
60. Goldenring JR, Mills JC (2022) Cellular plasticity, reprogramming, and regeneration: metaplasia in the stomach and beyond. Gastroenterology 162:415–430. https://doi.org/10.1053/j.gastro.2021.10.036
61. Carr NJ (2022) The pathology of healing and repair. Surg (United Kingdom) 40:13–19. https://doi.org/10.1016/j.mpsur.2021.11.003
62. Wlaschek M, Maity P, Makrantonaki E, Scharffetter-Kochanek K (2021) Connective tissue and fibroblast senescence in skin aging. J Invest Dermatol 141:985–992. https://doi.org/10.1016/j.jid.2020.11.010
63. Khan S, Hashmi GS (2015) Histology and functions of connective tissues. Univ J Dent Sci 1:1–2
64. Ariana R (2016) 済無
65. Journal H, Street RP, Brunswick N (1981) Connective tissue: an eclectic historical review with particular reference to the liver I. 396:341–396
66. Kamrani P, Jan A (2019) Anatomy, connective tissue blood supply and lymphatics. StatPearls Publ: 3–5
67. Nezwek TA, Varacallo M (2023) Physiology, connective tissue. 2–6
68. Schneider U, Murrell WD, Hollands P (2021) The regeneration of damaged connective tissue: wishful thinking or reality? CellR4: 1–11. https://doi.org/10.32113/cellr4
69. Forbes SJ, Rosenthal N (2014) Preparing the ground for tissue regeneration: from mechanism to therapy. Nat Med 20:857–869. https://doi.org/10.1038/nm.3653

70. Madhavan G (2003) Tissue and organ regeneration in adults
71. Rodrigues M, Kosaric N, Bonham CA, Gurtner GC (2019) Wound healing: a cellular perspective. Physiol Rev 99:665–706. https://doi.org/10.1152/physrev.00067.2017
72. Wilkinson HN, Hardman MJ (2020) Wound healing: cellular mechanisms and pathological outcomes: cellular mechanisms of wound repair. Open Biol 10. https://doi.org/10.1098/rsob.200223
73. Etc MCS, Das C, Lucia MS, HK, TJ (2019) 乳鼠心肌提取 HHS public access. Physiol Behav 176:139–148. https://doi.org/10.1016/j.addr.2018.09.010.Therapeutic
74. Takeo M, Lee W, Ito M (2015) Wound healing and skin regeneration. Cold Spring Harb Perspect Med 5:1–12. https://doi.org/10.1101/cshperspect.a023267
75. Wang PH, Huang BS, Horng HC et al (2018) Wound healing. J Chin Med Assoc 81:94–101. https://doi.org/10.1016/j.jcma.2017.11.002
76. Badiu D, Vasile M, Teren O (2011) Regulation of wound healing by growth factors and cytokines. Wound Heal Process Phases Promot: 73–93
77. Griffith CM, Wiley MJ, Sanders EJ (1992) The vertebrate tail bud: three germ layers from one tissue. Anat Embryol (Berl) 185:101–113. https://doi.org/10.1007/BF00185911
78. Sorg H, Tilkorn DJ, Hager S et al (2017) Skin wound healing: an update on the current knowledge and concepts. Eur Surg Res 58:81–94. https://doi.org/10.1159/000454919
79. Baron JM, Glatz M, Proksch E (2020) Optimal support of wound healing: new insights. Dermatology 236:593–600. https://doi.org/10.1159/000505291
80. Guillamat-Prats R (2021) The role of MSC in wound healing, scarring and regeneration. Cells 10. https://doi.org/10.3390/cells10071729
81. Martin P, Nunan R (2015) Cellular and molecular mechanisms of repair in acute and chronic wound healing. Br J Dermatol 173:370–378. https://doi.org/10.1111/bjd.13954
82. Werner S, Krieg T, Smola H (2007) Keratinocyte-fibroblast interactions in wound healing. J Invest Dermatol 127:998–1008. https://doi.org/10.1038/sj.jid.5700786
83. Tai K, Cockburn K, Greco V (2019) Flexibility sustains epithelial tissue homeostasis. Curr Opin Cell Biol 60:84–91. https://doi.org/10.1016/j.ceb.2019.04.009
84. Maclaren OJ, Byrne HM, Fletcher AG, Maini PK (2015) Models, measurement and inference in epithelial tissue dynamics. Math Biosci Eng 12:1321–1340. https://doi.org/10.3934/mbe.2015.12.1321
85. Blanpain C, Horsley V, Fuchs E (2007) Epithelial stem cells: turning over new leaves. Cell 128:445–458. https://doi.org/10.1016/j.cell.2007.01.014
86. Chen X, Reiter PL, McRee A-L (2011) 乳鼠心肌提取 HHS public access. Physiol Behav 176:139–148. https://doi.org/10.1016/j.cell.2021.05.036.Inflammatory
87. Debnath J, Brugge JS (2005) Modelling glandular epithelial cancers in three-dimensional cultures. Nat Rev Cancer 5:675–688. https://doi.org/10.1038/nrc1695
88. Horseman ND, Collier RJ (2014) Serotonin: a local regulator in the mammary gland epithelium. Annu Rev Anim Biosci 2:353–374. https://doi.org/10.1146/annurev-animal-022513-114227
89. Macias H, Hinck L (2012) Mammary gland development. Wiley Interdiscip Rev Dev Biol 1:533–557. https://doi.org/10.1002/wdev.35
90. O'Leary CE, Schneider C, Locksley RM (2019) Tuft cells-systemically dispersed sensory epithelia integrating immune and neural circuitry. Annu Rev Immunol 37:47–72. https://doi.org/10.1146/annurev-immunol-042718-041505
91. Alcolea MP, Jones PH (2014) Lineage analysis of epidermal stem cells. Cold Spring Harb Perspect Med 4:1–16. https://doi.org/10.1101/cshperspect.a015206
92. Shpichka A, Butnaru D, Bezrukov EA et al (2019) Skin tissue regeneration for burn injury. Stem Cell Res Ther 10:1–16. https://doi.org/10.1186/s13287-019-1203-3
93. Yang R, Wang J, Chen X et al (2020) Epidermal stem cells in wound healing and regeneration. Stem Cells Int 2020:9148310. https://doi.org/10.1155/2020/9148310
94. Kolios G, Moodley Y (2013) Introduction to stem cells and regenerative medicine. Respiration 85:3–10. https://doi.org/10.1159/000345615

95. Weatherbee BAT, Cui T, Zernicka-Goetz M (2021) Modeling human embryo development with embryonic and extra-embryonic stem cells. Dev Biol 474:91–99. https://doi.org/10.1016/j.ydbio.2020.12.010
96. Zakrzewski W, Dobrzyński M, Szymonowicz M, Rybak Z (2019) Stem cells: past, present, and future. Stem Cell Res Ther 10:68. https://doi.org/10.1186/s13287-019-1165-5
97. Blanpain C, Fuchs E (2014) Plasticity of epithelial stem cells in tissue regeneration. Science 344. https://doi.org/10.1126/science.1242281
98. Liu Y, Chen Y-G (2020) Intestinal epithelial plasticity and regeneration via cell dedifferentiation. Cell Regen (London, England) 9:14. https://doi.org/10.1186/s13619-020-00053-5
99. Lotfi L, Khakbiz M, Moosazadeh Moghaddam M, Bonakdar S (2019) A biomaterials approach to Schwann cell development in neural tissue engineering. J Biomed Mater Res Part A 107:2425–2446. https://doi.org/10.1002/jbm.a.36749
100. Lutzweiler G, Halili AN, Vrana NE (2020) The overview of porous, bioactive scaffolds as instructive biomaterials for tissue regeneration and their clinical translation. Pharmaceutics 12:1–29. https://doi.org/10.3390/pharmaceutics12070602
101. Alysandratos KD, Herriges MJ, Kotton DN (2021) Epithelial stem and progenitor cells in lung repair and regeneration. Annu Rev Physiol 83:529–550. https://doi.org/10.1146/annurev-physiol-041520-092904
102. Cotsarelis G (2006) Epithelial stem cells: a folliculocentric view. J Invest Dermatol 126:1459–1468. https://doi.org/10.1038/sj.jid.5700376
103. Houschyar KS, Borrelli MR, Tapking C et al (2020) Molecular mechanisms of hair growth and regeneration: current understanding and novel paradigms. Dermatology 236:271–280. https://doi.org/10.1159/000506155
104. Crosby LM, Waters CM (2010) Epithelial repair mechanisms in the lung. Am J Physiol Lung Cell Mol Physiol 298:1–19. https://doi.org/10.1152/ajplung.00361.2009
105. Heath DE (2019) A review of decellularized extracellular matrix biomaterials for regenerative engineering applications. Regen Eng Transl Med 5:155–166. https://doi.org/10.1007/s40883-018-0080-0
106. Kamimura J, Lee D, Baden HP et al (1997) Primary mouse keratinocyte cultures contain hair follicle progenitor cells with multiple differentiation potential. J Invest Dermatol 109:534–540. https://doi.org/10.1111/1523-1747.ep12336704
107. McKee CT, Last JA, Russell P, Murphy CJ (2011) Indentation versus tensile measurements of young's modulus for soft biological tissues. Tissue Eng Part B Rev 17:155–164. https://doi.org/10.1089/ten.teb.2010.0520
108. Rumpold H, Wolf D, Koeck R, Gunsilius E (2004) Endothelial progenitor cells: a source for therapeutic vasculogenesis? J Cell Mol Med 8:509–518. https://doi.org/10.1111/j.1582-4934.2004.tb00475.x
109. Baghdadi MB, Tajbakhsh S (2018) Regulation and phylogeny of skeletal muscle regeneration. Dev Biol 433:200–209. https://doi.org/10.1016/j.ydbio.2017.07.026
110. Dong R, Ma PX, Guo B (2020) Conductive biomaterials for muscle tissue engineering. Biomaterials 229:119584. https://doi.org/10.1016/j.biomaterials.2019.119584
111. Dumont NA, Bentzinger CF, Sincennes MC, Rudnicki MA (2015) Satellite cells and skeletal muscle regeneration. Compr Physiol 5:1027–1059. https://doi.org/10.1002/cphy.c140068
112. Zammit PS, Beauchamp JR (2001) Satellite_Stemcell_Zammit.Pdf: 193–204
113. Yang W, Hu P (2018) Skeletal muscle regeneration is modulated by inflammation. J Orthop Transl 13:25–32. https://doi.org/10.1016/j.jot.2018.01.002
114. Forcina L, Cosentino M, Musarò A (2020) Mechanisms regulating muscle regeneration: insights into the interrelated and time-dependent phases of tissue healing. Cells 9. https://doi.org/10.3390/cells9051297
115. Laumonier T, Menetrey J (2016) Muscle injuries and strategies for improving their repair. J Exp Orthop 3. https://doi.org/10.1186/s40634-016-0051-7
116. Fukada S, Akimoto T, Sotiropoulos A (2020) Role of damage and management in muscle hypertrophy: different behaviors of muscle stem cells in regeneration and hypertrophy.

Biochim Biophys Acta Mol Cell Res 1867:118742. https://doi.org/10.1016/j.bbamcr.2020.118742
117. Brown C, McKee C, Bakshi S et al (2019) Mesenchymal stem cells: cell therapy and regeneration potential. J Tissue Eng Regen Med 13:1738–1755. https://doi.org/10.1002/term.2914
118. Hashimoto H, Olson EN, Bassel-Duby R (2018) Therapeutic approaches for cardiac regeneration and repair. Nat Rev Cardiol 15:585–600. https://doi.org/10.1038/s41569-018-0036-6
119. Tenreiro MF, Louro AF, Alves PM, Serra M (2021) Next generation of heart regenerative therapies: progress and promise of cardiac tissue engineering. NPJ Regen Med 6. https://doi.org/10.1038/s41536-021-00140-4
120. Yi S, Zhang Y, Gu X et al (2020) Application of stem cells in peripheral nerve regeneration. Burn Trauma 8. https://doi.org/10.1093/burnst/tkaa002
121. Dalton PD, Harvey AR, Oudega M, Plant GW (2014) Tissue engineering of the nervous system. Tissue Eng Second Ed: 583–625. https://doi.org/10.1016/B978-0-12-420145-3.00017-1
122. Gordon T (2020) Peripheral nerve regeneration and muscle reinnervation. Int J Mol Sci 21:1–24. https://doi.org/10.3390/ijms21228652
123. Schmidt CE, Leach JB (2003) Neural tissue engineering: strategies for repair and regeneration. Annu Rev Biomed Eng 5:293–347. https://doi.org/10.1146/annurev.bioeng.5.011303.120731
124. Eltom A, Zhong G, Muhammad A (2019) Scaffold techniques and designs in tissue engineering functions and purposes: a review. Adv Mater Sci Eng 2019. https://doi.org/10.1155/2019/3429527
125. Roseti L, Parisi V, Petretta M et al (2017) Scaffolds for bone tissue engineering: state of the art and new perspectives. Mater Sci Eng C 78:1246–1262. https://doi.org/10.1016/j.msec.2017.05.017
126. Khalil HPSA, Jummaat F, Yahya EB et al (2020) A review on micro- to nanocellulose biopolymer scaffold forming for tissue engineering applications. Polymers (Basel) 12. https://doi.org/10.3390/polym12092043
127. Ambekar RS, Kandasubramanian B (2019) Progress in the advancement of porous biopolymer Scaffold: tissue engineering application. Ind Eng Chem Res 58:6163–6194. https://doi.org/10.1021/acs.iecr.8b05334
128. Xiong Y, Wang W, Gao R et al (2020) Fatigue behavior and osseointegration of porous Ti-6Al-4V scaffolds with dense core for dental application. Mater Des 195:108994. https://doi.org/10.1016/j.matdes.2020.108994
129. Bružauskaitė I, Bironaitė D, Bagdonas E, Bernotienė E (2016) Scaffolds and cells for tissue regeneration: different scaffold pore sizes—different cell effects. Cytotechnology 68:355–369. https://doi.org/10.1007/s10616-015-9895-4
130. Binette JS, Garon M, Savard P et al (2004) Tetrapolar measurement of electrical conductivity and thickness of articular cartilage. J Biomech Eng 126:475–484. https://doi.org/10.1115/1.1785805
131. Maganaris CN, Paul JP (1999) In vivo human tendon mechanical properties. J Physiol 521(Pt 1):307–313. https://doi.org/10.1111/j.1469-7793.1999.00307.x
132. Weber M, Rothschild MA, Niehoff A (2021) Anisotropic and age-dependent elastic material behavior of the human costal cartilage. Sci Rep 11:1–12. https://doi.org/10.1038/s41598-021-93176-x
133. Joines WT, Zhang Y, Li C, Jirtle RL (1994) The measured electrical properties of normal and malignant human tissues from 50 to 900 MHz. Med Phys 21:547–550. https://doi.org/10.1118/1.597312
134. Zieliński PS, Gudeti PKR, Rikmanspoel T, Włodarczyk-Biegun MK (2023) 3D printing of bio-instructive materials: toward directing thecell. Bioact Mater 19:292–327. https://doi.org/10.1016/j.bioactmat.2022.04.008
135. Hardarson T, Hanson C, Claesson M, Stenevi U (2004) Time-lapse recordings of human corneal epithelial healing. Acta Ophthalmol Scand 82:184–188. https://doi.org/10.1111/j.1600-0420.2004.00250.x

136. Tran RT, Thevenot P, Zhang Y et al (2010) Scaffold sheet design strategy for soft tissue engineering. Materials (Basel) 3:1375–1389. https://doi.org/10.3390/ma3021375
137. Lee JH, Yoon YC, Kim HS et al (2022) In vivo electrical conductivity measurement of muscle, cartilage, and peripheral nerve around knee joint using MR-electrical properties tomography. Sci Rep 12:1–10. https://doi.org/10.1038/s41598-021-03928-y
138. Spedden E, White JD, Naumova EN et al (2012) Elasticity maps of living neurons measured by combined fluorescence and atomic force microscopy. Biophys J 103:868–877. https://doi.org/10.1016/j.bpj.2012.08.005
139. Ren J, Jia L, Xu L et al (2009) Removal of autoantibodies by 4-mercaptoethylpyridine-based adsorbent. J Chromatogr B Anal Technol Biomed Life Sci 877:1200–1204. https://doi.org/10.1016/j.jchromb.2009.03.017
140. Cheng A, Schwartz Z, Kahn A et al (2019) Advances in porous scaffold design for bone and cartilage tissue engineering and regeneration. Tissue Eng Part B Rev 25:14–29. https://doi.org/10.1089/ten.teb.2018.0119
141. Collins MN, Ren G, Young K et al (2021) Scaffold fabrication technologies and structure/function properties in bone tissue engineering. Adv Funct Mater 31. https://doi.org/10.1002/adfm.202010609
142. Okada M, Matsumoto T (2016) Fabrication methods of hydroxyapatite nanocomposites. Nano Biomed 8:15–26. https://doi.org/10.11344/nano.8.15
143. Aras D (2015) 3.9.2015 1
144. Daskalakis E, Liu F, Cooper G, Weightman A (2021) Bio-materials and prototyping applications in medicine. Bio-Mater Prototyp Appl Med. https://doi.org/10.1007/978-3-030-35876-1
145. Ravarian R, Zhong X, Barbeck M et al (2013) Nanoscale chemical interaction enhances the physical properties of bioglass composites. ACS Nano 7:8469–8483. https://doi.org/10.1021/nn402157n
146. Farah S, Anderson DG, Langer R (2016) Physical and mechanical properties of PLA, and their functions in widespread applications—a comprehensive review. Adv Drug Deliv Rev 107:367–392. https://doi.org/10.1016/j.addr.2016.06.012
147. Souza IEP, Cambraia LV, Gomide VS, Nunes EHM (2022) Short review on the use of graphene as a biomaterial—prospects, and challenges in Brazil. J Mater Res Technol 19:2410–2430. https://doi.org/10.1016/j.jmrt.2022.05.170
148. Wei W, Qu X (2012) Extraordinary physical properties of functionalized graphene. Small 8:2138–2151. https://doi.org/10.1002/smll.201200104
149. Varghese S, Elisseeff JH (2006) Hydrogels for musculoskeletal tissue engineering. Adv Polym Sci 203:95–144. https://doi.org/10.1007/12_072
150. Martín C, Jun G, Schurhammer R et al (2019) Enzymatic degradation of graphene quantum dots by human peroxidases. Small 15. https://doi.org/10.1002/smll.201905405
151. Beri P, Matte BF, Fattet L et al (2018) Biomaterials to model and measure epithelial cancers. Nat Rev Mater 3:418–430. https://doi.org/10.1038/s41578-018-0051-6
152. Hamilton TD, Bučar DK, Baltrusaitis J et al (2011) Thixotropic hydrogel derived from a product of an organic solid-state synthesis: properties and densities of metal-organic nanoparticles. J Am Chem Soc 133:3365–3371. https://doi.org/10.1021/ja106095w
153. Vetrík M, Přádný M, Hrubý M, Michálek J (2011) Hydrazone-based hydrogel hydrolytically degradable in acidic environment. Polym Degrad Stab 96:756–759. https://doi.org/10.1016/j.polymdegradstab.2011.02.020
154. Gilpin A, Yang Y (2017) Decellularization strategies for regenerative medicine: from processing techniques to applications. Biomed Res Int 2017. https://doi.org/10.1155/2017/9831534
155. Choudhury D, Yee M, Sheng ZLJ et al (2020) Decellularization systems and devices: state-of-the-art. Acta Biomater 115:51–59. https://doi.org/10.1016/j.actbio.2020.07.060
156. Yi S, Ding F, Gong L, Gu X (2017) Extracellular matrix scaffolds for tissue engineering and regenerative medicine. Curr Stem Cell Res Ther 12:233–246. https://doi.org/10.2174/1574888x11666160905092513

157. Zhang X, Chen X, Hong H et al (2022) Decellularized extracellular matrix scaffolds: recent trends and emerging strategies in tissue engineering. Bioact Mater 10:15–31. https://doi.org/10.1016/j.bioactmat.2021.09.014
158. Lima EO, Ferrasi AC, Kaasi A (2019) Decellularization of human pericardium with potential application in regenerative medicine. Arq Bras Cardiol 113:18–19. https://doi.org/10.5935/abc.20190130
159. Kasravi M, Ahmadi A, Babajani A et al (2023) Immunogenicity of decellularized extracellular matrix scaffolds: a bottleneck in tissue engineering and regenerative medicine. Biomater Res 27:1–24. https://doi.org/10.1186/s40824-023-00348-z
160. Sitthisang S, Leong MF, Chian KS (2021) Perfusion decellularization of porcine esophagus: study of two processing factors affecting the folded mucosal structure of the esophageal scaffold. J Biomed Mater Res Part A 109:745–753. https://doi.org/10.1002/jbm.a.37060
161. Daskalakis E, Liu F, Cooper G, Weightman A, Koç B, Blunn G, Bártolo PJ (2021) Bioglasses for bone tissue engineering. In: Bio-materials and prototyping applications in medicine, pp 165–193. https://doi.org/10.1007/978-3-030-35876-1_9
162. Amani H, Kazerooni H, Hassanpoor H et al (2019) Tailoring synthetic polymeric biomaterials towards nerve tissue engineering: a review. Artif Cells Nanomed Biotechnol 47:3524–3539. https://doi.org/10.1080/21691401.2019.1639723

Chapter 3
Cell Sources for Tissue Engineering

Ayse Ceren Calikoglu-Koyuncu, Gozde Enguven, and Rumeysa Koyuncuoglu

Abstract One of the most essential elements of tissue engineering is the type of cell that is going to be used to regenerate a particular tissue. For this purpose, usually stem cells are preferred due to their remarkable features. They could be obtained from numerous tissues like bone marrow, adipose, and dental tissues. This chapter discusses cells utilized in general tissue engineering applications in terms of source, donor, potency, and engineered tissue type. Detailed classification of stem cells is reviewed, involving some basic methodology of stem cell isolation, characterization, and differentiation regarding the engineered tissue.

Keywords Tissue engineering · Cell culture · Cell sources · Stem cells · Regeneration

Introduction

In most tissue engineering studies, cells are essential to fully regenerate damaged or lost tissues and organs. Cells are categorized as *stem cells* or *differentiated cells* according to differentiation potential. They can be taken from autologous, allogenic, or xenogeneic sources. When a person's own cells are used for an engineering application it is called an autogenic source, whereas allogeneic sources indicate the transfer from another human. Animal sources other than humans are called xenogeneic. Xenogenic cells are currently considered an inconvenient source for cell transplantation because of the risk of zoonotic infection. Allogeneic transplants can be taken

A. C. Calikoglu-Koyuncu (✉) · G. Enguven · R. Koyuncuoglu
Center for Nanotechnology & Biomaterials Applications and Research (NBUAM), Marmara University, Istanbul, Turkey
e-mail: aceren@marmara.edu.tr

A. C. Calikoglu-Koyuncu
Faculty of Technology, Metallurgical and Materials Engineering Department, Marmara University, Istanbul, Turkey

Health Biotechnology Joint Research and Application Center of Excellence, Istanbul, Turkey

O. Gunduz et al. (eds.), *Biomaterials and Tissue Engineering*, Stem Cell Biology and Regenerative Medicine 74, https://doi.org/10.1007/978-3-031-35832-6_3

from brothers and sisters, parents, or unrelated donors. In the case of autogenic cell usage, the patient has no risk of developing an immune response because his own cells are used. However, there is the problem of generating morbidity in the healthy donor tissues because, in order to engineer tissues, we need a large number of cells, meaning that a significant amount of tissue should be taken from the healthy tissue. In allogenic and xenogeneic transplantation, there is also the risk of donor site morbidity, in addition to the possibilities of disease transfer and a severe immune response, which could result in the rejection of the transplant by the donor. Usually, the patient is supplied with antibiotics and immune suppressors after the organ or tissue transplant surgery to prevent infection and tissue incompatibility. Considering all these risk, it is safer to use autogenic cells. With the usage of stem cells as the autogenic cell source, the problem of tissue damage during cell collection from healthy tissues from the donor can be overcome more easily.

Cell Culture

Cells to be used in tissue regeneration could be taken directly from the relevant tissue by enzymatic or mechanical methods. After collecting the cells from their source, they must be grown and maintained in a laboratory under sterile conditions. This process is called *primary cell culture*. The two basic methods to obtain a primary culture are *explant* and *enzymatic isolation* [1]. In an explant culture, cell clusters are extracted from the tissues via mechanical ways, and then grown on glass or plastic plates. Few days later, cells in the tissue explant begin to migrate from the explant to the surface of the culture plate. Enzymatic extraction of cells involves the addition of proteolytic enzymes like trypsin or collagenase. These enzymes break down the chemical interactions between cells and their extracellular matrices, yielding single-cell suspension instead of tissue or cell clusters. The cell suspension should subsequently be spread on a culture plate so that individual cells can adhere and then the tissue debris can be discarded.

In order to cultivate the cells, chemically defined media containing essential micronutrients (amino acids, polysaccharides, etc.) have been produced. These nutrient media simulate the native tissue environment in vitro. Because different cell types will have different requirements, different media are required for each tissue engineering application. Replenishing the nutrient culture media at regular intervals provides cell proliferation and survival. After a couple of days of culturing, the cell number reaches a point where the nutrients in culture media are consumed. At this point, they need to be transferred to new larger culture dishes containing fresh media. This practice is called passaging or subculturing which is carried on until the proper amount of cells for application is achieved. The excess amount of cells can be further cultured or maintained in liquid nitrogen tanks for long-term storage. Primary cells can undergo limited number of cell divisions (approximately 40–50) because the telomere regions of chromosomes are shortened at each division, leading to cellular aging. The phenomenon of primary cells having a limited lifespan is called

Hayflick's limit [2]. After a particular number of passages, primary cells start to lose their phenotype and eventually die or become cancer cells. Though, some primary cells that have active telomerase can proliferate indefinitely, as the telomerase enzyme repairs the shortened telomeres of chromosomes. Some types of stem cells and many cancer cells have the intrinsic capability of unlimited cell division. Such immortal cells are called *cell lines*—homogeneous population of a particular cell. The best well-known example of a cell line is the cervical cancer line *HeLa*. These cells were derived from a female patient suffering from cervical cancer in 1951. The cell line was named "HeLa" after the death of Henrietta Lacks and is still being widely used in cancer research. Continuous cell lines can also be generated by transforming finite primary cells with viruses in vitro such that they do not undergo cellular aging. Both primary cells and cell lines can be produced by individual research laboratories with adequate cell culture facilities, or they can be purchased from cell banks, for instance American Type Culture Collection (ATCC)—a USA based company that holds the largest collection of cells (http://www.atcc.org) (Table 3.1).

Cells in culture can be either *adherent* or *suspension* according to their anchorage dependency in culture [5]. If the cells require a substrate to attach in order to survive, they are called adherent. Most of the mammalian somatic cells attach to the culture dish as monolayers. On the other hand, the cells that can grow as free-floating, such as blood cells, are suspension cells. Another classification is based on cell morphology, which divides adherent cells into categories as fibroblastic, epithelial, endothelial, and neuronal. In general, there are basically five groups of cells based on their shapes (see Table 3.2).

Table 3.1 Some of the most commonly used cell lines in mammalian tissue engineering research [3, 4]

Name	Origin
3T3	Mouse fibroblasts
A549	Human lung cancer cells
Caco-2	Human colorectal cancer epithelial cells
CHO	Chinese hamster ovary cells
COS	Monkey kidney cells
Cos-7	African green monkey kidney cells
HaCaT	Human keratinocytes
HeLa	Human cervical cancer epithelial cells
HEK-293	Human embryonic kidney cells
HUVEC	Human umbilical vein endothelial cells
L929	Murine fibroblasts
MCF-7	Human breast cancer cells
PC-12	Rat adrenal medullary tumor cells
Vero	African green monkey kidney epithelial cells

Table 3.2 Classification of in vitro cell morphology

Morphology	Shape	Attachment dependency
Fibroblastic	Bipolar, multipolar	Adherent
Epithelial	Polygonal	Adherent
Endothelial	Squamous	Adherent
Neuronal	Round, pyramidal, spindle-shaped	Adherent
Lymphoblastoid	Floating sphere	Non-adherent (suspension)

Stem Cells

Unspecialized cells which can develop into many cell types in the adult body, beginning from embryonic development, are called stem cells. Each cell in the body matures from a stem cell during development, which is known as the zygote. Fertilization of an egg cell with a sperm cell starts a cascade of cell division and reprogramming events. The zygote makes copies of itself through mitosis, forming a colony of identical cells. As the cell division continues, an 8-cell embryo is formed within 72 h. After 5–6 days, the embryo reaches the blastocyst stage, which contains hundreds of cells [6, 7]. Blastocyst formation is the first step of cell differentiation resulting in two different embryonic stem cell populations: the trophoblast and the inner cell mass (ICM, or embryoblast). The trophoblast forms the extraembryonic tissues like the placenta while all embryonic tissues are derived from the ICM [8].

Stem cells differ from adult somatic cells by two unique features. The first one is the capacity of unlimited cell division and self-renewal, meaning that the undifferentiated cells can proliferate to maintain the existing cell number in the pool of stem cells. The ability to divide limitlessly is determined by the small DNA chains known as "telomeres" located at the end of the chromosomes, and the enzyme called telomerase. Each time a stem cell divides, the telomeres are reconstituted by telomerase which is found in most of stem cells, but not in the terminally differentiated ones. Therefore, the more active the telomere enzyme is in a cell, the longer the telomeres remain. Stem cells owe their unlimited ability to divide to the intense activity of telomerases. There are two different ways of division in stem cells [9], which are greatly influenced by the cell's microenvironment:

1. **Symmetric division** Daughter cells, two cells having the identical characteristics of the parent cell, are formed as a result of symmetrical cell division. Thus, cells can proliferate without any genetic modification.
2. **Asymmetric division** One of the cells formed as a result of cell division remains as a backup for self-renewal, while the other one differentiates to form a progenitor cell for the following differentiation processes.

The second exclusive feature of a stem cell is the capability to specialize into different cells in the body. A stem cell can undergo a set of biochemical and phenotypic events to transform into a different type of cell, either through adaptation or the cell's own epigenetic program. The most important factors affecting the cell's fate of differentiation are growth factors which can be used to control gene expression patterns. Stem cells can differentiate into desired cell type when properly stimulated by different growth factors, repairing the tissue damage caused by diseases and injuries. This process is also called *directed differentiation.*

Stem cell plasticity defines the flexibility of a stem cell's phenotype, in other words its potential to differentiate. Stem cells can show totipotent, pluripotent, multipotent and unipotent properties according to their ability to transform into different cell lines. The potency of differentiation decreases from totipotent to unipotent, respectively.

- **Totipotent stem cells** have unlimited differentiation capacity, i.e. they can form any type of specialized cell. Zygote is the most well-known example of a totipotent stem cell, which has the total potency to form a whole organism.
- **Pluripotent stem cells** are able to differentiate into the cells of embryonic germ layers, namely endoderm, mesoderm, and ectoderm, e.g. the ICM of the blastocyst.
- **Multipotent stem cells** have limited differentiation capacity. Most of the adult stem cells show multipotency, with capacity of differentiation into similar cell lineages, e.g. bone marrow derived stem cells.
- **Unipotent stem cells** can only differentiate into one particular cell type. For instance, epidermal stem cells in the skin regenerate the somatic skin cells that are regularly being shed from the outermost skin layer and those that are lost when the skin is damaged. These cells also undergo self-renewal to maintain their population, which distinguishes them from terminally differentiated cells.

Stem cells can be investigated under two main categories with respect to plasticity: embryonic and adult stem cells.

Embryonic Stem Cells

During development, the zygote differentiates into every cell type in the body, including cells of extra-embryonic tissue. For research, embryonic stem cells (ESCs) are generally extracted from in vitro fertilized (IVF) embryos which means that the egg cell is fertilized outside of the body under cell culture conditions. The IVF cell is a totipotent stem cell, like zygote. ESC research began in 1964 with a study involving "embryonal carcinoma cells" [10]. They have discovered that these cells were embryonic cancer stem cells with multipotency of differentiation. The study led many scientists to eagerly study on ESCs in the following years [11]. In 1981, Evans and Kaufman published the first report on in vitro culturing of pluripotent ESCs from the ICM of mouse blastocysts [12]. The ICM is a pool of pluripotent ESCs. Human ESC colonies can be obtained by collecting the ICM from IVF embryos and then culturing them on embryonic mouse fibroblasts which are also known as the *feeder layer cells* [13]. They can also be incubated in a medium containing special growth factors, without using feeder layer cells [14]. The ESC colonies growing on fibroblasts are then transferred to larger cell culture plates every 2–3 days through subculturing to increase the number of viable cells. Like any type of cell, ESCs can be frozen for future use.

The pluripotent stem cells can produce 3D clusters in the culture dish, called embryoid bodies that have the ability differentiate into various cell types in vitro. They can form microtissues by exchanging signals among themselves, which can mimic many different tissues in the body (see Fig. 3.1). They are generally preferred in the pharmaceutical industry and clinical applications because of their ease of reproduction and ability to form microtissues.

Some pluripotent and multipotent stem cells can be extracted from the placenta, amniotic membrane or amniotic fluid during pregnancy or within the first few minutes of birth. Stem cells isolated from umbilical cord have gained attention of stem cell researchers in recent years, because of their pluripotency. There are also multipotent mesenchymal stem cells (MSCs) in the umbilical cord which develop from non-embryonic mesodermal tissue at 12–13 days of embryonic development. Within the first 10 min of birth, blood in the cord attached to the placenta is transferred into a sterile blood bag coated with anticoagulants. Then, the blood must be delivered to the cord blood bank in 24–48 h to preserve cell viability.

ESCs seem like a useful cell source for tissue engineering due to their unique properties, however, there are serious problems and difficulties in ESC selection and purification techniques, in addition to ethical concerns. Besides, there are many studies showing the possibility of teratoma formation after implantation, also known

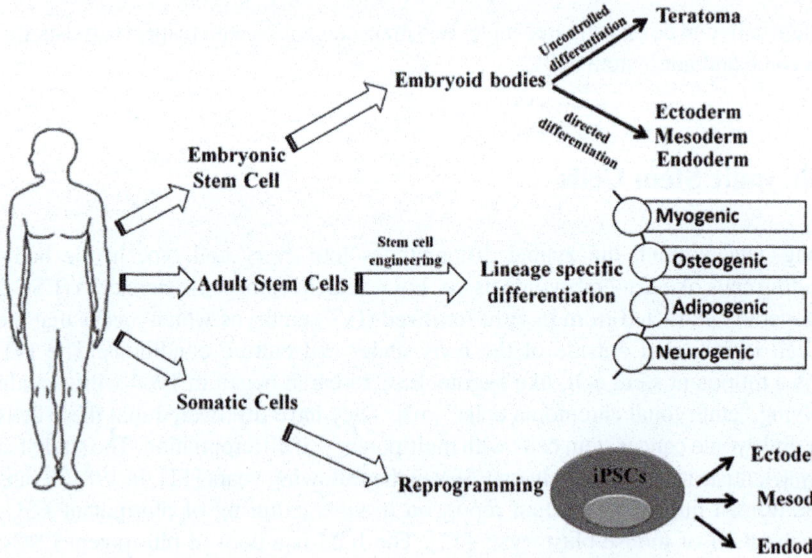

Fig. 3.1 Differentiation of ESCs, adult stem cells, and differentiated somatic cells [15]

as the tumor composed of cells from multiple germ layers. Therefore, the use of ESCs in tissue engineering is practically limited. On the other hand, differentiated cells in the adult body have lost their stemness properties; therefore, it is less likely for these cells to form tumors or to specialize into irrelevant types of cells upon administration, yet, the use of autologous differentiated cells can be disadvantageous due to their scarcity in the body. Accordingly, the most suitable cells for tissue engineering applications are suggested to be autologous adult stem cells.

Adult Stem Cells

Adult stem cells have been attractive candidates for tissue regeneration for many decades. They are naturally found in an organism that has finished its early embryonic development. Their ability to differentiate is rather limited than ESCs (see Fig. 3.1). Each adult stem cell resides in its specific microenvironment, or stem-cell niche, which defines specific locations in the body that house stem cells and regulate their fates [16]. When an adult tissue is damaged, adult stem cells can mature into tissue-specific cells to repair that particular tissue to assure its integrity.

Mesenchymal Stem Cells

The most promising and commonly used adult stem cells for tissue regeneration are the MSCs which can be extracted from many tissues like the bone marrow, adipose tissues, muscle, synovial membranes, umbilical cord, dental tissues, and even some terminally differentiated adult tissues like cartilage (see Fig. 3.2). They are pluripotent-multipotent stem cells that are able to detect and repair tissue injuries thanks to their immunoregulatory and immunosuppressive properties. Their ability to target tumors and metastatic sites make it possible to use MSCs as anti-tumor agents [17].

These MSCs, originating from the mesenchyme during development, can specialize into connective tissue cells (e.g. bone, cartilage, tendon) as well as into muscle, and nerve tissues. However, the proliferation and differentiation potency of MSCs cannot be easily ranked among the origins because cell characteristics differ from individual to individual. Besides, the difficulties in isolation vary among the tissues. For example, isolation of bone marrow is a very invasive process that involves a major surgery whereas cells can be easily collected from dental tissues and umbilical cord, by minimally invasive or even non-invasive protocols. Isolation procedures for MSCs are basically the same: first, the tissues are extracted, and then cells are isolated using specific enzymes that break down the extracellular components of the tissues. After that, the MSCs are selected among the cell population by using cell sieves or cell sorters or simply by using specific cell culture media that are defined for MSC isolation. Because MSCs are attachment-dependent cells

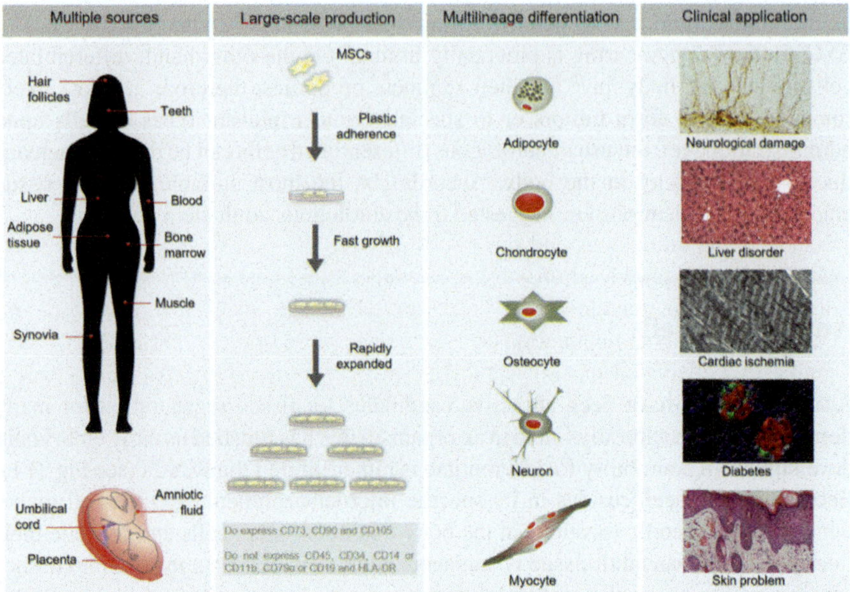

Fig. 3.2 Mesenchymal stem cells: sources, properties, differentiation potential, and application (Adapted from [18])

they can easily be expanded in the culture through standard subculturing, although their differentiation potential tends to decrease with increasing passage number. To validate the mesenchymal stemness of primary MSCs, the expression of certain cell surface antigens should be analyzed. CD90, CD73, and CD44 are among the most common MSC markers to test [19].

Bone Marrow The main source of MSCs and hematopoietic stem cells in an adult organism is the bone marrow. Especially bone marrow-derived MSCs have long been applied into numerous tissue engineering studies. Bone marrow allows the body to produce blood cells on a continuous basis for a lifetime. Stem cells in bone marrow play an extremely common and significant role in clinical use, especially in regeneration of bone and cartilage. The method of collecting bone marrow stem cells involves extraction of the bone marrow from the pelvic bone with special syringes under general anesthesia in an operation room [20]. Then, centrifugation is applied to the extract to separate the cells from tissue debris. Subsequently, the pellet which contains MSCs is spread on a culture plate containing a specific growth medium.

Umbilical Cord Recently, the umbilical cord has been used as another pool of MSCs; even the gene expression profile of umbilical cord-derived MSCs (UC-MSCs) is more similar to that of ESCs when compared to bone marrow-derived MSCs [21]. MSCs isolated from the umbilical cord are quite young relative to other adult stem cells; therefore, cellular aging might be stopped when properly stored. Under suitable conditions, UC-MSCs can differentiate into the cells of many tissues such as bone, cartilage, muscle, and nerve, similar to other MSC types. Their reproduction and colony formation rates are shown to be higher than the other sources of MSCs. In addition, the success rate of allogeneic tissue transplantation is high, although there is no complete matching of tissues between the donor and the recipient. The only negative side of using UC-MSCs is that you cannot obtain large amounts of tissues like you can get from bone marrow for instance. Other than that, the umbilical cord is a significantly promising source of cells in regenerative medicine.

Adipose Tissues There are two types of multipotent cells in adipose (or fat) tissues: white adipocytes and brown adipocytes. White adipocytes originate from MSCs whereas brown adipocytes are obtained from paraxial mesoderm. A gram of adipose tissue contains about 5000 stem cells with therapeutic properties [22], i.e. mesenchymal and hematopoietic stem cells. Similar to dental stem cells, MSCs can easily be isolated from discarded adipose tissue as a product of liposuction, a method to remove fat from areas where fat tissue is high, such as the abdomen, legs and hips. The properties of adipose-derived MSCs and bone marrow-derived MSCs are similar in terms of differentiation potential. They can easily be converted into bone and cartilage cells, and even neurons [23].

Dental Tissues Likewise, dental stem cells (DSCs) have gained attention in tissue engineering studies in recent decades because their isolation procedure is much less invasive and easier when compared to other sources. For instance, isolation of stem cells from bone marrow and adipose tissue require invasive surgical interventions. The proliferation and differentiation capacity of DSCs was found to be high and multiple like bone marrow-derived MSCs. Considering the umbilical cord, shed childhood teeth and third molars are already waste materials, it is favorable to evaluate the stem cells inside those tissues. Stem cells, mostly MSCs, can be extracted from many dental tissues [24].

- ***Dental pulp stem cells (DPSCs)*** These are multipotent MSCs with high proliferation rate and high plasticity. They can be obtained from the wisdom teeth, frequently. DPSCs can be extracted from the dental pulps of the deciduous incisors, which can differentiate into neurons, adipocytes, osteoblasts, and odontoblasts.
- ***Periodontal ligament stem cells (PDLSCs)*** They can be isolated from the surface of the roots of the extracted teeth and can acquire adipogenic, osteogenic, and chondrogenic phenotypes in culture. PDLSCs have been reported to form colonies but have a low in vitro osteogenic differentiation potential, but they provide tissue regeneration and periodontal repair when transplanted in mice.
- ***Stem cells derived from the apical papilla (SCAP)*** Obtained from the dental papillae of the impacted teeth or incisors at an early stage of dental development, SCAP can form more dentin than DPSCs. The number of stem cells in the apical papilla is higher than mature pulp, and when used together with periodontal ligament stem cells, they provide connective tissue formation. They can differentiate into osteogenic, odontogenic, neurogenic, adipogenic, chondrogenic, and hepatogenic cell types.

- ***Dental follicle cells*** Dental follicle, which can be easily isolated from third molar teeth, has progenitor cells that can mature into osteoblasts and cementoblasts, as well as chondrocytes and adipocytes.

Hematopoietic Stem Cells

Another adult stem cell population includes hematopoietic stem cells (HSCs) which have the capability to differentiate into blood cells, forming the blood tissue. Their primary functions are to transport oxygen, to maintain immune system function and to control bleeding [25]. HSCs can be obtained from bone marrow, peripheral blood, umbilical cord blood, and fetal liver, but the major source is bone marrow. Their applications are limited to the treatment of blood cancers, blood cell disorders, and autoimmune diseases, as well as liver tissue regeneration. They are divided into two classes as long-term HSCs and short-term HSCs. Those that proliferate within 12–16 weeks after transplantation are called long-lasting HSCs. Those that can give rise to all cells in blood serum after transplantation but have a limited life span are called short-lived HSCs. These cells are involved in the daily renewal of blood cells [26].

Induced Pluripotent Stem Cells

Induced pluripotent stem cells (iPSCs) have recently become popular in tissue engineering because they show similar characteristics to ESCs in terms of plasticity. The first iPS cells were produced in a study in 2006 in which mouse skin fibroblasts were converted into pluripotent stem cells as a result of induction with four transcription factors, namely Oct3/4, Sox2, c-Myc, and Klf4 [27]. Now, this group of four transcription factors are being called Yamanaka factors after Shinya Yamanaka. The findings of the study brought the scientist the Nobel Prize in Physiology or Medicine in the year of 2012. The prize was shared with another scientist named John Gurdon who discovered about 60 years ago that cellular reprogramming was possible [28]. Yamanaka factors are normally expressed in ESCs involved in mammalian development, and since that study in 2006, they are being used to generate pluripotent stem cells. IPSCs are capable of differentiation into cells of any origin without posing a teratoma formation risk like embryonic stem cells, as they are reprogrammed and differentiated simultaneously in culture. The technology of generating iPSCs from somatic cells provides that the resulting stem cells will match the immune system of the patient because they can be obtained from the patient himself (see Fig. 3.3). Consequently, an engineered tissue composed of autologous iPSCs would provide a reduced immune response. Techniques to stably produce and preserve iPSCs are still being established [29].

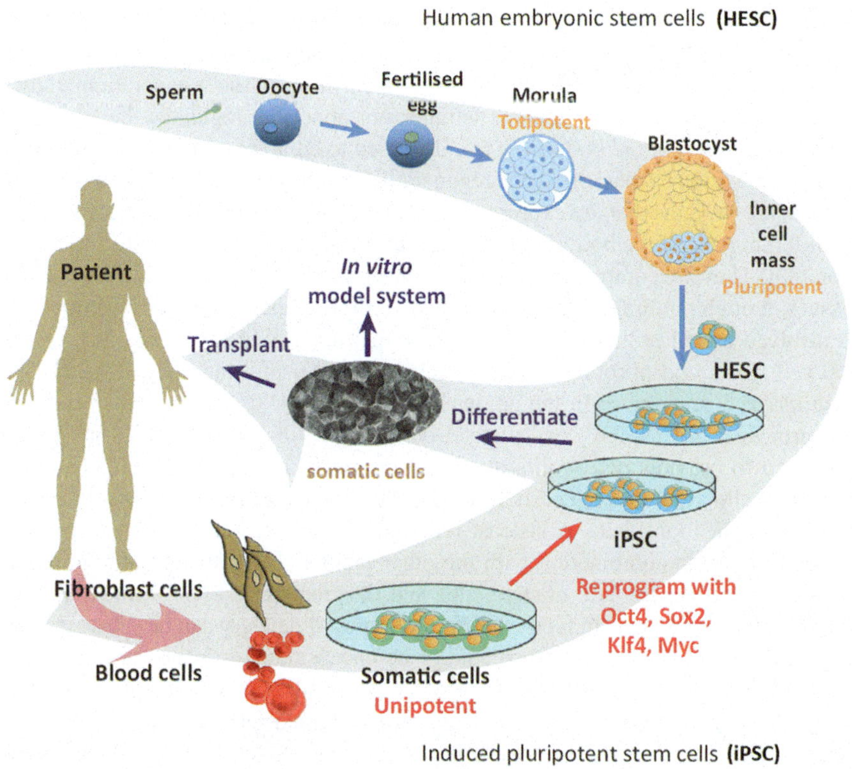

Fig. 3.3 Production and application of ESCs and iPSCs in vitro (Adapted from [30]) (licence number: 5537081036676). Fibroblasts and blood cells were given as examples

Differentiation of Adult Stem Cells

Isolation of cells from some tissues requires difficult extraction and expansion processes, and isolated cells can easily lose their specific phenotype in vitro; in other words, they could stop the expression of specific tissue markers. In order to promote re-differentiation, these cells should be cultured in three-dimensional scaffolds, or in an environment that includes growth and differentiation stimulating factors. In typical tissue engineering studies, various differentiation pathways can be activated by using particular stimulating factors. It is essential to maintain the tissue phenotype during the regeneration process; therefore, these molecular factors should be studied comprehensively according to the specific tissue engineering applications involving stem cells.

The transforming growth factor-beta (TGF-β) family of hormones are one of the most frequently investigated class of growth factors in engineering of tissues. Among the 20 members of TGF- β family, TGF-β1 is the most commonly used one that promotes the synthesis of muscle and connective tissue-specific proteins involved in extracellular matrix synthesis [31]. It can also activate cell proliferation, chemotaxis, and inflammatory cell recruitment, which are among the key steps in tissue healing. TGF-β3 is specifically used in cartilage tissue regeneration studies. Insulin-like growth factor (IGF) is another hormone that regulates the growth of numerous tissues in the body [32]. So, it is used to enhance cell growth and proliferation in various types of tissue regeneration studies. Fibroblast growth factors (FGFs) are proteins that are expressed by macrophages, and they have various roles in different parts of the body, including cell proliferation, differentiation, and migration [33]. They are commonly used for engineering blood vessels, skin, and nervous tissues, though not limited to these. Similar to FGF, epidermal growth factor (EGF) is involved in cell proliferation and survival, particularly of myoctyes [34], epithelial cells [35], and fibroblasts [36], which makes it useful mainly for muscle and skin tissue regeneration. In the engineering of vascular tissues, vascular endothelial growth factor (VEGF) and platelet-derived growth factor (PDGF) are mostly preferred to promote cell proliferation and migration as they are the regulators of wound healing and angiogenesis [36, 37]. All in all, for almost every tissue engineering application, there is at least one suitable growth factor that can be used either alone or in combination with the other factors. The selection of stimulatory factors determines the fate of stem cells and thus the tissue to be formed. To alter the effectiveness of growth factors, different routes of delivery can be selected. They can be directly added to culture media or be encapsulated in nano- or microparticles to obtain a more controlled release. Cells can also be genetically engineered to express a particular bioactive molecule for sustained release. Some tissues also require mechanical stimulation as well as molecular induction. For instance, cartilage regeneration can be improved by mechanical forces like compression, tension, and shear which can be applied by bioreactors, because the maintenance of cartilage tissue integrity depends on such mechanical forces [38].

A recently introduced cell culture technique called co-culture allows communication between different cells through molecular signals that promote differentiation [39]. It involves simultaneous culturing of different cell types in the same culture dish using various techniques (see Fig. 3.4). Because the technique provides reciprocal use of growth factors released naturally from the cells to culture media, it is favorable in engineering of complex tissues. A recent study in 2022 demonstrated that co-culturing of human vascular endothelial cells, fibroblasts, and adipose-derived stem cells (ADSCs) could promote osteogenic differentiation of ADSCs [40]. The differentiation of ADSCs into osteoblasts was induced by secretion of BMP-2, VEGF, and FGF from fibroblasts and vascular endothelial cells, eliminating the need for additional growth factors in the culture media.

Type	Diagram	Advantages	Disadvantages
Classical co-culture	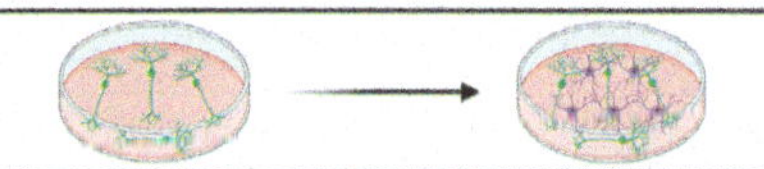	Easy to set up, easy to manipulate cell concentrations	Cannot differentiate between contact-dependent and independent interactions, media needs to sustain all cell types
Multi-layer co-culture	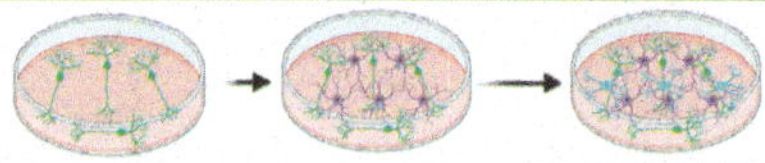	Easy to set up, affordable, more cell types than classic, so closer to *in vivo*	
Sandwich co-culture	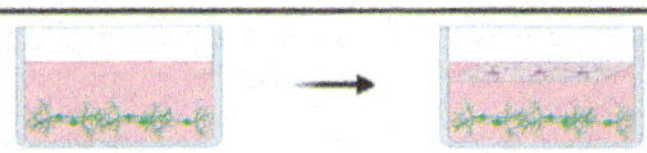	Can separate cell types for RNA/protein characterization after co-culture	Sandwich: more complicated set-up
Conditioned Media Transfer	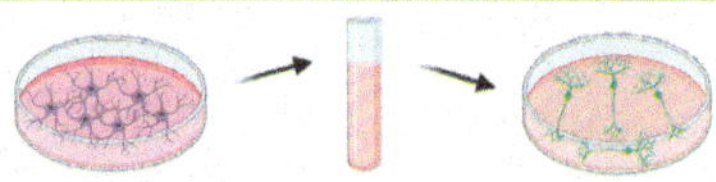	Easy to set up, well established, shows effects of soluble factors, unidirectional	Can only detect contact-independent interactions
Transwell Insert	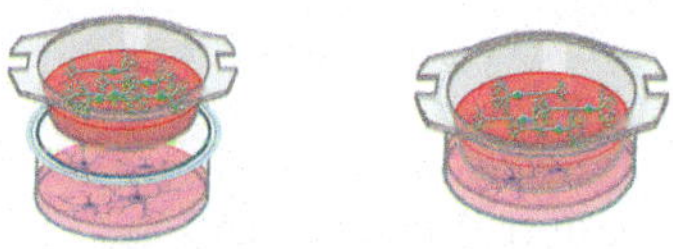	Easy to set up, can separate cell types for RNA/protein characterization	Expensive, usually not re-usable
Micro-fluidics chamber	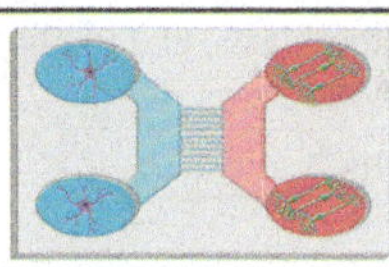	Can manipulate degree of contact of cell types, less cells and reagent needed, possible real time analysis	Complicated to set up, expensive

Fig. 3.4 Different co-culturing techniques, including the conventional and novel methods [39] (licence number: 5537081388754)

Tissue-Specific Differentiation

For some tissue engineering applications, differentiated cells from the healthy part of the particular tissue can be used. Because the cells are taken from the differentiated tissue, directed differentiation is unnecessary, unlike using the stem cells. However, since the process creates injury to the healthy tissue, the number of cells available for extraction from the healthy tissue is limited. For tissues like articular cartilage where the accessible tissue amount and inherent tissue regeneration are quite low, using autologous somatic cells from healthy part of the tissue can damage it permanently. In order to overcome these problems, pluripotent or multipotent adult stem cells can be used. By using certain growth and differentiation factors and cytokines, stem cells can be directed to differentiate towards a desired phenotype. Yet, most of the differentiation pathways are lineage specific, i.e. MSCs can differentiate into cells of mesodermal lineage. In tissue engineering, the selection of cell source primarily depends on tissue origin (see Table 3.3). Theoretically, only iPSCs can be used for all types of tissue engineering because they mainly possess embryonic stem cell features, such as the capacity of multi-lineage differentiation. Sometimes terminally differentiated cells could be activated to directly differentiate to another phenotype through a

Table 3.3 Potential cell sources for particular tissue engineering applications

Engineered tissue	Cell sources	References
Blood vessel	Endothelial cells (especially HUVEC), smooth muscle cells, fibroblasts	[42]
Bone	Osteoblasts, MSCs	[43]
Cornea	Corneal epithelial cells, corneal endothelial cells, fibroblasts, adipose-derived MSCs	[44]
Cartilage	Chondrocytes, fibroblasts, MSCs	[45]
Heart	Cardiomyocytes, cardiac stem cells, myoblasts, fibroblasts, MSCs	[46]
Intestine	Enterocytes, intestinal stem cells, MSCs	[47, 48]
Liver	Hepatocytes, hematopoietic stem cells, MSCs	[49]
Muscle	Myoblasts, fibroblasts, smooth muscle cells, muscle cell precursors, MSCs	[50]
Nerve	Neuroblasts, oligodendrocytes, astrocytes, microglial cells, Schwann cells, neural stem cells, adipose-derived MSCs	[51, 52]
Skin	Epidermal keratinocytes, dermal fibroblasts, endothelial cells, adipocytes, smooth muscle cells, MSCs	[53, 54]
Tooth	Odontoblasts, dental stem cells, MSCs	[55]

process similar to the generation of iPSCs, but without the stemness properties. The process of differentiating somatic cells into another cell type is called transdifferentiation in which the cell's gene expression pattern is completely changed by some growth factors and transcription factors supplied to the cell culture. For example, fibroblasts were induced to convert into cartilage cells when incubated in proper cell culture conditions. As a result of this transdifferentiation process, human dermal fibroblasts could produce cartilage-specific molecules like glycosaminoglycans and type II collagen in presence of IGF-1 [41].

Osteogenic Differentiation

In vitro osteogenic induction of MSCs can be possible by using a differentiation medium that contains combinations of various supplements such as dexamethasone, ascorbic acid-2-phosphate, β-glycerophosphate, vitamin D3, and transcription factors like TGF-β and BMPs [56, 57]. The effects of these molecules on transcription factors play an important role in the osteogenic induction of MSCs. Runx2, Osterix, and ß-catenin are the major transcription factors that regulate the osteogenic differentiation of MSCs. The Runx-2 transcription factor controls bone formation, acting as the key regulator of pre-osteoblastic differentiation by inhibiting adipogenesis and chondrogenesis [58].

Various signaling pathways involving Wnt, Bone Morphogenetic Protein (BMP), and Notch, regulate Runx2 expression. BMP has a crucial role in providing Runx2-Smad interaction, which is essentially required for osteogenic induction [59, 60]. Zhao et al. studied the role of Runx2 in osteogenic differentiation on mouse bone marrow derived MSCs that were transduced with Runx2 adenoviral vectors. The results demonstrated that Runx2 gene transfer increased osteocyte proliferation, indicating the crucial role of Runx2 in the osteogenic differentiation of MSCs [61].

Chondrogenic Differentiation

Due to the tight relationship between the processes of chondrogenesis and osteogenesis, many stimulating molecules that drive osteogenic differentiation are also partially associated with chondrogenic differentiation. The major stimulating factors that induce chondrogenic differentiation belong to the TGF-β family [62], such as TGF-β1, TGF-β3, BMP isoforms (BMP2, BMP4) [63], activin, osteogenic protein-1, and growth and differentiation factor-5 (GDF-5) [64–66]. However, using TGF-β1 alone was found insufficient in chondrogenic stimulation of periotic mesenchyme in vitro [67]. In order to overcome this problem, TGF-β1 and basic FGF were used in combination for chondrogenic induction of periotic mesenchyme [68]. FGF and its isoforms could contribute to chondrogenesis by transducing the mitogen-activated protein kinase (MAPK) signaling pathway, leading to an increase in the expression level of Sox9 which is a major transcription factor regulating chondrogenesis [69, 70].

In addition to TGF-β and FGF cytokine families, IGF-1 was shown to promote chondrogenic activity in adult dermal fibroblasts [41]. It was also reported that combining TGF-β1 with IGF-1 induced chondrogenesis of the periosteal MSCs [71].

Myogenic Differentiation

In vitro cardiomyogenic differentiation is based on mimicking the process that occurs naturally during embryonic cardiac development that can be achieved by regulating specific signaling pathways taking part in cardiomyogenesis, such as the WNT signaling pathway. The major growth factors used in the cardiogenic differentiation of MSCs are BMP4, FGF2, and activin A [72]. Further studies report that the use of BMP4 on mouse ESCs led to the formation of primitive hematopoietic precursors [73], and that TGF-β1 induced the differentiation of human cardiomyocyte progenitors into physiologically compatible cardiomyocytes [74].

Mishra et al. studied the myogenic differentiation capacity of umbilical cord blood (UCB)-derived MSCs and umbilical cord tissue (UCT)-derived MSCs. For the induction of myogenic phenotype, MSCs were cultured in HGF-2, IGF-1, and FGF-2 containing growth medium. Functional assays and transcriptomic analysis results revealed that UCT-derived MSCs expressed CD90, leading to faster myogenic differentiation than UCB-derived MSCs. As a result, UCT-derived MSCs present a more reliable and convenient usage than UCB-derived MSCs for myogenic lineage differentiation [75].

It has been suggested that hESCs-derived vascular progenitor cells were capable of differentiation into cells of endothelial and smooth muscle phenotype, based on the growth factor used. In an endothelial growth medium (EGM-2) containing VEGF-165, these vascular progenitor cells could obtain an endothelial-like phenotype, while adding platelet-derived growth factor-BB (PDGF-BB) in the EGM-2 could promote smooth muscle-like phenotype [76]. Likewise, human hair follicle stem cells (hHFSCs) could differentiate into functional smooth muscle cells when induced by PDGF-BB and TGF-β1 [77].

Ha et al. assessed the differentiation potential of DPSCs into smooth muscle-like cells. For smooth muscle-like differentiation, DPSCs isolated from wisdom teeth were cultured in two different commercially available culture plates, namely Norm-c and Gel-c (0.1% gelatin-coated), and in a culture medium supplemented with 5 ng/ml TGF-β1 and 2 ng/ml PDGF-BB. At different time points (7, 14, 21 days), smooth muscle-specific marker gene expression of differentiated cells was performed. Early, mid, and late smooth muscle cell-specific markers were observed in induced cells. According to the results, DPSCs cultured in both plates showed similar expression levels for smooth muscle cell early markers SM22-α, α-SMA, and SMTN, however, mRNA expression levels for mid and late markers (CALP and MHY-11) were higher in the Gel-c plate. Differentiated cells cultured in both plates exhibited high contractibility as determined by collagen gel contraction assay [78].

Epithelial and Endothelial Differentiation

In order to examine epithelial-like differentiation, MSCs from the sternum with various hematological disorders were first characterized to identify surface phenotype [79]. Differentiation of MSCs into epithelial cells was induced with 10 ng/ml keratinocyte growth factor (KGF), 20–30 ng/ml EGF, 60 ng/ml IGF-2, and 10 ng/ml hepatocyte growth factor (HGF) addition into the culture media. After 10–14 days following the induction, the fibroblastic-shaped MSCs achieved polygonal shape like epithelial cells. Immunohistochemistry and gene expression analysis demonstrated that the differentiated MSCs were expressing the simple epithelial markers cytokeratin 18 and 19 (CK18 and CK19) in contrast to undifferentiated MSCs.

Another study was performed for analyzing the in vitro endothelial differentiation ability of BMSCs [80]. Immediately after isolation, BMSCs tested positive for MSC surface markers (CD105, CD166, CD90, CD73, and CD44), but negative for endothelial and hematopoietic markers. However, after the confluent cells were induced with 50 ng/ml VEGF, the expression of endothelial markers KDR and FLT-1 remarkably increased. Similarly, the MSCs could form tube-like structures in a semi-solid medium in the presence of VEGF, as the result of in vitro angiogenesis assay for the analysis of capillary formation. In addition, differentiated MSCs exhibited mature endothelial cell-like expression patterns of VEGF-1/VEGF-2, von Willebrand factor (vWF), and VE-cadherin VCAM-1.

Neural Differentiation

Adipose tissue-derived stem cells (ADSCs) hold great promise for neural tissue engineering considering the increasing number of research in this field. ADSCs from the abdominal cavity of rats were able to exhibit Schwann-like spindle morphology in vitro when cultured in the presence of glial growth factor (GGF-2), basic FGF (bFGF), PDGF, and forskolin [81]. Co-culturing of ADSCs with motor neuron cells (NG108-15) resulted in the formation of neuron-like cells indicated by neurite growth.

In another research, an indirect co-culture of ADSCs and Schwann cells was tested for neural differentiation. ADSCs were extracted from testicular fat pads of Sprague–Dawley rats and then cultured in DMEM/F12 medium. Schwann cells from sciatic nerves of neonatal rats were cultured in the presence of heregulin1-ß1 extracellular domain (HRG1-ß1). Dorsal root ganglion (DRGN)/differentiated ADSCs co-culture systems were used to evaluate the myelination capacity of differentiated ADSCs. Schwann cell-like myelin formation was observed indicating that the indirect co-culture system of Schwann cells and ADSCs was effective in spindle-like morphology adoption of ADSCs [82].

Odontogenic Differentiation

Some research studies have focused on isolating stem cells from non-dental origin and differentiating them into odontogenic cells. Adipose tissue derived MSCs (AD-MSCs) could be differentiated into dental cells by providing them a proper medium that contains odontogenic signal molecules. For instance, Ferro et al. conducted a study to differentiate AD-MSCs into dental bud-like cells in vitro. For this purpose, AD-MSCs were isolated from human abdominal lipoaspirates and then transferred into culture media supplemented with various growth factors including PDGF-BB, EGF, IGF-1, and FGF-b. After four weeks, the induced cells exhibited structures like dental bud, expressing epithelial, mesenchymal, and basal lamina markers, which is a sign of dental morphogenesis [83].

Corneal Differentiation

Paired box 6 (PAX6) is a gene that is responsible for eye development and corneal limbal stem cell differentiation. In a recent study, rat AD-MSCs were transfected with recombinant PAX6 gene to examine the transdifferentiation ability of transfected PAX6 cells into corneal epithelial cells. PAX6-transfected cells expressed corneal-specific markers CK3/12 as well as the epithelial marker protein, E-cadherin, demonstrating that PAX6 transfection of AD-MSCs induced differentiation of AD-MSCs into corneal epithelial cells in vitro [84].

In another study, the co-culturing of human ADSCs with porcine-derived limbal epithelial stem cells induced ADSCs differentiation into the corneal epithelium. While the expression levels of corneal epithelial markers CK3 and CK12 increased at mRNA level in co-cultured cells, the expression of MSC markers (CD73, CD90, CD105) were decreased, indicating the transdifferentiation of ADSCs into limbal epithelial stem cells [85].

Intestinal Differentiation

The intestinal epithelium is structured in crypts and villi units. Continuous proliferation of intestinal adult stem cells leads to the renewal of epithelium completely. Differentiation of daughter cells occurs in the villus where they migrate from the bottom of the crypt. As a result of differentiation, daughter cells become effector cells, including niche factor secretory Paneth cells, nutrient absorbent enterocytes, hormone secretory enteroendocrine cells (EECs), mucus producer goblet cells, and mucosal immunity regulator tuft cells.

Culturing intestinal stem cells (ISCs) in vitro can generate three-dimensional organoids that have similar characteristics to in vivo intestinal epithelium with crypt and villus domains [47, 48]. In ISCs culturing, WNT signaling is the primary requirement because it is necessary for crypt proliferation, while EGF plays a role in regulating intestinal migration and proliferation [48, 86].

Hepatic Differentiation

Hepatocytes are the main cell type residing in the parenchymal (functional) tissue of the liver. To date, numerous experiments have been conducted to manipulate cell culture conditions in vitro to provide hepatic differentiation of stem cells. ESCs and human MSCs were shown to be induced to possess a functional hepatocyte-like phenotype [87, 88]. Growth factors and cytokines with hepatic growth and differentiation effects are HGF, EGF, TGF, IGF, and bFGF. Differentiated cells can perform characteristic functions of the liver cells in vitro, such as producing albumin, secreting urea and storing glycogen [89].

Conclusion and Future Prospects

The choice of cells for tissue engineering is one critical step to regenerating a fully functional tissue construct. The cells, and growth factors necessary for their directional differentiation should be selected according to the type of tissue. In recent years, co-cultures have been investigated for tissue repair in vitro and in vivo. Given that co-culturing involves the combinational use of two or more different cell types to produce multifunctional tissues, it holds great promise for whole organ engineering someday.

References

1. Yoon JH, Roh EY, Shin S et al (2013) Comparison of explant-derived and enzymatic digestion-derived MSCs and the growth factors from Wharton's jelly. Biomed Res Int 2013
2. Shay JW, Wright WE (2000) Hayflick, his limit, and cellular ageing. Nat Rev Mol Cell Biol 1:72–76
3. Singh J, Goswami A (2014) Applications of cell lines as bioreactors and in vitro models. October 2012
4. Verma A (2013) Animal tissue culture: principles and applications. Elsevier
5. Malik P, Mukherjee S, Mukherjee TK (2023) Mammalian cell culture types and guidelines of their maintenance. Pract Approach Mamm Cell Organ Cult 1–27
6. Clift D, Schuh M (2013) Restarting life: fertilization and the transition from meiosis to mitosis. Nat Rev Mol Cell Biol 14:549–562
7. Firmin J, Maître JL (2021) Morphogenesis of the human preimplantation embryo: bringing mechanics to the clinics. Semin Cell Dev Biol 120:22–31
8. Vázquez-Diez C, Fitzharris G (2018) Causes and consequences of chromosome segregation error in preimplantation embryos. Reproduction 155:R63–R76
9. Evano B, Khalilian S, Le Carrou G et al (2020) Dynamics of asymmetric and symmetric divisions of muscle stem cells in vivo and on artificial niches. Cell Rep 30:3195-3206.e7
10. Kleinsmith LJ, Pierce GB Jr (1964) Multipotentiality of single embryonal carcinoma cells. Can Res 24:1544–1551
11. Armstrong L, Lako M, Buckley N et al (2012) Editorial: our top 10 developments in stem cell biology over the last 30 years. Stem Cells 30:2–9
12. Evans MJ, Kaufman MH (1981) Establishment in culture of pluripotential cells from mouse embryos. Nature 292:154–156
13. Khan FA, Almohazey D, Alomari M, Almofty SA (2018) Isolation, culture, and functional characterization of human embryonic stem cells: current trends and challenges. Stem Cells Int 2018
14. Brimble SN, Zeng X, Weiler DA, Luo Y, Liu Y, Lyons IG, Freed WJ, Robins AJ, Rao MS, Schulz TC (2004) Karyotypic stability, genotyping, differentiation, feeder-free maintenance, and gene expression sampling in three human embryonic stem cell lines derived prior to August 9, 2001. Stem Cells Dev 13(6):585–597

15. Singh RK, Gaikwad S (2014) Stem cells: the holy grail of regenerative medicine. In: Cai W (ed) Engineering in translational medicine (XIX, Issue February). Springer, London. https://doi.org/10.1007/978-1-4471-4372-7
16. Jones DL, Wagers AJ (2008) No place like home: anatomy and function of the stem cell niche. Nat Rev Mol Cell Biol 9(1):11–21
17. Ding DC, Shyu WC, Lin SZ (2011) Mesenchymal stem cells. Cell Transplant 20:5–14
18. Hmadcha A, Martin-Montalvo A, Gauthier BR et al (2020) Therapeutic potential of mesenchymal stem cells for cancer therapy. Front Bioeng Biotechnol 8:1–13
19. Scott S, Clouser C, Aho J (2015) Methods to validate mesenchymal stem cell quality. Bio-Techne 5–6
20. Bianco P, Riminucci M, Gronthos S, Robey PG (2001) Bone marrow stromal stem cells: nature, biology, and potential applications. Stem Cells 19:180–192
21. Nagamura-Inoue T, He H (2014) Umbilical cord-derived mesenchymal stem cells: their advantages and potential clinical utility. World J Stem Cells 6(2):195
22. Ozen A, Gül Sancak I, Ceylan A, Özgenç O (2016) Isolation of adipose tissue-derived stem cells. Turkish J Vet Anim Sci 40:137–141
23. Huang SJ, Fu RH, Shyu WC et al (2013) Adipose-derived stem cells: Isolation, characterization, and differentiation potential. Cell Transplant 22:701–709
24. Calikoglu Koyuncu AC, Dogan E, Uzun M, Gunduz O (2021) An overview of the use of dental stem cells and polycaprolactone scaffolds in tissue engineering. Eng Mater Stem Cell Regen 139–162
25. Lee JY, Hong SH (2020) Hematopoietic stem cells and their roles in tissue regeneration. Int J Stem Cells 13:1–12
26. Liu L, Papa EF, Dooner MS, et al (2012) Homing and long-term engraftment of long- and short-term renewal hematopoietic stem cells. PLoS One 7
27. Takahashi K, Yamanaka S (2006) Induction of pluripotent stem cells from mouse embryonic and adult fibroblast cultures by defined factors. Cell 126:663–676
28. Gurdon JB (1962) The developmental capacity of nuclei taken from intestinal epithelium cells of feeding tadpoles. J Embryol Exp Morphol 10:622–640
29. Horiguchi I, Kino-oka M (2021) Current developments in the stable production of human induced pluripotent stem cells. Engineering 7:144–152
30. Carr AJF, Smart MJK, Ramsden CM et al (2013) Development of human embryonic stem cell therapies for age-related macular degeneration. Trends Neurosci 36:385–395
31. Xu X, Zheng L, Yuan Q et al (2018) Transforming growth factor-β in stem cells and tissue homeostasis. Bone Res 6
32. Ren X, Zhao M, Lash B et al (2020) Growth factor engineering strategies for regenerative medicine applications. Front Bioeng Biotechnol 7:1–9
33. Yun YR, Won JE, Jeon E et al (2010) Fibroblast growth factors: biology, function, and application for tissue regeneration. J Tissue Eng 1:1–18
34. Wroblewski OM, Vega-Soto EE, Nguyen MH et al (2021) Impact of human epidermal growth factor on tissue-engineered skeletal muscle structure and function. Tissue Eng—Part A 27:1151–1159
35. Fatimah SS, Tan GC, Chua KH et al (2012) Effects of epidermal growth factor on the proliferation and cell cycle regulation of cultured human amnion epithelial cells. J Biosci Bioeng 114:220–227
36. Yu A, Matsuda Y, Takeda A et al (2012) Effect of EGF and bFGF on fibroblast proliferation and angiogenic cytokine production from cultured dermal substitutes. J Biomater Sci Polym Ed 23:1315–1324
37. Banfi A, Degenfeld G, Gianni-Barrera R et al (2012) Therapeutic angiogenesis due to balanced single-vector delivery of VEGF and PDGF-BB. FASEB J 26:2486–2497
38. Solem RC, Eames BF, Tokita M, Schneider RA (2011) Mesenchymal and mechanical mechanisms of secondary cartilage induction. Dev Biol 356(1):28–39. https://doi.org/10.1016/j.ydbio.2011.05.003

39. Whiteley JT, Fernandes S, Sharma A et al (2022) Reaching into the toolbox: stem cell models to study neuropsychiatric disorders. Stem Cell Rep 17:187–210
40. Liao X, Zhong R, Zhang H, Wang F (2022) Effects of coculture fibroblasts and vascular endothelial cells on proliferation and osteogenesis of adipose stem cells. Comput Math Methods Med 2022
41. French MM, Rose S, Canseco J, Athanasiou KA (2004) Chondrogenic differentiation of adult dermal fibroblasts. Ann Biomed Eng 32:50–56
42. Song HHG, Rumma RT, Ozaki CK et al (2018) Erratum: vascular tissue engineering: progress, challenges, and clinical promise. Cell Stem Cell 22(3):340–354, S1934590918300717. https://doi.org/10.1016/j.stem.2018.02.009; Cell Stem Cell 22:608
43. Orciani M, Fini M, Di Primio R, Mattioli-Belmonte M (2017) Biofabrication and bone tissue regeneration: cell source, approaches, and challenges. Front Bioeng Biotechnol 5:1–15
44. Ghezzi CE, Rnjak-Kovacina J, Kaplan DL (2015) Corneal tissue engineering: recent advances and future perspectives. Tissue Eng—Part B Rev 21:278–287
45. Vinatier C, Guicheux J (2016) Cartilage tissue engineering: from biomaterials and stem cells to osteoarthritis treatments. Ann Phys Rehabil Med 59:139–144
46. Nguyen AH, Marsh P, Schmiess-Heine L et al (2019) Cardiac tissue engineering: state-of-the-art methods and outlook. J Biol Eng 13:1–21
47. Sato T, van Es JH, Snippert HJ, Stange DE, Vries RG, van den Born M, Barker N, Shroyer NF, van de Wetering M, Clevers H (2011) Paneth cells constitute the niche for Lgr5 stem cells in intestinal crypts. Nature 469(7330):415–418
48. Sato T, Vries RG, Snippert HJ, van de Wetering M, Barker N, Stange DE, van Es JH, Abo A, Kujala P, Peters PJ, Clevers H (2009) Single Lgr5 stem cells build crypt-villus structures in vitro without a mesenchymal niche. Nature 459(7244):262–265
49. Mirdamadi ES, Kalhori D, Zakeri N et al (2020) Liver tissue engineering as an emerging alternative for liver disease treatment. Tissue Eng—Part B Rev 26:145–163
50. Carnes ME, Pins GD (2020) Skeletal muscle tissue engineering: biomaterials-based strategies for the treatment of volumetric muscle loss. Bioengineering 7:1–39
51. Bellamkonda RV, Clements IP, Munson JM (2013) Neuronal tissue engineering. Third Edit, Elsevier
52. Doblado LR, Martínez-Ramos C, Pradas MM (2021) Biomaterials for neural tissue engineering. Front Nanotechnol 3
53. Vig K, Chaudhari A, Tripathi S et al (2017) Advances in skin regeneration using tissue engineering. Int J Mol Sci 18
54. Yu JR, Navarro J, Coburn JC et al (2019) Current and future perspectives on skin tissue engineering: key features of biomedical research, translational assessment, and clinical application. Adv Healthc Mater 8:1–19
55. Bluteau G, Luder HU, De Bari C, Mitsiadis TA (2008) Stem cells for tooth engineering. Eur Cells Mater 16:1–9
56. Friedenstein AJ, Chailakhyan RK, Gerasimov UV (1987) Bone marrow osteogenic stem cells: in vitro cultivation and transplantation in diffusion chambers. Cell Prolif 20(3):263–272
57. Okamoto T, Aoyama T, Nakayama T, Nakamata T, Hosaka T, Nishijo K, Nakamura T, Kiyono T, Toguchida J (2002) Clonal heterogeneity in differentiation potential of immortalized human mesenchymal stem cells. Biochem Biophys Res Commun 295(2):354–361
58. Komori T (2006) Regulation of osteoblast differentiation by transcription factors. J Cell Biochem 99(5):1233–1239
59. Bandyopadhyay A, Tsuji K, Cox K, Harfe BD, Rosen V, Tabin CJ (2006) Genetic analysis of the roles of BMP2, BMP4, and BMP7 in limb patterning and skeletogenesis. PLoS Genet 2(12):2116–2130
60. Fujii M, Takeda K, Imamura T, Aoki H, Sampath TK, Enomoto S, Kawabata M, Kato M, Ichijo H, Miyazono K (1999) Roles of bone morphogenetic protein type I receptors and SMAD proteins in osteoblast and chondroblast differentiation. Mol Biol Cell 10(11):3801–3813
61. Zhao Z, Zhao M, Xiao G, Franceschi RT (2005) Gene transfer of the Runx2 transcription factor enhances osteogenic activity of bone marrow stromal cells in vitro and in vivo. Mol Ther: J Am Soc Gene Ther 12(2):247–253

62. Wozney JM (1998) The bone morphogenetic protein family: multifunctional cellular regulators in the embryo and adult. Eur J Oral Sci 106(1 Suppl):160–166
63. Kramer J, Hegert C, Guan K, Wobus AM, Müller PK, Rohwedel J (2000) Embryonic stem cell-derived chondrogenic differentiation in vitro: activation by BMP-2 and BMP-4. Mech Dev 92(2):193–205
64. Asahina I, Sampath TK, Nishimura I, Hauschka PV (1993) Human osteogenic protein-1 induces both chondroblastic and osteoblastic differentiation of osteoprogenitor cells derived from newborn rat calvaria. J Cell Biol 123(4):921–933
65. Jiang TX, Yi JR, Ying SY, Chuong CM (1993) Activin enhances chondrogenesis of limb bud cells: stimulation of precartilaginous mesenchymal condensations and expression of NCAM. Dev Biol 155(2):545–557
66. Spiro RC, Liu LS, Heidaran MA, Thompson AY, Ng CK, Pohl J, Poser JW (2000) Inductive activity of recombinant human growth and differentiation factor-5. Biochem Soc Trans 28(4):362–368
67. Frenz DA, Galinovic-Schwartz V, Flanders KC, Van De Water TR (1992) TGF-b1 is an epithelial-derived signal peptide that influences otic capsule formation. Dev Biol 153:324–336
68. Frenz DA, Liu W, Williams JD, Hatcher V, Galinovic-Schwartz V, Flanders KC, van de Water TR (1994) Induction of chondrogenesis: requirement for synergistic interaction of basic fibroblast growth factor and transforming growth factor-beta. Dev (Camb, Engl) 120(2):415–424
69. Murakami S, Kan M, McKeehan WL, de Crombrugghe B (2000) Up-regulation of the chondrogenic Sox9 gene by fibroblast growth factors is mediated by the mitogen-activated protein kinase pathway. Proc Natl Acad Sci USA 97(3):1113–1118
70. Ornitz DM, Marie PJ (2002) FGF signaling pathways in endochondral and intramembranous bone development and human genetic disease. Genes Dev 16(12):1446–1465
71. Fukumoto T, Sperling JW, Sanyal A, Fitzsimmons JS, Reinholz GG, Conover CA, O'Driscoll SW (2003) Combined effects of insulin-like growth factor-1 and transforming growth factor-β1 on periosteal mesenchymal cells during chondrogenesis in vitro. Osteoarthritis Cartilage 11(1):55–64
72. Yang L, Soonpaa MH, Adler ED, Roepke TK, Kattman SJ, Kennedy M, Henckaerts E, Bonham K, Abbott GW, Linden RM, Field LJ, Keller GM (2008) Human cardiovascular progenitor cells develop from a KDR+ embryonic-stem-cell-derived population. Nature 453(7194):524–528
73. Kattman SJ, Witty AD, Gagliardi M et al (2011) Stage-specific optimization of activin/nodal and BMP signaling promotes cardiac differentiation of mouse and human pluripotent stem cell lines. Cell Stem Cell 8:228–240
74. Goumans MJ, de Boer TP, Smits AM, van Laake LW, van Vliet P, Metz CHG, Korfage TH, Kats KP, Hochstenbach R, Pasterkamp G, Verhaar MC, van der Heyden MAG, de Kleijn D, Mummery CL, van Veen TAB, Sluijter JPG, Doevendans PA (2008) TGF-β1 induces efficient differentiation of human cardiomyocyte progenitor cells into functional cardiomyocytes in vitro. Stem Cell Res 1(2):138–149
75. Mishra S, Sevak JK, Das A, Arimbasseri GA, Bhatnagar S, Gopinath SD (2020) Umbilical cord tissue is a robust source for mesenchymal stem cells with enhanced myogenic differentiation potential compared to cord blood. Sci Rep 10(1):1–12
76. Ferreira LS, Gerecht S, Shieh HF, Watson N, Rupnick MA, Dallabrida SM, Vunjak-Novakovic G, Langer R (2007) Vascular progenitor cells isolated from human embryonic stem cells give rise to endothelial and smooth muscle like cells and form vascular networks in vivo. Circ Res 101(3):286–294
77. Xu ZC, Zhang Q, Li H (2013) Human hair follicle stem cell differentiation into contractile smooth muscle cells is induced by transforming growth factor-β1 and platelet-derived growth factor BB. Mol Med Rep 8(6):1715–1721
78. Ha J, Bharti D, Kang YH, Lee SY, Oh SJ, Kim SB, Jo CH, Son JH, Sung IY, Cho YC, Rho GJ, Park JK (2021) Human dental pulp-derived mesenchymal stem cell potential to differentiate into smooth muscle-like cells in vitro. BioMed Res Int, 2021

79. Păunescu V, Deak E, Herman D, Siska IR, Tănasie G, Bunu C, Anghel S, Tatu CA, Oprea TI, Henschler R, Rüster B, Bistrian R, Seifried E (2007) In vitro differentiation of human mesenchymal stem cells to epithelial lineage. J Cell Mol Med 11(3):502–508
80. Oswald J, Boxberger S, Joergensen B, Bornhaeuser M, Ehninger G, Werner C (2004) Mesenchymal stem cells can be differentiated into endothelial cells in vitro. Stem Cells (Dayton, Ohio) 22(3):506
81. Kingham PJ, Kalbermatten DF, Mahay D, Armstrong SJ, Wiberg M, Terenghi G (2007) Adipose-derived stem cells differentiate into a Schwann cell phenotype and promote neurite outgrowth in vitro. Exp Neurol 207(2):267–274
82. Wei Y, Gong K, Zheng Z, Liu L, Wang A, Zhang L, Ao Q, Gong Y, Zhang X (2010) Schwann-like cell differentiation of rat adipose-derived stem cells by indirect co-culture with Schwann cells in vitro. Cell Prolif. 2010 Dec 43(6):606–616. https://doi.org/10.1111/j.1365-2184.2010.00710.x. PMID: 21039999; PMCID: PMC6496385
83. Ferro F, Spelat R, Falini G, Gallelli A, D'Aurizio F, Puppato E, Pandolfi M, Beltrami AP, Cesselli D, Beltrami CA, Ambesi-Impiombato FS, Curcio F (2011) Adipose tissue-derived stem cell in vitro differentiation in a three-dimensional dental bud structure. Am J Pathol 178(5):2299–2310
84. Sun J, Liu WH, Deng FM, Luo YH, Wen K, Zhang H, Liu HR, Wu J, Su BY, Liu YL (2018) Differentiation of rat adipose-derived mesenchymal stem cells into corneal-like epithelial cells driven by PAX6. Exp Ther Med 15(2):1424–1432
85. Sikora B, Skubis-Sikora A, Kimsa-Furdzik M, Ciszek W, Kostrzewski M, Stojko J, Mazurek U, Gola J (2019) Adipose-derived stem cells undergo differentiation after co-culture with porcine limbal epithelial stem cells. Stem Cell Res 41:101609
86. Holmberg FE, Seidelin JB, Yin X, Mead BE, Tong Z, Li Y, Karp JM, Nielsen OH (2017) Culturing human intestinal stem cells for regenerative applications in the treatment of inflammatory bowel disease. EMBO Mol Med 9(5):558–570
87. Snykers S, de Kock J, Rogiers V, Vanhaecke T (2009) In vitro differentiation of embryonic and adult stem cells into hepatocytes: state of the art. Stem Cells 27(3):577–605
88. Taléns-Visconti R, Bonora A, Jover R, Mirabet V, Carbonell F, Castell JV, Gómez-Lechón MJ (2007) Human mesenchymal stem cells from adipose tissue: differentiation into hepatic lineage. Toxicol Vitro 21(2):324–329
89. Chivu M, Dima SO, Stancu CI, Dobrea C, Uscatescu V, Necula LG, Bleotu C, Tanase C, Albulescu R, Ardeleanu C, Popescu I (2009) In vitro hepatic differentiation of human bone marrow mesenchymal stem cells under differential exposure to liver-specific factors. Transl Res 154(3):122–132

Chapter 4
Biomaterials

Angela Spoială, Cornelia-Ioana Ilie, Denisa Ficai, and Anton Ficai

Abstract Even since antiquity, and thanks to their ingenuity, doctors have benefited from using natural materials "biomaterials" to substitute tissue, organs or other body parts. Over time, traditional and rudimentary biomaterials have been replaced with improved and targeted synthetic materials specially designed for specific applications. This chapter provides an overview of natural and synthetic biomaterials, used in developing innovative devices for tissue engineering applications, from polymers and ceramics to composite materials (metals were discussed in more details in Chap. 1). The first part of this chapter presents a comprehensive overview of the first biomaterials used in ancient medicine. This part highlights the essential role that primitive biomaterials brought in today's medicine. In the second part, natural and synthetic biomaterials are very thoroughly presented. The main aspect of these biomaterials is related to their physicochemical and mechanical properties, which must be considered when featured for tissue engineering applications. The focus of the third part will significantly illustrate the interconnection and combination between medicine and scientific research to develop new platforms for tissue engineering applications. Over the past few decades, researchers and scientists have demonstrated considerable progress in developing novel biomaterials as substitutes for replacing and

A. Spoială · C.-I. Ilie · D. Ficai · A. Ficai (✉)
Department of Science and Engineering of Oxide Materials and Nanomaterials, Faculty of Chemical Engineering and Biotechnologies, University Politehnica of Bucharest, 1-7 Gh Polizu Street, 011061 Bucharest, Romania
e-mail: anton.ficai@upb.ro

D. Ficai
e-mail: denisaficai@yahoo.ro

A. Spoială · C.-I. Ilie · A. Ficai
National Centre for Micro and Nanomaterials and National Centre for Food Safety, University Politehnica of Bucharest, 313 Splaiul Independentei, 060042 Bucharest, Romania

D. Ficai
Department of Inorganic Chemistry, Physical Chemistry and Electrochemistry, Faculty of Chemical Engineering and Biotechnologies, University Politehnica of Bucharest, 1-7 Gh. Polizu Street, 011061 Bucharest, Romania

A. Ficai
Academy of Romanian Scientists, 3 Ilfov Street, 050045 Bucharest, Romania

O. Gunduz et al. (eds.), *Biomaterials and Tissue Engineering*, Stem Cell Biology and Regenerative Medicine 74, https://doi.org/10.1007/978-3-031-35832-6_4

repairing diverse damaged tissues. Lastly, conclusions and future trends review the most important and complex aspects that tissue engineering provides in combing diverse materials to develop suitable substitutes to alleviate patients' lives.

Keywords Biomaterials · Biocompatibility · Composites · Tissue engineering

An Overview of Biomaterials

Humankind's use of materials to repair the body has dated since antiquity. Doctors benefit from natural materials such as wood to substitute-lost tissues or bone parts followed by disease, injury, or trauma. Historically, doctors used these materials based on their availability and ingenuity in designing and implementing this prosthetic. To enhance performance, increased reproducibility, and functionality of natural materials, synthetic polymers, ceramics, and metals (alloys) have begun to be extensively used. This approach increases the usage limits and efficiency of biomaterials, resulting in life savings and improved devices in applications such as vascular stents, dental restoration or implants, artificial hips, skin or nerve substitutes, contact lenses [1].

The appreciation of traditional biomaterials has been consistently growing since their inception and has rapid development in various applications. Biomaterials played an essential role in the ancient past and even more in today's medicine. Primitive biomaterials date since antiquity when Egyptians used animal sinew to manufacture sutures. The British ophthalmologist Harold Ridley made the first modern medical implants in the 1940s. When analyzing the pilot's eyes, Harold Ridley accidentally saw that fragments of plastic canopy seemed to heal their eyes. After that, he developed his first lens implant in 1949. His breakthrough in cataract problems had a tremendous impact on developing modern intraocular lenses (IOLs), which helped millions of people. When Harold Ridley was developing IOLs, other innovators had remarkable disclosures in diverse fields of medicine. For instance, Charnley designed a novel prosthetic implant, Vorhees fabricated a vessel graft, Kolff invented dialysis, and Hufnagel manufactured the first artificial heart valve. Even with rudimentary medical materials, these pioneers revolutionized medicine and saved lives. Until the 1960s, engineers, chemists, and biologists, in association with physicians, created new concepts and opened new frontiers for biomaterials design, which was the path to understanding the reaction of a living organism to an implant [2].

Biomaterials have been used to substitute body parts, such as joints, legs, knees, hips, and other tissues or organs in the human body. Tathe et al. [3] define "biomaterial as a material that comprises the whole or part of a living structure or biomedical device, which performs, augments, or replaces a natural function". These biomaterials might be designed from natural or synthetic materials. Several authors and institutions have comprehensively defined biomaterials in the biomedical sector [4–6].

Natural and synthetic, organic and inorganic, as well as hybrid materials, are used for biomedical purposes. Usually, materials used in biomedical applications are divided into four categories according to their nature: metals, ceramics, polymers, and composites (Fig. 4.1). Several factors are considered when designing these materials for specific applications. Nevertheless, the key factor is biocompatibility, which specifies the materials' chemical, biological, and physical appropriateness, especially in terms of resorption rate, flexibility or mechanical properties [7, 8]. Besides biocompatibility, integration rate, adherent or non-adherent surfaces, electrical, optical, magnetic properties, etc., are other major factors that must be considered when choosing biomaterials for specific applications [9, 10]. Considering the compositional similitude, it can conclude that biomaterials based on pure polymers have great potential for all soft tissue engineering. In contrast, materials based on metals and ceramics, especially calcium phosphates and bioactive glasses alone or in the composite form, should be the first-choice material in hard tissue engineering. Unfortunately, due to the many challenges in the field of tissue engineering, mainly caused by various diseases or diverse factors inducing the need for such biomaterials, compositional and morphological design is more and more clear to be needed.

Figure 4.2 highlights the repartition of the annual needs of grafting materials over the major classes used in tissue engineering [11]. By far, the need for bone grafting materials is the largest, 49% of the total amount of grafts for bone tissue engineering. Here, a major reason is associated with fractures and congenital diseases. It is worth knowing that just the need for blood is larger than the need for bone grafting materials. The need for blood vessels, nervous and skin substitutes is quite similar and together, ~30% of the global market of grafting materials. Indeed, the compositional and morphological requirements are different because of the various functions, which is why the materials used in these classes can differ very much.

As presented above, the biomaterials from the four classes are the best option that could be successfully used in tissue engineering. Advancements in the field

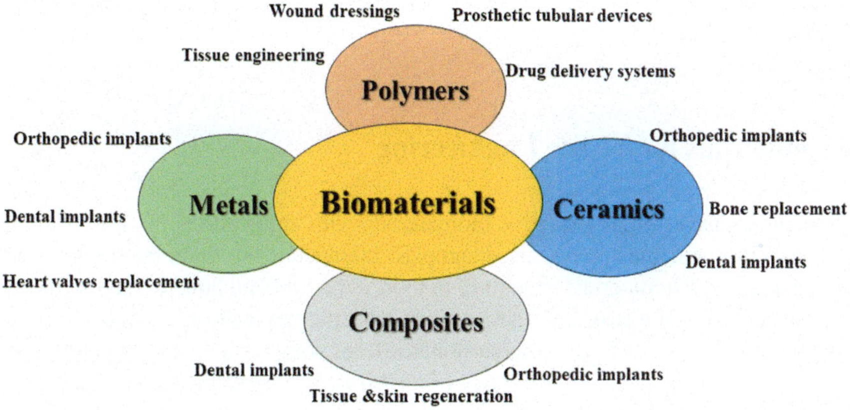

Fig. 4.1 Graphical display of the classification and applications of the biomaterials

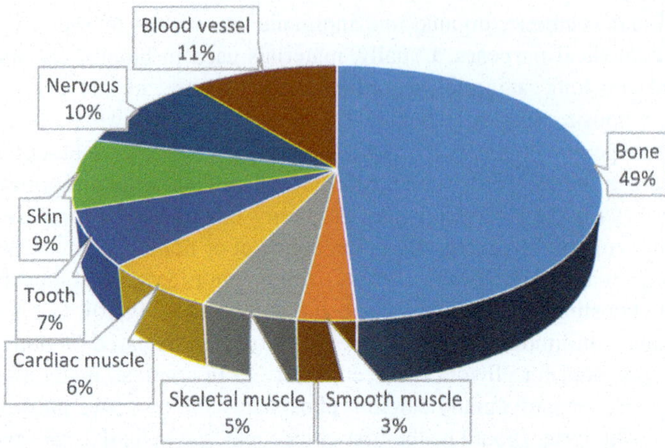

Fig. 4.2 Prospects of using biomaterials in tissue engineering (with the kind permission of [11])

of biomaterials for tissue engineering have enabled their usage in soft and hard tissue engineering. Therefore, scientists have focused their attention on the progress of tissue engineering applications by using these biomaterials in bone, skin and nerve tissue engineering, development of blood vessel substitutes, and in dentistry or even synthetic muscle [12, 13]. Natural and synthetic polymers are used for soft tissue engineering alone or in association with ternary components used in a specific ratio. Among these ternary components, some synthetic materials such as carbon-based materials (including graphene and graphene oxide and carbon nanotubes), oxides materials (such as bioactive glasses, zinc oxide, magnetite, etc.) but also metal nanoparticles (especially silver, gold and cooper nanoparticles) have become attractive for inducing the desired properties and performances. Therefore, graphene nanocomposites are considered the next generation of soft tissue engineering biomaterials, especially if conductive materials are needed or if electric triggering delivery is requested [11].

Biomaterials for Tissue Engineering

Researchers have found attractive biomaterials with applications in several industries and biomedicine (prosthetic, diagnosis, therapeutic, storage, etc.) to improve patients' lives. Biomaterials in nature, both in terrestrial and aquatic environments, are particularly interesting and are used in specific applications. These biomaterials are spider silk, silk, calcium carbonates from corals, shells and eggshells, fish bones, etc., which can be used directly or turned into valuable biomedical materials. Remarkable physicochemical and mechanical properties allow them to be used in various fields of medicine, such as orthopaedics, neurology, dentistry, and tissue engineering. These biomaterials have multiple applications, such as carriers, healing agents, antimicrobial agents, and fillers. Because they have an extremely long life within the host, there is a minimal need for replacement or repair [14].

Natural Biomaterials

The most natural polymers used in tissue engineering belong to proteins and polysaccharides. Protein-based biomaterials include collagen and gelatine (collagen derivatives), fibrin, silk fibroin, and polysaccharide-based biomaterials include chitosan, alginate, cellulose and hyaluronic acid [15]. Natural polymers such as collagen, matrigel, chitosan and alginate are extensively used as scaffolds.

Collagen

Many scientists have shown in their work that collagen biomaterials have biomimetic properties, providing great applicability in tissue engineering and cosmetic applications (especially as gelatine). Collagen is an important polymer for bone, skin, blood vessels, cartilage, nerve repair, and bladder engineering. Collagen has poor mechanical and swelling properties in vivo environments due to hydrophilicity; therefore, its physicochemical and mechanical properties must be tailored [16]. Using natural collagen has a great advantage due to its abundance in nature, low immunological response, and ability to form fibres. Collagen, a resorbable polymer, has a high swelling ability, low antigenicity, cytocompatibility, and tissue regeneration potential [17–20].

Fibrin

Because of its excellent biocompatibility, biodegradability, and transport of cells and biomolecules, fibrin has gained interest in skin and bone tissue engineering applications. Furthermore, fibrin has been a preferred biomaterial in tissue engineering applications due to its precursors, fibrinogen and thrombin, which can enable autologous scaffolds [21].

Alginate

Alginate biopolymers have been used in diverse tissue engineering, including cartilage and cardiac tissues [22]. Due to its biocompatibility, low toxicity, cost-effectiveness, and easy gelation, alginate is a hydrogel widely used in many biomedical applications. Alginate's chemical structure presents extended features, making it preferable for 3D printing applications [23].

Silk Fibroin

Due to its excellent biocompatibility, nontoxicity, physical features, adhesion and proliferation capability, silk fibroin (SF) is a natural protein with potential biomedical applications. Also, it has excellent biological properties and applicability in tissue engineering [24]. Researchers found that the paramagnetic behaviour of SF/CS-based magnetic scaffolds decrease the phosphate-buffered saline (PBS) uptake but are still used in tissue engineering [25]. Recently, authors were inspired by natural silk to develop a biomimetic meso-assembly processing engineering (MAPE) technology to synthesize biomaterials that mimic the structure and mechanics of biological tissues. Thanks to this approach, it could manufacture silk fibroin-based biomaterials with exceptional mechanical properties for tissue engineering applications [26–28].

Hyaluronic Acid

Another natural biomaterial used in tissue engineering as a scaffold is hyaluronic acid (HA), which could promote healing and induce chondrogenesis. Because of its excellent biocompatibility and structure versatility, hyaluronic acid (HA) has applicability in the biomedical field. Alginate-hyaluronic acid composites are attractive biomaterials used in tissue engineering due to their physical, mechanical, and biological features [29].

Elastin

The protein responsible for the elasticity of the body's various tissues is elastin, which has a soluble precursor, tropoelastin. Also, elastin is an abundant biopolymer along with collagen, which has been used in developing elastic tissues. The incorporation of elastin in biomaterials promotes elasticity and biological effects. In association with collagen, elastin has usage in tissue engineering applications [30].

Agarose

Agarose, a natural polymer extracted from agar by removing agaropectin, is considered a potential candidate for neural and cartilage scaffolds in tissue engineering [31].

Chitosan

Chitosan hydrogels can be useful in soft tissue engineering applications as biomaterials, with great potential in developing scaffolds. Chitosan hydrogels can create different body tissue types, such as skin, muscles, blood vessels, and nerves. Due to its biological effects, such as antibacterial, antitumor, antioxidant, and tissue regeneration effects, chitosan can be used as a dressing to prevent and treat soft tissue diseases [32–34]. Chitosan (CS) is derived from the deacetylation of chitin, it is a biomaterial extracted from crustacean exoskeletons. The major limitation of CS in tissue engineering applications could be overcome by adding biocompatible polymers or suitable cross-linkers. Blending it with polymers or bioactive-based materials could improve its physicochemical and biological properties and applicability in tissue engineering [35].

Silk Spider

Lately, biomaterial research has become paramount in modern medicine. For ages, *spider silk fibres* have captivated scientists' attention, primarily due to their outstanding mechanical properties. Their strength and elasticity give that toughness that another natural or synthetic fibre could not achieve. Moreover, the preferable properties for using biomaterials such as spider silk in biomedical applications are biocompatible, biodegradable, and hypoallergenicity. The nucleus of spider silk contains numerous parts of amino acids (containing glycine and polyalanine), providing an impressive structure responsible for their mechanical properties. The development of spider silk's technological process made it possible to be used as an implanted biomaterial with many applications (coatings, drug systems, tissue engineering, and more) [36–38].

Calcium Carbonate

Another important inorganic biomaterial with a natural origin is represented by calcium carbonate, especially calcite. It can be found in eggshells, corals, shells, etc. The *eggshell* consists of three layers: a foamy cuticle on the outside with ceramic looks, a spongy layer in the middle, and the inside presents lamellar stripes. The membrane and the eggshell represent ~10% of the entire weight, with ~90% calcite $CaCO_3$. Besides its unique microstructure and interesting features, it has applications as a fertilizer, a sorbent for heavy metals and dyes, and a source of calcium for synthesising hydroxyapatite, which is very popular in hard tissue engineering [39–41]. Corals were found to be promising materials that can be easily transformed into HA with a special microstructure, making them suitable especially for hard tissue engineering [42, 43].

Synthetic Biomaterials

Synthetic polymers can be personalized to have more extensive mechanical and chemical properties than their natural equivalents. Use synthetic polymers as biomaterials must be either resorbable or non-resorbable. Poly(tetrafluoroethylene) (Teflon), a non-degradable polymer that is biocompatible and widely used in vascular grafts and hip implant applications.

Poly-L(lactic acid) (PLLA), poly(glycolic acid) (PGA), and poly(lactic-co-glycolic acid) (PLGA) are synthetic, biodegradable polymers with a large range of applications in tissue engineering. They can be used as pure scaffolds and medical devices, including for drug delivery purposes [22, 44].

The synthetic polymers poly-4-hydroxybutyrate (P4HB) and polyhydroxyalkanoate (PHA) has been used for heart valve development due to their excellent elastomeric properties [45, 46]. Because of their high biocompatibility and processability, hydrogels have drawn great interest in tissue engineering [47].

Synthetic ceramics such as calcium phosphates–CaPs, zirconium dioxide–ZrO_2, alumina–Al_2O_3, etc., are especially used as a ceramic powder or ceramic body but also as coatings–especially on metallic implants. Most of these ceramic materials in the medical field are related to bone tissue engineering, alone or as composite materials, but also as drug delivery systems. These materials can be prepared to start from natural resources. In this case, they can retain some of the features of the precursors (porosity, oligo-elements content, etc.) [48–51].

Carbon-based materials are increasingly used in tissue engineering. Since the development of carbon nanotubes and later graphene-related materials, these special materials have gained more and more interest in many fields, including tissue engineering. It is worth mentioning that with high electrical and mechanical performances, the tuneable hydrofil/hydrophobe ratio and thus the ability to specifically absorb/desorb specific biological active agents make these materials very attractive

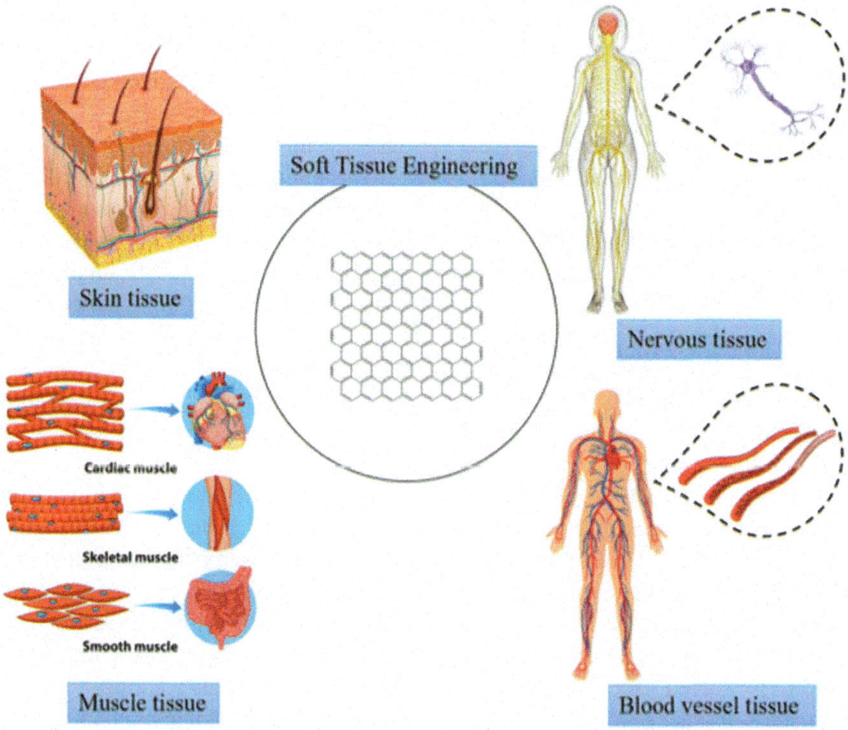

Fig. 4.3 Applications of graphene biomaterials in soft tissue engineering applications [11] (reprint kindly granted by MDPI)

in medical applications, especially as an additive in polymeric, ceramic or composite materials. Some of the most important applications are schematically presented in Fig. 4.3.

Advances in Tissue Engineering

The modern area of biomaterials has increased significantly due to the interconnection of medicine with scientific research, which has greatly influenced the tissue engineering field. In the past few decades, implantable materials and devices (heart valves, stents, grafts, joints, ligaments, tendons, dental, and nerves) have evolved considerably due to advances in material science. This is mainly because of the advancement in materials processing and design. Scientists have promoted different techniques for tissue healing, such as sutures, clips, staples, and resorbable dressings. Combining biomaterial-based scaffolds, cells, and bioactive molecules are displayed as a modern tissue engineering solution. Researchers have made a great breakthrough in using nanoparticles in biomedicine, especially in imaging and cancer therapy.

Some biosensors can detect specific carriers and register data for monitoring diverse activities within the body or brain. Also, drug delivery systems, carrying and/or applying meds to a targeted site, are displayed as implantable stents for cancer therapy [52].

A multidisciplinary field involving cell biology, materials science, device engineering, and clinical research has developed tissue engineering to create new tissues and organs. According to Statistics, the most transplanted organ worldwide is the kidney, followed by the liver and heart. In the past four years, ~155.000 organ transplants have occurred worldwide. Even if organ transplantation can be challenging and complex, it is still the best and only treatment for end-stage organ failure [53]. Enhancing the stem cell control within biological, chemical and physical tools has offered significant insights into tissue engineering [22].

Biomaterials have evolved since their first use in antiquity, and their degree of complexity (compositional, morphological, etc.) has increased significantly. Even gradient-like devices and grafts were developed to mimic the natural tissue better or to design better the functional properties required for the final applications [54]. In the past years, research in this direction has involved complex and unparallel approaches, including drug delivery functionalization, micropatterning, microfluidics, and other technologies. The timeline of using biomaterials in clinical use has shown that the evolution road continuously surprises us in developing new biomaterials with impressive performances. However, still, there are many challenges which have to be solved. Before their first development, biomaterials were used without awareness of sterilization, inflammation, or biodegradation problems and limitations. However, their use in various surgical procedures was successful, proving the human body's impressive capability to adapt and accommodate foreign objects. Therefore, their evolution and interaction with the human body have increased scientists' and physicians' interest in the issues related to biocompatibility [6].

Biocompatibility issues are fundamental when using diverse biomaterials to develop new platforms for tissue engineering applications. Biocompatibility assessments are a crucial and complex procedure that aims to verify the material's ability to avoid adverse reactions and function perfectly when in contact with the biological environment. Short- and long-term evaluations must be considered when referring to the biocompatibility of a particular material. Also, it has to consider other aspects related to the chemical composition and mechanical and physical properties of the parts involved in the process, especially concerning the surrounding tissues and organs [55].

To develop or create the conditions to manufacture a specific tissue, it must consider the nature of the material and its chemical and physical features. Different materials for tissue engineering have been used, but *metals* and *ceramics* were first used in *orthopaedic* applications [56, 57]. Metals and ceramics (hydroxyapatite and bioglass) have been successfully used in orthopedic applications. Polymer materials closely match most biological tissues' chemical and mechanical properties. They are especially suitable as a scaffold and drug delivery systems in soft tissue engineering [58]. Combining the proper cells with a particular material under a specific environment leads to tissue foundation, the ground rule of tissue engineering.

Foreign implanted metals show diverse reactions due to the body's exposure to those implants. Most metals used in surgical implants are oxidized, releasing ions, sometimes toxic ions, which degrade the implanted device making it impossible to perform for a long period of time–usually, metal implants require revisions within 10–15 years. Therefore, ceramic and polymer materials have started to gain interest in biomaterials. The world of polymers has continuously developed and includes numerous different substances with customized features to extend their biomedical applications [59]. Some materials are used as biomaterials, especially ceramics, with applications in dentistry and tissue engineering. Ceramics were recommended as an alternative for metals and polymers to enhance bone fixation/integration. Generally, they are biocompatible, inert, or bioactive and have adverse reactions. Thus, they have limitations regarding rigidity and fragility. Both represent essential disadvantages for many practical applications; still, they could perform a direct bond with living tissues [60].

The synthesis and microfabrication of biomaterials play an important role in developing new devices/scaffolds for diverse fields of tissue engineering. The conventional approach to designing engineered tissues begins with incorporating cells within a scaffold-based resorbable material. A matrix consisting of natural or synthetic materials acts as an immuno-isolation barrier and supports the cells. The main goal is that the cells integrate with the host, remodel the scaffold (by resorbing it and developing new tissue, these two processes coincide), and ultimately develop the desired tissue for a specific application. Polymeric biomaterials synthesis strategies are important in developing potential biomedical applications, including tissue engineering scaffolds [61].

Natural or synthetic biomaterials are often used in tissue engineering applications. Their development has used several methods: conventional versus additive manufacturing methods. Material design is proven to bring important benefits to the materials, which is normal and can be easily verified just by analysing the differences between compact and spongious bones. Even though the composition is quite the same, the properties are very different. For instance, the mechanical performances of compact bone tissue are much higher. In contrast, the ion released from these tissues are opposite (a major role in homeostasis). Traditional methods may imply solvent casting, gas foaming, freeze-drying, melt moulding, and electrospinning are some of the most utilized. Additive manufacturing techniques use software such as CAD/CAM to develop pre-designed structures. There are cases when CT scans and MRIs on natural tissues are used to create the required CAD/CAM technique system. Several methods based on additive manufacturing have been developed, such as 3D printing, stereolithography, and fused deposition modelling. The most used biomaterials in 3D printing are natural polymers. They are widely used to fabricate engineered skin, nerve, and bone tissue substitutes and organs [23, 62, 63]. Considering the classes of biomaterials, Table 4.1 illustrates some of their most relevant clinical applications and synthesis methods.

Table 4.1 Classes of biomaterials and their synthesis methods and applications

Biomaterial		Methods of synthesis and processing	Applications	References
Metals	Stainless steel, titanium, tantalum, tungsten, platinum	Vacuum casting, casting, melting, thin-film deposition, spin coating method; 3D printing	Orthopaedic implants, artificial hip & dental implants, bone implants, medical implants, and chip heart implants	[64–68]
Ceramics	Alumina, zirconia, hydroxyapatite	Solvothermal synthesis, sol–gel synthesis, uniaxial pressing and sintering, spark-plasma sintering, coating technique, casting method, and 3D printing	Ceramic powders, dental and hip implants, dental prostheses, oral implants, orthopaedic & dental implants, and drug delivery systems	[69–71]
Polymers	Collagen, fibrin, agarose, chitosan, hyaluronic acid, polyglycolic acid	Crosslinking, surface modification, von Kossa's method, polymerization, phase inversion technique, and melt-spinning process; 3D printing	Tissue engineering (Skin, neural, cardiac, ligaments, cartilage, bone), regenerative medicine, and drug delivery systems	[16, 72, 72–75]
Composites	Fibroin/chitosan/magnetite; silk fibroin/calcium polyphosphate, collagen/hydroxyapatite; ZrO_2/HA composite materials; peptide/alginate, microfluidic alginate fibre	Freeze-casting method, polymerization, electrospinning, micro-spinning method; 3D printing	Bone tissue engineering, bone regeneration, cardiac tissue regeneration, tissue reconstruction, and drug delivery systems	[25, 76–78]

For more than three decades of history since the first synthesis of coral-derived hydroxyapatite, researchers have used natural corals to develop human bone alternatives for different medical purposes: orthopaedic, craniofacial, dental, and neurosurgery, having in mind the special microstructure but also the presence of specific oligo-elements. Most of the corals used in studies are the genus Porites, Goniopora, Alveo-pore, and Acropora. Figure 4.4 illustrates some biomaterials found in nature with tissue engineering applications.

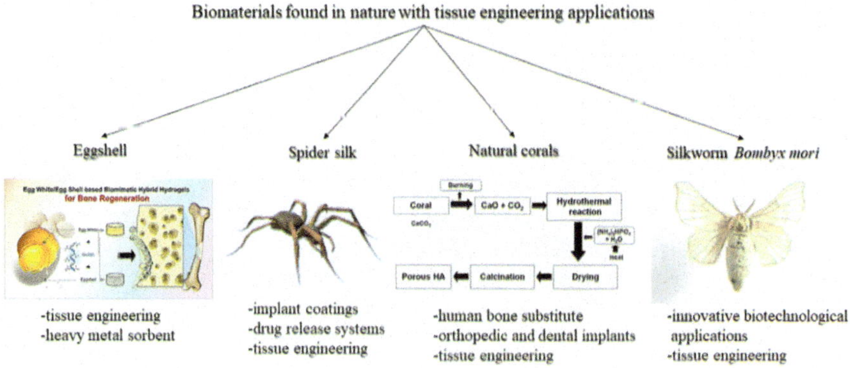

Fig. 4.4 Natural biomaterials used in tissue engineering applications

Bone Tissue Engineering

Bone is almost 80% of the human body and is responsible for movement, ion homeostasis, production of blood cells, protection and support of the organs, etc. Bone is mainly based on collagen and hydroxyapatite. Its composition is ~70% mineral phase, mainly doped hydroxyapatite, ~21% organic phase, mainly collagen and about 9% water. The special microstructure of these composite materials can lead to compact/cortical as well as cancellous or trabecular (also known as spongious) bone tissue. The bone is especially known for its special mechanical properties, well adapted to the needs of the individuals, but also because it assures the homeostasis of several elements, especially Ca, P, Mg, and so on.

Figure 4.5 illustrates the bone structure from macro to sub-nanostructure [79]. As it is depicted, the hierarchical structure of the collagen molecules and hydroxyapatite crystals is essential to generate the requested properties. So, starting from the elementary components, hydroxyapatite crystals and collagen molecules, a first step of assembling occurs, and mineralized collagen molecular array is obtained. Looking at the figure, these represent the sub-nanostructure at the bone level. Starting from these arrays, the assembling process continues, and mineralized fibrils and fibers are developed. These two stages represent the nanostructure of the bone tissue. At the microstructure level, it can see that the assembling structures can be more compact, resulting in osteons–corresponding to compact tissue or laxer, resulting in trabeculae–corresponding to trabecular tissue. At macrostructure, these trabeculae and osteons are assembled, forming bones and the entire skeleton. It is important to mention that bones may contain both specific bonny tissue in a different ratio; for instance, it is represented in the femoral bone where the median structure is a more complex structure with gradient distribution, the outer side being compact while the inner side being trabecular.

The imperious need for developing bone graft biomaterials has been focused on manufacturing various materials with impressive properties such as strength,

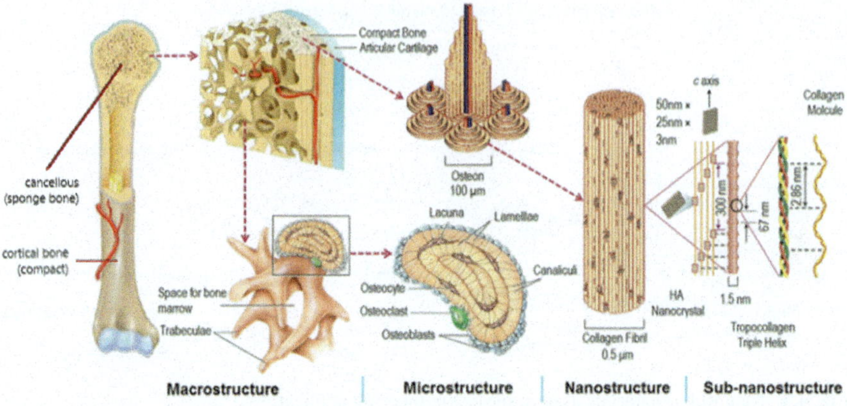

Fig. 4.5 Bone structure from macro to sub-nano structure (with the kind permission of [79])

and mechanical and biological properties, which are just some of them. Figure 4.6 illustrates the biomaterials progression in bone tissue engineering and their usage starting from 1950 until 2020 ~ 2030 [80].

From an evolutionary point of view, since 1950, metals and alloys have started to be used. Even if they are not bioresorbable or bioactive, they are still used nowadays because they can be easily implanted and have good mechanical properties. Ceramics and polymers represent the second generation of biomaterials used in bone tissue engineering; they can be bioresorbable or bioactive. The third generation of biomaterials tried to combine the properties of the ceramics and polymers and is represented by the composite materials and was further improved by the addition of the tissue-related components (growth factors or even cells) representing the fourth generation of the bone grafting materials. Nowadays, special attention is paid to Materials Design. Even if not yet recognized by all, it seems that the next milestone

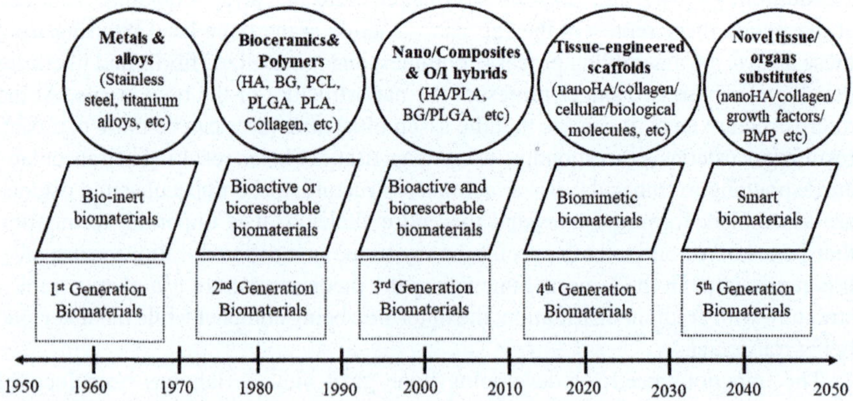

Fig. 4.6 Biomaterials progression in bone tissue engineering. Figure adapted from [80]

(or generation) will be represented by the materials obtained by Materials Design, including 3D printing.

Among the biomaterials presented above, metal is the strongest, with high-tensile strength (600–1085 MPa) and elastic modulus (120–210 GPa), depending on its type [8]. "Vanadium steel" was the first metal alloy specially designed for bone fractures (Sherman plates) and screws. Due to their biocompatibility with the host body (human and animal), implant-based metals (Fe, Cr, Co, Ni, Ti, Ta, Nb, Mo, and W) were manufactured. The naturally occurring forms of these metallic elements are vital to red cells (Fe) and vitamin B12 (Co). However, they are still dangerous in large quantities [81]. When considering these metallic implants in vivo environments, they must be biocompatible with the host [68]. When corrosion weakens the implant, the surrounding tissues and organs are damaged [10, 82].

Numerous types of metal are used for medical applications, such as amalgam, tungsten, tantalum, titanium alloys, stainless steel, and aluminium alloys. Metal has been applied in the medical sector to make dental materials, prosthetic devices, knee joints, hip joints, bone devices, and others. The metals used in medical implants are selected based on their mechanical properties: high load-bearing resistance, wear resistance, and good fatigue limit. These properties are suitable in artificial joints exposed to repeated loading and unloading; they can also hold good mechanical loads. Other than that, metal-based materials have good corrosion resistance and low cost [66]. Stainless steel is a standard metal that forms the backbone of medical device manufacturing, in which 316L stainless steel is the most typically used metal in all implants. Tantalum is a metal that encourages bone development and is also used in dentistry and tissue regeneration [65]. Ryan et al. [83] published an extensive review of porous metals used in orthopaedic applications. The porosity is important, especially on the surface, thus, entirely porous implants but also porous coatings were reported, and in this case, a good surface adherence can be achieved, even cells can migrate inside the porous surface, and new bone tissue can be obtained even inside the porous metal surface, and thus a good connection/osteointegration is achieved.

Ceramics are mainly used in orthopedics and dentistry as a skeletal and dental materials. However, in terms of mechanical properties, ceramics have some limitations compared to metal. To fulfil its lifetime in the host, a ceramic implant must satisfy the following properties (a) non-toxicity, (b) non-carcinogenicity, (c) non-allergenicity, (d) non-inflammatory, (e) biocompatible, (f) biofunctional, and thus they are extensively used as monolith, coatings, or composites [10, 84]. The hardness of ceramics is measured against ceramic materials and directed towards developing ceramic-related implants. Ceramics are classified as non-absorbable (relative inert), bioactive or reactive (semi-inert) [85], and biodegradable or resorbable (non-inert) [86]. Examples of inert bioceramics include alumina, zirconia, silicon nitrides, and carbons. Semi-inert (bioactive) ceramics are calcium phosphates and especially hydroxyapatite. Still, calcium sulfate, aluminate, and bioactive glasses are resorbable ceramics (non-inert) [87]. Ceramics are widely used as prosthetic hips, prosthetic knees, and replacements for jawbones, bone grafts, dental implants, ear implants, heart valves, and skeletal system parts. Sarkar and Banerjee explained the applications and types of ceramic-based materials in biomedicine [88].

By far, calcium phosphates–CaPs and especially hydroxyapatite are the most used bioceramic materials being followed, by far, by bioactive glasses, zirconia, and alumina as a result of the ISI Clarivate/Thompson database (Fig. 4.7). As a general observation, 4816 entries were found using Bioceramics as a keyword (using a restriction over Title/Keywords/Abstract). In both major cases, doping is an interesting topic, bioactive glasses as well as CaPs are improved, or new properties are induced by doping adequately. It is worth mentioning that bioactive glasses are increasingly also used for skin tissue engineering alone or associated with polymers [89–91].

Polymers are less used in developing bone grafts. Looking into Scopus, 18.900 records involve polymers for bone engineering, from which over 1700 records are related to polylactic acid, followed by collagen (>1400 records) and polyesters (>1300 records). Chitosan (>1200 records), polycaprolactone (>1150 records), etc. Collagen and chitosan-based polymers are used in bone tissue regeneration as porous scaffolds, alone or in association with other polymers, but also entrapping osteoblasts or loaded with active agents, including medium-cross-linked recombinant collagen peptides, bone morphogenetic proteins, growth factors, etc. [92–95]. PEG-based materials are especially used, alone or in association with other polymers in tissue engineering, being well-suitable as materials for 3D printing [96]. CS's attractive properties in bone-forming cells and mineralization of the bone matrix make it a viable candidate for tissue engineering [75]. Modifying CS with imidazole and methyl pyrrolidinone makes it osteoconductive in vivo, promoting bone defect regeneration [74]. Also, polymer hydrogels loaded with mesenchymal stem cells (MSCs) enhance bone formation, and their mixture with different materials improves mechanic ability and bone-matrix interface strength. Undoubtedly, 3D polymer scaffolds loaded with various stem cells can be developed in tissue engineering applications, including skin, cartilage, cardiac, nerve, etc. [97].

Composite materials in biomedicine include orthopedic applications, dental applications, prosthetic devices, and skin regeneration. Extensive reviews on the use of these types of materials for biomedical applications have been presented by Mehboob

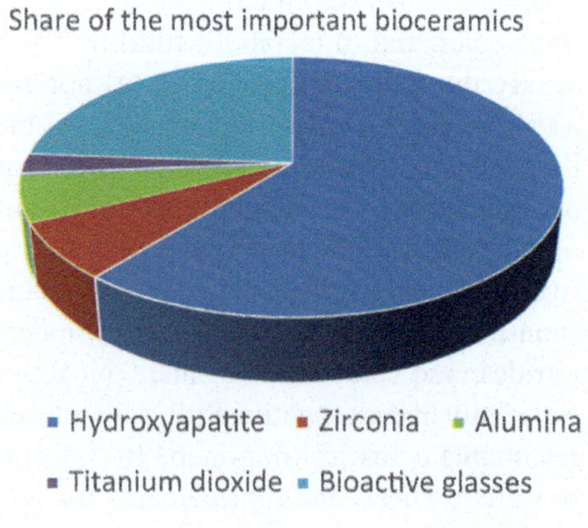

Fig. 4.7 The share of the most abundant 5 materials, according to ISI Clarivate, over the bioceramics records available (using a preliminary restriction of title/keywords/abstract

and Chang [98], Iftekhar [99], and Salernitano and Migliaresi [100]. As a general definition, a composite material combines two or more physical and chemically different materials (matrix and reinforcing materials) to produce materials with enhanced properties. According to the design, composition and fabrication methodology, the properties of the composite materials can be tuned accordingly to the needs. It is important to mention that the best materials for medical applications have to be designed considering also the final applications and neighbouring tissues. It is well known that animals and human beings are working in perfect harmony/equilibrium and thus, any imbalances can lead to diseases. For instance, the use of stronger materials as a substitute can alterate the neighbouring tissues or, if their mechanical properties are inferior, they will be destroyed, and additional intervention will be necessary to replace them.

Considering the composition of the bone, some of the most suitable materials used in bone tissue engineering, particularly bone replacement, are collagen/hydroxyapatite composite materials. But, to obtain the best bone substitute, first must overcome disadvantages such as crystal shape, size, and orientation for collagen and hydroxyapatite. After studying the literature with proper synthesis methods and other conditions, it has been shown that the prepared nanocomposite materials present great potential for being used as bone grafts [101–103]. The improvements concerned compositional and morpho-structural design. Figure 4.8 illustrates some representative morphologies of the COLL/HA composites obtained using specific conditions of synthesis and orientation. The oriented assemblies were obtained by self-assembling [102] or using an appropriate electric field [103]. In both cases, good orientation (>95%) and dense structures were obtained when started from collagen gel. Porous composite materials can be obtained starting from gels, matrices or fibres, the porosity and pore sizes being controlled. Intermediate porosity can also be obtained by combining controlled air drying with freeze drying, as presented by Ficai et al. [104]. It can see that the porosity can be easily tuned within a large range, from 27.5 to 96.5%, porosity which is suitable for tissue engineering, including drug delivery.

The literature presents numerous research works on bone tissue engineering grafts. However, still, there is a challenge to mimic and reproduce a bone using COLL/HA composite materials because of problematic clinical implantation, fixation and integration. The collagen/hydroxyapatite composite materials lead to enhanced osteogenic differentiation of the mesenchymal stem cells, and thus better performances are assured. Unfortunately, at this moment, only the metals/alloys are suitable for being designed and executed adequately for a sufficiently good fixation into the human body, so they do not require additional fastening systems. But such composite materials seem to be suitable grafting materials, as filler, or even as an implantable body even in high-loading bones. Still, metallic components, especially screws, rods, plates or wires, are required in these cases. Nevertheless, better grafting would be achieved if vascularization and regeneration could coincide. But, in most cases, the grafting material is entirely replaced by the regenerated tissue. Therefore, there is still debatable which method could be used to develop the best biomaterial for bone replacement [105].

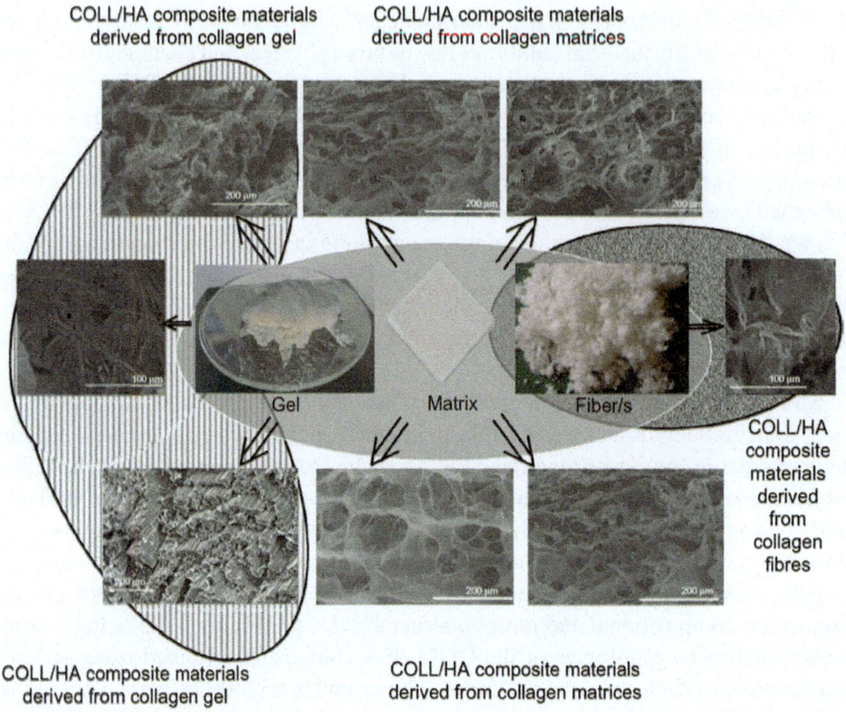

Fig. 4.8 Morfo-structural design of the COLL/HA composite materials (with the kind permission of the [106–109])

Studies have shown that pure CS scaffolds are improper for bone tissue engineering, but combining them with other materials, such as ceramics and polymers, enhances their mechanical and biological properties and has become promising biomaterials for bone regeneration [106]. Incorporating more functional biomaterials into CS could significantly improve bone tissue engineering scaffolds' functionalities. For instance, incorporating nanohydroxyapatite (nHAp) in the CS matrix in situ could enhance the mechanical strength of the final composite. In the case of obtaining fibrous scaffolds, adding fillers or synthetic polymers helps easily incorporate nHAp into the solution [107]. Also, heparin immobilized on CS scaffolds could exploit the natural interaction between growth factors and extracellular matrix; therefore, it has been proven that CS scaffolds based-biocompatible polymers or other materials have much better properties than pure CS scaffolds. Even if natural polymers present great potential, they still have drawbacks and concerns about their usage. One includes challenges in controlling their mechanical properties and degradation in time [108]. Secondly, exists the possibility that natural-derived materials provoke a severe immune response and store microbes or viruses. Due to these concerns, natural materials must be carefully characterized and screened before their use in biomedical applications [109].

Researchers have proven that collagen-bioactive ceramic composites have great potential in developing improved scaffolds [110–112]. It has been shown that loading poly lactic-co-glycolic acid (PLGA) microspheres into collagen/CS/β-TCP for developing scaffolds improve the regeneration properties of the obtained implant [113]. It has been shown that collagen gels reinforced with bioactive glass nanoparticles have great potential in bone tissue engineering applications [114]. Faster osteointegration and healing can be obtained if composite materials are loaded with mesenchymal stem cells and growth factors.

To augment, repair and replace damaged skin or other body parts, tissue engineering is one of the most effective "tools". Approaches in vivo and in vitro techniques are used for artificial tissue regeneration applications. Such practices are needed to support cell attachment, growth, and proliferation. Polymers, metals, and ceramics were used to develop 2D and 3D scaffold structures manufactured through electrospinning and 3D printing. Natural and synthetic polymers have been used in diverse tissue replacement. Polymers that include chitosan, collagen, polyethylene oxide, polylactic acid, and ceramics, especially hydroxyapatite, were used for bone replacement. Even if there have been significant breakthroughs in biomimetic development, tissues and organs with natural properties have been minimal. Since polymer engineering and technology development continuously evolve throughout specific applications, the biomimicking approach has been limited [115, 116]. The word "biomimetics" derives from the Greek "bios" (life) and "mimesis" (to imitate). Understanding the biomimetics of nature or natural phenomena and obtaining ideas that could benefit science, engineering, and medicine [117], this concept has attracted considerable attention in tissue regeneration applications. Mimicking the development of skeletal defects is challenging, and vascularization has an essential role in bone regeneration. 3D scaffolds have been extensively used in bone repair, even if they must overcome angiogenesis and osteogenesis impediments. Researchers were inspired by lotus plants and successfully developed root-like lotus biomaterials via a novel 3D printing technique (Fig. 4.9). The obtained biomimetic materials can considerably enhance cell attachment and proliferation by promoting cell delivery and bone regeneration applications [118].

The bio-inspired or biomimetic design provides novel opportunities for manufacturing biocompatible implant devices. To improve the capability of an implant to integrate with bone tissue, it must enhance osteoblastic functions. An essential breakthrough in laser technology has been made. Yttria-stabilized with tetragonal zirconia was considered an allergy-free implantable, biocompatible material appropriate for hard tissue engineering. By solid-state laser etching, novel zirconia was created with mesoscale cactus-like spikes and nanoscale bone-like trabecular morphology. Further, novel technologies and platform-design strategies for biomaterial surfaces have been optimized to enhance the integration and regeneration of the implanted bone [119]. The biomimetic approach has lent its application in architecture, where novel solutions are essential. Architects are expected to deliver and develop creative solutions better than before. Engineers, designers, and architects were inspired by nature to innovate and improve architectural quality. The University of Technology from Vienna has an innovative "Biomimetics design exercise"

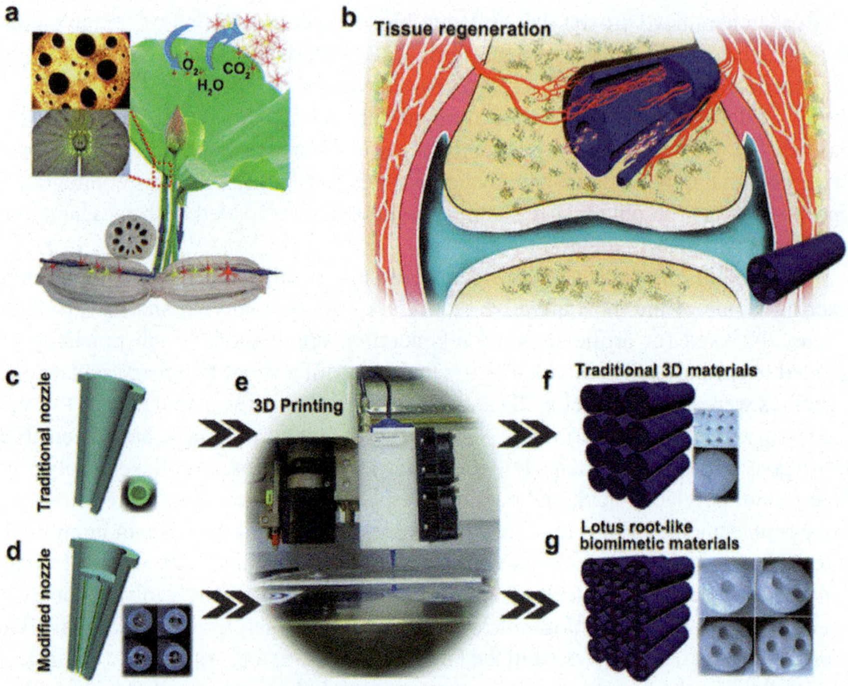

Fig. 4.9 Illustration of a 3D printing of root-like lotus structures for tissue regeneration applications, **a** lotus root microstructure, **b** application in tissue regeneration, **c** traditional nozzle, **d** modified nozzle, **e** 3D printing method, **f** traditional 3D materials, **g** lotus root-like biomimetic materials (with the kind permission of [118])

programme involving natural phenomena in architecture. This architectural approach of using plants for buildings, and being inspired by nature, has presented inspirational ideas and collaboration within interdisciplinary fields in future projects [120].

It was observed that pore size and porosity are key factors in developing materials for hard tissue engineering. It is worth mentioning that materials with pores within 100–300 μm are optimal for bone grafting because (i) the osteoblasts have a size of about 20–30 μm. They need slightly larger pores to be able to penetrate inside. An "in-depth osteointegration" and (i) larger pores induce a significant decrease in the mechanical properties of these grafts, and the risk of secondary fracture is relatively high. On the other size, Karageorgiou and Kaplan showed that larger pores can allow direct bone formation without an intermediate osteochondral formation [121], with vascularization being faster. They also show that a gradient is important because, in this case, multiple tissues and interfaces can be obtained. Thus, overall better integration with the bone and fibrous tissue is obtained. Such structures can be easily obtained using 3D printing when large channels can be developed (allowing the transport of high amounts of nutrients and oxygen) and, at the same time, the strands can have controlled porosity (Fig. 4.10).

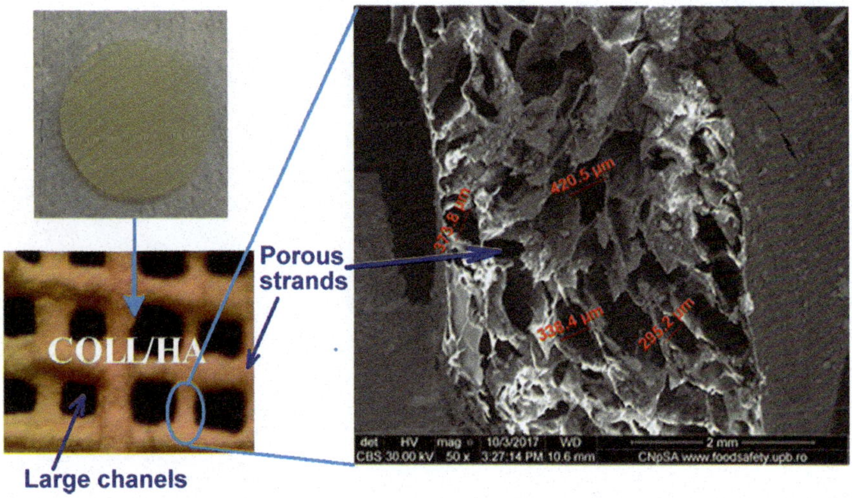

Fig. 4.10 3D-printed COLL/HA composite materials highlighting the large channels and strands but also the porous nature of the strands (adapted according to [18])

Skin Tissue Engineering

The skin is considered the largest organ of the human body, having almost 10% of the body mass. As the figure shows (up-left part), the skin is composed of three distinct layers, starting from the top to down: epidermis, dermis, and hypodermis. Studies have shown that the most common problem with the skin is associated with burns. The challenge of skin tissue engineering involves working relentlessly with other interdisciplinary fields to develop bioengineered skin substitutes to improve the victim's life. According to a report from 2018 by the WHO, there are approximately 11 million burns worldwide every year, which cause almost 180.000 deaths. In Fig. 4.11, we can see the three major classes of burns according to their characteristics. The main difference is that 3rd-degree burns are destroyed mostly from hair follicles and sweat glands deep into the tissue and many times also affect the fibrous tissue/bone, not only the skin [122, 123].

It is worth mentioning that many biomaterials are available for skin grafts, starting from films to matrices and hydrogels. The previous section presents the most used polymers for skin tissue engineering. They include natural (collagen, alginate, chitosan, etc.) and synthetic polymers (PLA, PLGA, PVA, PEG, etc.). The process is based on implanting the biocompatible scaffold into the site of injury, these materials acting as pure regenerative but also as drug release supports. Many classes of skin grafting materials are available on the market, and even more, are researched at the preclinical and clinical levels. It is worth mentioning that acellular but also cell-based grafting materials are available. These skin grafts can be loaded with specific cells: keratinocytes (particular to the epidermis) and fibroblasts (specific to the dermis). The most complex grafts mimic the skin's structure; thus, these skin

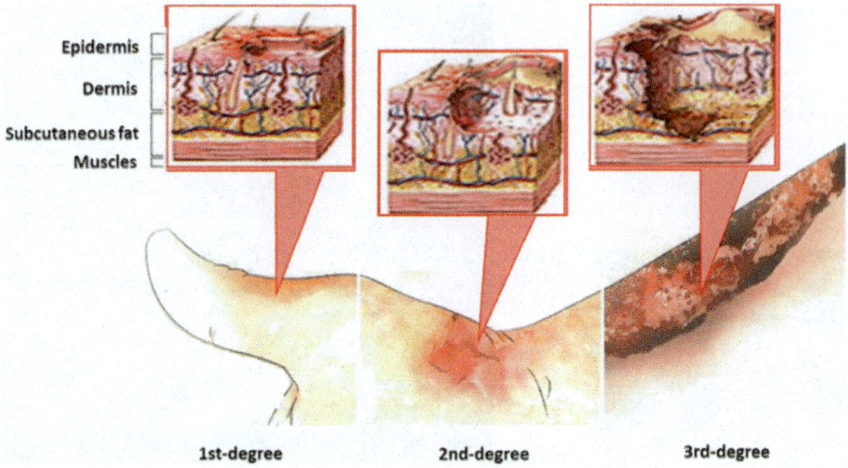

Fig. 4.11 Skin burn classification (with the kind permission of [123])

substitutes are called dermo-epidermal substitutes containing both types of cells, as seen in Fig. 4.12.

Many acellular skin substitutes can be used, including soluble gels (especially collagen-based) but also matrices and films which, depending on the wounds, are adapted to adsorb more or less exudate, the proper humidity being essential in healing.

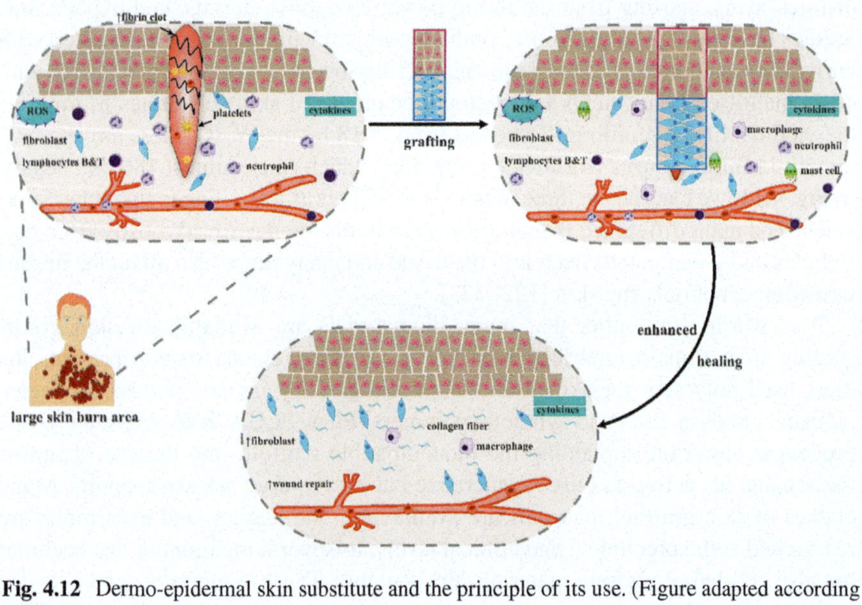

Fig. 4.12 Dermo-epidermal skin substitute and the principle of its use. (Figure adapted according to [124])

Raghunath et al. [125, 126] first reported the importance of the graft's mature and functional dermo-epidermal junction with the skin. The first trial used a coculture of keratinocytes and fibroblasts embedded in a collagen matrix using normal human serum instead of foetal calf serum [127]. These living skin equivalents have shown a regular stratification of the epidermis and the dermal–epidermal junction after 3 weeks of coculture. This approach uses autologous serum and patients' cells (keratinocytes and fibroblasts), which is beneficial because the risks of transmission of infections are decreased considerably. They also proved that the proliferation and survival of the fibroblasts were much better in collagen/fibrin hydrogel compared with pure collagen. This support is recommended to further seed keratinocytes on the surface.

In 2011, Brazilius et al. [128] realized engineered autologous porcine skin grafts for transplantation in a large animal model using epidermal keratinocytes and dermal fibroblasts harvested from the pig's abdomen. More complex dermo-epidermal skin analogues were obtained by loading fibroblast cells into the collagen type I dermal layer. Further epidermal cells were seeded on them [129]. According to the cells used for the epidermal layer, three different skin substitutes were obtained: (i) keratinocytes only, (ii) keratinocytes and melanocytes, and (iii) sweat gland cells. In that study, after 8 weeks, all the grafts were found to develop nerves because these dermo-epidermal skin substitutes can attract nerve fibres from adjacent host tissues assuring proper healing. Biedermann et al. [130] proved that human autologous tissue-engineered skin grafts with a dermo-epidermal structure, including keratinocytes and melanocytes seeded onto a fibroblast-loaded collagen type I hydrogel, can develop blood vessels within 3 weeks but no nerves. This group also proved the advantages of using polymeric net-like meshes embedded into the collagen hydrogel containing fibroblasts and seeded onto the top with keratinocytes [131].

Recent work [132] showed the benefits of 3D printing in designing and fabricating advanced large pigmented and vascularized human dermo-epidermal skin substitutes.

Advanced Nerve Substitute

The purpose of these substitutes is to simulate the structure and functions of the neural tissue to sustain, support and improve the functions of the affected tissue. Even though scientists have made great advances in nerve tissue engineering, clinical repair of a nerve defect remains one of the most challenging surgical problems. Therefore, acellular nerve allografts and artificial nerve repair are the most promising substitutes for nerve autografts. According to the FDA, there are a few products already approved, such as Neurotube (PGA), NeuoMatrix, Neuroflex, Neuragen (Collagen, type I), and SaluBridge are some of them [133].

Synthetic nerves include non-degradable and degradable structures, mainly based on collagen, poly-L-lactic acid, polyglactin, chitosan, polycaprolactone, polyglycolic acid, polyesters, hyaluronic acid, etc. Considering the functions of the nerves, increasing attention is devoted to electric conductive polymers, such as polypyrrole or composites that contain conductive fillers, especially carbon-based materials [134–142].

A novel biocomposite material used for nerve regeneration is polypyrrole/collagen/strontium substituted bioactive glass composite, which has been proven to be a good substitute because it can fulfil both features of the based materials but still require electric conditions to achieve faster regeneration [143].

As has already been presented, graphene-related materials (GRMs) have shown great potential in promoting excellent conductivity and biocompatibility between the biomaterial and host tissue. Conduction can be assured using carbon-based biomaterials such as graphene and graphene oxide. The healing process is enabled through electrical-assisted methods. Figure 4.13 illustrates various methods for fabricating nerve grafts based on graphene and graphene oxide biomaterials [11].

It is important to mention that using the right material is very important, especially when combining proper material with proper physicochemical stimulation. For instance, in peripheral nerve injury, surgical suturing or implants are required but not consistently enough. In such cases, when tissue-engineered grafts are not inducing the necessary therapeutic effect, several solutions are available. Using biological (nerve growth) factors could be a suitable solution, especially when combined with physical factors such as electrical, mechanical, light, and magnetic stimulation, as presented mainly by Zeng et al. [139].

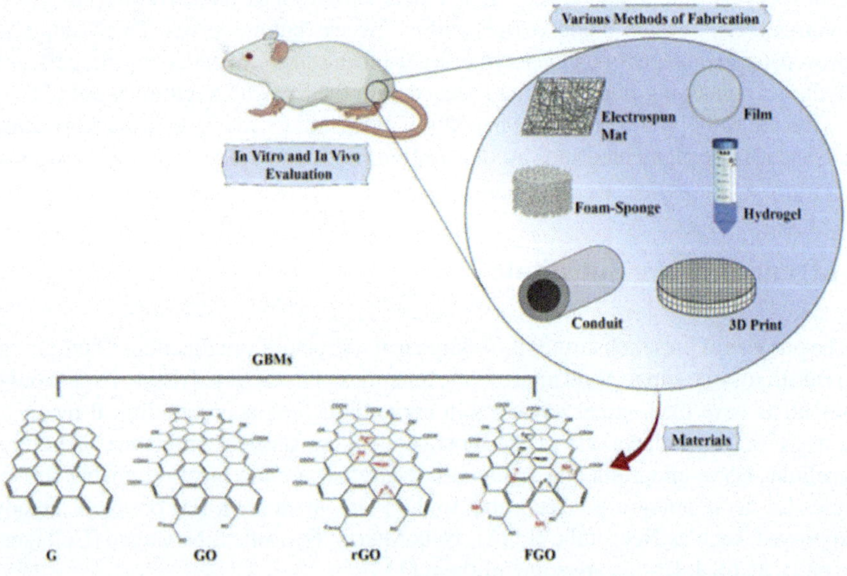

Fig. 4.13 Methods of fabrication for nerve grafts based on graphene oxide-based biomaterials (with the kind permission of [11])

Blood Vessels' Substitutes

The soft tissue engineering of biological substitutes for the replacement or/and repair of blood vessels and other body components presents an important and challenging interest for many scientists and engineers. Even if there has been considerable progress in the clinical trial approach, FDA approval is still far in the future. Over the past 40 years, scientists have focused on revolutionizing the implant industry to repair/regenerate diverse soft tissues [144]. Among all diseases, cardiovascular diseases are the leaders in the majority of mortality and choosing proper materials for blood vessels substitute has become a provocation. After studying various biomaterials for replacing or repairing soft tissue, it has been proven that biodegradable polymers were the first fit [145]. Therefore, scientists focused their interest on developing collagen-based blood vessel substitutes due to their biological properties [146]. Important biopolymers used for blood vessels are chitosan hydrogels. It has been proven that combining collagen with chitosan is beneficial for developing body tissues, such as blood vessel substitutes [34]. Another biopolymer-based biomaterial used for creating blood vessel substitutes is alginate-based hydrogels. Like other natural polymers, alginate is widely used due to its biocompatibility and gelling properties in developing vascularized scaffolds for blood vessel substitutes [147, 148]. It has been demonstrated that combining collagen with other natural polymers improves outcomes when developing tissue-engineered blood vessels [149]. Another promising material used for blood vessel grafts is bacterial cellulose (BC) due to its mechanical and biological properties. These BC grafts exhibit attractive perspectives in using them in future cardiovascular surgeries as blood vessel substitutes [150, 151].

Cellular vascular substitutes can also be obtained in specific (perfusion) bioreactors using 3D vascular graft templates, as presented by Michael et al. [152]. They could design and manufacture 3D electrospun scaffolds based on gelatine and polycaprolactone with appropriate characteristics (fibre diameters and alignment but also porosity) to assist the biomimetic cell organization of the cocultured cells: human vascular endothelial and human fibroblasts.

Conclusions and Future Trends

Tissue engineering is a complex field involving multidisciplinary strategies to develop new tissues and organs. Combining materials, drugs and cells that could imitate specific environments and promote tissue and organ development has significantly impacted tissue engineering [22]. Considering using biomaterials as promising in tissue engineering, smart biomaterials should be considered. Moreover, smart biomaterials could lead to impressive progress in tissue engineering applications by providing better precision in clinical treatments, with relevant development towards advances in diagnosis, promoting the evolution of invasive therapy without side effects. Besides, novel tissue or organ substitutes might substantially minimize the demand for implants and facilitate modern treatments that can help patients with revivable organ failure [2, 153]. Researchers and scientists have shown that biocompatibility, biodegradability, and bioavailability are important features of natural and synthetic materials when using them in developing platforms for tissue engineering. Even though natural polymers have promising results, new technologies must overcome their limitations. Materials design is the coming age in developing materials for medical applications. The properties and performances can be tuned by design, and improving fabrication techniques can allow better experimental models. Additive manufacturing techniques will enable the fabrication of 3D scaffolds from natural/synthetic materials (including metals, ceramics, polymers or even composites), bioactive molecules, and proper living cells. By understanding the interaction mechanisms between the materials and host tissues, representative progress will be made in the tissue engineering area. It has been proven that natural materials can be ideal biomaterials and can be used alone or in association with other materials to fabricate composite structures for tissue engineering. Collagen, chitosan, cellulose, and alginate are suitable matrices for soft tissue engineering. Depending on the final application, different characteristics have to be induced. For instance, an essential characteristic in nerve grafting is electric conductivity, which is why many nerve substitutes contain electro-conductive polymers or carbon-based materials; in synthetic blood vessels, mechanical properties are important, etc. Still, combined with inorganic materials such as calcium phosphates, bioglasses, etc., these could also act as promising solutions for bone tissue engineering. Moreover, combining natural with synthetic polymers could become a viable solution. Thus, for these particular cases, FDA-approved polymers such as poly-L-lactic acid (PLA) and polyglycolic acid (PGA). It has been proven that polylactic-co-glycolic acid (PLGA) may amplify some physical properties of the scaffolds. Also, there are materials such as graphene and graphene oxides and carbon nanotubes that are not yet approved by Food and Drug Administration–FDA (US) or European Medicines Agency–EMEA and could have a real impact in the future and could get approval for being used in medical applications. It is also worth mentioning that additional properties can be desirable depending on the aetiology, such as antimicrobial, antitumoral, and antiosteoporotic efficiency. These properties can be achieved by loading these substitutes with proper biological active agents.

Also, concerns related to the impact of these biomaterials on the environment and the awareness of the need to develop materials with a low carbon footprint which implies less and less waste and energy consumption, are playing an increasing role. Along with these, the design of the more complex architectures, which better mimics the natural tissue/organ that will be replaced or augmented, is expected to lead to better materials and devices, in general. Understanding how nature works may help scientists and researchers develop and design novel materials that mimic nature by replicating exact biological structures and reproducing the perfect behaviour of the biological materials to obtain biomaterials for specific applications. This direction will probably get towards researching new generations of biomaterials, imprinting their understanding in novel applications to revolutionize modern biomedicine.

References

1. Huebsch N, Mooney DJ (2009) Inspiration and application in the evolution of biomaterials. Nature 462(7272):426–432
2. Ratner BD, Bryant SJ (2004) Biomaterials: where we have been and where we are going. Annu Rev Biomed Eng 6:41–75
3. Tathe A, Ghodke M, Nikalje AP (2010) A brief review biomaterials and their application. Int J Pharm Pharm Sci 2(4):19–23
4. Helmus MN, Gibbons DF, Cebon D (2008) Biocompatibility: meeting a key functional requirement of next-generation medical devices. Toxicol Pathol 36:70–80
5. Patel NR, Gohil PP (2012) A review on biomaterials: Scope, applications and human anatomy significance. Int J Emerg Technol Adv Eng 2(4):91–101
6. Ratner BD, Hoffman AS, Schoen FJ, Lemons JE (2020) Biomaterials science: an introduction to materials in medicine. Elsevier Academic Press, London
7. Namvar F, Jawaid M, Md Tahir P, Mohamad R, Azizi S, Khodavandi A, Rahman HS, Nayeri MD (2014) Potential use of plant fibres and their composites for biomedical applications. BioResources 9(3):5688–5706
8. Ramakrishna S, Mayer J, Wintermantel E, Leong KW (2001) Biomedical applications of polymer-composite materials: a review. Compos Sci Technol 61:1189–1224
9. Kutz M (2003) Standard handbook of biomedical engineering and design. In McGraw-Hill (ed)
10. Parida P, Behera A, Mishra SC (2012) Classification of biomaterials used in medicine. Int J Adv Appl Sci 1(3):125–129
11. Chen C, Xi Y, Weng Y (2022) Progress in the development of graphene-based biomaterials for tissue engineering and regeneration. Mater (Basel) 15(6)
12. Chong ETJ, Ng JW, Lee P-C (2022) Classification and medical applications of biomaterials—a mini review. BIO integration
13. Eldeeb AE, Salah S, Elkasabgy NA (2022) Biomaterials for tissue engineering applications and current updates in the field: a comprehensive review. AAPS PharmSciTech 23(7):267
14. Rubežić MZ, Krstić AB, Stanković HZ, Ljupković RB, Ranđelović MS, Zarubica AR (2020) Different types of biomaterials—structure and applications—a short review. Adv Technol 9(1):69–79
15. Anil S, Chalisserry EP, Nam SY, Venkatesan J (2019) Biomaterials for craniofacial tissue engineering and regenerative dentistry. In: Advanced dental biomaterials, pp 643–674
16. Ferreira AM, Gentile P, Chiono V, Ciardelli G (2012) Collagen for bone tissue regeneration. Acta Biomater 8(9):3191–3200
17. Abou Neel EA, Salih V, Revell PA, Young AM (2012) Viscoelastic and biological performance of low-modulus, reactive calcium phosphate-filled, degradable, polymeric bone adhesives. Acta Biomater 8(1):313–320
18. Ardelean IL, Gudovan D, Ficai D, Ficai A, Andronescu E, Albu-Kaya MG, Neacsu P, Ion RN, Cimpean A, Mitran V (2018) Collagen/hydroxyapatite bone grafts manufactured by homogeneous/heterogeneous 3D printing. Mater Lett 231:179–182
19. Marques C, Ferreira JM, Andronescu E, Ficai D, Sonmez M, Ficai A (2014) Multifunctional materials for bone cancer treatment. Int J Nanomedicine 9:2713–2725
20. Pien N, Pezzoli D, Van Hoorick J, Copes F, Vansteenland M, Albu M, De Meulenaer B, Mantovani D, Van Vlierberghe S, Dubruel P (2021) Development of photo-crosslinkable collagen hydrogel building blocks for vascular tissue engineering applications: a superior alternative to methacrylated gelatin? Mater Sci Eng C Mater Biol Appl 130:112460
21. Noori A, Ashrafi SJ, Vaez-Ghaemi R, Hatamian-Zaremi A, Webster TJ (2017) A review of fibrin and fibrin composites for bone tissue engineering. Int J Nanomedicine 12:4937–4961
22. Lavik E, Langer R (2004) Tissue engineering: current state and perspectives. Appl Microbiol Biotechnol 65(1):1–8
23. Sun J, Tan H (2013) Alginate-based biomaterials for regenerative medicine applications. Materials (Basel) 6(4):1285–1309

24. Ribeiro VP, Pina S, Oliveira JM, Reis RL (25 April, 2018) Silk fibroin-based hydrogels and scaffolds for osteochondral repair and regeneration. In: Advanced experimental medicine biology (ed), vol 1058. Springer, Cham, pp 305–325
25. Aliramaji S, Zamanian A, Mozafari M (2017) Super-paramagnetic responsive silk fibroin/chitosan/magnetite scaffolds with tunable pore structures for bone tissue engineering applications. Mater Sci Eng C Mater Biol Appl 70(Pt 1):736–744
26. Li G, Sun S (2022) Silk fibroin-based biomaterials for tissue engineering applications. Molecules 27(9)
27. Yao X, Zou S, Fan S, Niu Q, Zhang Y (2022) Bioinspired silk fibroin materials: from silk building blocks extraction and reconstruction to advanced biomedical applications. Mater Today Bio 16:100381
28. Zhang Y, Ye S, Cao L, Lv Z, Ren J, Shao Z, Yao Y, Ling S (2022) Natural silk spinning-inspired meso-assembly-processing engineering strategy for fabricating soft tissue-mimicking biomaterials. Adv Func Mater 32(27):2200267
29. Ganesh N, Hanna C, Nair SV, Nair LS (2013) Enzymatically cross-linked alginic–hyaluronic acid composite hydrogels as cell delivery vehicles. Int J Biol Macromol 55:289–294
30. Muiznieks LD, Keeley FW (2013) Molecular assembly and mechanical properties of the extracellular matrix: a fibrous protein perspective. Biochim Biophys Acta 1832(7):866–875
31. Zarrintaj P, Manouchehri S, Ahmadi Z, Saeb MR, Urbanska AM, Kaplan DL, Mozafari M (2018) Agarose-based biomaterials for tissue engineering. Carbohydr Polym 187:66–84
32. Ressler A (22, 2022) Chitosan-based biomaterials for bone tissue engineering applications: a short review. Polymers (Basel), 14(16)
33. Xia Y, Wang D, Liu D, Su J, Jin Y, Wang D, Han B, Jiang Z, Liu B (2022) Applications of chitosan and its derivatives in skin and soft tissue diseases. Front Bioeng Biotechnol 10:894667
34. Yang Y, Campbell Ritchie A, Everitt NM (2021) Recombinant human collagen/chitosan-based soft hydrogels as biomaterials for soft tissue engineering. Mater Sci Eng C Mater Biol Appl 121:111846
35. Balagangadharan K, Dhivya S, Selvamurugan N (2017) Chitosan based nanofibers in bone tissue engineering. Int J Biol Macromol 104(Pt B):1372–1382
36. Nova A, Keten S, Pugno NM, Redaelli A, Buehler MJ (2010) Molecular and nanostructural mechanisms of deformation, strength and toughness of spider silk fibrils. Nano Lett 10(7):2626–2634
37. Sponner A, Vater W, Monajembashi S, Unger E, Grosse F, Weisshart K (2007) Composition and hierarchical organisation of a spider silk. PLoS ONE 2(10):e998
38. Vollrath F, Barth P, Basedow A, Engström W, List H (2002) Local tolerance to spider silks and protein polymers in vivo. In vivo (Athens, Greece) 16(4):229–234
39. Flores-Cano JV, Leyva-Ramos R, Mendoza-Barrón J, Guerrero-Coronado RM, Aragón-Piña A, Labrada-Delgado GJ (2013) Sorption mechanism of Cd(II) from water solution onto chicken eggshell. Appl Surf Sci 276:682–690
40. Gergely G, Weber F, Lukacs I, Toth AL, Horvath ZE, Mihaly J, Balazsi C (2010) Preparation and characterization of hydroxyapatite from eggshell. Ceram Int 36(2):803–806
41. Park HJ, Jeong SW, Yang JK, Kim BG, Lee SM (2007) Removal of heavy metals using waste eggshell. J Environ Sci 19(12):1436–1441
42. Ebaretonbofa E, Evans JRG (2002) High porosity hydroxyapatite foam scaffolds for bone substitute. J Porous Mater 9(4):257–263
43. Xu Y, Wang DZ, Yang L, Tang HG (2001) Hydrothermal conversion of coral into hydroxyapatite. Mater Charact 47(2):83–87
44. Dang W, Davlau T, Ylng P, Zhao Y, Nowotmk D, Clow CS, Tyler B, Brern H (1996) Effects of GLIADEL wafer initial molecular weight on the erosion of wafer and release of BCNU. J Control Release 42:89–92
45. Martin DP, Williams SF (2003) Medical applications of poly-4-hydroxybutyrate: a strong flexible absorbable biomaterial. Biochem Eng J 16(2):97–105
46. Sodian R, Sperling JS, Martin DP, Egozy A, Stock U, Mayer JE, Vacanti JP (2000) Fabrication of a trileaflet heart valve scaffold from a polyhydroxyalkanoate biopolyester for use in tissue engineering. Tissue Eng

47. Anseth KS, Metters AT, Bryant SJ, Martens PJ, Elisseeff JH, Bowman CN (2002) In situ forming degradable networks and their application in tissue engineering and drug delivery. J Control Release 78(1):199–209
48. Andronescu E, Nastase M, Ghitulica C, Stefan E (2002) Composite bioceramics. Euro Ceramics Vii, Pt 1–3(206–2):1591–1594
49. Balazsi C, Weber F, Kover Z, Horvath E, Nemeth C (2007) Preparation of calcium-phosphate bioceramics from natural resources. J Eur Ceram Soc 27(2–3):1601–1606
50. Filip N, Radu I, Veliceasa B, Filip C, Pertea M, Clim A, Pinzariu AC, Drochioi IC, Hilitanu RL, Serban IL (2022) Biomaterials in orthopedic devices: current issues and future perspectives. Coatings 12(10)
51. Zhou YL, Wu CT, Chang J (2019) Bioceramics to regulate stem cells and their microenvironment for tissue regeneration. Mater Today 24:41–56
52. Biomaterials (2019) National institutes of health, national institute of biomedical imaging and bioengineering
53. Elflein J (2021) Statistica—global number of organ transplantations 2019
54. Kosowska K, Domalik-Pyzik P, Sekula-Stryjewska M, Noga S, Jagiello J, Baran M, Lipinska L, Zuba-Surma E, Chlopek J (2020) Gradient chitosan hydrogels modified with graphene derivatives and hydroxyapatite: physiochemical properties and initial cytocompatibility evaluation. Int J Mol Sci 21(14)
55. Moizhess TG (2008) Carcinogenesis induced by foreign bodies. Biochem (Mosc) 73(7):763–775
56. Ducheyne P, Qiu Q (1999) Bioactive ceramics—the effect of surface reactivity on bone formation and bone cell function. Biomaterials 20:2287–2303
57. Hench LL (1998) Bioceramics. J Am Ceram Soc 8:1705–1728
58. Seala BL, Oterob TC, Panitcha A (2001) Polymeric biomaterials for tissue and organ regeneration. Mater Sci Eng R 34:147–230
59. LeVeen HH, Rarberio JR (1949) Tissue reaction to plastics used in surgery with special reference to teflon. Ann Surg 129:74–84
60. Hench LL (1988) Bioactive ceramics. Ann NY Acad Sci 523:54–71
61. Bettinger CJ (2009) Synthesis and microfabrication of biomaterials for soft-tissue engineering. Pure Appl Chem 81(12):2183–2201
62. Guo L, Liang Z, Yang L, Du W, Yu T, Tang H, Li C, Qiu H (2021) The role of natural polymers in bone tissue engineering. J Control Release 338:571–582
63. Todros S, Todesco M, Bagno A (2021) Biomaterials and their biomedical applications: from replacement to regeneration. Processes 9(11)
64. Grosberg A, Alford PW, McCain ML, Parker KK (2011) Ensembles of engineered cardiac tissues for physiological and pharmacological study: heart on a chip. Lab Chip 11(24):4165–4173
65. Mohandas G, Oskolkov N, McMahon MT, Walczak P, Janowski M (2014) Porous tantalum and tantalum oxide nanoparticles for regenerative medicine. Acta Neurobiolgiae Experimentalis (Wars) 74(2):5688–5706
66. Niinomi M (2002) Recent metallic materials for biomedical applications. Metall Mater Trans A 33(3):477
67. Shah Idil A, Donaldson N (2018) The use of tungsten as a chronically implanted material. J Neural Eng 15(2):021006
68. Yoo YR, Cho HH, Jang SG, Lee KY, Son HY, Kim JG, Kim YS (2007) Effect of co-content on the corrosion of high performance stainless steels in simulated bio-solutions. Key Eng Mater 342–343:585–588
69. Kolos E, Ruys A (2015) Biomimetic coating on porous alumina for tissue engineering: characterisation by cell culture and confocal microscopy. Materials 8(6):3584–3606
70. Pieralli S, Kohal RJ, Jung RE, Vach K, Spies BC (2017) Clinical outcomes of zirconia dental implants: a systematic review. J Dent Res 96(1):38–46
71. Rahman SU (2020) Hydroxyapatite and tissue engineering. In: Handbook of ionic substituted hydroxyapatites, pp 383–400

72. Freed LE, Vunjak-Novakovic G, Biron RJ, Eagles DB, Lesnoy DC, Barlow SK, Langer R (1994) Biodegradable polymer scaffolds for tissue engineering. Bio/Technol 12:689–693
73. Kim BS, Sung HM, You HK, Lee J (2014) Effects of fibrinogen concentration on fibrin glue and bone powder scaffolds in bone regeneration. J Biosci Bioeng 118(4):469–475
74. Kim HD, Valentini RF (2002) Retention and activity of BMP-2 in hyaluronic acid-based scaffolds in vitro. J Biomed Mater Res 59(3):573–584
75. Seol YJ, Lee JY, Park YJ, Lee YM, Young K, Rhyu IC, Lee SJ, Han S-B, Chung C-P (2004) Chitosan sponges as tissue engineering scaffolds for bone formation. Biotechnol Lett 26:1037–1041
76. Kang E, Choi YY, Chae SK, Moon JH, Chang JY, Lee SH (2012) Microfluidic spinning of flat alginate fibers with grooves for cell-aligning scaffolds. Adv Mater 24(31):4271–4277
77. Shachar M, Tsur-Gang O, Dvir T, Leor J, Cohen S (2011) The effect of immobilized RGD peptide in alginate scaffolds on cardiac tissue engineering. Acta Biomater 7(1):152–162
78. Xie H, Gu Z, Li C, Franco C, Wang J, Li L, Meredith N, Ye Q, Wan C (2016) A novel bioceramic scaffold integrating silk fibroin in calcium polyphosphate for bone tissue-engineering. Ceram Int 42(2):2386–2392
79. Naresh K, Suyash N, Behzad F (2022) Additively manufactured porous Ti6Al4V for bone implants: a review. Metals 12(4)
80. Allo BA, Costa DO, Dixon SJ, Mequanint K, Rizkalla AS (2012) Bioactive and biodegradable nanocomposites and hybrid biomaterials for bone regeneration. J Funct Biomater 3(2):432–463
81. Anupam S (2011) An overview of metallic biomaterials for bone support and replacement, biomedical engineering, trends in materials science. In Laskovski A (ed). InTech China, pp 153–168
82. Mudali UK, Sridhar TM, Raj B (2003) Corrosion of bio implants. Sadhana 28(3&4):601–637
83. Ryan G, Pandit A, Apatsidis DP (2006) Fabrication methods of porous metals for use in Orthopaedic applications. Biomaterials 27(13):2651–2670
84. Behera A (2012) Classification of biomaterials used in medicine. Int J Adv Appl Sci 1(3):125–129
85. Billotte WC (2006) Ceramic biomaterials, pp 339.1–39.34
86. Tateish T (2007) Biomaterials in Asia: in commemoration of the 1st Asian biomaterials (Vol. ISBN: 13-978-981-283-574-1). World Scientific
87. Atala A (2008) Principles of regenerative medicine. Academic Press
88. Sarkar R, Banerjee G (2010) Ceramic based bio-medical implants. Interceram 59(2):98–102
89. Workie AB, Shih SJ (2022) Mesoporous bioactive glasses: synthesis, characterization, and their medical applications. Surf Rev Lett
90. Wu JJ, Wang SX, Zheng Z, Li JB (2022) Fabrication of biologically inspired electrospun collagen/silk fibroin/bioactive glass composited nanofibrous scaffold to accelerate the treatment efficiency of bone repair. Regenerative Ther 21:122–138
91. Yang ML, Yu S, Zhao P, Shi GF, Guo Y, Xie LW, Lyu G, Yu JJ (2022) Fabrication of biologically inspired electrospun collagen/silk fibroin/bioactive glass composited nanofibrous to accelerate the treatment efficiency of wound repair. Int Wound J
92. Akiyama Y, Ito M, Toriumi T, Hiratsuka T, Arai Y, Tanaka S, Futenma T, Akiyama Y, Yamaguchi K, Azuma A, Hata KI, Natsume N, Honda M (2021) Bone formation potential of collagen type I-based recombinant peptide particles in rat calvaria defects*. Regenerative Ther 16:12–22
93. Gassling V, Hedderich J, Acil Y, Purcz N, Wiltfang J, Douglas T (2013) Comparison of platelet rich fibrin and collagen as osteoblast-seeded scaffolds for bone tissue engineering applications. Clin Oral Implant Res 24(3):320–328
94. Geiger M, Li RH, Friess W (2003) Collagen sponges for bone regeneration with rhBMP-2. Adv Drug Deliv Rev 55(12):1613–1629
95. Santos TD, Abuna RPF, de Almeida ALG, Beloti MM, Rosa AL (2015) Effect of collagen sponge and fibrin glue on bone repair. J Appl Oral Sci 23(6):623–628

96. Lutolf MR, Weber FE, Schmoekel HG, Schense JC, Kohler T, Muller R, Hubbell JA (2003) Repair of bone defects using synthetic mimetics of collagenous extracellular matrices. Nat Biotechnol 21(5):513–518
97. Schneider RK, Puellen A, Kramann R, Raupach K, Bornemann J, Knuechel R, Pérez-Bouza A, Neuss S (2010) The osteogenic differentiation of adult bone marrow and perinatal umbilical mesenchymal stem cells and matrix remodelling in three-dimensional collagen scaffolds. Biomaterials 31(3):467–480
98. Mehboob H, Chang S-H (2014) Application of composites to orthopedic prostheses for effective bone healing: a review. Compos Struct 118:328–341
99. Iftekhar A (2004) Biomedical composites. Standard handbook of biomedical engineering and design. McGraw-Hill Companies
100. Salernitano E, Migliaresi C (2018) Composite materials for biomedical applications: a review. J Appl Biomater Funct Mater 1(1):3–18
101. Ficai A, Andronescu E, Ghitulica C, Voicu G, Trandafir V, Manzu D, Ficai M, Pall S (2009) Collagen/hydroxyapatite interactions in composite biomaterials. Materiale Plastice 46(1):11–15
102. Ficai A, Andronescu E, Trandafir V, Ghitulica C, Voicu G (2010) Collagen/hydroxyapatite composite obtained by electric field orientation. Mater Lett 64(4):541–544
103. Ficai A, Andronescu E, Voicu G, Ghitulica C, Vasile BS, Ficai D, Trandafir V (2010) Self-assembled collagen/hydroxyapatite composite materials. Chem Eng J 160(2):794–800
104. Andronescu E, Voicu G, Ficai M, Mohora IA, Trusca R, Ficai A (2011) Collagen/hydroxyapatite composite materials with desired ceramic properties. J Electron Microsc (Tokyo) 60(3):253–259
105. Valtanen RS, Yang YP, Gurtner GC, Maloney WJ, Lowenberg DW (2021) Synthetic and bone tissue engineering graft substitutes: what is the future? Injury 52(Suppl 2):S72–S77
106. Levengood SKL, Zhang M (2014) Chitosan-based scaffolds for bone tissue engineering. J Mater Chem B 2(21):3161–3184. https://doi.org/10.1039/C4TB00027G
107. Wang F, Zhang YC, Zhou H, Guo YC, Su XX (2014) Evaluation of in vitro and in vivo osteogenic differentiation of nano-hydroxyapatite/chitosan/poly(lactide-co-glycolide) scaffolds with human umbilical cord mesenchymal stem cells. J Biomed Mater Res A 102(3):760–768
108. Lee CH, Singla A, Lee Y (2001) Biomedical applications of collagen. Int J Pharm 221:1–22
109. Schmidt CH, Baier JM (2000) Acellular vascular tissues: natural biomaterials for tissue repair and tissue engineering. Biomaterials 21:2215–2231
110. Andronescu E, Voicu G, Ficai M, Mohora IA, Trusca R, Ficai A (2011) Collagen/hydroxyapatite composite materials with desired ceramic properties. J Electron Microsc 60(3):253–259
111. Ficai A, Albu MG, Birsan M, Sonmez M, Ficai D, Trandafir V, Andronescu E (2013) Collagen hydrolysate based collagen/hydroxyapatite composite materials. J Mol Struct 1037:154–159
112. Nitti P, Kunjalukkal Padmanabhan S, Cortazzi S, Stanca E, Siculella L, Licciulli A, Demitri C (2021) Enhancing bioactivity of hydroxyapatite scaffolds using fibrous type I collagen. Front Bioeng Biotechnol 9:631177
113. Bian S, He M, Sui J, Cai H, Sun Y, Liang J, Fan Y, Zhang X (2016) The self-crosslinking smart hyaluronic acid hydrogels as injectable three-dimensional scaffolds for cells culture. Colloids Surf B Biointerfaces 140:392–402
114. Moreira CD, Carvalho SM, Mansur HS, Pereira MM (2016) Thermogelling chitosan-collagen-bioactive glass nanoparticle hybrids as potential injectable systems for tissue engineering. Mater Sci Eng C Mater Biol Appl 58:1207–1216
115. Prianka TR, Subhan N, Reza HM, Hosain MK, Rahman MA, Lee H, Sharker SM (2018) Recent exploration of bio-mimetic nanomaterial for potential biomedical applications. Mater Sci Eng C Mater Biol Appl 93:1104–1115
116. Reddy R, Reddy N (2018) Biomimetic approaches for tissue engineering. J Biomater Sci Polym Ed 29(14):1667–1685
117. Benyus JM (2009) Biomimicry: innovation inspired by nature. Pymble, NSW, Harper Collins

118. Feng C, Zhang W, Deng C, Li G, Chang J, Zhang Z, Jiang X, Wu C (2017) 3D printing of lotus root-like biomimetic materials for cell delivery and tissue regeneration. Adv Sci (Weinh) 4(12):1700401
119. Saruta J, Ozawa R, Okubo T, Taleghani SR, Ishijima M, Kitajima H, Hirota M, Ogawa T (2021) Biomimetic zirconia with cactus-inspired meso-scale spikes and nano-trabeculae for enhanced bone integration. Int J Mol Sci 22(15)
120. Gruber P (2009) Biomimetics in architecture—inspiration from plants 6th plant biomimetics conference—cayenne, November 16–21
121. Karageorgiou V, Kaplan D (2005) Porosity of 3D biomaterial scaffolds and osteogenesis. Biomaterials 26(27):5474–5491
122. Dearman BL, Boyce ST, Greenwood JE (2021) Advances in skin tissue bioengineering and the challenges of clinical translation. Front Surg 8:640879
123. Kuznetsova TA, Andryukov BG, Besednova NN (2022) Modern aspects of burn injury immunopathogenesis and prognostic immunobiochemical markers (Mini-Review). BioTech (Basel) 11(2)
124. Spoială A, Ilie C-I, Ficai D, Ficai A, Andronescu E (2022) Synergic effect of honey with other natural agents in developing efficient wound dressings. Antioxidants 12(1), Article 34
125. Raghunath M, Hopfner B, Aeschlimann D, Luthi U, Meuli M, Altermatt S, Gobet R, BrucknerTuderman L, Steinmann B (1996) Cross-linking of the dermo-epidermal junction of skin regenerating from keratinocyte autografts—anchoring fibrils are a target for tissue transglutaminase. J Clin Investig 98(5):1174–1184
126. Raghunath M, Meuli M (1997) Cultured epithelial autografts: diving from surgery into matrix biology. Pediatr Surg Int 12(7):478–483
127. Hinterhuber G, Marquardt Y, Diem E, Rappersberger K, Wolff K, Foedinger D (2002) Organotypic keratinocyte co-culture using normal human serum: an immunomorphological study at light and electron microscopic levels. Exp Dermatol 11(5):413–420
128. Braziulis E, Biedermann T, Hartmann-Fritsch F, Schiestl C, Pontiggia L, Bottcher-Haberzeth S, Reichmann E, Meuli M (2011) Skingineering I: engineering porcine dermo-epidermal skin analogues for autologous transplantation in a large animal model. Pediatr Surg Int 27(3):241–247
129. Biedermann T, Bottcher-Haberzeth S, Klar AS, Pontiggia L, Schiestl C, Meuli-Simmen C, Reichmann E, Meuli M (2013) Rebuild, restore, reinnervate: do human tissue engineered dermo-epidermal skin analogs attract host nerve fibers for innervation? Pediatr Surg Int 29(1):71–78
130. Biedermann T, Klar AS, Bottcher-Haberzeth S, Schiestl C, Reichmann E, Meuli M (2014) Tissue-engineered dermo-epidermal skin analogs exhibit de novo formation of a near natural neurovascular link 10 weeks after transplantation. Pediatr Surg Int 30(2):165–172
131. Hartmann-Fritsch F, Biedermann T, Braziulis E, Luginbuhl J, Pontiggia L, Bottcher-Haberzeth S, van Kuppevelt TH, Faraj KA, Schiestl C, Meuli M, Reichmann E (2016) Collagen hydrogels strengthened by biodegradable meshes are a basis for dermo-epidermal skin grafts intended to reconstitute human skin in a one-step surgical intervention. J Tissue Eng Regen Med 10(1):81–91
132. Pontiggia L, Van Hengel IAJ, Klar A, Rutsche D, Nanni M, Scheidegger A, Figi S, Reichmann E, Moehrlen U, Biedermann T (2022) Bioprinting and plastic compression of large pigmented and vascularized human dermo-epidermal skin substitutes by means of a new robotic platform. J Tissue Eng 13
133. Qiang A (2016) Progress of nerve bridges in the treatment of peripheral nerve disruptions. J Neurorestoratology 4:107–113
134. Abdelbasset WK, Jasim SA, Sharma SK, Margiana R, Bokov DO, Obaid MA, Hussein BA, Lafta HA, Jasim SF, Mustafa YF (2022) Alginate-based hydrogels and tubes, as biological macromolecule-based platforms for peripheral nerve tissue engineering: a review. Ann Biomed Eng 50(6):628–653
135. Gu XK, Yi S, Deng AD, Liu H, Xu L, Gu JH, Gu XS (2022) Combined use of chitosan-PGLA nerve grafts and bone marrow mononuclear cells to repair a 50-mm-long median nerve defect

combined with an 80-mm-long ulnar nerve defect in the human upper arm. Curr Stem Cell Res Ther 17(4):389–397
136. Khan HM, Liao XX, Sheikh BA, Wang YX, Su ZX, Guo C, Li ZY, Zhou CC, Cen Y, Kong QQ (2022) Smart biomaterials and their potential applications in tissue engineering. J Mater Chem B 10(36):6859–6895
137. Nune M, Bhat M, Nagarajan A (2022) Design of ECM functionalized polycaprolactone aligned nanofibers for peripheral nerve tissue engineering. J Med Biol Eng 42(2):147–156
138. Wang B, Lu CF, Liu ZY, Han S, Wei P, Zhang DY, Kou YH, Jiang BG (2022) Chitin scaffold combined with autologous small nerve repairs sciatic nerve defects. Neural Regeneration Res 17(5):1106-+
139. Zeng ZP, Yang YJ, Deng JY, Rahman MSU, Sun CM, Xu SS (2022) Physical stimulation combined with biomaterials promotes peripheral nerve injury repair. Bioeng-Basel 9(7). https://doi.org/10.3390/bioengineering9070292
140. Zhang FS, Zhang M, Liu SY, Li C, Ding ZT, Wan T, Zhang PX (2022) Application of hybrid electrically conductive hydrogels promotes peripheral nerve regeneration. Gels 8(1)
141. Zhang LL, Zheng TT, Wu LL, Han Q, Chen SY, Kong Y, Li GC, Ma L, Wu H, Zhao YH, Yu YX, Yang YM (2021) Fabrication and characterization of 3D-printed gellan gum/starch composite scaffold for Schwann cells growth. Nanotechnol Rev 10(1):50–61
142. Zhang XD, Meng YX, Gong BW, Wang T, Lu YL, Zhang LQ, Xue JJ (2022) Electrospun nanofibers for manipulating soft tissue regeneration. J Mater Chem B 10(37):7281–7308
143. Lin B, Dun G, Jin D, Du Y (2019) Development of polypyrrole/collagen/nano-strontium substituted bioactive glass composite for boost sciatic nerve rejuvenation in vivo. Artif Cells Nanomed Biotechnol 47(1):3423–3430
144. Nerem RM, Ensley AE (2004) The tissue engineering of blood vessels and the heart. Am J Transplant 4(Suppl 6):36–42
145. Bertanha M, Moroz A, Almeida R, Alves FC, Acorci Valerio MJ, Moura R, Domingues MA, Sobreira ML, Deffune E (2014) Tissue-engineered blood vessel substitute by reconstruction of endothelium using mesenchymal stem cells induced by platelet growth factors. J Vasc Surg 59(6):1677–1685
146. Copes F, Pien N, Van Vlierberghe S, Boccafoschi F, Mantovani D (2019) Collagen-based tissue engineering strategies for vascular medicine. Front Bioeng Biotechnol 7:166
147. Antunes M, Bonani W, Reis RL, Migliaresi C, Ferreira H, Motta A, Neves NM (2022) Development of alginate-based hydrogels for blood vessel engineering. Biomater Adv 134:112588
148. Idaszek J, Volpi M, Paradiso A, Nguyen Quoc M, Górecka Ż, Klak M, Tymicki G, Berman A, Wierzbicki M, Jaworski S, Costantini M, Kępczyńska A, Chwalibóg ES, Wszoła M, Święszkowski W (2021) Alginate-based tissue-specific bioinks for multi-material 3D-bioprinting of pancreatic islets and blood vessels: a step towards vascularized pancreas grafts. Bioprinting 24:e00163
149. Du J, Hu X, Su Y, Wei T, Jiao Z, Liu T, Wang H, Nie Y, Li X, Song K (2022) Gelatin/sodium alginate hydrogel-coated decellularized porcine coronary artery to construct bilayer tissue engineered blood vessels. Int J Biol Macromol 209(Pt B):2070–2083
150. Brewster L, Brey EM, Greisler HP (2014) Blood vessels, pp 793–812
151. Scherner M, Reutter S, Klemm D, Sterner-Kock A, Guschlbauer M, Richter T, Langebartels G, Madershahian N, Wahlers T, Wippermann J (2014) In vivo application of tissue-engineered blood vessels of bacterial cellulose as small arterial substitutes: proof of concept? J Surg Res 189(2):340–347
152. Michael PL, Yang NJ, Moore M, Santos M, Lam YT, Ward A, Hung JC, Tan RP, Wise SG (2022) Synthetic vascular graft with spatially distinct architecture for rapid biomimetic cell organisation in a perfusion bioreactor. Biomed Mater 17(4)
153. Amukarimi S, Ramakrishna S, Mozafari M (2021) Smart biomaterials—a proposed definition and overview of the field. Curr Opin Biomed Eng 19

Chapter 5
Micro and Nanotechnology

Chukwuka Bethel Anucha and Erwann Guénin

Abstract Scale measurement is one of the frequently used approaches and tools for classifying micro, nano, and even macro technological application domains, despite the fundamentally significant variations in the chemical and physical mechanisms. Unit scales of macro, micro, and nano are used to understand and interpret better the size of objects in an associated and related manner at different level of scales in order to be able to understand the boundary-evading nature of materials fabricated at different levels for general and specialized fit-for-purpose applications. This book chapter focused on a relatively small portion of the larger and far-reaching field of micro/nanotechnology, which currently encompasses every aspect of science and engineering as well as anything we could possibly dream of or envision.

Keywords Technological advancements · Nanofabrication · Microfabrication · Scale measurement · Biosensing

The History of Micro and Nanotechnologies

Historically, materials always have existed at different dimensional scales, and humans have used different particles and structures since the first ancient civilization. Whether they had been aware or not, the Romans and the Damascans demonstrated one of the most interesting technologies in the ancient world by respectively using nanoparticles to create iridescent glasswares and exceptionally sharp edge swords. Developmental traces of nanoscience in the time of ancient Greek philosophers like Democritus in the fifth century led scientist asked whether matter is a continuum, and therefore, indefinitely divisible into smaller portions, or made up of tiny, indivisible, and indestructible components which today's scientists refer to as atoms [1]. This can only consolidate the belief that archaic artisans were nanotechnologists who had practically used nanoscience to develop historic artefacts [1]. As early as

C. B. Anucha · E. Guénin (✉)
Centre de Recherche Royallieu, Université de Technologie de Compiègne, ESCOM, TIMR (Integrated Transformations of Renewable Matter), CS 60 319, 60 203 Compiègne Cedex, France
e-mail: erwann.guenin@utc.fr

O. Gunduz et al. (eds.), *Biomaterials and Tissue Engineering*, Stem Cell Biology and Regenerative Medicine 74, https://doi.org/10.1007/978-3-031-35832-6_5

the fourth century, Roman artists had discovered that adding gold and silver to glass developed a remarkable effect of colour change with the glass appearing slate green when lit from outside and glowing red when lit from within [2–4]. The ceremonial Lycurgus cup is the most famous surviving example of this technique [1, 2, 4]. While nanoscience intermarries nanotechnology for its practical application, the same goes with the relationship between microtechnology and nanotechnology. Over the course of modern historical evolution of scientific and technological advancements, by scale measurement, micro and nanotechnological materials have progressed hand in hand and used in various application domains. In essence, nowadays, we can buy available precision technological processes workup materials at microscale or work them down to nanoscale, thus, the interwoven nature of both technologies exist at scale boundary line.

In a historical precedence, groundbreaking discoveries and events continue to emerge in an evolutionary timeline fashion resonating the witnessed progress advancement of nanoscience and technology. For instance: Stained glass windows of cathedrals (500–1450 BC) [5], Iridescent/metallic clusters of the Derutta Poetry (1450–1600 BC) [5], colloidal rugby gold nanoparticles synthesis [6], light scattering of nanoparticles (Mie 1908) [7], near field optical microscope [8], cathode ray oscilloscope (CRO) [9], field-ion electron microscope reporting the first surface atom visualization [10], Watson and Crick discovery of DNA [11], electron tunneling [12], etc.; are in a random mix, few of the huge achievements already recorded within the sphere of nanoscience and nanotechnology and has offered highly innovative solutions to various scientific and technological domains tackling human challenges. Table 5.1 gives a more comprehensive overview of the evolutionary trend timeline marking the events that has shaped historical progress in the advancement of nanoscience and nanotechnology to date. This timeline depicts premodern examples of nanotechnology, and in like manner, modern era discoveries so far reached as milestones.

Heralding the era of what today is referred to as modern nanotechnology, American Physicist and nobel laurate Richard Feynman, came up first with ideas and concepts behind nanotechnology in 1959 [1]. At this time and during the annual meeting of the American Physical Society held at California Institute of Technology (Caltech), Feynman presented a lecture titled "There's Plenty of Room at the Bottom" [25]. Though not mentioning the word 'nanotechnology' in his lecture, Feynman posed the question: "why can't we write the entire 24 Encyclopedia Britannica on the head of a pin"?, laying down the possibility of using machines to fabricate and construct smaller machines to molecular level scale [31]. From this point onward, a revolutionary trend in the manipulation of materials at different scale range levels of "micro" and "nano" erupted giving birth to the traverse being witnessed today in the diverse domains of science, engineering and technology for the development of best result outcome materials as a function of size, shape and functions entirely different from their bulk state.

Table 5.1 Evolutionary trend of progress advancement in nanoscience and nanotechnology

Year	Event/discovery & milestone	Brief introduction about finding	References
4th century	The Lycurgus cup (Rome)	An example of a dichroic glass of colloidal gold and silver that makes the glass look opaque green when lit from outside and translucent red when light shines through the inside	[1]
6th–15th centuries	Vibrant stained glass windows (Seen in European cathedrals)	Their rich colours owed to nanoparticles of gold chloride ($AuCl_4$), and other metal oxides and chlorides	[5]
15th–16th centuries	The Italian Renaissance pottery	The Deruta and Gubbio centre became the most famous centre for the production of lustred glazed majolica containing nanoparticle components	[13, 14]
9th–17th centuries	Glowing, glittering "luster" ceramic glazes	Used in Islamic world and later in Europe contains silver or copper or other metallic nanoparticles	[15]
13th–18th centuries	"Damascus" saber blades	Contain carbon nanotube (CNT), and cementite nanowires: an ultra-high carbon-steel formulation that bequeathed them with strength, resilience, ability to hold a well defined edge, and pronounced moiré pattern in the steel from where the blades name originated	[16]
1857	Colloidal "ruby" gold	Michael Faraday demonstrated that nanostructured gold under certain lightening conditions produces different coloured solutions	[5]
1897	Cathode ray oscilloscope (CRO)	Ferdinand Braun developed the CRO for measuring and displaying different forms of electrical signals	[9]
1895	X-ray diffraction (XRD)	Wilhelm Conrad Rontgen stumbled across effects that could create images on fluorescent screens and subsequently on photographic plates at a distance from an optically hidden crooks or similar structured tubes	[17]
1908	Light scattering of nanoparticles	Gustav Mie demonstrated the scattering of nanoparticles by light	[7]

(continued)

Table 5.1 (continued)

Year	Event/discovery & milestone	Brief introduction about finding	References
1928	Near field optical microscope	Edward Hutchinson Synge laid out and published an incredibly complete theoretical description of near field microscope: an instrument used in nanotechnology several decades before being experimentally developed	[8]
1928	Raman spectroscopy	Sir C.V. Raman discovered a change that beckoned in the frequency and wavelength of deflected light when passed through a spectrograph	[18]
1931	Transmission electron microscopy (TEM)	Max Knoll and Ernest Ruska invented the technique where beam of electrons is transmitted through a specimen to form an image	[19]
1936	Field electron microscope (FEM)	Erwin Müller while working at Siemens research laboratory invented the field emission microscope that would allow near atomic resolution images of materials	[10]
1937	Scanning electron microscopy (SEM)	Manfred von Ardenne made an invention that utilized electromagnetic lenses to focus scanning electron beam on the target surface and then collect scattered electrons giving back information on the sample topography and structure	[20]
1947	Semiconductor transistor	John Barden, William Shockley, and Walter Brattain at Bells Laboratory made the discovery of semiconductor transistor and expanded greatly the scientific knowledge of semiconductor interfaces, which today has turned out to be the foundation for electronic devices and information age	[9]
1950	Growth process for monodispersed colloidal materials	Victor La Mer & Robert Dinegar developed the theory and process for growing monodispersed colloidal materials which immensely contributed to the ability of controlling and fabricating myriad of materials for diverse industrial applications like specialized papers, paints, thin films, and even for dialysis treatments	[21]

(continued)

Table 5.1 (continued)

Year	Event/discovery & milestone	Brief introduction about finding	References
1951	Field Ion microscope (FIM)	Erwin Müller pioneerd the field ion microscope- a means to image the arrangement of atoms at the surface of a sharp tip. He first recorded the image of tungsten atoms	[10, 22]
1953	DNA discovery	James Watson & Francis Crick discovered the molecular structure of Deoxyribose Nucleic Acid (DNA) laying the foundation and paving the way on interdisciplinary research involving scientist from diverse background of physics, chemistry, material science, computer science and medicine in finding solutions to challenges bordering on related issues with this intricate and complex building block of human cells and tissues	[12]
1956	Molecular engineering	Arthur von Hippel introduced many concepts bordering on nanotechnology and coined the term "Molecular Engineering" as applied to dielectrics, ferroelectrics, piezoelectrics, etc.	[23]
1958	Electron tunneling	Leo Esaki first observed and interpreted the negative resistance phenomena in degenerate germanium semiconductor p-n junctions	[12]
1958	Integrated circuit	Jack Kilby of Texas Instruments devised the concept, designed and built the first integrated circuit	[24]
1959	"There is plenty of room at the bottom"	Richard Feynman of the California institute of technology during an American physical society meeting at Caltech delivered what is widely considered to be the first lecture on technology and engineering at atomic scale	[25]
1960	Zeolites and catalysis	Charles Plank and Edward Rosinki developed and patented crystalline zeolite composite for the catalytic cracking of hydrocarbons	[26]

(continued)

Table 5.1 (continued)

Year	Event/discovery & milestone	Brief introduction about finding	References
1963	Ferrofluids	Stephen Papell achieved and patented low viscosity magnetic fluid he had obtained by the colloidal suspension of magnetic particles	[27]
1965	Moore's law	Intel's co-founder Gordon E. Moore described in electronics Magazine several trends he foresaw in the field of electronics. One of such trends known today as Moore's law was that the density of transistors on an integrated chip (IC) will double every 12 months (later amended to 2 years). Moore also saw chip sizes and costs shrinking with their growing functionality as they show transformational effect on the way people live and work	[28]
1969	X-ray photoemission spectroscopy (XPS)	A technique for analyzing material's surface chemistry was discovered	[29]
1970	C60 existence as icosahedron	Eiji Osawa was first to have predicted the existence of C60 as icosahedron structure	[30]
1974	First time the term "Nanotechnology" was used	Prof. Norio Taniguchi was the first to coin the term "Nanotechnology" describing the precision of manipulating materials to within atomic scale dimensional limits	[31]
1974	Molecular electronics	Mark A. Ratner & Arieh Avram constructed a rectifier as a simple electronic device based on the use of a simple organic molecule- methylene	[32]
1977	Surface enhanced Raman spectroscopy (SERS)	Richard P. van Dunye and Co-workers verified that by Raman spectroscopy, adsorbed pyridine on a silver surface showed a remarkable sensitivity, which by extension was applicable to other nitrogen heterocycles and amines	[33]
1980	Self assembly monolayers (SAMs)	Jacop Sagip formulated oleophobic mixed layer structures on solid surfaces	[34]

(continued)

Table 5.1 (continued)

Year	Event/discovery & milestone	Brief introduction about finding	References
1981	Scanning tunneling microscope (STM)	Gerd Binning & Heinrich Rohrer discovered the technique that makes possible the imaging of surfaces at atomic scale, and facilitate the creation of atoms and molecules for the creation of structures	[35, 36]
1981	Nanocrystalline quantum dots in a glass matrix	Russian Scientist Alexei Ekimov discovered nanocrystalline semiconducting quantum dots in a glass matrix and conducted pioneering studies of their electrical and optical properties	[37]
1981	Molecular engineering	Eric Drexler developed method equipped with general capabilities for the realization of various manipulated molecules	[38]
1982	DNA concept of nanotechnology	Nadrian Seeman shaded light on the structural junctions and lattices of DNA, paving way for more opportunities and related challenges of DNA nanotechnology	[39, 40]
1983	Colloidal quantum dot	Louis Brus discovered colloidal quantum dots studying the electronic and optical properties of cadmium selenide (CdS)	[41, 42]
1985	Buckminsterfullerene C60 discovery	Harold Kroto, Sean O'Brien, Robert Curl, and Richard Smalley- researchers from Rice university discovered the buckminsterfullerene commonly known as Buckyball: a molecule of the shape of a ball composed entirely of carbon as graphite and diamond	[43]
1986	Atomic force microscopy (AFM) or the scanning force microscope (SFM)	IBM scientists-Gerd Binnig, Calvin Quate, and Christoph Gerber co-invented the first experimental AFM apparatus allowing the use of an atomic size tip to probe the surface of a material, facilitating the creation of a 3D map of the surface to be constructed, down to the scale of individual atoms	[44, 45]

(continued)

Table 5.1 (continued)

Year	Event/discovery & milestone	Brief introduction about finding	References
1987	Single-electron tunneling (SET) transistor	Dimitri Averin and Konstantin Likharev developed a macroscopic approach to the theoretical examination of small, current-biased tunnel junctions used to account for the secondary quantization of both the single electron (quasiparticle), and Cooper-pair (Josephson) current components	[46]
1990	Manipulation of the 35 individual Xenon (Xe) atoms to inscribe and spell out the IBM logo	Don Eigler & Erhard Scweizer at IBM's Almaden research center demonstrated the ability to precisely manipulate atoms: a turning point that ushered in the applied use of nanotechnology	[47]
Early '90s	Pioneer nanotechnology companies start to be operational	Nanotechnology companies like Nanophase technologies (1989, Helix energy Solutions group (1990), Zyvex (1991), Nano-Tex (1998), etc.; commenced operations	(Initiative)
1991	Multi-wall carbon nanotube discovery	Though there were early observations of tubular carbon structures, Sumio Iijima of Japanese multinational information technology and electronics corporation (NEC), is credited with the discovery of Carbon Nanotube (CNT): a material like the buckyball but tubular in shape boasting of extraordinary properties of strength, electrical and thermal conductivity amongst other unique features	[48]
1992	Nanostructured catalytic materials MCM-41 & MCM-48	Charles T. Kresge and colleagues at Mobil oil discovered nanostructured catalytic materials: MCM-41 & MCM-48 -mesoporous molecular sieve silica materials that have found wide applications in crude oil refining, drug delivery, water treatment, etc.	[49, 50]
1993	Single-wall carbon nanotube discovery	Sumio Iijma & Donald Bethune grew by cobalt catalyzed reaction; carbon nanotubes with single atomic layer walls and shell of diameter 1 nm	[51, 52]
1993	Method for controlled synthesis of nanocrystals (Quantum dots)	Moungi Bawendi of MIT, and contribution works of other researchers like Louis Brus, and Chris Murray resulted to methods for synthesizing quantum dots for applications ranging from computing to biology to high-efficiency photovoltaics and lighting	[53]
1996	Self assembly molecule (SAM) of DNA + gold colloids	Chad Mirkin & Robert Letsinger developed DNA-based method for the rational manipulation of nanoparticles into macroscopic materials	[54]

(continued)

Table 5.1 (continued)

Year	Event/discovery & milestone	Brief introduction about finding	References
1998	Nanotechnology Research Directions: Vision for the Next Decade (1999)	The Interagency working group on nanotechnology was formed under the auspices of the national science and technology council to investigate the state of the art in nanoscale science and technology to forecast possible future developments	[48]
1998	Transistor Creation with Carbon Nanotube	Cees Dekker developed a room temperature operating Carbon nanotube-based transistor	[55]
1999	Invention of Dip-Pen Nanolithography	Chad Mirkin at Northwestern university developed dip-pen nanolithography leading to the manufacturing, and reproducibility of electronic circuits, biomaterial patterning for cell biology, nanoencryption, and other applications	[56]
1999	Molecule Assemblage	Cornell university researchers Wilson Ho and Hyojune Lee investigated the secrets of chemical bonding by assembling iron carbonyl ($Fe(CO)_2$ molecule built from its components of iron (Fe) and carbon (ii) oxide (CO_2) using a scanning tunneling microscope (STM)	[57]
2000	Feedback-Controlled Lithography (FCL)	Mark Hersam & Joseph Lyding developed reliable technique for making individual dangling bond templates on Si(100)-2 × 1:H surface, detecting individual H desorption events while patterning for tip structure variation compensation	[58]
1999–early 2000's	Consumer Products derived by Nanotechnology starts to make Market Entry	consumer products like: light weight nanotechnology -enabled automobile bumpers with denting and scratching resistances, straighter fly golf balls, stiffer tennis rackets, better flex and kick baseball bats, nano-silver antibacterial socks, clear sunscreens, wrinkle and stain resistant clothing, deep-penetrating therapeutic cosmetics, scratch resistant glass coatings, faster-recharging batteries for cordless electric tools, improved TV, cell phones, and digital camera displays etc. made their ways into the market	[48]

(continued)

Table 5.1 (continued)

Year	Event/discovery & milestone	Brief introduction about finding	References
2000	United States National Nanotechnology Initiative (US NNI)	President Bill Clinton announced the US NNI for the coordination of the Federal R&D efforts and the positioning of the US competitiveness in nanotechnology	[59]
2001	Molecular Nanomachines: molecular motor (rotor) with nanoscale silicon devices	Carlo Montemagno explored vision of realizing molecular nanomachines through the development of molecular motor powered by nanosilicon components	(CD [60])
2002	DNA functionalized carbon nanotube	Cees Dekker and co-workers conducted the study of unifying single walled carbon nanotube (SW-CNT) by coupling to peptide nucleic acid (PNA)- an uncharged DNA analogue and hybridized the macromolecular wires with complementary DNA thereby unravelling a pathway for achieving versatilized SWCNTs as probes in biological systems	[61]
2003	21st century nanotechnology and development act (P.L. 108–153) enacted by the US congress and officially signed into law	The US congress enacted the 21st century nanotechnology and development Act (P.L. 108–153), providing a statutory foundation for the NNI, established programs, assigned agency responsibilities, authorized funding levels, and promoted research to address key issues and was finally signed into law by United States President, President George W. Bush	[48]
2003	Development of gold nanoshells	Naomi Halas, Jennifer West, and Renata Pasqualin at Rice university developed gold nanoshells tunned in size to absorb near infrared light, serving as integrated discovery, diagnosis, and breast cancer treatment platform with no invasive biopsies, surgery, systemic destructive radiation or chemotherapy	[62, 63]
2004	The EU commission adoption of "towards a European strategy for nanotechnology"	The EU commission adopted the communication "Towards a European strategy for nanotechnology" COM (2004) 338, with the proposal of intitutionalizing European nanoscience and nanotechnology Research and Developments efforts within an integrated and responsible strategy triggering European action plans and progressive funding for nanotechnology	[48]

(continued)

Table 5.1 (continued)

Year	Event/discovery & milestone	Brief introduction about finding	References
2004	Nanoscience and nanotechnology: opportunities and uncertainties published by Britain's royal society and royal society of engineering	In this publication, the Britain's royal society and royal society of engineering advocated the need to address potential health, environmental, social, ethical, and regulatory issues associated with nanotechnology	[48]
2004	College of nanoscale Science and nanotechnology engineering launched	SUNY Albany launched the first college level education programme in Nanotechnology in the US	[48]
2004	Graphene discovery	Adre Geim & Konstantin Novoselov discovered graphene material as atomically thin carbon film	[64]
2004	Fluorescent carbon dot discovery	Xu et al.; discovered single-walled fluorescent carbon dots	[65]
2005	Nanocar on buckyball wheels	James Tour and colleagues at Rice university built nanoscale car turning on buckyball wheel	[66, 67]
2005	DNA -based computation and algorithmic self assembly	Erik Winfree and Paul Rothemund from the California institute of technology (CIT) developed theories for DNA -based computation and algorithmic self assembly in which computations are embedded in the processes of nanocrystal growth	[48]
2006	DNA origami	By his own coined out term @"scaffolded DNA origami, Paul Rothemund was able to assemble six different shapes of DNA strands, e.g., squares, triangles, five-pointed stars etc.	[68]
2007	Type of bacteria lithium ion battery	Angela Belcher and Colleagues at MIT built lithium-ion battery with a common type of bacteria that is unharmful to humans	[48]
2007	Artificial molecular machines	J. Fraser Stoddart developed artificial molecular machines that is like a pH -triggered muscle	[69]

(continued)

Table 5.1 (continued)

Year	Event/discovery & milestone	Brief introduction about finding	References
2008	Development of green fluorescent protein (GFP) won a nobel	Osamu Shimomura, Martin Chalfie, and Roger Y. Tsien were awarded Nobel prize in Chemistry for their collaboration on the development of green fluorescent protein (GFP)	[70]
2008	NNI strategy for nanotechnology-related environmental, health, and safety (EHS) research published	The NNI strategy for Nanotechnology related EHS was published after 2-year process of NNI-sponsored investigations and public dialogs. The strategy was further updated in 2011 after series of workshops and public review	[48]
2009–2010	DNA-like robotic nanoscle assembly devices	Nadrian Seeman and colleagues at New York University developed several DNA-like robotic nanoscale assembly device, and also created DNA assembly line for which he shared the prestigious Kavli Prize in Nanoscience 2010	[71]
2010	Development of an ultra-fast lithography to create 3D nanoscale textured surface	IBM developed the method of creating nanoscale patterns and structures as small as 15 nm at a reduced cost and complexity opening new prospects in the field of electronics, optoelectronics, and medicine	[72]
2011	Electro-mechano properties of individual molecules and the polymer chains	Leonhard Grill used scanning tunneling microscope (STM) for the description of electrical, and mechanical properties of individual molecules and polymer chain	[73]
2011	Updating of the NNI strategic plan and NNI environmental, health, and safety research strategy	The nanoscale science, engineering and technology -NSET subcommittee of the NNI updated the NNI strategic plan and the EHS strategy drawing extensive input from public workshops, and online dialog with stakeholders from Government, academia, NGOs, the public, and others	[48]
2012	Launching of two more nanotechnology signature initiatives (NSIs)	The NNI launched nanosensors, and the nanotechnology knowledge infrastructure (NKI)	[48]

(continued)

Table 5.1 (continued)

Year	Event/discovery & milestone	Brief introduction about finding	References
2013	NNI strategic planning	NNI starts next round of strategic planning commencing with stakeholder workshop	[48]
2013	Carbon nanotube computer	Standford researchers developed first carbon nanotube computer	[48]
2014	Release of updated strategic planning	NNI releases the updated 2014 strategic plan, progress review on the coordinated implementation of the NNI 2011 EHS research startegy	[48]
2016	Nobel for design and synthesis of molecular machines	Jean-Pierre Sauvage, Sir J. Fraser Stoddart, and Bernard L. Feringa won the 2016 nobel prize in chemistry for their work on design and synthesis of molecular machines	[74]
2017	Nobel for gravitational waves	Nobel prize in physics for work on gravitational waves	[31]
2018	World smallest game board with DNA	World smallest tic-tac-toe game board was developed with DNA	[75]
2018	Object shrinkage	Object shrinkage to nanoscale was demonstrated	[76]

Definition of Microtechnology

Micro and nanotechnologies are roughly related on scale basis; however do have significant differences underlying their physical and chemical mechanisms [77–79]. However, no hardline differences between these technologies, scientific and technology communities regimentally approach them differently [77–81]. Hence, the challenge of arriving to an all-inclusive and sufficient definition of these concepts transcending multiple disciplinary frontiers with no fit-in-one category exists. As particularly pervasive as they are, microtechnology extends into nanotechnology. Just as science and technology have always being heavily intertwined, impossible to discuss and indeed advance independently, such connection exists between micro and nanotechnology. In essence, both micro and nanotechnologies are about miniaturizing and scaling materials down to produce other material products. Majority of material manufacturing processes which earlier were classified as microtechnological processes have now further being scaled down and falling within the scope of nanotechnology. Microtechnology is the application of technological processes in the manufacture of miniaturized objects or systems at microscale [77–79]. Matter manipulation at this scale is within the range of 1micron (μm) or 10^{-6} m (in the equivalent of meter) or 10^{-3} mm (in the equivalent of millimeter) or even 10^{3} nm (in the equivalent of nanometer scale).

Definition of Nanotechnology

Fifteen years later following the exploration of Feyman, the word "Nanotechnology" was first used and defined as "a process comprising mainly of separation, consolidation, and deformation of materials by one atom or one molecule" by Norio Tanuguchi, a Japanese scientist [31]. In recognition of his inspiring contributions, Richard Feynman was credited and widely regarded today as the father of modern nanotechnology [1]. Following the discovery and insight into this new research field, scientists and researchers were highly illuminated with a trigger of huge interest, focus and concerted efforts. Such scientific research community approach has advanced findings with tremendous impact translating nanoscience capability of manipulating materials at nano and molecular scales to observing, measuring, manipulating, assembling, controlling and manufacturing matter at the nanometer scale [80]. In the light of the huge potential prompting the declaration of nanotechnology as one of the most promising technologies of the twenty-first century, the National Nanotechnology Initiative (NNI) in the United States of America (USA) defined nanotechnology as the science, engineering, and technology of manipulating materials at the nanoscale range of (1–100 nm) size, making it possible to leverage on their exhibited outstanding properties of sizes and shapes at this scale; for extensive novel application to variety of fields ranging from chemistry, physics, and biology to medicine/life sciences, engineering, and electronics [48]. Enshrined in the NNI definition of nanotechnology are the structural manipulation of materials based on sizes and shapes at nanoscale, and the inherent novelty taking advantage of materials manipulation at this scale where unique physical, chemical, and biological properties not obtainable at either smaller scales like atoms nor larger scales of millimeters/ inches frequently used in everyday life emerge [48].

Increased information capabilities, system miniaturization, new material science development with increased functionality and autonomy are some of natural consequences in the advances of micro and nanotechnologies resultant effect of scaling to small size [77–81]. The definition of microtechnology do not completely border on categorized microscale dimension materials but could however also be classified based on manufacturing composition for instance microelectrionic material devices originally fabricated with microscale development processes [77, 81]. The transitional means in fabrication of micro and nanotechnologies offers material workability at both scales to meet need for specific applications. This interplay has led to revolutioned material advancement with robust and ultimate properties emerging at nanoscale [81]. Typically, apparent trends in microelectronics with component miniaturization, increased capability as per information density, reduced cost function, increased reliability and ruggedness etc., are improved features at the micro-nanotechnologies interface [77, 81].

In absolute sense, the two particularly pervasive themes of microtechnology with extension into nanotechnology lie interfacially at the convergence of science disciplines of chemistry, biology, physics, and material science and subsequently lay foundation for nanoscience. While nanoscience interwove with these principal core science disciplines, the technological application of the knowledge emerging from the involved nanoscience interconnectivity delivers the capability of measurement observation, manipulation, assemblage, control, and matter manufacturing at the scale of nanometer and micrometer. In accounting for a more regulated and laid out definition of micro and nanotechnologies, several organizations and regulatory bodies and agencies have provided explanation to scientific terms associated with research areas focusing on the understanding and manipulation of matter at atomic and molecular scales. The associative defined terms as relayed in Table 5.2, complements the capacity for the observation, measurement, manipulation, production and manufacturing of materials at nano scale to facilitate their integration and incorporation into miniaturized microsystems, components and subcomponents fit for variety of applications.

Differences Between Macro, Micro and Nano Scale

Despite the underlying significant differences in the chemical and physical mechanisms of micro, nano, and even macrotechnological application domains, one of the widely used approach and means for their classification is scale measurement [96]. In order to be able to come to terms with across the boundary evading nature of materials fabricated at different levels for general and specialized fit –for-purpose applications;

Table 5.2 Nanotechnology and related terms definition by international regulatory bodies

Regulatory body/ committee/organization	Term	Definition	References
British Standards Institution (BSI)	Nanoscale	Nanometer size measurement of range ~1 to 100 nm	[82]
British Standards Institution (BSI)	Nanoscience	Study of the understanding of size and structure dependent features of nano-scaled matter and their comparison for related differences with individual atoms, molecules or bulk materials	[82]
Technical committee of the international standard organization- ISO/TC 229: Nanotechnologies	Nanotechnology	Available scientific data/ information for the manipulation and control of nano-matter (100 nm) needed for the harnessing of its size- and specific structural properties different from its individually existing atoms/molecules, or bulk materials	[83]
European Patent Office (EPO)	Nanotechnology	Description of entities controlled at geometrical sizes of < 100 nm displayed by atleast one functional component of one or more dimensional exhibition with resultant effect to size-dependent physical, chemical, and biological responses	[84]
American National Standard Institute (ANSI)- Nanotechnology standards panel	Nanotechnology	Manipulation and control of matter in nanoscale (~1 to 100 nm), at which point new properties emerge of materials and for new applications. A composition of nanoscale science, engineering and technology, in which nanomaterials are observed and imaged, measured, modelled and simulated, and manipulated	[85]

(continued)

Table 5.2 (continued)

Regulatory body/ committee/organization	Term	Definition	References
European Commission-EC	Nanomaterials	Natural, incidental, or manufactured material including unbound, aggregated, or agglomerated particles with ≥ 50% of those population present in the 1–100 nm size range. The ≥ 50% threshold maybe replaced exactly at 1–50% by the reason of specified related environmental, health, safety concerns, or issues of competition with such materials. Fullerenes, graphene flakes, and single-walled carbon nanotubes (SWCNT) with one or even more external dimensions < 1 nm are by the EC regulations classified as nanomaterials	[86]
European commission for novel foods (Amending regulation regulation No 258/97 (under harmonization)	Nanomaterials	Intensionally produced material within the dimension range (≥1– ≤ 100 nm), or made up of distinct functional parts at the internal or surface of the material, many of which are of ≥ 1 nm to ≤ 100 nm in dimension. Aggregate or agglomerate structures of size > 100 nm but with nanoscale level specific properties are also included	[86]
European Commission -EC: cosmetic product regulation	Nanomaterials	Intensionally produced insoluble or bio-persistent material with one or more external dimension or internal structure at the nanoscale size range of 1–100 nm	[87, 88]

(continued)

Table 5.2 (continued)

Regulatory body/ committee/organization	Term	Definition	References
European Union scientific committee on consumers products	Nanomaterials	Nanoscale materials of one or more external dimension or internal structure within the ~1 to 100 nm nanoscale size range with possible exhibition of newly acquired novel properties upon manipulation different from the same material not at the nanoscale	[89]
American chemistry council	Engineered nanomaterials	Intentionally produced materials with possible existence at 1, 2, or 3 dimensions (1D, 2D, or 3D) lying within 1–100 nm size range. However, such materials might- (i) not exhibit any acquired new or novel properties compared to bulk materials, (ii) be soluble in water or relevant biological solvent media at the molecular size level, but with the exception of micelles and single-polymer compounds	[90]
British Standards Institution (BSI)	Nanostructured materials	Internal or surface nanostructured exhibiting materials	[91]
British Standards Institution (BSI)	Nanostructures	Materials with interconnected structural constituent parts within the nanoscale region	[92]
British Standards Institution (BSI)	Nanocomposites	Multiphase materials with atleast one phase within the nanoscale region	[93]
British Standards Institution (BSI)	Nanofibres	Nanomaterials with similar exterior nanoscale dimension different from the third dimension which is larger	[94]

(continued)

Table 5.2 (continued)

Regulatory body/ committee/organization	Term	Definition	References
British Standards Institution (BSI)	Nanoparticles	Nano-objects with 3 different external nanoscale dimensions. Terms like nanorods, nanoplates, nanosheets are used to describe such nanoparticles with different dimensional lengths of the longest and shortest axis	[95]
British Standards Institution (BSI)	Nano-objects	Materials with atleast one or more external parts displaying nanoscale dimensions	[96]

unit scales of macro, micro and nano are used to understand and interpret better size of objects in associated and related manner at different level of scales [77, 81]. The connection between size and scale in the understanding of the differences between macro, micro, and nano terminologies of size and scale can for a better insight be comparatively illustrated. Therefore, to draw a clear-cut line and make differences between objects within the macro, micro, and nano range, scale and size needs to be associated by correlation. While sizes of objects can be comparatively illustrated in a scale, scale itself defines the relationship of what object is being compared and how that relationship can be represented either numerically or visually. While objects within the macro, micro, and nano brackets can be numerically defined by scale, visual control has the restriction function of what the human naked eye can see, and objects at both the nano and to a point micro scale can only be visualized by machine-aided visual devices. As a classic example, the human bone tissue for instance, is an open material constituent of different components that exist at different scale of measurement and sizes for its components [97].

Figure 5.1 shows the macro, micro, and nano scale measurements and the corresponding sizes within those three scale levels where the cortical and cancellous bone, osteons with Haversian systems, lamellae, the collagen fibrous assemblages of collagen fibrils, bone mineral crystals (HA), collagen molecules, and non-collageneous protein layers lie [98]. Figure 5.1 example is a demonstration of translational manipulation of different sizes and at different scales by nature of the human bone tissue material composed of five-level hierarchical structures at the macro (10 mm to several cm), micro (10–500 μm), sub-nano (1–10 μm), and nano (<1 μm) scales [99, 100]. In a practical approach of the application of micro and nanotechnologies for instance in Biomedical/Tissue Engineering related deployments, bone tissue therapies effortlessy adopt this kind of scale-size related bone architecture to fabricate bone-mimicking scaffolds in the development of bone support materials. For a more generalized example, Fig. 5.2, depicts a comparative scale and size chart of various materials from minute objects at femto level to pico level to nano to micro, millimeter and to meter level at which point macro scale materials reside. By a conversion factor of X1000, the next scale level is obtained from the preceding one for scale region material size assignment (Fig. 5.2).

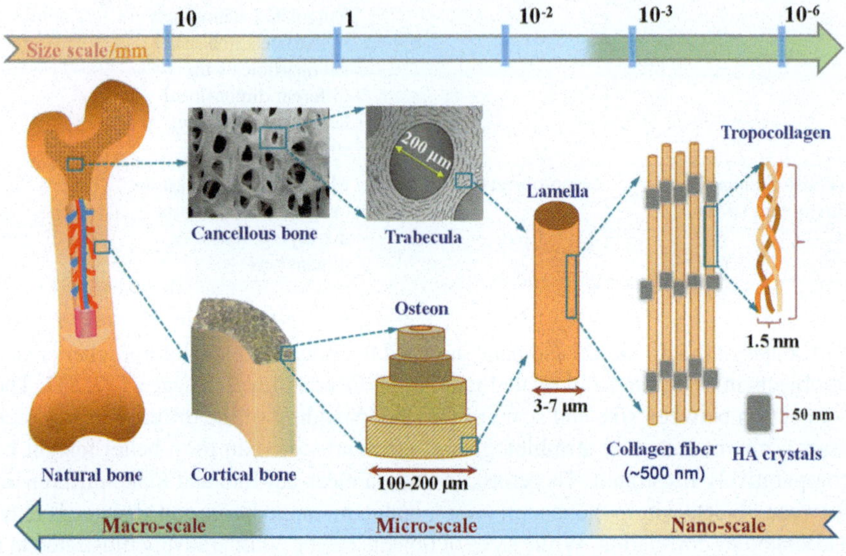

Fig. 5.1 Hierachical Bone Tissue Structure Components showing: cortical and cancellous bone, osteons with Haversian systems, Lamellae, the collagen fibrous assemblages of collagen fibrils, bone mineral crystals (HA), collagen molecules, and non-collageneous protein layers at their different scale measurement residences and sizes [98]

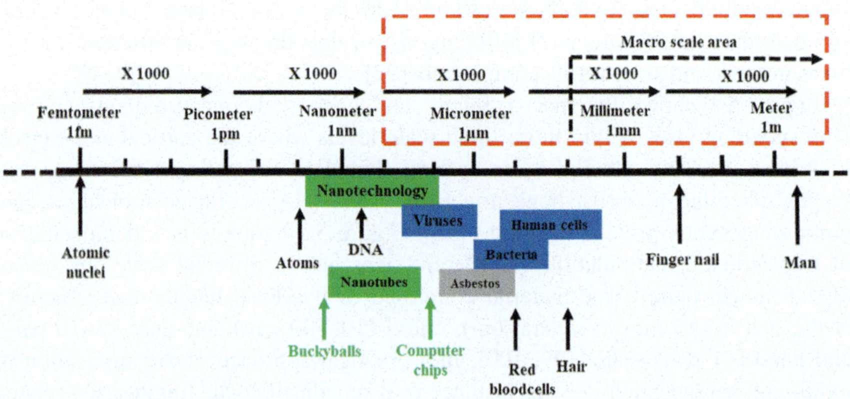

Fig. 5.2 Comparative size chart of materials from femtometer to meter scale range

The exploration of the existing borderlines as a matter of differences between materials in the macro, micro, and nano scale and size measurements of materials will continue to find a balance in the combined application of micro and nanotechnologies for the now and future perspective. The miniaturization of things in effect of application of micro and nanotechnologies has for instance led to transistors going from macroscopic (~1-mm) junction length-devices to, (~90-nanometeter) gates in the recent commercial chips and to ~10-nm gates in laboratory devices in a linear fashion and will continue to reduce approaching the size of an atom ~ (0.2 nm) [96]. On the other hand, volumetric approach had equally resulted to new materials with reliance on different computing strategies allowing for the advancement of increased capabilities and function of electronic and storage devices at weight/volume/power of electronics. In scale progression, there is the appreciation of the challenge in the control of the applicative technological tolerances at different levels. Scales at macro, micro and nano levels by virtue of material sizes have differences in application. Comparatively, macroscopic devices are smaller, lighter, more energy efficient, and fabricated with fewer materials than microscopic devices while by equivalence of application micro devices have edge over macro devices in terms of reliability, efficiency, selectivity, response time and energy consumption [77, 78, 81, 96]. Applicatively, micro scale exist in microtechnology and in the production of microsystems and microsystem components. Microsystems are miniaturized integrated systems in a small package or more specifically, micro sized components functioning together as a unit system and assembled into a package that fits on a pinhead [81, 96]. They are referred to as microsystems (MST) in Europe or microelectromechanical systems (MEMS) in the US and have been interchangeably used. Microsystems are microscopic scale level, integrated, self-aware, stand-alone products with sensory, thoughts, communication and action capabilities and have found application in areas of accelerometers, micro fluidic pumps, pressure sensor, spatial light modulators, lab on a chip, radio frequency (RF), mass storage devices etc. [77–81, 96]. The downscaling of microsystems where as they get smaller and smaller, their components correspondingly does eventually creates a meeting point between microtechnology and nanotechnology. Despite the transient tiny line difference that exist between macro, micro, and nano scale, their fabrication is another parameter that can be used to separate them. While microtechnology generally uses "top down" method type of fabrication, nanotechnology more often uses what is referred to as "bottom up approach" method of fabrication [20].

Micro and Nanotechnology: Size and Properties

The manipulation of materials at the micro and or nano level of scale measurement have led to the creation and the use of structures, devices, and systems that have novel properties and functions because of their small and or intermediate size. Classically, dimensions as well as composition and structure impact material properties in micro and nano scale. The decrease in the dimension of an object from the macroscopic to micrometric scale properties is not the same. For instance, at the microscale and at the surface of a liquid, an insect is able to stay afloat as gravitational influence becomes negligible relative to surface tension. Scaling law dependence properties can also be seen in microfluidics where transition from laminar to turbulent flow given by the Reynolds number (R_e) for the prediction of flow patterns, depend on the size of the tubing [101]. Low R_e numbers favour flows dominated by laminar flow, while at high R_e values turbulent flows dominate. However, due to their size in microfluidic systems, turbulence flow dominance disappears as flow properties become particular. The gecko's ability to stand on a wall is due to micro and nano structuration of its leg. This micro and nano scale level structuration is a scale dependence property of the material found in the gecko's leg which facilitates it's locomotry function as friction and adhesion forces become prominent over gravity [102]. Properties and functions of micro and nanotechnological systems have not only being regulated by scale sizes but shape as well relatively. Enhanced surface to volume ratio as seen in materials for example tiny nanoparticles of single crystal structure have exhibited drastically different shape- and size dependent features (e.g.; thermal decomposition, melting, electrical/thermal conductivity, magnetism, optical behaviour, and catalytic and bioactivity properties, sensing and plasmonic features, steric features etc.) [103–105]. The size and shape control of materials have offered the needed leverage in the identification of critical sizes below which target properties of interest differs from the bulk material to be able to achieve simple, cost-effective, environmentally begin, and easily scalable production methods and processes [97]. Various synthesis methods and procedures have been used to adjust the properties of nanomaterials as per their target application of interest [97]. While Fig. 5.3 show size property dependence of micro/nano scale level systems structural function reliance, Fig. 5.4 depicts combination of size/shape dependence for target application functions.

The size and property phenomenon in the functionality of micro/nanoscale materials is as well predominant in the development processes of fabricating microelectronics, and microscale devices, which are not only classified based on dimensional

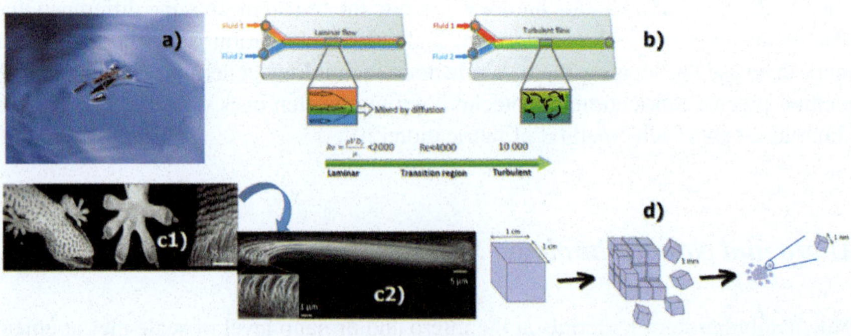

Fig. 5.3 Image representation of size-dependent properties and features of micro/nano objects: **a** microscale effect on an afloat water insect, **b** laminar and turbulent flow restrictions in microfluidics system [101], **c1, c2** micro and nano structuration of gecko's leg [102], **d** a demonstration of surface interface display for enhanced surface/volume ratio

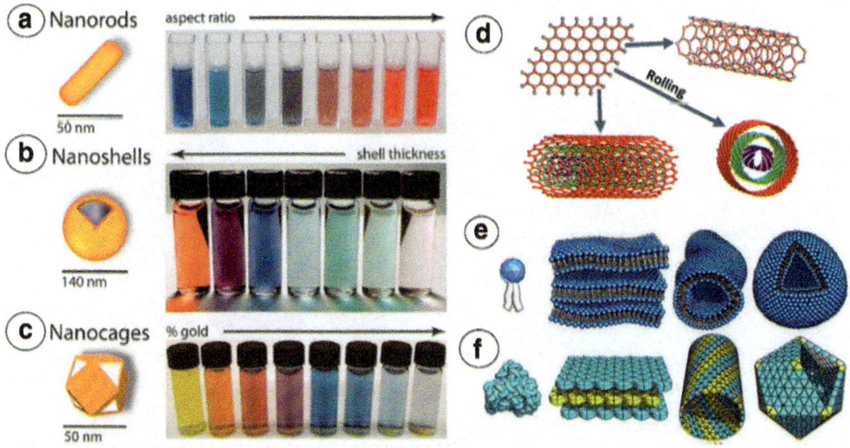

Fig. 5.4 Size- and shape dependent properties of nanomaterials: (a–c) Changes in optical properties (colour) [106], (d) Integration by rolling graphite layer into single-walled and multi-walled carbon nanotubes [107], Self-assemblage of (e) and (f) proteins into complex nanostructures [108]

scale, but also their composition and manufacturer. The two distinct yet overlapping fields of microelectromechanical systems (MEMS) and nanosystems or nanotechnology share a common set of engineering design considerations unique from other more typical engineering systems distinguishing their existence, effectiveness, and development like in the area of micro-scale and nanoscale transducers from those of conventional scale [81, 96]. To achieve application target of these systems, physics of scaling and the suitability of manufacturing techniques and processes largely convey their function-based link of size property dependence. Just as molecular machines involve macrobiolgical molecules (e.g. Proteins, DNA etc.,), nanotechnology has equally played considerable role in the down sizing of these micro-scale devices for biomedical applications down to assembly of individual molecules to fabricate molecular machines. Nanoscale and its implication on medicine has equally showed that biological molecules are in size range of nanomaterials. Once again, the size property related function is at play here as representation of comparison existing between particularly the involved biological molecule and nanoparticle for a particular target application functionalization achieved via size property based interaction (Fig. 5.5) [109].

Preparation of Micro-nano Objects

Unlike natural nanomaterials, micro and nano material objects are designed, fabricated, and or processed in the laboratory/industry. During the manipulation and engineering of these objects, their size, morphology and composition are controlled

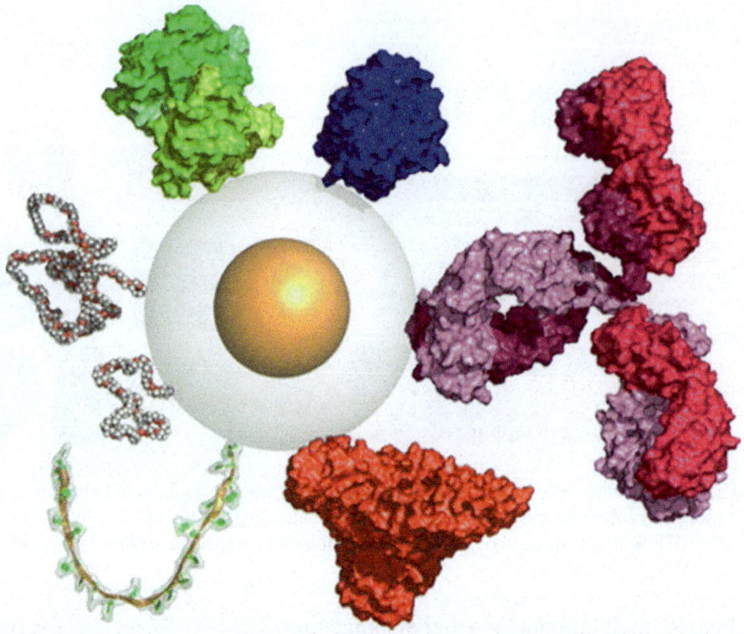

Fig. 5.5 Relative size of nanoparticles and biomolecules, drawn to scale. Schematic representation of a nanoparticle with 5 nm core diameter, 10 nm shell diameter, with PEG molecules of 2000 and 5000 g mol^{-1} (on the left, light grey), streptavidin (green), transferrin (blue), antibody (IgG, purple), albumin (red), single-stranded DNA (20mer, cartoon and space filling). Proteins are crystal structures taken from the Protein Data Bank (http://www.rcsb.org) and displayed as surfaces; PEG and DNA have been modelled from their chemical structure and space filling. Reused with permission of Sperling and Parak [109]

with high level of precision. The complexity involved in the fabrication of micro/nano scale level devices, have led to the progress sophistication of the processes and procedures used in the realization of micro and nano material objects. Notwithstanding, nanomanufacturing/nanofabrication still adopt chemical processes that have been around for centuries as variety of archived nanomaterials under microscopic observation have shown their micro and or nano scale manifestation for expected or unexpected reasons of production at that time [110, 111]. Several techniques for the fabrication of nanomaterials in general have been reported, however the two main strategies used irrespective of the origin of the nanoparticle are "Bottom up and Top down approaches" [112, 113].

Bottom Up Method

This method fabricates materials building blocks through the assembling of individual atoms or molecules [112]. It involves the manipulation of atoms, ions, molecules, or engage the unique property of nanoparticles self-assembly ability, their physico-chemical interactions (e.g. hydrogen and ionic bonds, van der Waals forces, and water-media generated hydrogen bond) to assemble fundamental building blocks into macroscopic structures [112, 114]. Theoretically, it ensures the perfect control over the production of nanomaterials with well defined size, shape and highly homogeneous size dispersion. As a chemical synthesis route, liquid synthesis (also known as wet synthesis) widely used for the preparation of inorganic nanoparticles is a typical example of bottom up approach [115]. This methodology also used in the industry nevertheless often suffer from a poor scalability. Several techniques can be employed in liquid synthesis depending on the precursors, the reactants and the nature of the expected end nanomaterial. The most common liquid syntheses abundantly described in the literature are chemical precipitation synthesis [116], solvothermal synthesis [117], and sol–gel synthesis [118]. However, the mechanisms of the nano-object formation are widely studied such as the nucleation and growth mechanism described by Lamer [119]. The main challenges of these syntheses is the inability to effectively control the growth of the nanocrystals and afford a stable and homogeneous suspension in solution, and lack of the needed tools to effectively handle the seed atoms and molecules [1].

Top Down Method

The top down procedure primarily involves the breaking down of bulk materials to get down to nanoparticles or the solid compartmentalization of uniformly distributed materials into smaller fractions for the formation of nanoparticles [1, 97]. Techniques like advanced industry precision engineering and lithography, etching, etc.; are some of the reported suitable approaches of top down method for the production of nanostructures [1, 97]. Others are ball milling, flame synthesis, laser ablation methods, and plasma technology. The superiority of application of top down method in the fabrication of electronic circuitry for their effect of integration and network interconnectivity has been highly recommended and thus, widely used in microelectronic industry [113, 120]. Thin film synthesis and nanoparticles creation of size > 100 nm with unique properties acquisition relative to their bulk parent material analogues, are good credits endorsement for top down approach [1, 97]. Though the precision engineering support of top down approach has made it an established preferred fabrication choice technique for the creation of nanomaterials within the electronic industry, limitation in resolution remains a big challenge to overcome [121]. In tackling this challenge, the strategy of combining both the bottom up and top down fabrication methods have been adopted for improved product outcome. This combinative approach of bottom up and top down fabrication method has led to a third type of fabrication approach known as 'hybrid approach' developed and optimized to weather off bottom up and top down challenges and deliver fabrication improvements [1, 97].

In the hybrid fabrication administration, there is the concurrent application of bottom up and top down methods where representative particles from these two methods can complement their fabrication process material outcome either as desired nanomaterial process result outcome or precursor fragment fit as a new source raw material for bottom up creation [97]. Lithographic methods e.g. photolithography, electron beam lithography, focused ion beam, soft lithography, neutral atomic beam lithography, nanoimprint lithography etc.; that can be used to realize 1D or 2D nanostructures, microtechnological materials etc.; is considered a hybrid fabrication process when combined with etching as the complementary top down technique, or with ion growth layering as the bottom up complementary method [122, 123]. Other improvement approach geared towards better created product outcome include the use of advanced nanostructured diamond or boron nitride-based sensors capable of controlling size, coupled with numerical control and advanced servo-drive technologies [1]. Though the introduction of the hybrid system has dealt with fabricated product outcome resolution; cost and the use of non-environmentally friendly chemicals, is still a setback and needs addressing.

Figure 5.6, displays the fundamental concepts involved in the bottom up and top down fabrication methods for micro and or nanotechnological material development.

Objects in the Micro and Nano Scales

Long before 1959 when the concept of nanotechnology was introduced by Richard Fenyman (Table 5.1, [25], and subsequently its definition in 1974 [31], variety of materials of either organic or inorganic origin and source that can be characterized by the concept terminology of nanotechnology were already in existence. The nano scale (nm) unit measurement of materials placed such a restriction in the sense that on the premise of a material size falling within 1–100 nm, it can only be classified a nanomaterial, otherwise it cannot. By this narrow definition, several biological materials of organic components fits within this nano scale range and have a long established history of medical benefits. For instance, the development and application Cowpox virus developed by Edwar Jenner in 1798 that helped with the saving of millions of lives from small pox and facilitate the eradication of such a deadly disease, underlines the impact nanotechnologies have already had in the history of the world [124]. Outside the walls of nanotechnology, the range of these biomedical materials and components is huge. As the concepts related to them continue to widen, the definition of nanotechnology in this area has been ceiled further a top and in the range of +100 nm to accommodate them. Therefore, from the viewpoint of biomedical applications and platforms, advances in nanomedicine including the emerging field of biologics where cellular size consideration in the range of (1–25 + μm) exist, ropes in the entire biomedical platforms and promote their inclusivity within the advancing micro-and nanotechnology.

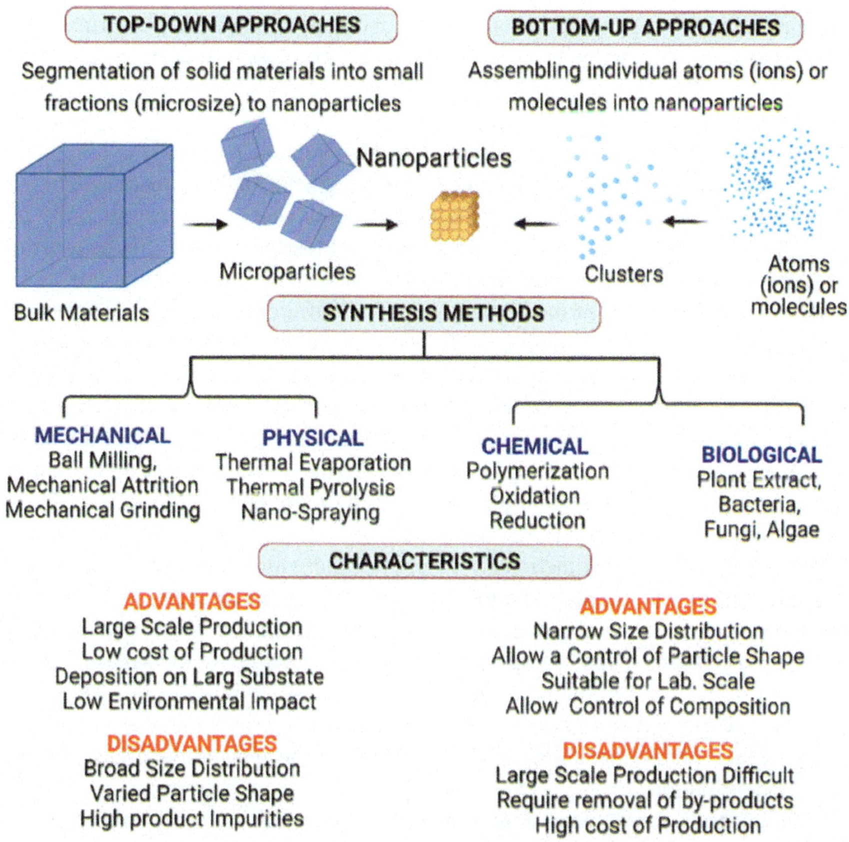

Fig. 5.6 Schematic illustration of top down and bottom up approaches applied in the fabrication of materials showing their synthesis methods and pros and cons [97]

Nanomedicine

The development of nanotechnology gave rise to the nanomedicine concept where nanomaterials having new properties owing to their nanosize are able to be used in several medical science related domains. Micro and nanomaterial unique properties of increased surface to volume ratio, charge density, surface chemistry etc.; are among the various acquired new properties that facilitate desired interaction between the cell environment and the administered micro and nanotechnology platforms [125]. For example, nobel metal plasmonic properties can be used for sensors or phototherapy. Superparamagnetism obtained of several metal oxides at the nanoscale can be used for magnetic resonance imaging (MRI) or even developed as therapeutic tools for magnetic hyperthermia. There has been evaluation of vast majority of nanomaterials for drug delivery and vaccine with several of such systems already present, for instance COVID vaccines. Micro and nano objects application within the veils of nanomedicine are expanding into the micro and nano biomedical platform and are familiar to areas of imaging, drug delivery and vaccine therapy, biosensing etc. [126].

- Imaging

Medical imaging has become a fundamental component in the field of biomedical research, clinical practise and diagnosis. Advancement made in medical imaging techniques have led to important breakthrough researches in clinical anatomy and forensic pathology. The emergence of new medical imaging technologies have served well and instrumentally too healthcare professionals with proper vision and understanding of the anatomy of the human body from different perspectives. Following the discovery of X-ray radiography in the nineteenth century, several progress advancement in diverse imaging modalities have been achieved. Prominent amongst the most advanced medical imaging modalities are magnetic resonance imaging (MRI), computed tomography (CT), positron emission tomography (PET), single photon emission computed tomography (SPECT), fluorescence, ultrasound imaging (sonography), etc. Spatial resolution, sensitivity, field strength, pulse sequence speed, absence of ionizing radiation, etc.; are among the various imaging capabilities used in the specification of the functional efficiency of these modalities and which can by combination present bundled features to leverage high performance level imaging applications. In order to obtain accurate imaging information that is representative of analysed cell tissue, contrast agents are utilized, to facilitate and make easy during imaging test the recognition by difference of a healthy cell tissue from a pathogenic one. Generally, contrast agents are small molecules or organometallic complexes, which by their presence manifest high toxicity, short life span blood circulation, and poor biodistribution.

Nanomaterial acquisition of new and novel properties upon manipulation at nanoscale have presented huge opportunity for enhanced performance contrast agents for medical imaging lately [127]. Contrast agents of nanoparticulate nature as a benefit; deliver various advantages that can enhance their application as imaging media. Ability to manipulate and control the size features via synthesis parameteric variations ensure and guarantee uniformity in biodistribution. Additionally, blood circulation enhancement through size conjugation by enhanced permeability and retention (EPR) effect promotes passive targeting [128]. The possibility in regulating the surface of the nanomaterials permits its tuning for the addition of new properties and features designed for improved functionalities of the contrast agents. The practicality in active targeting is effected by the addition of targeting molecules for increased spatial localization of pathologic tissue or the targeting of a specific type of cell with receptor suppression. Strategic modification of the chemical nature of the surface by charge alteration, covalent polymeric attachment with organic compounds like polyethylene glycol (PEG) possibly effects biodistribution, retention, and clearance or blood-circulation half-life modification for a balanced stealth and echogenic contrast agent. Summarized introduction and mode of operation of the key medical imaging techniques as already mentioned will be focused in the below section.

- Magnetic Image Resonance (MRI)

Magnetic Resonance Imaging (MRI) is a nonivasive modality that supports the observation of anatomic structures, physiological functions and molecular tissue compositions. It functions based on Nuclear Magnetic Resonance (NMR) with production of multiplanar and true 3D invivo subject datasets, excellent soft tissue contrast of high spatial image resolution, and with no harmful ionizing radiation effect. It is commonly used for stroke and cancer pathology diagnosis. MRI are of two different modalities and the contrast agents are categorized into T1 or T2 contrast agents in relation to the time of relaxation of the water proton. Bright contrast paramagnetic gadolinium (Gd^{3+}) chelates belongs to T1 contrast agents and in the modern MRI used as the gold standard. However, concerns of low sensitivity and perception of toxicity has widen search for options. Iron oxide nanoparticle as an alternative with superparamagnetic behaviour and as a T2 black contrast agent have attracted an overwhelming interest towards the end of the last century [129]. Majority of such contrast agents approved by Food and Drug Administration (FDA) have already had market entry and administered for years. Market available ones with trademark names: Feridex®, Resovist®, Combidex®, Lumiren® are however, trailing with limitation issues of signal intensity identified with T1 contrast agents [130]. Considerably, manganese oxide nanoparticles are feasibly T1 contrast agent alternatives [131], as well as gadolinium oxide nanoparticles that boast a dual T1/T2 modality. Recent advancement in size control, aqueous media stability, composition, and coating thickness permitted iron oxide nanoparticle as a MRI T1 contrast agent [132].

- Computed Tomography (CT)

Computed tomography also referred to as computerized tomography or computed axial tomography (CAT) is a non-invasive medical examination or procedure that relies on specialized X-ray equipment to generate cross-sectional images of the body for variety of reasons ranging from diagnosis, treatment, screening, and interventional purposes for outpatient procedures. It is credited for its low cost of operation, fast result delivery and high image resolution. High x-ray attenuation need for contrast agents used in CT analysis have placed huge interest of consideration on some metallic nanoparticles like gold, tantalum or zirconium oxide [133]. The interest over gold (Au) nanoparticles hinges back on its ease of production and functionalization that facilitates targeting and enhanced biocompatibility. Commercialzed Au nanoparticle radiosensitizer for instance AuroVist™ though with no market entry yet, is potentially for invivo research. Though with long track record in clinical imaging, iodine based contrast agents with diverse limitation issues of lack of fast clearance, renal toxicity potential, blood pool distribution nonspecificity, and adverse effects/anaphylaxis documentations has been upgraded by the complementary vector platform support of nanoparticles, and therefore, continue to be administered for biomedical imaging applications.

- Positron Emission Tomography (PET)/Single Photon Emission Computed Tomography (SPECT)

Positron emission tomography (PET) is a nuclear medicine technology overly used and credited with high tissue penetration, high-sensitivity, and real-time quantitative imaging analysis. Single photon emission computed tomography (SPECT), similarly, is another imaging tool of nuclear medicine technology for the detection of abnormal biochemical functions prior to body anatomical alterations. Despite limitations of high cost, radioactive exposure, and less image resolution in comparison to MRI, PET/SPECT continue to find application in molecular and early diagnostic imaging with the use of target specific radionuclide nanomaterials. Common radionuclides like ^{68}Ga, ^{64}Cu, for PET, and ^{67}Ga, ^{99m}Tc for SPECT have been applied for the advantages of reasonably having longer half-life over the likes of Flourine-18 with a half-life of about 109.8 min [133]. Direct introduction of radionuclides during nanoparticle preparations and incorporation of radionuclide tracers post synthetically are such methodologies adopted to circumvent short half-life issues during preparation and thus, for the improvement of cellular tissue material uptake for rapid diagnosis [134].

- Fluorescence

Nanoparticle fluorescence imaging has been applied in research and the monitoring of real time therapeutic effects including gene detection, protein analysis, enzyme activity evaluation, element tracing/cell tracking, early stage disease diagnosis etc. Equipped with near-infrared fluorescence capability, fluorescence imaging technology offer highest spatial resolution for disease diagnosis on macroscopic level taking advantages of its near-infrared responsiveness for deeper tissue penetration, and less non-specific tissue auto-fluorescence in comparison to visible light operation mode [133]. Notwithstanding, low penetration depth, and auto-fluorescence with related scattering properties in some cellular tissues continues to hamper its outright clinical utility [135]. In this aspect, nanoparticles properties have been seen useful to subdue the limiting challenges of fluorescence imaging. The use of large number of dye molecules loaded onto nanoparticles for extra signal provision, structural modifications for counter potential quenching of near-infrared (NIR) when necessary, nanoparticle concentration lesion increase via active/passive protocols, upgraded circulation time for uptake enhancement in lesion regions, and lower energy to high energy conversion design for blink and photobleaching reduction effects, etc.; are some of the strategies directed towards the development of fluorescent material-based nanoparticle [136–138]. Noteworthy still is the possibility in the combination of different imaging techniques for the execution of a hyphenated multimodal imaging agents beckoned on the versatility of nanomaterials to leverage coupled techniques (e.g.; PET/CT, PET/MRI, PET/US, CT/MRI, etc.;) for enhanced imaging therapeutic capabilities. Therapeutic capabilities of drug delivery systems (DDS), typified intrinsic physical property like photothermal therapy display of Au nanoparticles or magnetic hyperthermia of iron oxide nanoparticles, etc., are amongst the many accessible benefits of nanomaterials in such coupled/hyphenated medical imaging systems and have already received attention for potential dual exploitation of therapeutics and diagnosis in the recent emerging field of theragnostics [139].

- Ultrasound Imaging (Sonography)

Ultrasound imaging is a well-established imaging technique for clinical analysis. Its mode of operation is based on the detection of reflected sound waves in the range of 2–12 MHz frequency. The degree of reflection largely depends of the material nature whereby molecules in the gas phase return higher image contrast than the ones residing in the biological tissue media. For a long time, gas core microbubbles and lipids, protein or polymer shell have continuously been deployed as contrast agents for disease diagnosis in the vascular space. Limitation in the extravasation of these microbubble contrast agents due to inherent issues of size (1–3 μm), and instability has hindered their extended use as contrast agents for diagnostic ultrasound imaging applications, and in particular for vascular anatomy and gastrointestinal tract (GIT) [140]. Nanosized formulations of hollow silica nanoparticles, carbon nanotubes (CNT), nanobubbles based natural polymers or lipids, and gas vesicles of microbial origin has been studied as potential alternatives for ultrasound contrast agent fit for biomedical applications [141]. Dose-dependent ultrasound demonstration outcome so far has revealed microbial originated gas vesicles ability to diffuse into extracellular and intracellular environments due to their small size and waterproof-like gas-permeable protein shells. This biocompatibility feature, through their gene-coded form has given them a huge exploitable potential as ultrasound gene reporters in the same manner fluorescent proteins have served as optical reporters [141].

- Biosensing

Biosensors are analytical devices operating on biological recognition element with the capability of generating short time interval quantitative or semi-quantitative information through the transduction of chemical reactions into quantifiable physical response. Categorically, they fall into two groups of: biological recognition elements such as DNA, enzymes, antibodies, microorganisms, tissues, cell receptors, etc.; and transduction principle system group such as optical, electrochemical and mass-based biosensors. There is high specificity and sensitivity in the mechanism of biosensors of the category of biologically recognized elements as macromolecules are for instance used to match antibody-antigen or enzyme substrate pairs. Biosensor development is overached by physiological modifications in biological fluids or the apparition of pathogenic molecules as early signal indicators for the detection and subsequent follow up on various disease outcomes. High costs, and long signal data treatment time are amongst some of the noticeable drawbacks limiting wider application of biosensors in majority of fields. Thus, the incorporation of nanomaterials into biosensors has led to the emergence of bionanosensors delivering the possibility of new physico-chemical properties at nanoscale and increasing the possibility of surface coupling of biomolecules to nanomaterials [142]. Notwithstanding, the fabrication of diverse restructured material composites including nanomaterials open up the development of wider spectrum window of medical health and services including point of personal care medicine and other improvement such as for example wearable electronic devices for healthcare monitoring [143].

Majority of nanomaterials based biosensing devices; compose of gold nanoparticles [144]. As has been earlier stated, at the nanoscale, materials inherit new properties that are unique and different from its bulk material state. There is the exhibition of plasmonic effect by gold (Au) nanoparticle as a consequence of its small size acquisition at nanoscale, and that is completely absent with Au material at macroscopic level. This phenomenon is referred to as localized surface plasmon resonance (LSPR): an effect generated due to the collective oscillations of particle's free electron at the surface of the metal nanoparticles. The enhancement of the near surface electromagnetic field by the surface plasmon resonance (SPR) of the plasmonic metal nanoparticle improves the detection capabilities of deployed bionanosensor and have been extended to detection application techniques such as surface enhanced Raman spectroscopy (SERS). Optical extinction display at maximum absorption wavelength of the plasmon resonance frequency is also a registered effect linked to the size, form or agglomeration state of the nanoparticles [145]. These properties are consequential effects of nanoscale material manipulation underlining the functional mode of operation in gold and other plasmonic metal biosensor nanomaterials. Many nanomaterial biosensing applications have also been witnessed in the area of colorimetrics, Qdot tagged fluorescence, Förster resonance energy transfer (FRET), etc. [146, 147]. Comparatively, surface enhanced Raman scattering (SERS) spectroscopy have recently gained more attention as one of the most effective detection techniques particularly due to its capability of single molecule detection.

SERS is an analytical technique that functions based on plasmonic metallic colloids like Au, Ag, Pd, Pt etc., and using their surface roughness to detect target analyte samples. Plasmonic phenomenon properties and charge transfer between particle and the analyte generate Raman signal intensification signal of up to 10^6 response in sensitivity value. Interest on SERS application has grown over the detection of small molecules, proteins, DNA [148], or even viruses [149], giving it a competitive advantage over other detection/sensing method protocols like dynamic light scattering and hyper-Rayleigh scattering-based sensing, Two-photon photoluminescence (TPPL)-based sensing, chiroplasmonic activity-based sensing etc. [150].

Property features applications in biosensor operations have equally found their way into the lateral flow Immunoassay strips test, which has been clinically, tested for different detection measurements [151, 152]. Amongst those test, human chorionic gonadotrophin (HCG) pregnancy test is well documented. The pink appearing colour for the pregnancy positivity test assurance is often due to the presence of gold nanoparticle in the sensor material. Lateral flow immunoassay based gold nanoparticles kit analyzers have also been utilized for HIV detection and lately COVID tests. As a basic functioning principle, these tests are generally employing a detection strategy derived from the most used clinical technique for biomarker sample presence evaluation: enzyme-linked immunosorbent assay (ELISA). Usually, ELISA strips are coated with antibodies for the selected antigen with the incubation of the analyte performed with other antibodies bound to Horse Radish peroxidase (HRP) for the oxidization of the colour changing substrates, thereby helping to facilitate the manifestation. Nanomaterial-based architectural instruments have been used to enhance the detection capabilities of ELISA. Antibodies decorated nanoparticles are utilized for the elaboration of surface binding sites, which in turn allows improved selectivity for the detection of analyte. Catalytic properties of the nanoparticles can leverage visualization in combination with colour changing substrates. However, nanoparticles can also be directly used as the detected object owing to their plasmonic properties for example gold nanoparticle or even due to their magnetic properties, and for example in the case of iron oxide-based immunoassays [153]. In the later scenario, such developed dedicated detector have been reported of high sensitivity possession [154]. Displayed in Fig. 5.7 is the timeline evolution indicating first reported instances of micro and nanotechnology nanomedical therapeutic, vaccine, and imaging applications but to mention a few.

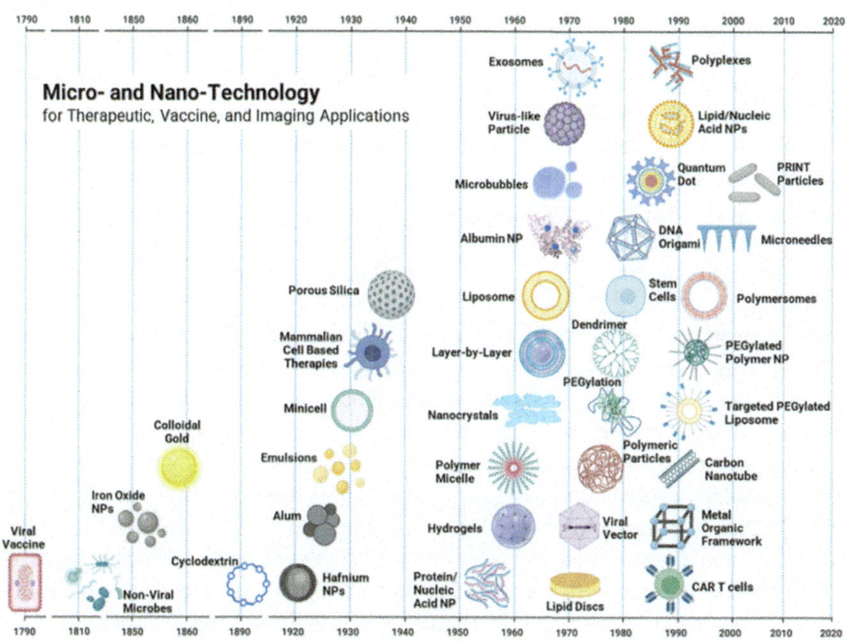

Fig. 5.7 First report timeline of micro and nanotechnology used for therapeutic, vaccine and imaging applications [155]. *CAR T cells: chimeric antigen receptor T cells; NP: nanoparticle; PEG: polyethylene glycol; PEGlyation/PEGlyated: PEG covalently bound; PRINT: particle replication in non wetting templates

Micro-nano Technology and 3D Printing

The development of micro-electro-mechanical systems (MEMS): thanks to the cross-border overlapping technological advancement at nano and micro scale of material and devices; have now unleashed the huge potential of this technology in the area of life science (biology and medicine) for diverse of applications. Interest on the extension and application of MEMS to applied biological domain has created a link and built a bridge across two words of biology or biomedical sciences and micro-electronic and mechanical systems to set us up with what is now referred to as "BioMEMS". The huge potential embodiment of BioMEMS is unrivalled starting all the way from the two areas of biomedical and life sciences, stretching and encompassing diagnostics, therapeutics, drug delivery, biosensors, as well as tissue engineering [156]. The transformation witnessed now in those two areas of biology and medicine of the life sciences has drastically changed the way one is able to see and detect substances in the biological world [101, 157]. This tremendous contribution has been registered in bioassay and instrumentation through micro and nanotechnology, and thus, bringing such an overwhelming development to the domain of biomedical sciences. This revolution of biological assays and instrumentation is made possible by the miniaturization of the current biological tests by combining different microtechnological tools like microfluidics, microactuators, and other micromachine processes [101]. These novel and advanced biomedical devices, now guarantee high performance throughput methodologies by running on robust and efficient biomedical device platforms like microarrays, organ-on-a-chip, etc. [101, 157]. With all the outstanding progress achievements recorded so far in the wide domain world of science, engineering, and medicine through breakthrough advancements in micro and nanotechnology; the contribution of additive manufacturing cannot but be commended. The many capabilities of additive manufacturing processes e.g. design alteration, rapid manufacturing, industry scale up cost effectiveness, ease of material integration, structure and functions, which are readily amenable to diverse technology processes are its uncommon blessing that has left it positioned in such a central indispensable role within the ever-dynamic micro and nanotechnology fabrication processes world (Grandviewresearch n.d.). 3D Printers (3DPs) are continuously evolving in terms of their capabilities and specification as the extrusion of multiple material is now possible with wide range printable materials due to diverse presence of specialized printers, and material science evolution. By the advances made in printing processes, scale of precision has been minimized to nano and micro scale level which relatively have effect on the material properties as well the application that can be targeted with such materials [158]. On this premise, precision fabrication type of micro/nano scale printing to achieve range of biomedical and electronic devices have diffused. Stereolithography printing, Two Photon Polymerization (TPP), Dip-Pen Nanolithography (DPN), Inkjet printing, Piezoelectric Inkjet Printing, Thermal Inkject Printing, Electrohydrodynamic (EHD) printing are all diverse kinds of 3D printing processes of different control/process parameters and other capabilities adopted for the micro/nano fabrication of various material target

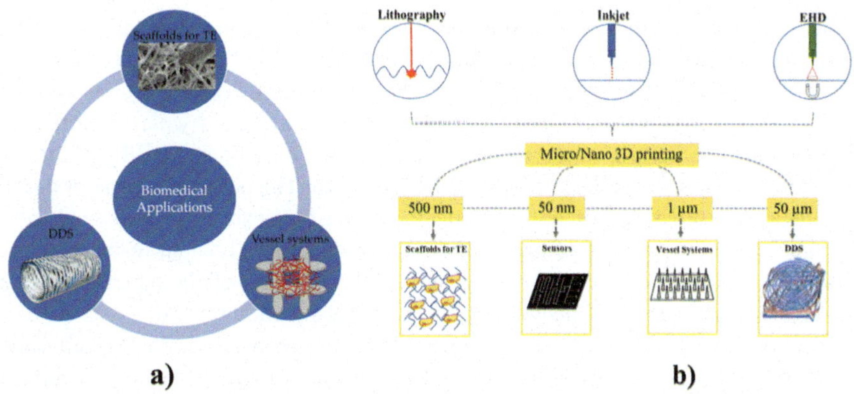

Fig. 5.8 A representative outlook of the many evolving printable materials for biomedical application (**a**) [163, 164], and a narrow spectrum of the most deployed 3D printing process for the realization of different biomedical application materials at micro/nano level (**b**) [160]

application purposes from biomedical, to electronics, to environmental, etc. [159, 160]. Differently from the conventional 3D printing processes, lithography printing, ink-jet printing, and electrohydrodynamic (EHD) printing processes are specifically deployed for printing operations where precision is highly needed. They are suitable for biomedical and electronic device printing application where they transfer their capabilities of offering physical and chemical properties to printed materials, introduce with ease micro/nano structures, which remain unattainable with conventional fabrication processes [158, 161, 162]. Figure 5.8 shows the possibility of arriving to micro/nano level by 3D printing processes to execute materials fit for different biomedical/other applications from tissue engineering, sensor, cell vessels to biomaterial for drug delivery system (DDS) [159].

Conclusion

In this book chapter, we have covered five different sections of- the history of micro and nanotechnology where we looked at the definitions of both worlds and x-rayed their closely knitted differences. This section laid the foundation of this book chapter by looking at the dynamics and ever evolving world of micro and nanotechnology, while at the same time traced back to historic times, the origin of what today is driving innovation in science and engineering and that will continue to dominate in the coming years. In doing this, we spurred the curiosity, in the mind of the reader to ask questions of how emerging nanotechnology and ofcourse the microtechnology counterpart is; as historic evidence of fabricated archived premedivial, materials show nanometric proofs. The line difference between scale levels of materials is thin and transient. In differentiating nanomaterials or materials at nano and micro level, it should not only be a question of what the present scale is as per the materials in view at a point in time, but rather a holistic understanding of the material composition, target of application in mind before the fabrication and the manufacturer of such a material. As it is possible for a nanomaterial to exist at microscale theoretically, it is practical to build a micro matter from nano scale components like in micro-electronics and mechanical systems (MEMS) devices. The pervasive world of micro and nanotechnology leaves no clear-cut line along the scale for instance where a certain device constituting of nano building unit in a micro content resides. In this case, various modalities in the conveyance of the function of the material and its composition gives the best insight into its level classification by scale.

Section 5.2: Looked at the new acquired properties of materials due to their scale size level and established in a narrowed down fashion of what is a broad spectrum of diverse field of applications, a connection between different sizes and shapes of materials for particular and target material functions.

Section 5.3: was about the two main fabrication protocols of micro and nanomaterials and the meaning of those approaches. For a more specialized material function, a combination of the bottom up and top down approach could result to improved material property for specialized function.

Section 5.4: drew insight about the application of micro and nanotechnology in medical science, an overached field, presently designated as nanomedicine. In this chapter, we took a random look of the application of nanomedicine using micro and nano technological prowess to drive several medical diagnosis and treatments.

Finally, Sect. 5.5 focused on the connection between micro and nanotechnology and 3D printing (3DPs). We looked at the best available technology (BAT) practices in printing fabrication processes to realize materials, which serve purpose in several application domains of biomedicine and electronics.

Micro and nanotechnology is a huge and diversified field within nanoscience and nanotechnological application. The intricacies of both areas yet one for the sake of their service of purpose lies in their overlapping interlocking nature especially in systems where materials of differences in scale level occupy the same suit to serve target-particularized functions.

This book chapter explored just about a narrow overlay of the wider and far-reaching world of micro/nanotechnology which now occupies everything and anything that we could ever think or imagine possible within science and engineering enclave.

References

1. Bayda S, Adeel M, Tuccinardi T et al (2020) The history of nanoscience and nanotechnology: from chemical–physical applications to nanomedicine. mdpi.com 25:112. https://doi.org/10.3390/molecules25010112
2. Barber D, Archaeometry IF (1990) An investigation of the origin of the colour of the Lycurgus Cup by analytical transmission electron microscopy. academia.edu

3. Carrapiço A, Martins M, Caldeira A et al, Biosynthesis of metal and metal oxide nanoparticles using microbial cultures: mechanisms, antimicrobial activity and applications to cultural heritage. mdpi.com
4. Freestone I, Meeks N, Sax M, Higgitt C (2008) The Lycurgus Cup—A Roman nanotechnology. Gold Bull 40:270–277. https://doi.org/10.1007/BF03215599
5. Sciau P, Sciau P (2012) Nanoparticles in ancient materials: the metallic lustre decorations of medieval ceramics. https://doi.org/10.5772/34080
6. Faraday M (1857) Experimental relations of gold (and other metals) to light. Philos Trans R Soc London 147:145–181. https://doi.org/10.1098/RSTL.1857.0011
7. Mie G (1908) Beiträge zur Optik trüber Medien, speziell kolloidaler Metallösungen. Ann Phys 330, 377–445
8. Synge E (1928) XXXVIII. A suggested method for extending microscopic resolution into the ultra-microscopic region. Taylor Fr 6:356–362. https://doi.org/10.1080/14786440808564615
9. Shockley W (1951) Circuit element utilizing semiconductive material. Google Patents 25:347
10. Müller EW (1951) Das Feldionenmikroskop. Zeitschrift für Phys 131:136–142. https://doi.org/10.1007/BF01329651
11. Watson J, Crick FHC (1953) Molecular structure of nucleic acids: a structure for deoxyribose nucleic acid. Nature 171:737–738
12. Esaki L (1958) New phenomenon in narrow germanium p-n junctions [3]. Phys Rev 109:603–604. https://doi.org/10.1103/PHYSREV.109.603
13. Padeletti G, Fermo P (2003) How the masters in Umbria, Italy, generated and used nanoparticles in art fabrication during the Renaissance period. Springer 76:515–525. https://doi.org/10.1007/s00339-002-1935-1
14. Poole C, Owens F (2003) Introduction to nanotechnology
15. Pradell T, Climent-Font A, Molera J et al (2007) Metallic and nonmetallic shine in luster: an elastic ion backscattering study. J Appl Phys 101. https://doi.org/10.1063/1.2734944
16. Reibold M, Paufler P, Levin A, Kochmann W (2006) Carbon nanotubes in an ancient Damascus sabre. Nature 444:286
17. Kevles B (1997) Naked to the bone: medical imaging in the twentieth century, pp 19–22
18. Singh R (2002) C. V. Raman and the discovery of the Raman effect. Phys Perspect 4:399–420. https://doi.org/10.1007/S000160200002
19. Ruska E (1980) The early development of electron lenses and electron microscopy—PubMed. App Opt 5:820
20. Ford BJ (2001) The royal society and the microscope. Notes Rec R Soc 55:29–49. https://doi.org/10.1098/RSNR.2001.0124
21. Lamer VK, Dinegar RH (1950) Theory, production and mechanism of formation of monodispersed hydrosols. J Am Chem Soc 72:4847–4854. https://doi.org/10.1021/JA01167A001
22. Müller EW, Bahadur K (1956) Field ionization of gases at a metal surface and the resolution of the field Ion microscope. Phys Rev 102:624–631. https://doi.org/10.1103/PHYSREV.102.624
23. Von Hippel A (1956) Molecular engineering. Science (80-) 123:315–317. https://doi.org/10.1126/SCIENCE.123.3191.315
24. Melliar-Smith CM, Haggan DE, Troutman WW (1997) Key steps to the integrated circuit. Wiley Online Libr 2:15–28. https://doi.org/10.1002/bltj.2081
25. Feynman RP (1960) There's plenty of room at the bottom. Eng Sci 23:22–36
26. Plank C, Rosinski EJ (1964) US Patent 3 140,249, 1964 undefined Catalytic cracking of hydrocarbons with a crystalline zeolite catalyst composite. Google Patents
27. Papell S (1965) Low viscosity magnetic fluid obtained by the colloidal suspension of magnetic particles Patent
28. Moore G (1965) Cramming more components onto integrated circuits. Electronics 38:114–117
29. Moffitt C (2011) Fundamental aspects of XPS and the development of XPS imaging. Micros Today 19:16–21

30. Mulvaney P (2015) Nanoscience vs nanotechnology-defining the field. ACS Nano 9:2215–2217. https://doi.org/10.1021/ACSNANO.5B01418
31. Kang G (2017) Gravitational waves and the 2017 Nobel Prize in Physics. Nobel Prize Phys
32. Aviram A, Ratner M (1974) Molecular rectifiers. Chem Phys Lett
33. Jeanmaire DL, Van Duyne RP (1977) Surface Raman spectroelectrochemistry: Part I. Heterocyclic, aromatic, and aliphatic amines adsorbed on the anodized silver electrode. J Electroanal Chem Interfacial Electrochem 84:1–20
34. Sagiv J (1980) Organized monolayers by adsorption. 1. Formation and structure of oleophobic mixed monolayers on solid surfaces. J Am Chem Soc 102:92–98. https://doi.org/10.1021/JA00521A016
35. Binnig G, Rohrer H, Gerber C et al (1982) Tunneling through a controllable vacuum gap. aip.scitation.org 40:178. https://doi.org/10.1063/1.92999
36. Binning G, Rohrer H, Gerber C, Weibel E (1982) Surface studies by scanning tunneling microscopy. Phys Rev Lett 49:57–61. https://doi.org/10.1103/PHYSREVLETT.49.57
37. Ekimov A (1982) Quantum size effect in the optical-spectra of semiconductor micro-crystals. Sov Phys Semicond 16:775–778
38. Drexler KE (1981) Molecular engineering: an approach to the development of general capabilities for molecular manipulation. Proc Natl Acad Sci U S A 78:5275–5278. https://doi.org/10.1073/PNAS.78.9.5275
39. Pinheiro A, Han D, Shih W, Yan H (2011) Challenges and opportunities for structural DNA nanotechnology. Nat Nanotechnol 6:763–772
40. Seeman N (1982) Nucleic acid junctions and lattices. J Theor Biol 99:237–247
41. Rossetti R, Nakahara S, Physics LB-TJ of C (1983) Quantum size effects in the redox potentials, resonance Raman spectra, and electronic spectra of CdS crystallites in aqueous solution. aip.scitation.org 79:1086. https://doi.org/10.1063/1.445834
42. Steigerwald ML, Alivisatos AP, Gibson JM et al (1988) Surface derivatization and isolation of semiconductor cluster molecules. J Am Chem Soc 110:3046–3050. https://doi.org/10.1021/JA00218A008
43. Kroto H, Heath J, O'Brien S et al (1985) C60: Buckminsterfullerene. Nature 318:162–163
44. Gerber C, Quate C, Binning G (1986) Atomic force microscope. Phys Rev Lett 56:930–933
45. Quate C (1994) The AFM as a tool for surface imaging. Surf Sci 980–995
46. Averin DV, Likharev KK (1986) Coulomb blockade of single-electron tunneling, and coherent oscillations in small tunnel junctions. J Low Temp Phys 62:345–373. https://doi.org/10.1007/BF00683469
47. Eigler D, Schweizer E (1990) Positioning single atoms with a scanning tunnelling microscope. Nature 344:524–526
48. Iijima S (1991) Helical microtubules of graphitic carbon. Nature 354:56–58
49. Beck JS, Vartuli JC, Roth WJ et al (1992) A new family of mesoporous molecular sieves prepared with liquid crystal templates. J Am Chem Soc 114:10834–10843. https://doi.org/10.1021/JA00053A020
50. Kresge C, Leonowicz M, Roth W, Vartuli J (1992) Ordered mesoporous molecular sieves synthesized by a liquid-crystal template mechanism. Nature 359:710–712
51. Bethune D, Kiang C, De Vries M, Gorman G (1993) Cobalt-catalysed growth of carbon nanotubes with single-atomic-layer walls. Nature 363:605–607
52. Iijima S, Ichihashi T (1993) Single-shell carbon nanotubes of 1-nm diameter. Nature 363:603–605
53. Murray CB, Nirmal M, Norris DJ, Bawendi MG (1993) Synthesis and structural characterization of II-VI semiconductor nanocrystallites (quantum dots). Zeitschrift für Phys D Atoms, Mol Clust 26:231–233. https://doi.org/10.1007/BF01425674
54. Mirkin C, Letsinger R, Mucic R, Storhoff JJ (1996) A DNA-based method for rationally assembling nanoparticles into macroscopic materials. Nature 382:607–609
55. Tans S, Verschueren A, Dekker C (1998) Room-temperature transistor based on a single carbon nanotube. Nature 393:49–52

56. Piner RD, Zhu J, Xu F et al (1999) "Dip-pen" nanolithography. Science (80-) 283:661–663. https://doi.org/10.1126/SCIENCE.283.5402.661
57. Lee HJ, Ho W (1999) Single-bond formation and characterization with a scanning tunneling microscope. Science (80-) 286:1719–1722. https://doi.org/10.1126/SCIENCE.286.5445.1719
58. Hersam MC, Guisinger NP, Lyding JW (2000) Isolating, imaging, and electrically characterizing individual organic molecules on the Si(100) surface with the scanning tunneling microscope. J Vac Sci Technol A Vacuum, Surfaces, Film 18:1349–1353. https://doi.org/10.1116/1.582352
59. Lok C (2010) Nanotechnology: small wonders. Nature 467:18–21
60. Montemagno CD (2001) Nanomachines: a roadmap for realizing the vision. J Nanoparticle 3:1–3
61. Williams K, Veenhuizen P, de la Torre B, Eritja R (2002) Carbon nanotubes with DNA recognition. Nanotechnology 420:761
62. Hirsch LR, Stafford RJ, Bankson JA et al (2003) Nanoshell-mediated near-infrared thermal therapy of tumors under magnetic resonance guidance. Proc Natl Acad Sci U S A 100:13549–13554. https://doi.org/10.1073/PNAS.2232479100
63. Loo C, Lin A, Hirsch L et al (2004) Nanoshell-enabled photonics-based imaging and therapy of cancer. Technol Cancer Res Treat 3:33–40. https://doi.org/10.1177/153303460400300104
64. Novoselov KS, Geim AK, Morozov S V et al (2004) Electric field in atomically thin carbon films. Science (80-) 306:666–669. https://doi.org/10.1126/SCIENCE.1102896
65. Xu X, Ray R, Gu Y et al (2004) Electrophoretic analysis and purification of fluorescent single-walled carbon nanotube fragments. J Am Chem Soc 126:12736–12737. https://doi.org/10.1021/JA040082H
66. Morin JF, Shirai Y, Tour JM (2006) En route to a motorized nanocar. Org Lett 8:1713–1716. https://doi.org/10.1021/OL060445D
67. Shirai Y, Osgood A, Zhao Y et al (2005) Directional control in thermally driven single-molecule nanocars. ACS Publ 5:2330–2334. https://doi.org/10.1021/nl051915k
68. Rothemund P (2006) Folding DNA to create nanoscale shapes and patterns. Nature 440:297–302
69. Du G, Moulin E, Jouault N et al (2012) Muscle-like supramolecular polymers: integrated motion from thousands of molecular machines. Angew Chemie 124:12672–12676. https://doi.org/10.1002/ANGE.201206571
70. Sanders J, Jackson S (2009) The discovery and development of the green fluorescent protein. GFP. Chem Soc Rev 38:2821
71. Zheng J, Birktoft J, Chen Y et al (2009) From molecular to macroscopic via the rational design of a self-assembled 3D DNA crystal. Nature 461:74–77
72. Knoll AW, Pires D, Duerig U et al (2010) Probe-based 3-D nanolithography using self-amplified depolymerization polymers. Wiley Online Libr 22:3361–3365. https://doi.org/10.1002/adma.200904386
73. Lafferentz L, Ample F, Yu H et al (2009) Conductance of a single conjugated polymer as a continuous function of its length. Science (80-) 323:1193–1197. https://doi.org/10.1126/SCIENCE.1168255
74. Richards V (2016) Molecular machines. Nature. https://doi.org/10.1038/nchem.2687
75. Petersen P, Tikhomirov G, Qian L (2018) Information-based autonomous reconfiguration in systems of interacting DNA nanostructures. Nat Commun 9:5362
76. Oran D, Rodriques SG, Gao R et al (2018) 3D nanofabrication by volumetric deposition and controlled shrinkage of patterned scaffolds. Science (80-) 362:1281–1285. https://doi.org/10.1126/SCIENCE.AAU5119
77. Council NR (2001) Physics in a new era: an overview. Natl Acad Press, Washington, DC
78. Reid T (2001) The chip: how two Americans invented the microchip and launched a revolution. NY Random House Inc., New York
79. Shermer M (2001) Nano nonsense and cryonics. Sci Am 285:29

80. Allhoff F (2007) On the autonomy and justification of nanoethics. NanoEthics 1:185–210. https://doi.org/10.1007/S11569-007-0018-3
81. Roco MC (2001) International strategy for nanotechnology research and development. J Nanoparticle Res 3:353–360. https://doi.org/10.1023/A:1013248621015
82. Kallinger C, Veffkind V, Michalitsch R, Verbandt Y (2008) Patenting nanotechnology: a European patent office perspective. Nanotechnol Law Bus 5:95
83. Hammani S, Moulai-Mostefa N, Samyn P et al (2020) Morphology, rheology and crystallization in relation to the viscosity ratio of polystyrene/polypropylene polymer blends. Materials (Basel) 13:926. https://doi.org/10.3390/ma13040926
84. Bowman D, May N, Maynard A (2018) Nanomaterials in cosmetics: regulatory aspects. Anal Cosmet Prod 289–302
85. Sharma V, Kumar A, Dhawan A (2012) Nanomaterials: exposure, effects and toxicity assessment. Proc Natl Acad Sci India Sect B Biol Sci 82:3–11. https://doi.org/10.1007/S40011-012-0072-7
86. Atanassov A, Trifonova S, Saraivanova J et al (2017) Assessment of the administrative burdens for businesses in Bulgaria according to the national legislation related to the European Union internal market. J Contemp Manag Issues 22:21–49
87. Hischier R, Walser T (2012) Life cycle assessment of engineered nanomaterials: State of the art and strategies to overcome existing gaps. Sci Total Environ 425:271–282
88. Kazakov S, Ruiz-Alba JL, Muñoz MM (2020) The impact of information and communication technology and internal market orientation blending on organisational performance in small and medium enterprises. Eur J Manag Bus Econ 30:129–151. https://doi.org/10.1108/EJMBE-04-2020-0068/FULL/HTML
89. Rauscher H, Rasmussen K, Sokull-Klü B (2017) Regulatory aspects of nanomaterials in the EU. Wiley Online Libr 89:224–231. https://doi.org/10.1002/cite.201600076
90. Chen X, Ke C (2016) Structural and physical properties of boron nitride nanotubes and their applications in nanocomposites. Boron Nitride Nanotub Nanomedicine
91. Jeevanandam J, Barhoum A, Chan Y (2018) Review on nanoparticles and nanostructured materials: history, sources, toxicity and regulations. Beilstein J Nanotechnol 9:1050–1074
92. Bathi J, Moazeni F, Upadhyayula V (2021) Behavior of engineered nanoparticles in aquatic environmental samples: current status and challenges. Sci Total Environ 793
93. Zidan S, Silikas N, Alhotan A et al (2019) Investigating the mechanical properties of ZrO_2-impregnated PMMA nanocomposite for denture-based applications. Materials (Basel) 12:1344
94. Dahm MM, Schubauer-Berigan MK, Evans DE et al (2018) Exposure assessments for a cross-sectional epidemiologic study of US carbon nanotube and nanofiber workers. Int J Hyg Environ Heal 221:429–440
95. Slezakova K, Morais S, Do M, Pereira C (2013) Atmospheric nanoparticles and their impacts on public health. Curr Top Public Heal
96. Council NR (2002) Implications of emerging micro-and nanotechnologies
97. Barhoum A, Garcia-Betancourt ML, Jeevvanandam J et al (2022) Review on natural, incidental, bioinspired, and engineered nanomaterials: history, definitions, classifications, synthesis, properties, market, toxicities, risks and regulations. Nanomaterials 12:177
98. Suzan-Monti M, La SB, Raoult D (2006) Genomic and evolutionary aspects of Mimivirus. Virus Res 117:145–155
99. Gao C, Peng S, Feng P, Shuai C (2017) Bone biomaterials and interactions with stem cells. Bone Res 5:1–33
100. Wang X, Xu S, Zhou S et al (2016) Topological design and additive manufacturing of porous metals for bone scaffolds and orthopaedic implants: a review. Biomaterials 83:127–141
101. Azizipour N, Avazpour R, Rosenzweig D, Sawan M (2020) Evolution of biochip technology: a review from lab-on-a-chip to organ-on-a-chip. Micromachines 11:599. https://doi.org/10.3390/mi11060599
102. Autumn K, Gravish N (2008) Gecko adhesion: evolutionary nanotechnology. Philos Trans R Soc A Math Phys Eng Sci 366:1575–1590. https://doi.org/10.1098/RSTA.2007.2173

103. Cremers V, Rampelberg G, Barhoum A et al (2018) Oxidation barrier of Cu and Fe powder by atomic layer deposition. Surf Coatings Technol 349:1032–1041
104. Dreaden EC, Alkilany AM, Huang X et al (2012) The golden age: gold nanoparticles for biomedicine. Chem Soc Rev 41:2740–2779
105. Shin W, Cho J, Kannan A et al (2016) Cross-linked composite gel polymer electrolyte using mesoporous methacrylate-functionalized SiO_2 nanoparticles for lithium-ion polymer batteries. Sci Rep 6:26332
106. Yang S, Yan Y, Huang J et al (2017) Giant capsids from lattice self-assembly of cyclodextrin complexes. Nat Commun 8:15856
107. Collegium JR (2011) The impacts of nanotechnology. researchgate.net 7:28–66. https://doi.org/10.1016/B978-0-323-39311-9.00018-2
108. Bahrami F, Panahi F, Daneshgar F et al (2016) Synthesis of new α-aminophosphonate derivatives incorporating benzimidazole, theophylline and adenine nucleobases using l-cysteine functionalized magnetic. RSC Adv 6:5915–5924
109. Sperling RA, Parak WJ (2010) Surface modification, functionalization and bioconjugation of colloidal Inorganic nanoparticles. Philos Trans R Soc A Math Phys Eng Sci 368:1333–1383. https://doi.org/10.1098/RSTA.2009.0273
110. Bai W, Ross CA (2016) Functional nanostructured materials based on self-assembly of block copolymers. MRS Bull 41:100–107. https://doi.org/10.1557/MRS.2016.1
111. Keller AA, McFerran S, Lazareva A, Suh S (2013) Global life cycle releases of engineered nanomaterials. J Nanoparticle Res 15:1–17. https://doi.org/10.1007/S11051-013-1692-4
112. Iqbal P, Preece JA, Mendes PM (2012) Nanotechnology: the "top-down" and "bottom-up" approaches. Supramol Chem. https://doi.org/10.1002/9780470661345.SMC195
113. Patel JK, Patel A, Bhatia D (2021) Introduction to nanomaterials and nanotechnology. Emerg Technol Nanoparticle Manuf 3–23. https://doi.org/10.1007/978-3-030-50703-9_1
114. Nnaji CO, Jeevanandam J, Chan YS et al (2018) Engineered nanomaterials for wastewater treatment: current and future trends. Fundam os Nanoparticles 129–168
115. Cushing BL, Kolesnichenko VL, O'Connor CJ (2004) Recent advances in the liquid-phase syntheses of inorganic nanoparticles. Chem Rev 104:3893–3946. https://doi.org/10.1021/CR030027B
116. LaGrow A, Besenhard M, Hodzic A, Sergides A (2019) Unravelling the growth mechanism of the co-precipitation of iron oxide nanoparticles with the aid of synchrotron X-Ray diffraction in solution. Nanoscale 11:6620–6628
117. Shah SI, Hector AL, Li X, Owen JR (2015) Solvothermal synthesis and electrochemical charge storage assessment of Mn 3 N 2. J Mater Chem 3:3612–3619
118. Bokov D, Jalil AT, Chupradit S et al (2021) Nanomaterial by sol-gel method: synthesis and application. Adv Mater Sci Eng 21:
119. Vreeland EC, Watt J, Schober GB et al (2015) Enhanced nanoparticle size control by extending LaMer's mechanism. ACS Publ 27:6059–6066. https://doi.org/10.1021/acs.chemmater.5b02510
120. Schmidt OG, Deneke C, Nakamura Y et al (2007) Nanotechnology—Bottom-up meets top-down. Adv Solid State Phys 231–240. https://doi.org/10.1007/3-540-45618-X_18
121. Ranjan S, Dasgupta N, Rajendran B et al (2016) Microwave-irradiation-assisted hybrid chemical approach for titanium dioxide nanoparticle synthesis: microbial and cytotoxicological evaluation. Environ Sci Pollut Res 23:12287–12302. https://doi.org/10.1007/S11356-016-6440-8
122. D VH-TEPJ (2011) Atomic mechanism of vitrification process in simple monatomic nanoparticles. Springer 61:627–635. https://doi.org/10.1140/epjd/e2011-10651-1
123. Marcelo G, Lodeiro C, Capelo J et al (2020) Magnetic, fluorescent and hybrid nanoparticles: From synthesis to application in biosystems. Mater Sci Eng 106
124. Proceedings SR-BUMC (2005) Edward Jenner and the history of smallpox and vaccination. Taylor Fr 18:21–25. https://doi.org/10.1080/08998280.2005.11928028

125. Genito CJ, Batty CJ, Bachelder EM, Ainslie KM (2021) Considerations for size, surface charge, polymer degradation, co-delivery, and manufacturability in the development of polymeric particle vaccines for infectious. Wiley Online Libr 1:2000041. https://doi.org/10.1002/anbr.202000041
126. Huang J, Xie J, Chen K et al (2010) HSA coated MnO nanoparticles with prominent MRI contrast for tumor imaging. Chem Commun 46:6684–6686. https://doi.org/10.1039/c0cc01041c
127. Kitrungrotsakul T, Han X, Iwamoto Y, Lin L (2019) VesselNet: a deep convolutional neural network with multi pathways for robust hepatic vessel segmentation. Comp Med Imag 75:74–83
128. Alasvand N, Urbanska A, Rahmati M (2017) Therapeutic nanoparticles for targeted delivery of anticancer drugs. Biosens Diagn, pp 245–259
129. Stephen Z, Kievit F, Zhang M (2011) Magnetite nanoparticles for medical MR imaging. Mater Today 14:330–338
130. Starsich FHL, Eberhardt C, Keevend K et al (2018) Reduced magnetic coupling in ultrasmall iron oxide T1 MRI contrast agents. ACS Publ 1:783–791. https://doi.org/10.1021/acsabm.8b00244
131. Mauri M, Collico V, Morelli L et al (2020) MnO nanoparticles embedded in functional polymers as T1 contrast agents for magnetic resonance imaging. ACS Appl Nano Mater 3:3787–3797. https://doi.org/10.1021/ACSANM.0C00474
132. Latràs JE, Estelrich J, Sánchez-Martín MJ (2015) Nanoparticles in magnetic resonance imaging: from simple to dual contrast agents. diposit.ub.edu 10:1727–1741. https://doi.org/10.2147/IJN.S76501
133. Han X, Xu K, Taratula O et al (2019) Applications of nanoparticles in biomedical imaging. Nanoscale 11:799–819
134. Lee SB, Kim HL, Jeong H-J et al (2013) Mesoporous silica nanoparticle pretargeting for PET imaging based on a rapid bioorthogonal reaction in a living body. Wiley Online Libr 125:10743–10746. https://doi.org/10.1002/ange.201304026
135. Foucault-Collet A, Gogick KA, White KA et al (2013) Lanthanide near infrared imaging in living cells with Yb^{3+} nano metal organic frameworks. Proc Natl Acad Sci U S A 110:17199–17204. https://doi.org/10.1073/PNAS.1305910110
136. Cai Z, Ye Z, Yang X et al (2011) Encapsulated enhanced green fluorescence protein in silica nanoparticle for cellular imaging. pubs.rsc.org 3:1974–1976
137. Genovese D, Bonacchi S, Juris R et al (2013) Prevention of self-quenching in fluorescent silica nanoparticles by efficient energy transfer. europepmc.org 52:5965–5968. https://doi.org/10.1002/anie.201301155
138. Grebenik EA, Nadort A, Generalova AN et al (2013) Feasibility study of the optical imaging of a breast cancer lesion labeled with upconversion nanoparticle biocomplexes. J Bio Opt 18
139. Abd Elkodous M, El-Sayyad GS, Abdelrahman IY et al (2019) Therapeutic and diagnostic potential of nanomaterials for enhanced biomedical applications. Colloids Surf B 180:411–428
140. Alphandéry E (2022) Ultrasound and nanomaterial: an efficient pair to fight cancer. J Nanobiotechnology 20:1–30. https://doi.org/10.1186/S12951-022-01243-W
141. Martinez H, Kono Y, Blair S, Sandoval S (2010) Hard shell gas-filled contrast enhancement particles for colour Doppler ultrasound imaging of tumors. pubs.rsc.org 1:266–270
142. Datta LP, Manchineella S, Govindaraju T (2019) Development of biomolecule integrated materials and their biological applications. World Sci 173–198. https://doi.org/10.1142/9789811211331_0008
143. Peng C, Xi N, Hong Z et al (2022) Acceptance of wearable technology: a meta-analysis. osuva.uwasa.fi 5101–5110. https://doi.org/10.24251/HICSS.2022.621
144. Aldewachi H, Chalati T, Woodroofe M, Bricklebank N (2018) Gold nanoparticle-based colorimetric biosensors. Nanoscale 10:18–33. https://doi.org/10.1039/C7NR06367A
145. Szunerits S, Spadavecchia J, Boukherroub R (2014) Surface plasmon resonance: signal amplification using colloidal gold nanoparticles for enhanced sensitivity. Rev Anal Chem 33:153–164. https://doi.org/10.1515/REVAC-2014-0011

146. Chern M, Kays JC, Bhuckory S, Dennis AM (2019) Sensing with photoluminescent semiconductor quantum dots. Methods Appl Fluoresc 7:012005. https://doi.org/10.1088/2050-6120/AAF6F8
147. Medintz IL, Mattoussi H (2009) Quantum dot-based resonance energy transfer and its growing application in biology. Phys Chem Chem Phys 11:17–45. https://doi.org/10.1039/B813919A
148. Kahraman M, Mullen ER, Korkmaz A, Wachsmann-Hogiu S (2017) Fundamentals and applications of SERS-based bioanalytical sensing. Nanophotonics 6:831–852. https://doi.org/10.1515/NANOPH-2016-0174/PDF
149. Ambartsumyan O, Gribanyov D, Kukushkin V et al (2020) SERS-based biosensors for virus determination with oligonucleotides as recognition elements. Int J Mol Sci 21:3373. https://doi.org/10.3390/IJMS21093373
150. Polavarapu L, La PA, Novikov SM et al (2014) Pen-on-paper approach toward the design of universal surface enhanced Raman scattering substrates. Small 10:3065–3071. https://doi.org/10.1002/SMLL.201400438
151. Ding Y, Shang H, Wang X, Chen L (2020) A SERS-based competitive immunoassay for highly sensitive and specific detection of ochratoxin A. Analyst 145:6079–6084. https://doi.org/10.1039/D0AN01220C
152. Serebrennikova KV, Byzova NA, Zherdev AV et al (2021) Lateral flow immunoassay of SARS-CoV-2 antigen with SERS-based registration: development and comparison with traditional immunoassays. Biosensors 11:510. https://doi.org/10.3390/BIOS11120510
153. Gao Y, Zhou Y, Chandrawati R (2020) Metal and metal oxide nanoparticles to enhance the performance of enzyme-linked immunosorbent assay (ELISA). ACS Appl Nano Mater 3:1–21. https://doi.org/10.1021/ACSANM.9B02003/ASSET/IMAGES/MEDIUM/AN9B02003_0018.GIF
154. Orlov AV, Khodakova JA, Nikitin MP et al (2013) Magnetic immunoassay for detection of staphylococcal toxins in complex media. Anal Chem 85:1154–1163. https://doi.org/10.1021/AC303075B
155. Stiepel RT, Duggan E, Batty CJ, Ainslie KM (2023) Micro and nanotechnologies: the little formulations that could. Bioeng Transl Med 8:e10421. https://doi.org/10.1002/BTM2.10421
156. Grayson ACR, Shawgo RS, Johnson AM et al (2004) A BioMEMS review: MEMS technology for physiologically integrated devices. Proc IEEE 92:6–21. https://doi.org/10.1109/JPROC.2003.820534
157. Pereiro I, Aubert J, Kaigala GV (2021) Micro-scale technologies propel biology and medicine. Biomicrofluidics 15:021302. https://doi.org/10.1063/5.0047196/5.0047196.MM.ORIGINAL.V1.MPG
158. Ru C, Luo J, Xie S, Sun Y (2014) A review of non-contact micro- and nano-printing technologies. J Micromechanics Microengineering 24:053001. https://doi.org/10.1088/0960-1317/24/5/053001
159. Wang L, Luo Y, Ahmad Z et al (2018) Fabrication of stacked-ring netted tubular constructs via 3D template electrohydrodynamic printing. J Mater Sci 53:11943–11950. https://doi.org/10.1007/S10853-018-2468-0
160. Wu S, Li JS, Mai J, Chang MW (2018) Three-dimensional electrohydrodynamic printing and spinning of flexible composite structures for oral multidrug forms. ACS Appl Mater Interfaces 10:24876–24885. https://doi.org/10.1021/ACSAMI.8B08880
161. Initiative NN Nanotechnology Timeline | National Nanotechnology Initiative (2023)
162. 3D Printing Market Size, Share & Trends Analysis Report, 2030. https://www.grandviewresearch.com/industry-analysis/3d-printing-industry-analysis. Accessed 18 March 2023
163. Li X, Zhang C, Wu S et al (2019) Precision printing of customized cylindrical capsules with multifunctional layers for oral drug delivery. ACS Appl Mater Interfaces 11:39179–39191. https://doi.org/10.1021/ACSAMI.9B13568/SUPPL_FILE/AM9B13568_SI_003.MP4
164. Wu S, Wang B, Ahmad Z et al (2017) Surface modified electrospun porous magnetic hollow fibers using secondary downstream collection solvent contouring
165. Soltani A, Pouypouy H (2019) Standardization and regulations of nanotechnology and recent government policies across the world on nanomaterials. Adv Phytonanotechnology 419–446

166. Can gold melt at room temperature? Melting temperature depression!—Sustainable Nano. https://sustainable-nano.com/2018/02/14/melting-temperature-depression/. Accessed 18 March 2018

Chapter 6
Bioceramics

Tuba Bedir, Eray Altan, Kubra Aranci-Ciftci, and Oguzhan Gunduz

Abstract Bioceramics commonly employed materials for the restoration, replacement and recovery of unhealthy and impaired pieces of the muscle and skeletal system, as well as periodontal anomalies. According to the host tissue interactions, bioceramics can be graded as nearly bioinert, bioactive, and bioresorbable. Most of the clinical applications of bioceramics comprise orthopedic and dental surgery and also have potential in the field of tissue engineering. This chapter aims to introduce a concise and accessible overview of the past of bioceramics to the present. From bioinert to bioactive and bioabsorbable bioceramics, the classification of materials is discussed and bioceramics characteristics such as biodegradability, bioactivity, biocompatibility, porosity, mechanical and surface properties, as well as osteoconductivity and osteoinductivity emphasized in depth. Production processes of bioceramics are also considered herein. At the end of this chapter, the biomedical applications of bioceramics including orthopedic, dental, surface coatings, and bone tissue engineering, challenges, and future research expectations in the area of bioceramics are also highlighted.

T. Bedir · E. Altan · K. Aranci-Ciftci · O. Gunduz (✉)
Center for Nanotechnology and Biomaterials Application and Research (NBUAM), Marmara University, 34722 Istanbul, Turkey
e-mail: ucemogu@ucl.ac.uk

T. Bedir
e-mail: tubabedir@marun.edu.tr

E. Altan
e-mail: eray.altan@marmara.edu.tr

T. Bedir · E. Altan · O. Gunduz
Department of Metallurgical and Materials Engineering, Faculty of Technology, Marmara University, 34722 Istanbul, Turkey

K. Aranci-Ciftci
Department of Bioengineering, Faculty of Engineering, Marmara University, 34722 Istanbul, Turkey
e-mail: kciftci@marun.edu.tr

O. Gunduz
Health Biotechnology Joint Research and Application Center of Excellence, 34220 Istanbul, Türkiye

O. Gunduz et al. (eds.), *Biomaterials and Tissue Engineering*, Stem Cell Biology and Regenerative Medicine 74, https://doi.org/10.1007/978-3-031-35832-6_6

Keywords Bioceramics · Bioactive glasses · Hydroxyapatite · Calcium phosphates

Background and Scope of Bioceramics

Ceramic materials represent a significant source of biomaterials for biomedical applications. A class of ceramics called bioceramics, which can be of synthetic or natural sources, are used to restore and regenerate distorted and unhealthy pieces of the muscle and skeletal system, as well as periodontal anomalies [100]. Bioceramic materials can be categorized into three main classes based on how tissues react to them: nearly bioinert (alumina and zirconia), bioactive (hydroxyapatite (HA) and bioactive glass, and bioresorbable [tricalcium phosphate (TCP)] [65]. These ceramics, which are meant to contact live tissues, have seen serious advancements in the past 50 years. Due to their favorable interactions with tissues, bioceramics are preferred in many clinical applications, such as the replacement of tendons, ligaments, hips, knees, teeth and spinal fusion, stabilization and enhancement of the jaw bones, maxillofacial reconstruction, also as bone cavity fillers and as carbon coatings [35, 44].

The history of ceramics in biomedical applications extends to the late 1960s, with considerable research by Hulbert and his colleagues [61, 62, 130]. The advancement of the usage of bone tissue growth in porous ceramics as a technique for mechanically interlocking prostheses was presented together with the biocompatibility of oxide ceramics. Oxide ceramics, which have been intensively researched starting in the late 1960s, are nearly inert bioceramics. Hench and colleagues thoroughly assessed surface reactive bioceramics in the early 1970s, which led to the development of an intermediate strategy [53, 101, 112]. Hench et al. presented 45S5 Bioglass®, which demonstrated multi-stage and complicated surface reactions for the development of biologically active hydroxy-carbonate apatite layers resembling the natural bone minerals [54]. Synthetic HA and β-TCP became commercially available in the early 1980s as replacement bone materials for use in dentistry and medicine [26]. Apatite-wollastonite, a novel glass–ceramic material, was created in the 1980s and has the highest fracture toughness, bending strength, and elasticity modulus among various bioceramics [71]. Studies on the application of various calcium phosphate ceramics for in vivo implantation considerably increased in the 1980s and 1990s [17]. The clinical use of bioceramics is restricted to non-load-bearing locations because of their poor mechanical characteristics. As a result, various techniques or materials have been tried to enhance the mechanical characteristics of bioceramics [46]. In order to enhance the plasticity, toughness, strength, and graded mechanical stiffness of polymeric composites, Bonfield et al. suggested using bioceramics as fillers at these dates [22] (Fig. 6.1). Various bioceramic materials are now employed as granules, nanopowders, metallic implant coatings, polymeric scaffold fillers, and bone cement that hardens on its own [93]. There are different types of hybrid composite scaffolds and research in this area is very active [9, 121]. Targeted delivery of biomolecules

Fig. 6.1 An overview of the history of bioceramics

embedded in biodegradable nanoparticles can enhance the functionality of these structures as a novel approach [110, 122].

Characteristics of Bioceramics

Biodegradability

The ability of a material to break down biologically is known as biodegradability. When used as scaffolds, biodegradable materials temporarily replace a hole while degrading over time as new bone grows in its place. On the other hand, non-biodegradable materials block the stimulation for the surrounding bone to grow (stress shielding). During the degradation of the material, the room is formed for the growing tissue and the load progressively shifts from the implant to the bone. In the end, the defect is substituted by healthy bone tissue, and degradation and elimination from the body of biomaterial occurs safely [119]. Bioceramics reveal various degradation behaviour according to their composition. The degree of dissolution of bioceramics will vary according to the HA:β-TCP content, taking into account the particle size and macroporosity: since β-TCP is more soluble than HA [74]. Thus, a higher amount of HA causes less dissolution. The synthesis and sintering methods of the material have also an impact on dissolution. Certain microporosity, mechanical characteristics, surface area, and mode of particle release can affect the bioceramic's ability to degrade, dissolve, and be absorbed by the tissue and cells around it. Processing factors (temperature and sintering time) that may have an impact on the entire macroporosity and microporosity may be the origin of this occurrence. The degree of dissolution increases with both macro- and microporosity. A decline in crystal size and an enhancement in macro- and microporosity accompany in vivo dissolution of bioceramics [26]. In addition, the properties of the physiological solution like temperature, pH, ion concentration, protein concentration and buffer capacity also affect dissolution [72].

Bioactivity

The ability of a substance to interact with a living system or tissue is known as bioactivity. It was suggested to measure a material's bioactivity using the in vivo bioactivity index I_B, which has the formula IB = $100/t_{50bb}$. The amount of time needed for an interface to bind more than 50% is t_{50bb} [50]. The nature and microstructure of bioactive materials affect both the ratio at which bone bonds to implants and the bond's strength and stability. In general, higher bioactivity index values of the material result in faster apatite production on the surface and stronger attachment to bone. Adsorption of the medium's protein altered how Bioglass® interacted with the solution. In the protein-containing solutions, the formation of hydroxyl carbonate apatite (HCA) crystals on the Bioglass® surface grew complicated or took longer [104]. Conversely, the media containing serum proteins dramatically slowed down or even prevented the time required for the production of apatite on HA. Bioactive materials' surface reactivity has an impact on how bone cells connect, proliferate, differentiate, and mineralize [33]. The biological processes by which cells perceive and react to implant materials through this protein layer hinge on the adsorption of proteins on material surfaces [128].

Biocompatibility

The capability of a material to function with a suitable host reaction in a particular application is the definition of biocompatibility [131]. It is a significant characteristic of a biomaterial. In vitro and in vivo are often involved in the biological evaluation of a biomaterial. Even though it would be ideal to directly observe the host tissue's reaction to a biomaterial in vivo, this may be problematic due to the in vivo processes' high level of intricacy. As a result, the in vitro evaluation of the isolated cells' reactions to a biomaterial is carried out rather than researching the complicated in vivo response. This enables a controlled investigation into a particular cellular reaction to a test material. While beneficial for customizing material to the living tissue, information gained in vitro cannot substitute in vivo evaluation. Particularly, a conflict is present for a bioactive glass that has been assessed in vitro and in vivo [59].

For in vitro modelling of biological responses, numerous different cell lines are employed. They are obtained from humans or animals (mice, rats), and can be stem cells or primary cells that have transformed. Different cell types make up tissues and organs, so interactions between various cell types are crucial for how cells behave physiologically.

A biomaterial's in vitro biological responses can be evaluated qualitatively, such as by looking at how cytoskeletal proteins are organized, or quantitatively, such as by performing cytotoxicity tests and measuring the phenotype, gene expression, and growth, proliferation, and differentiation of cells. The physicochemical properties of the material, like surface roughness and crystallinity, have an impact on the cells grown in contact with it. Additionally, it was discovered that inorganic ions with micromolar concentrations, such as Si, might promote the growth, differentiation, and gene expression of osteoblasts [109]. Appropriate in vivo investigations are needed to evaluate the biocompatibility/bioactivity of scaffolds. Mice or rats seldom show lamellar cortical bone remodelling, while bigger animals demonstrate human-like bone development and remodelling [56]. For example, sheep are suitable animals for examining load-bearing effects on bone repair, as they can produce larger defect sizes. Calcium phosphate implants' bioactivity has been satisfactorily assessed using the sheep model [97].

Porosity

Crucial factors for porous ceramics and glasses materials include porosity and pore size, distribution, and interconnectivity, which impact their mechanical characteristics and biocompatibility [31]. For blood vessels and nearby bone to develop into the scaffold and for implanted or moving cells to deposit bone inside of it, there must be sufficient porosity [118]. According to the studies, pore sizes between 300 and 600 μm are optimum for tissue development [127]. The intrinsic weakness brought on by porosity is the typical restriction of all porous materials. Therefore, porous scaffolds are not preferred in non-load-bearing applications like fractured skulls [29].

Interconnected macroporosity and suitable microporosity are significant for the ideal biological performance of bioceramics, including bioceramics resorption, bioceramics-cell interactions, bioceramic tissue interface, and new bone development [20, 76, 79]. Macroporosity in calcium phosphate (CaP) ceramics is achieved by the inclusion of evaporative materials (porogens such as hydrogen peroxide, polymers, sugar etc.), heating to sublimation or calcination-inducing temperatures (80–500 °C) followed by sintering at higher temperatures [26, 76]. The temperature and period of sintering have an impact on microporosity: the specific surface area and microporosity decrease with increasing temperature. According to some research, a higher micropore concentration is associated with higher osteogenic characteristics in bony regions with the same macroporosity [55, 86]. The majority of the commercial CaP bioceramics that are now on the market have a porosity of 70%, according to analysis, albeit there are significant differences across proprietary bioceramics [26].

Mechanical Properties

The structural applications of a material are significantly influenced by its mechanical properties. Several mechanical characteristics of bioceramics are presented in Table 6.1. Compressive strength, which is frequently assessed for ceramics, is the capacity of the material to withstand crushing. Bending testing, also known as flexural testing, is a common method of determining the strength of ceramics. One of the most important indicators of the mechanical characteristics of ceramics is fracture toughness. The load applied during the bending and tensile tests are vertical to the crack. K_{IC} measures the level of the stress intensity at which a crack will begin to spread and result in fracture toughness. It is more challenging to start and spread a crack when the fracture toughness is higher. Resistance of the material to localized deformation caused by scratching or indentation is referred to as hardness. To determine a material's hardness, a tiny indenter is applied to its surface, and the size of the resulting indentation is recorded. The wear behaviour of ceramic is influenced by its fracture toughness and hardness. For use in joint replacement, the wear characteristics of the bioceramics are crucial. Because wear debris may promote inflammatory reactions in the tissue around an implant, which could ultimately result in implant failure [49, 52, 59, 60, 69].

The distinctive mechanical characteristics of bone are a result of the interplay between elastic collagen fibers and rigid HA crystals in the nanoscale structure of bone. Because of this, replacing the bone with synthetic materials is challenging [19, 131]. Mechanical stimulation is necessary for bone tissue regeneration and remodelling. The majority of these stimulating pressures are absorbed by permanent metal implants; this is known as the stress shielding effect, which causes bone resorption around the implant [19, 59]. A tissue-engineered scaffold should ideally be created to closely resemble the mechanical characteristics of the particular bone it will replace. The mechanical performance of the scaffold must be enough to sustain the implantation process and support the loads and stresses that the new tissue will eventually be subjected to [11].

Table 6.1 Mechanical properties of different bioceramic materials [60, 116, 131]

Materials	Density ($g\ cm^{-3}$)	Hardness (Vickers, HV)	Young's modulus (GPa)	Bending strength (MPa)	Compressive strength (MPa)	Fracture toughness K_{IC} ($MPa\ m^{1/2}$)
Alumina	3.98	2400	380–420	595	4000–4500	4–6
Zirconia (TZP)	6.05	1200	150	1000	2000	7
Zirconia (Mg-PSZ)	5.72	1120	208	800	1850	8
Bioglass 45S5	2.66	458	35	40–60	~500	0.4–0.6
A-W glass–ceramic	3.07	680	118	215	1080	2.0
Sintered HA	3.156	500–800	70–120	20–80	100–900	0.9–1.3

Surface Properties

In general, cellular response is affected by the physicochemical characteristics of implant materials, like hydrophilicity or hydrophobicity, surface energy and charge, which can alter protein absorption and cell attachment. Thereof, cellular reactions can be greatly influenced by the surface characteristics of materials. The bioceramic surface modification, which results from local environment interaction, is increasingly accepted as a crucial component. The loss of soluble silica and the emergence of Si–OH (silanols) at the glass-solution interface is caused by interchanges of Na^+ and K^+ with H^+ and H_3O^+ ions from the solution at the glass surface. Following this phase, an amorphous CaP layer forms when calcium and phosphate ions move through the silica-rich layer to the surface. When carbonate, hydroxyl, and fluoride ions are added, they can crystallize to form an apatite layer [51]. Another strategy for obtaining a desired tissue response to implant materials is to improve the interaction between tissue and biomaterials. Contact guidance develops due to cells' recognition and response to surface properties [25]. Material topography is a crucial design factor for obtaining maximum cell responses [28]. In an effort to increase cell activity, electrohydrodynamic spraying and print-patterning have been used to create micro- and nano-scaled surface topography utilizing HA [2, 94].

Osteoinduction, Osteoconduction, and Osseointegration

Inducing osteogenesis is done by a process called osteoinduction. It is a phenomenon that is observed regularly in all kinds of bone healing processes. Osteoinduction is the process of collecting immature cells and encouraging them to develop into preosteoblasts. In a situation like a fracture, bone healing depends on this term. Osteoconduction refers to the growth of bone over a surface. This phenomenon occurs regularly with bone implants. Copper, silver, and bone cement, which are implant materials with minimal biocompatibility, exhibit little to no osteoconduction [4]. Osseointegration is regarded as a requirement for implant loading and the longtime clinical success of end-bone dental implants. It is described as the direct structural and functional link between healthy, live bone and the surface of a load-bearing implant [96]. A major problem in clinics is how to use bone transplants to achieve the best possible regeneration healing of massive bone lesions. The length of the defect, the state of the soft tissue, the surroundings, age, and other factors are only a few that can affect the outcome of treatment [113]. The majority of artificial bone products in use only exhibit osteoconductivity. Hence, it is challenging to achieve fully fresh bone regeneration for significant bone abnormalities [57]. So, it is a great advance to make artificial bone grafts that have good osteoconduction and osteoinduction to mend.

Habibovic et al. produced bioceramic brushite and monetite implants with different geometries using low-temperature direct 3D printing. The design of the bioceramic implants (open or closed pores) allowed osteoinduction and osteoconduction in bone repair [47]. Moreover, other comparative studies that assess the osteoinductive qualities of bioceramic, coral, and processed graft alternatives have been conducted. Different bone graft substitutes were used in rats. For instance, a demineralized rat bone graft showed slight fibrovascular invasion and there was little clue of new bone formation around the implant. Also the coral graft, due to its extreme hardness, wasn't observed in some sections. The thin layer of the dense mineral was seen in contact with the remaining surface of implanted coral. It can be said that all types of bioceramics had different impacts on the host tissue [16]. In another study, Frayssinet et al. studied the osseointegration levels between calcium phosphate as the control group and composite grafts. They improved the strength of macroporous CaP by filling the pores with a highly soluble, CaP cement made of TCP and dicalcium phosphate (DCPD). After 20, 60, and 120 days, cylinders of the finished product were implanted in sheep and examined. At 20 days, composite materials exhibited developing in the soft tissue surrounding the implants while the control group also had some fibrous tissue invasion of the pores. In the pores of the composite implants, where the cement was gradually being replaced by bone tissue, a progressive ingrowth was observed at 60 days. For the control group, there was only seen some partial integration. Finally, at 120 days, all the pores in bone tissue were totally filled, and they showed severe signs of deterioration. Control ceramics had been slowly integrated compared with the composite implant and exhibited signs of resorption after 120 days [40].

Classifications of Bioceramics

Bioceramics are classified by considering factors such as their origin and composition, tissue response, and crystallinity (Fig. 6.2) [45]. As in the general classification of biomaterials, starting with the classification of bioceramics according to their interaction with living tissue may be more appropriate. Thus, synthetic ceramics are

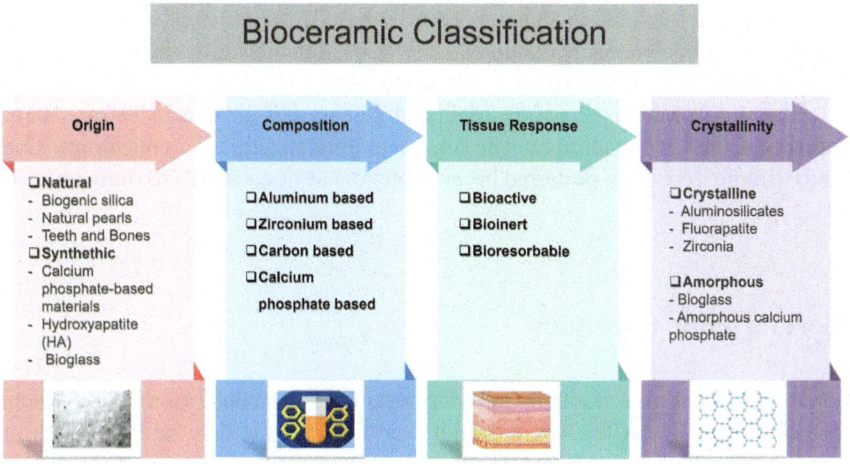

Fig. 6.2 Factors for bioceramic classification

sub-categorized as bioinert, bioactive and biodegradable ceramics according to the reaction that occurs in the tissue after implantation [37].

Bioinert Ceramics

Bioinert ceramics do not stimulate any tissue responses when interacting with biological systems. Since the absence of interaction with the surrounding tissues, they are highly resistant to corrosion in the physiological environment. Generally, they are classified into two categories, oxide and non-oxide bioinert ceramics, according to whether they contain oxide or not. The oxide bioinert is composed of alumina (Al_2O_3), zirconia (ZrO_2), and titania (TiO_2) [45, 123]. The carbon- and nitride-based ceramics, such as silicon nitride, titanium nitride, and zirconium nitride, are referred to as non-oxide bioinert ceramics [83].

Bioactive Ceramics

Bioactive ceramics induce tissue response by producing strong biochemical bonds with the surrounding tissue when they are interacting with the physiological system. This interaction with body fluids and tissues provides the initiation of many biological processes, such as osteoblast adhesion, differentiation of stem cells, and enzyme activity [39]. Bioactive ceramics are especially utilized as coating materials to increase the mechanical and corrosion resistance of bone graft implants owing to their osteoconductive behaviours [83]. The most widely known bioactive ceramics are glasses, glass ceramics, calcium phosphate-based (HA and tricalcium phosphate-based), and carbon-based bioceramics. They have high biocompatibility. Their most important feature is supporting the formation of strong bones. In addition to bone tissue applications, they have a significant role in the advancement of metallic implants, toothpaste, and composite materials [39]. In surgical applications, CaP-based ceramics are often preferred because they act as tissue scaffolds during healing [102].

Bioresorbable Ceramics

Ceramics that dissolve gradually in living organisms are called resorbable (or bioresorbable) ceramics. After they are implanted in the body, they integrate with the tissue without any toxic effect and are constantly replaced by the host tissue [37]. The most popular bioresorbable ceramics are TCP, porous HA, CaPs and salts and some bioglasses. Bioresorbable and bioactive ceramics can sometimes confuse due to their interaction with tissue. For example, HA in dense form is not resorbable and can remain in living tissue for a long time (approximately 5–7 years). Therefore, bioactive ceramics show similar properties to both bioinert and bioresorbable ceramics [45]. The most important factors that distinguish bioactive ceramics and bioresorbable ceramics are their chemical compositions and microstructures.

After the implanted resorbable ceramics create a suitable environment for tissue regeneration, the resorption process begins. The ceramic's dissolution rate and resorption rate are related. One of their most important advantages of them is that the resorption rate can be easily controlled by modifying the chemical bonds in the structure. For example, the resorption rate increases with the decrease in the size of the particles and the crystallinity in the structure [37]. However, the resorption rate must be compatible with the tissue healing process. In addition, resorbable ceramics during healing should provide the necessary mechanical performance for the tissue. Otherwise, it may cause failures and difficulties in the application of bioceramics [37, 45]. Thus, these parameters should be taken into consideration when designing and selecting bioresorbable ceramics.

Types of Bioceramics

Alumina and zirconia

Alumina has a lot of uses in engineering because of its durability, insulating qualities, chemical stability, and thermal stability. This material is fragile and challenging to process despite having desirable qualities [124]. Biomedical-grade zirconia is possibly the orthopaedic material that has generated the most discussion among researchers, industrialists, and medical professionals. To address the problem of alumina fragility and possible failure of implants, biomedical-grade zirconia was developed twenty years ago [23]. Lately, zirconia implants have been suggested for individuals with metal sensitivity and soft tissue injury in particularly difficult cosmetic scenarios [1].

Glasses and glass-ceramics

Glass–ceramics are microcrystalline solids synthesized with the devitrification of glass in a controlled manner. Glass is melted, moulded to be shaped, and then, by heat treatment, it is transformed into a primarily crystalline ceramic. Effective internal nucleation, the foundation of controlled crystallization, enables the creation of small, randomly oriented grains devoid of voids, microcracks, or other porosity [13]. A glass–ceramic typically has a microstructure that is between 50 and 95% crystalline by volume while the remaining material is glass. Heat treatment causes the formation of one or more crystalline phases. The composition of the residual glass differs from the lead (main) glass because the position is often different from the main glass [107]. Crystalline phases in particular forms mixed with the glassy phase make up the basic microstructure of glass ceramics [138]. Synthetic glass–ceramics were accidentally found in 1953 by Stanley Donald Stookey. Corning Inc. created and commercialized two novel glass–ceramics based on Li-aluminosilicates (LAS) and Mg-aluminosilicates between 1953 and 1963, following the discovery of lithium disilicate glass–ceramic (MAS). As LAS glass–ceramic has a very low coefficient of thermal expansion, it has been utilized as cookware (CTE). Researchers attempted to create transparent, nanocrystalline glass–ceramics between 1963 and 1980 [114]. Glass–ceramics have several advantages over conventional powder-processed ceramics. In addition to the flexibility to form in the glassy state, glass–ceramics have a microstructure homogeneity followed by the reproducibility of properties resulting from the homogeneity of the starting glass [14]. Dental glass–ceramics are now mostly employed as coatings for teeth. More durable and resilient glass ceramics must be created in order to broaden their uses beyond dental coatings to include implant bodies, dental abutments, and even one-piece implants. The issues associated with dental implants made of titanium or titanium alloys, like titanium allergy, accumulation of metal ions, and subpar aesthetics, can be efficiently avoided by using glass–ceramic implants [114].

Calcium phosphate bioceramics

Chemical compounds very similar to the inorganic portion of the main normal and pathological calcified tissues of mammals are known as CaPs [32]. These bioceramics exhibit exceptional biological performance, including biocompatibility, bioresorbability, and osteoconductivity. This provides them with the ability to be integrated into living tissue via the identical mechanisms used in bone remodelling. CaPs are also affordable to buy and reasonably easy to certify as medical grade. However, CaPs are mostly utilized as coatings and fillers in the biomedical industry and are only appropriate for load-bearing applications owing to their weak mechanical qualities, particularly in terms of fatigue resistance and strength [21]. Moreover, CaPs are also offered as dense or porous blocks, particles, injectable formulations, coatings for implants, and composites with polymers [34].

According to their solubility and acidity, the most well-known CaPs have Ca/P molar ratios that range from 0.5 to 2. Lower Ca/P molar ratios cause CaP to become more acidic and water-soluble [78]. Most CaPs are only poorly soluble in water, however, all of them are easily soluble in acidic solutions but insoluble in alkaline ones. CaPs can be rated in order of decreasing in situ degradation rate based on solubility as follows: HA > β-TCP > OCP > DCPD > α-TCP ≈ TTCP > MCPM [100]. Because implants consisting of calcined HA can remain in the defected bone for a very long time, biomedical interest has generally centred on HA, α- and β-TCP, CDHA, and biphasic CaPs [27].

Hydroxyapatite

The expression 'apatite' refers to a group of materials with close structures but different compositions. $Ca_{10}(PO_4)_6(OH)_2$ makes up hydroxyapatite, often known as HA or HAp, and its Ca/P stoichiometric ratio is 1.67 [15]. With a space group of $P6_3/m$, it possesses a hexagonal crystal structure [67] and typical unit cell parameters of $a = 9.422$ and $c = 6.883$ Å [132]. The HA crystal structure is represented in Fig. 6.3. [92]. Spectral characteristics, lattice parameters, morphology and crystal size of apatite are all impacted by substitutions in its structure, this, in turn, influences its thermal stability and chemical stability (solubility) [135]. The size and quantity of the replacing ions determine the level of impact [77]. A robust bone-implant interface can be created using HA, a bioactive substance. Additionally, it promotes bone formation away from the interface region, making it osteoconductive [72].

Mineral or inorganic phases of typical calcified tissues, like teeth and bone, are created by biological apatites; also, they are present in several pathological calcifications, including urinary stones, cardiac calcifications, dental calculi, and soft tissue calcifications. Contrary to pure HA, biological apatites contain minor substituents like CO_3^{2-}, Na^+, and Mg^{2+} and are hence better called carbonated apatite [75, 136]. Fluoride (F^-) and carbonate (CO_3^{2-}) are both present in other biologic apatites like shark and other fish enameloids [80].

Products obtained from bovine bone, marine algae or coral are examples of biologically originated commercial HA products. By hydrothermally transforming coral Porites, composed of calcium carbonate in the form of aragonite, in the existence of

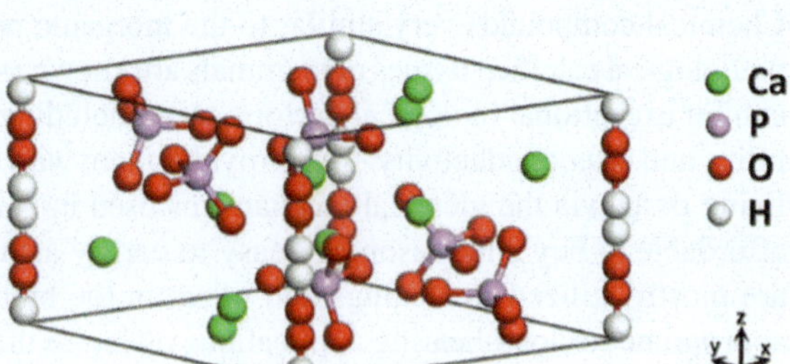

Fig. 6.3 The crystal structure of HA [92]

ammonium phosphate, coralline HA is produced [133]. Around 275 °C and 1200 psi are used for the procedure. Coralline HA, such as ProOsteon®, Interpore®, CA, and Irivine is distinct from bone and synthetic HA in terms of its composition, dissolving characteristics, and crystal size. Products made from bovine are divided into (a) bovine bone apatite produced by deproteinizing bovine bone or (b) bovine bone produced at around 1000 °C. Biologic apatite loses carbonate during sintering, and crystal size also grows [75].

The compression and sintering conditions, the characteristics of the apatite powder, as well as macro- and microporosity, all affect the dense HA's mechanical properties. A few mechanical properties, such as compressive strength, decreases when microporosity increases. Grain size, compressive, density, flexural, and torsional strength, as well as elastic moduli in compression, all rise at higher sintering temperatures. When compared to cortical bone, which has an elastic modulus of 12–18 GPa, HA has a 40–117 GPa range [77].

Currently, granules, blocks, and scaffolds made of synthetic or biological HA are employed for the regeneration and repair of bone; it is used on its own or as a compound with polymers or other ceramics, or as coatings on orthopaedic or dental implants [77].

Tricalcium phosphate-based ceramics

The chemical formula of tricalcium phosphate (TCP) is $Ca_3(PO_4)_2$ and it is bioactive and resorbable bioceramic [34]. The Ca/P ratio of stoichiometric TCP is 1.5, it comes in four different, more soluble forms compared to stoichiometric HA [12]. The first two types of TCP are low-temperature, unstable phases called amorphous (am-TCP) and apatitic (ap-TCP). The final two types, crystalline high-temperature phases include α-TCP and β-TCP. TCP's crystalline phases resemble bone's inorganic component in relation to composition and degree of crystallinity even if they are not totally crystalline themselves [34].

In comparison to HA, TCP dissolves more quickly. The TCP members with the highest dissolution rates are am-TCP and α-TCP. Following these, the dissolving rate of ap-TCP is lower, while β-TCP has the lowest dissolving rate of all. Therefore, TCP has been employed for the filling of bones in orthopaedic and dentistry applications. By using TCP with smaller particle size and greater microporosity, the process of filling the bones can be regulated and improved to osteoconduction level. Bone remodelling and filling are regulated by TCP's osteoclastic activity as well as its in vivo resorption rate [30]. Also, according to the basic idea that the solubility of calcium phosphates determines their biological resorbability, they are all bioresorbable [100]. Along with being bioresorbable, am-TCP and α-TCP in particular demonstrate extremely strong reactivity in aqueous conditions and can rapidly change into apatite. In physiological settings, am-TCP and α-TCP are unstable, they either hydrolyze forming weakly crystalline apatite similar to bone minerals or spontaneously dissolve. Ap-TCP should react similarly to bone minerals because it is most likely within the spectrum of bone solubility. Theoretically, β-TCP cannot hydrolyze into apatite in the lack of cell activity or dissolve spontaneously in vivo [70].

It was discovered that sintering circumstances affected the mechanical behaviour of β-TCP ceramics [126]. It was observed that elasticity modulus and hardness increased at a temperature below 1300 °C, however above that point, a decline in elasticity modulus was noted. Above 1400 °C, it was proved that a portion of the β-TCP changed into α-TCP. This transformation helped to lower the elasticity modulus [70].

Fabrication Process of Bioceramics

In the last 50 years, there have been quite new developments in bioceramic production. It is possible to produce bioceramics in more than one shape by more than one method. However, in the most general sense, the fabrication of bioceramics consists of 4 steps: powder preparation, forming of the ceramic green body, sintering and machining [102, 115].

Powder Preparation

The properties and preparation methods of ceramic powders are very important for general ceramic production. In the most general form, powder preparation methods are examined under two titles; mechanical and chemical methods [105].

Mechanical Methods

The process of bringing large pieces into smaller pieces by using mechanical forces is called comminution. There are processes such as crushing and grinding in comminution. The milling method is most commonly used to obtain powder by reducing its size. In the milling, mechanical stresses are applied to the particles, creating elastic and inelastic deformations and breaking the particles. However, the surface chemical properties of the particle also change with the mechanical energy applied for a very long time. For this reason, the milling device to be used and the duration of the applied mechanical energy should be chosen carefully [105].

Chemical Methods

The chemical methods are processes for producing advanced ceramics using raw materials of natural or synthetic origin. In the last 30 years, the use of chemical methods in powder preparation processes has become very popular. The chemical methods are divided into 3 titles; dry chemical process (solid-state), wet-chemical process and vapour-phase reactions (Fig. 6.4). In this study, only dry and wet-chemical processes will be discussed because of their importance. The dry chemical process is a method in which chemical reactions are carried out between solid-state sources using high temperatures. The high temperature facilitates the diffusion of ions and accelerates chemical reactions, resulting in powder production. The solid-state process is sometimes called a reaction of decomposition. In this method, a solid reactant is heated and decomposed into a new solid plus gas form. The decomposition is mostly used in the preparation of oxides and simple thermodynamic laws, reaction kinetics play a role here. Although the solid-state chemical process is cost-effective, there are many parameters to be considered such as the structure of the reactants, the nature of the product, size and size distribution, reaction atmosphere, temperature and time etc. Incomplete and poorly mixed chemical reactions cause the formation of undesirable phases and it becomes very difficult to control the size and shape of the obtained ceramic powders [115].

The wet-chemical processes are methods used to produce crystalline bioceramics at low temperatures. Starter sources are in aqueous form. The major advantage of using liquid precursors is mixing at the atomic level. There are two methods of obtaining powder ceramics from liquid initiators. The first is the evaporation of the liquid. The second way is precipitation by adding chemical agents to the solution. Crystal precipitation is a method consisting of two steps; crystal nucleation and crystal growth. The nucleation process is the chemical process that begins in the

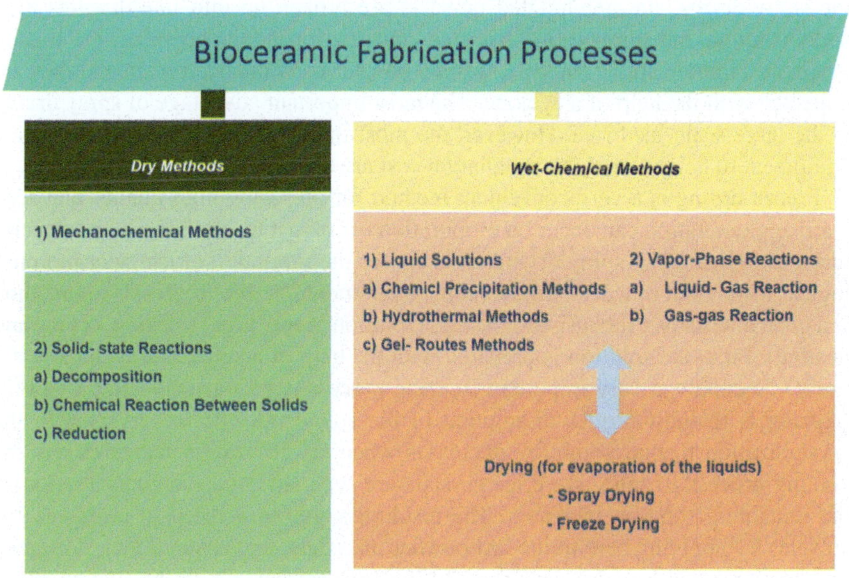

Fig. 6.4 Bioceramic fabrication process

aqueous phase, leading to the formation of molecular aggregates, which can then transform into macroscopic crystals in the crystalline growth phase [42, 115].

For several bioceramics, the chemical precipitation method has been chosen as the best method. For example, the precipitation method is frequently used in the production of commercialized HA. However, co-precipitation, hydrolysis and sol–gel methods are used in some ceramics such as Y-TZP. In the sol–gel method, the monomers are exposed to chains of hydrolysis/condensation/drying/calcination processes to form a colloidal suspension (sol) that turns into a gel under low temperature. The most important advantage is that ultrafine ceramic powders with particle sizes from nanometer to micrometer scale can be obtained. The sol–gel method's drawbacks include the pricey beginning materials and the time-consuming nature of the procedure [73, 115].

In addition to precipitation and sol–gel techniques, hydrothermal techniques are favourable. Hydrothermal synthesis can be defined as chemical precipitation performed at high temperatures and pressure in an autoclave and pressure vessel in the presence of organic solvents. The high temperature and pressure increase reactivity. It is particularly effective in the preparation of components with poor solubility. However, high temperatures and pressure require expensive equipment, so this process is more expensive than other methods [42, 73].

Drying may be required after wet-chemical synthesis methods. Upon heating, the solution is supersaturated. The drying process is divided into two heated and unheated. While the most commonly used method in heated drying is spray drying, the freeze-drying method is most frequently used in unheated drying methods. In the spray drying, the precipitated powders are broken up into fine droplets by a spray solution and sprayed into a heated chamber. Inside the chamber, the solvent evaporates immediately. Thus, the desired powder components remain and they are collected with the help of a cyclone. The most important advantage of spray drying is the large-scale feed rate. However, the most important disadvantage is that the equipment to be used and the installation cost are expensive [105, 115, 120].

Freeze-drying is a very convenient method for preparing high-quality powders. Hard agglomeration can occur on evaporation of solvent by applying high temperatures, as indicated by spray drying. This causes the formation of irregular morphological features in powders. To overcome agglomeration, the freeze-drying method is a good alternative. It consists of 4 steps (solution preparation, freezing, drying and heating). First, the solution is prepared with the help of a suitable solvent (usually water). Secondly, a very important freezing process is performed. The purpose of freezing is to limit further dissolution of the solute into the ice crystals. In the crystallization theory, solubility decreases when crystallization increases. As the cooling process is done, more ice crystals are separated and the concentration of the remaining solution increases. The resulting solution is dried to evaporate the solvent. By applying heat to the last powder materials, it is ensured that more pure and homogeneous powders are formed [88]. In accordance with the studies, it has been proven that the morphologies of freeze-dried powders are more regular and smaller in size [58, 129, 139]. The production of ceramic powders with high purity, regular morphology and desired dimensions is very important for the second stage, green body production.

Forming of the Ceramic Green Body

The process of combining ceramic powders to create the green body structure (unsintered solid blank) is called forming. It is one of the significant steps in the production of bioceramic products. The ceramic powders can be brought into the desired shape by different shaping methods that are roughly examined under 3 titles; dry forming, wet forming and solid-free form production. Each of these methods has its advantages and disadvantages. However, the dry forming method is most frequently used in the production of advanced ceramic powders. The wet forming method is more advantageous in shaping ceramic powders in low molecular solvents. In the dry and wet forming methods, shaping moulds (tools) are used to shape powders. However, the shaping mould is not necessary for the solid free-form method, and the green body is printed layer by layer using computer-aided programs (CAD). Therefore, they are very fast methods. Regardless of the method, the structures of the obtained green bodies should be homogeneous and the defects in the structure should be minimal. The green body's microstructure will affect the sintering stage and the engineering-processing properties of the fired body [105, 115, 120].

Sintering

The mechanical characteristics of bioceramics are largely defined by the microstructure. Fine-grained structures are harder and stronger than large-pored structures. The sintering process is critical in creating microstructures with desired properties. The sintering process involves heating green ceramic bodies to high temperatures below their melting point in order to harden them. At high temperatures, the surface energy of the system drops and the ions of the system enter an irreversible process to compensate for the energy deficit. In this process, strong chemical bonds are formed and ceramic materials are hardened. There are two classes of sintering processes: pressurized (divided into sub-headings) and pressureless methods [105, 120, 141].

Machining

Correct shaping in bioceramics depends on the last step, machining. Traditional processing methods (mechanical methods-milling, grinding, drilling) are not useful due to the fragile nature of ceramics. Although chemical processing techniques can be considered as an option, it should not be forgotten that chemicals will harm the environment. For this reason, non-contact studies have been carried out on processing methods without contacting ceramics. For example, laser processing, plasma processing, electrochemical processing, electron beam processing and hybrid processing methods are some non-contact methods [18, 36]. Each has different advantages and disadvantages. For example, electrochemical machining is a very expensive process, while plasma machining is inexpensive. Surface quality is lower in laser machining, chemical machining and electron beam machining, but higher in plasma machining. In the selection of the machining method, the choice should be made according to the practical usability and industrially applied criteria [18, 36, 38].

Biomedical Applications of Bioceramics

Bioceramics have a wide range of applications. Though they are primarily used in medical fields such as orthopaedics and bone tissue engineering, they are also of great importance in dental applications, replacing jaws, hips, and tendons (Fig. 6.5) [45, 111, 140].

Orthopedic applications

Orthopaedic reconstructive surgery is a branch of medicine that typically deals with bone tumours, congenital abnormalities, scoliosis, arthritis, and osteoporosis. In general, patients' complaints are minimized or diseased musculoskeletal tissue is largely repaired by using prosthetic biomaterials. In the twentieth century, with the improvement of new biomaterials and anaesthesia methods, prosthetic repair of bone defects and replacement of joints became practical. In particular, Sir John Charnley's total hip arthroplasty in 1960 is known as the most successful surgery of the twentieth century, and its success has pioneered many other studies. In this progression, ceramic materials have played a crucial role [87]. It was thought that they could be used as femoral heads in hip reconstructive surgery prostheses. However, many of them were broken in vivo because they did not have sufficient mechanical strength and toughness. Orthopedists used polymeric and metallic materials until 1970. However, the utilization of ceramics in orthopaedic reconstructive surgery became widespread again due to the improvement of their characteristics like durability, biocompatibility, stiffness, and corrosion resistance of ceramics in the 1970s [45, 87].

Calcium phosphates, calcium silicates and calcium sulfates are examples of bioactive materials [41]. In particular, calcium phosphate-based HA and TCP ceramics exhibit properties similar to the structure of bone and have various applications in the field of orthopaedics [111]. For example, HA is used in the bioactive coating of orthopaedic implants due to its high osteoconductivity. In another example, TCP is utilized as a filling material in the treatment of bone defects owing to its weak mechanical characteristics. However, both HA and TCP are not suitable for utilization in bone grafts owing to their poor mechanical characteristics. Therefore, calcium phosphate-based ceramics cannot be used alone. To improve both mechanical and

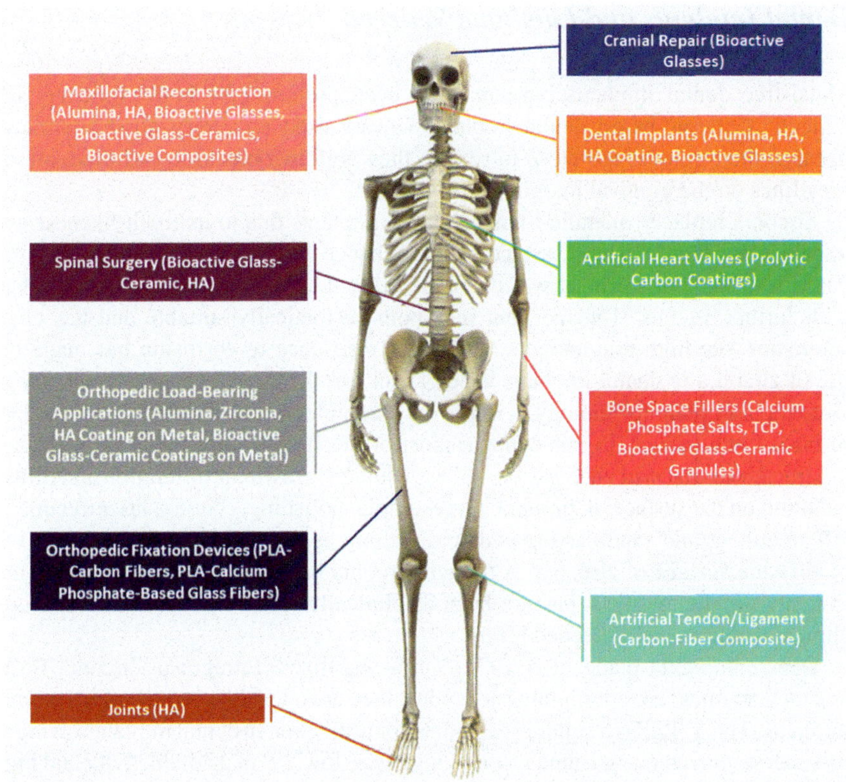

Fig. 6.5 Several applications of bioceramics in the body

biodegradation properties, formulations are controlled, and reinforcing compounds are added to the structure and can be made ready for use in bone implants [64].

Dental Applications

Materials to be used in dental applications should be durable and aesthetically acceptable. In recent years, biomaterials consisting of the main components of the tooth and exhibiting very similar properties to the structure of the tooth have been developed. The biomaterials are designed to accelerate remineralization in dental problems by using biomimetic approaches [68]. Bioceramics are generally used in crowns, implants, bridges, inlays/onlays, veneers and implant forms [10].

Dental Implants and Laminate Veneers

Metal-free dental implants have recently been preferred by both clinicians and patients. Ceramic-based dental implants can be used as a very good alternative because their colours are close to reality, they are biocompatible and there are no grey lines on the gingival margin [99].

Zirconia replaces metallic-titanium-based implants due to its tooth-likeness and translucency. It was first used as a coating to enhance the osseointegration of titanium-based implants. However, it was concluded that it can be used as a direct implant with further studies. The fact that it is both aesthetically suitable and has characteristics like high toughness, strength and resistance to corrosion has made the use of zirconia in dental implants valid. Thanks to the superior biocompatibility of zirconia in implant technology, it improves osseointegration and minimizes the loss of bone-tooth tissues. In this context, most of the studies have focused on surface modifications that will enable more cell attachment, growth, proliferation and differentiation on the surface of zirconia. For example, polishing, coating, laser treatment, UV treatment and sandblasting and acid etching are some of the methods used to modify the surface of zirconia. Although zirconia has many advantages for its use in dental implantology, it has not been as clinically tested as titanium. Therefore, clinical studies are still needed [103].

Dental implants made of Y-TZP (Yttria-Stabilized Tetragonal Zirconia Polycrystal Ceramics) are commonly used today, they are considered standard in ceramic implants. The Y-TZP is a suitable candidate for utilization in dental implants owing to its superior corrosion resistance, good toughness, low thermal conductivity and high biocompatibility [7, 103]. It is usually a one-piece implant. However, many manufacturers have started to develop two-piece dental implants. The design of them is a hard issue due to the tiny overall diameter of the instrument and the reduced wall thickness in the connection area. However, their designs can be improved by using composite materials [99].

Recently, studies on silicon nitride (advanced bioceramic material) and its use in dental implants have been increased owing to its high durability, high corrosion coefficient, low friction coefficient and improved imaging properties [48]. According to the study results, its mechanical properties are better than titanium alloys [108]. However, clinical studies have not yet been conducted on the consequences of using silicon nitride as a dental implant material [48].

Increasing demands for dental restorations in terms of aesthetics and durability have popularized the use of ceramic laminate veneers (feldspathic porcelain) for the treatment of misaligned, broken, worn, and discoloured tooth disorders. For protection or aesthetic reasons, the anterior teeth may be covered with a thin coating of material called a veneer [90]. To date, glass–ceramic (leucite glass and fluorapatite glass) materials have been used as laminate veneers. The most important features are their longevity and ability to show low complications. However, it has been reported that the most common problems of ceramic (porcelain) veneers are related to microleakage and fractures [137].

Recently, new veneering materials with different properties and categories have been developed, such as thick monochromatic veneers, leucite-enhanced feldspathic veneers, lithium disilicate veneers and lumineers. The key features of these coatings are that they are all ultra-thin, offering excellent aesthetics and maximum mechanical properties [90].

Ceramics in Endodontics

The area of dentistry that studies the physiology, morphology and pathology of the soft tissues inside the tooth and human dental pulp is known as endodontics [8]. The root canal and vital pulp treatment, and repair of damage to the dental pulp and surrounding tissues due to causes like infection, fractures, caries and trauma are some of the endodontic procedures. Progress in endodontic materials science has been cited as the first reason for the rise in the field of endodontics in recent years. Bioceramics is one of the most frequently used biomaterials in the field of endodontics [6]. They are generally applied as cement, root repair materials, root canal sealers, coating and filling materials [106].

The first bioceramic-based endodontic material is MTA cement (Mineral Trioxide Aggregate) which was applied in the early 90s. The MTA is a hydrophilic, biocompatible endodontic cement composed of tricalcium oxide, silicon oxide and bismuth oxide. The parent component of MTA is Portland cement, which makes up 75%. Although MTA and Portland cement have similar properties in relation to chemical composition and microscopic properties, there are differences in their physical and biological properties [5]. The FDA-approved MTA contains bismuth oxide as a radio pacifier. However, Portland cement cannot enter the ISO radiopacity standard due to its insufficient radiopacity [125].

The biggest advantage of the use of MTA in endodontics is that it hardens in a humid environment. MTA placed in a humid environment, turns into calcium hydroxide, which is the main component in its structure when it comes into contact with moisture. Together with this conversion, the pH of the environment rises and the antibacterial activity increases. In addition, the solubility of MTA is lower than calcium hydroxide. Therefore, it maintains its physical integrity in a humid environment [117].

In addition to MTA, there are two advantages of using bioceramics as cement and sealers in endodontics. Firstly, due to their high biocompatibility and bone-like structure, they are not rejected by the surrounding tissues. Secondly, it can properly close the gaps between the dentinal walls and the occlusive material. Therefore they can easily interact with periapical tissues and aid neoformation [6].

Surface Coatings

Metallic biomaterials constitute 70–80% of implants utilized in hard tissue applications. However, the most important disadvantages of inert metallic implants with poor osteointegration are that they form a fibrous collagenous tissue in the area where they are implanted and cause the release of systemic or locally toxic metals over time. Moreover, bioinert ceramics induce bacterial and chronic inflammation and physical discomfort. Hereof, the use of a bioactive coating is a successful tactic to enhance the function of metallic implants while providing substantial advantages to the patient. When the surface coating is placed, the metallic implants' surface will be protected from corrosion and will not discharge dangerous metal ions into the body [3, 66]. Furthermore, surface coating is a practical and affordable method for separating the substrates (metallic implant's surface) and the corrosive medium [81].

Ceramics are one of the most popular materials used for surface coatings, as they are more chemically stable than metals and alloys. Oxides (Al_2O_3, TiO_2, ZrO_2), silicates and calcium phosphate (CaP) and salts are often used in ceramic coatings. However, in a more specific sense, ceramic coating is divided into two bioinert and bioactive. Bioinert ceramic surface coatings have better mechanical qualities and proper biocompatibility than bioactive coatings [95]. This coating method is generally used for dental and ophthalmic implants for increasing mechanical properties and corrosion resistance [82]. However, the application of the bioinert ceramic coating is limited due to its weak interaction capability with the surrounding living tissues and high elastic modulus [95]. On the other hand, adhesion between implant and tissue is increased and osseointegration increases by using bioactive ceramic surface coatings. Because the bioactive ceramics (calcium phosphate and glass/glass ceramics) used in this coating method show very similar properties with the structure of bone and tooth tissue. Thus, human cell growth is promoted and recovery time is accelerated thanks to bioactive ceramic coatings [91]. In addition to increasing bioactivity and mechanical properties, bioceramic surface coating can also help modify the degradation properties of materials with high degradation rates [95].

Bone Tissue Engineering

Due to the growing economic burden associated with bone injury and illness, the creation of biomaterials for bone regeneration is the most active research subject in tissue engineering. The biggest obstacle in bone tissue engineering is to produce and develop a material that is mechanically strong and also biodegradable [24]. Ideally, the scaffold should be degradable therefore this biodegradation avoids the detrimental effects of a persistent foreign material and permits it to be gradually replaced by new bone. In terms of a biological view, it is a common approach to unite polymers and ceramics to produce scaffolds [89]. Natural bone is the combination of a naturally occurring polymer and biological apatite. Also, only one material type does not always provide the required mechanical and chemical characteristics for this application [85].

3D Printed Bioceramic Scaffolds for Bone Tissue Engineering

Scaffolds are essential in providing a 3D environment for cell adhesion and growth. Conventional fabrication techniques like gas foaming, freeze drying etc. cannot obtain adequate pore size, architecture, or interconnectivity of the scaffolds therefore they can not provide tissue repair and cell growth [84]. In order to prevent these unwanted situations, 3D printing technologies have been improving towards the years. According to the fabrication procedure, there are numerous sorts of 3D printing techniques [63]. Scaffolds built by stereolithography (SLA) are produced in slices from the bottom of a liquid polymer to the top. Afterwards, hardened with exposure to radiation from an ultraviolet laser [43]. Rather than using a UV laser to produce materials, selective laser sintering (SLS) uses sintering methods. The laser is used in SLS to sinter the powder and make the particles adhere to one another [134]. In fused deposition modelling (FDM), scaffolds are made by melting and extruding layers of filament from a heated nozzle [98]. In 3D printing, ceramics are layered in liquid binders to achieve particle adhesion and build scaffolding for the shape-forming process [142].

Conclusions and Future Perspectives

Bioceramics are highly effective materials for bone repair and regeneration in the human body. These biomaterials are commonly utilized in biological applications including dental and orthopaedic implants as well as porous scaffolds for tissue engineering. Superior mechanical performance for load-bearing applications can be found in bioinert ceramics like alumina and zirconia. Besides, bioactive glasses and calcium phosphate ceramics exhibit osteoconductive properties. In this context, it is essential to comprehend the material and therapeutic requirements to enable the fabrication of customized scaffolds. In this chapter, a concise and accessible overview of the past of bioceramics to the present was addressed. Materials ranging from bioinert to bioactive and bioabsorbable bioceramics were described. Moreover, the properties that bioceramic materials should have and their production processes were emphasized. Finally, recent biomedical applications of bioceramics including orthopaedic and dental applications, surface coatings and bone tissue engineering were highlighted.

Despite significant progress in recent years regarding the number of bioceramics employed in clinical practice and the potential of bone regeneration they provide, major developments are still required in this field. These include enhancing the mechanical capabilities of current bioactive ceramics, increasing bioactivity in relation to the enhancement of mechanical stability and biological agent delivery properties of biomedical coatings, gene activation, creating intelligent materials that can combine sensing and bioactivity, and the improvement of biomimetic composites.

Bioceramic materials have been used in a variety of clinical settings; the two most prevalent are skin grafting and bone generation. New advances in stem cells, tissue engineering and 3D printing technology hold promise for future transplantation necessities, the preparation of bioinks, and the development of different scaffolds and coating materials for teeth and gums. Different organs like the skin, ear, liver, heart and kidney are tried to be developed under laboratory conditions by using bioceramics at different stages, but this has not been successful yet. The key difficulty is determining the precise organ size while ensuring optimal vascular system channelling. This is a visionary effort to address the demands of the medical sector in the supply of various bones and organs to treat different ailments and unintentional organ loss.

Even if there have been substantial advancements in the engineering of novel tissues, future research should be concentrated on in order to make significant improvements and turn it into a therapeutically useful approach. A thorough comprehension of the interactions between bioceramics and tissue, the longtime use of a hierarchical structure, and the corresponding mechanical strength, particularly the fatigue threshold when exposed to repeated external load, should be a focus of strategies.

References

1. Afrashtehfar KI, Del Fabbro M (2020) Clinical performance of zirconia implants: a meta-review. J Prosthet Dent 123:419–426
2. Ahmad Z, Huang J, Edirisinghe MJ et al (2006) Electrohydrodynamic print-patterning of nano-hydroxyapatite. J Biomed Nanotechnol 2:201–207
3. Al Mugeiren OM, Baseer MA (2019) Dental implant bioactive surface modifiers: An update. J Int Soc Prev Community Dent 9:1
4. Albrektsson T, Johansson C (2001) Osteoinduction, osteoconduction and osseointegration. Eur Spine J 10:S96–S101
5. Ali MS, Kano B (2019) Endodontic materials: from old materials to recent advances. In: Advanced dental biomaterials. Elsevier, pp 255–299
6. Aljifan M, Almakrami A, Al Mutairi B et al (2022) Advantages and types of bioceramics in endodontics
7. Apratim A, Eachempati P, Salian KKK et al (2015) Zirconia in dental implantology: a review. J Int Soc Prev Community Dent 5:147
8. Aubeux D, Peters OA, Hosseinpour S et al (2021) Specialized pro-resolving lipid mediators in endodontics: a narrative review. BMC Oral Health 21:276
9. Aydogdu MO, Mutlu B, Kurt M et al (2019) Developments of 3D polycaprolactone/beta-tricalcium phosphate/collagen scaffolds for hard tissue engineering. J Aust Ceram Soc 55:849–855
10. Babu PJ, Alla RK, Alluri VR et al (2015) Dental ceramics: Part I-An overview of composition, structure and properties. Am J Mater Eng Technol 3:13–18
11. Baino F, Novajra G, Vitale-Brovarone C (2015) Bioceramics and scaffolds: a winning combination for tissue engineering. Front Bioeng Biotechnol 3:1–17

12. Barrère F, van Blitterswijk CA, de Groot K (2006) Bone regeneration: molecular and cellular interactions with calcium phosphate ceramics. Int J Nanomedicine 1:317–332
13. Beall G (1992) Design and properties of glass-ceramics. Annu Rev Mater Res 22:91–119
14. Beall GH, Pinckney LR (1999) Nanophase glass-ceramics. J Am Ceram Soc 82:5–16
15. Beevers CA, McIntyre DB (1946) The atomic structure of fluor-apatite and its relation to that of tooth and bone material. (With Plates XVI-XVIII.). Mineral Mag J Mineral Soc 27:254–257
16. Begley CT, Jo Doherty M, Mollan RAB, Wilson DJ (1995) Comparative study of the osteoinductive properties of bioceramic, coral and processed bone graft substitutes. Biomaterials 16:1181–1185
17. Best SM, Porter AE, Thian ES, Huang J (2008) Bioceramics: past, present and for the future. J Eur Ceram Soc 28:1319–1327
18. Bilal A, Jahan MP, Talamona D, Perveen A (2019) Electro-discharge machining of ceramics: a review. Micromachines 10
19. Bobyn JD, Mortimer ES, Glassman AH et al (1992) Producing and avoiding stress shielding. Laboratory and clinical observations of noncemented total hip arthroplasty. Clin Orthop Relat Res 79–96
20. Bohne W, Pouezat JA, Peru L, Daculsi G (1993) Heating of calcium phosphate crystals: morphological consequences and biological implications. Cells Mater 3:377–382
21. Bohner M (2001) Physical and chemical aspects of calcium phosphates used in spinal surgery. Eur Spine J 10:S114–S121
22. Bonfield W, Grynpas MD, Tully AE et al (1981) Hydroxyapatite reinforced polyethylene–a mechanically compatible implant material for bone replacement. Biomaterials 2:185–186
23. Brunski JB (1988) Biomaterials and biomechanics in dental implant design. Int J Oral Maxillofac Implants 3:85–97
24. Chen Q, Zhu C, Thouas GA (2012) Progress and challenges in biomaterials used for bone tissue engineering: bioactive glasses and elastomeric composites. Prog Biomater 1:2
25. Curtis A, Wilkinson C (1997) Topographical control of cells. Biomaterials 18:1573–1583
26. Daculsi G (2014) Handbook of bioceramics and biocomposites
27. Daculsi G, Laboux O, Malard O, Weiss P (2003) Current state of the art of biphasic calcium phosphate bioceramics. J Mater Sci Mater Med 14:195–200
28. Dalby MJ, Gadegaard N, Tare R et al (2007) The control of human mesenchymal cell differentiation using nanoscale symmetry and disorder. Nat Mater 6:997–1003
29. Deville S, Saiz E, Tomsia AP (2006) Freeze casting of hydroxyapatite scaffolds for bone tissue engineering. Biomaterials 27:5480–5489
30. Dorozhkin S (2016) Calcium orthophosphate-based bioceramics and biocomposites
31. Dorozhkin SV (2010) Bioceramics of calcium orthophosphates. Biomaterials 31:1465–1485
32. Dorozhkin SV (2009) Calcium orthophosphates in nature biology and medicine. Materials (Basel) 2:399–498
33. Ducheyne P, Qiu Q (1999) Bioactive ceramics: the effect of surface reactivity on bone formation and bone cell function. Biomaterials 20:2287–2303
34. Eliaz N, Metoki N (2017) Calcium phosphate bioceramics: a review of their history, structure, properties, coating technologies and biomedical applications. Mater (Basel, Switzerland) 10
35. Eslami H, Tahriri M, Moztarzadeh F et al (2018) Nanostructured hydroxyapatite for biomedical applications: from powder to bioceramic. J Korean Ceram Soc 55:597–607
36. Fan L, Zhang Y, Peng L et al (2020) Crystal growth, finite element analysis and oxygen annealing of ZnO bulk crystal grown by low-temperature-gradient chemical vapor transport. Ceram Int 46:23107–23115
37. Farid SBH (2018) Bioceramics: for materials science and engineering. Woodhead Publishing
38. Ferraris E, Vleugels J, Guo Y et al (2016) Shaping of engineering ceramics by electro, chemical and physical processes. CIRP Ann 65:761–784
39. Ferraris S, Yamaguchi S, Barbani N, et al (2020) Bioactive materials
40. Frayssinet P, Mathon D, Lerch A et al (2000) Osseointegration of composite calcium phosphate bioceramics. J Biomed Mater Res 50:125–130

41. Gao C, Peng S, Feng P, Shuai C (2017) Bone biomaterials and interactions with stem cells. Bone Res 5:17059
42. Gomes DS, Santos AMC, Neves GA, Menezes RR (2019) A brief review on hydroxyapatite production and use in biomedicine. Cerâmica 65:282–302
43. Griffith M, Freeform JH, Fabrication of ceramics via stereolithography
44. Gul H, Khan M, Khan AS (2019) Bioceramics: types and clinical applications. Elsevier Ltd
45. Gul H, Khan M, Khan AS (2020) 3—Bioceramics: types and clinical applications. In: Khan AS, Chaudhry AA (eds) Handbook of ionic substituted hydroxyapatites. Woodhead Publishing, pp 53–83
46. Gunduz O, Yetmez M, Ekren N et al (2018) Microstructural and mechanical properties of nano-yttria-oxide doped hydroxyapatite composites. Mater Sci Forum 923 MSF:89–92
47. Habibovic P, Gbureck U, Doillon CJ et al (2008) Osteoconduction and osteoinduction of low-temperature 3D printed bioceramic implants. Biomaterials 29:944–953
48. Heimann RB (2021) Silicon nitride, a close to ideal ceramic material for medical application. Ceramics 4:208–223
49. Hench L, Best S (2013) Ceramics, glasses, and glass-ceramics. Biomater Sci An Introd to Mater Med 153–170
50. Hench LL (1991) Bioceramics: from concept to clinic. J Am Ceram Soc 74:1487–1510
51. Hench LL (2006) The story of bioglass. J Mater Sci Mater Med 17:967–978
52. Hench LL, Andersson O (1993) Bioactive glasses. In: An introduction to bioceramics. World Scientific, pp 41–62
53. Hench LL, Paschall HA (1973) Direct chemical bond of bioactive glass-ceramic materials to bone and muscle. J Biomed Mater Res 7:25–42
54. Hench LL, Splinter RJ, Allen WC, Greenlee TK (1971) Bonding mechanisms at the interface of ceramic prosthetic materials. J Biomed Mater Res 5:117–141
55. Hing KA, Annaz B, Saeed S et al (2005) Microporosity enhances bioactivity of synthetic bone graft substitutes. J Mater Sci Mater Med 16:467–475
56. Hing KA, Revell PA, Smith N, Buckland T (2006) Effect of silicon level on rate, quality and progression of bone healing within silicate-substituted porous hydroxyapatite scaffolds. Biomaterials 27:5014–5026
57. Ho-Shui-Ling A, Bolander J, Rustom LE et al (2018) Bone regeneration strategies: engineered scaffolds, bioactive molecules and stem cells current stage and future perspectives. Biomaterials 180:143–162
58. Hu W, Yu L, Ma Z, Liu Y (2019) W-Y_2O_3 composite nanopowders prepared by freeze-drying method and its sintering characteristics. J Alloys Compd 806:127–135
59. Huang J, Best S (2014) Ceramic biomaterials for tissue engineering. In: Tissue engineering using ceramics and polymers, 2nd edn, pp 3–34
60. Hulbert SF (1993) The use of alumina and zirconia in surgical implants. In: An introduction to bioceramics. World Scientific, pp 25–40
61. Hulbert SF, Hench LL, Forbers D, Bowman LS (1982) History of bioceramics. Ceram Int 8:131–140
62. Hulbert SF, Young FA, Mathews RS et al (1970) Potential of ceramic materials as permanently implantable skeletal prostheses. J Biomed Mater Res 4:433–456
63. Hwa LC, Rajoo S, Noor AM et al (2017) Recent advances in 3D printing of porous ceramics: a review. Curr Opin Solid State Mater Sci 21:323–347
64. Ielo I, Calabrese G, De Luca G, Conoci S (2022) Recent advances in hydroxyapatite-based biocomposites for bone tissue regeneration in orthopedics. Int J Mol Sci 23:9721
65. Jones JR, Gibson IR (2020) 1.3.4—Ceramics, Glasses, and Glass-Ceramics: Basic Principles. In: Wagner WR, Sakiyama-Elbert SE, Zhang G, Yaszemski MJBT-BS (eds) Biomaterials science, 4th edn. Academic Press, pp 289–305
66. Joy-anne NO, Su Y, Lu X et al (2019) Bioactive glass coatings on metallic implants for biomedical applications. Bioact Mater 4:261–270
67. Kay MI, Young RA, Posner AS (1964) Crystal structure of hydroxyapatite. Nature 204:1050–1052

68. Khan AS, Syed MR (2019) A review of bioceramics-based dental restorative materials. Dent Mater J 38:163–176
69. Kokubo T (1991) Bioactive glass ceramics: properties and applications. Biomaterials 12:155–163
70. Kokubo T (2008) Bioceramics and their clinical applications, 1st edn. Woodhead Publishing
71. Kokubo T, Shigematsu M, Nagashima Y et al (1982) Apatite- and wollastonite-containg glass-ceramics for prosthetic application. Bull Inst Chem Res Kyoto Univ 60
72. Kumar P, Dehiya BS, Sindhu A (2018) Bioceramics for hard tissue engineering applications: a review
73. Kumar R, Mohanty S (2022) Hydroxyapatite: a versatile bioceramic for tissue engineering application. J Inorg Organomet Polym Mater 1–17
74. LeGeros RZ (1993) Biodegradation and bioresorption of calcium phosphate ceramics. Clin Mater 14:65–88
75. LeGeros RZ (1991) Calcium phosphates in oral biology and medicine. Monogr Oral Sci 15:1–201
76. LeGeros RZ (2002) Properties of osteoconductive biomaterials: calcium phosphates. Clin Orthop Relat Res 81–98
77. LeGeros RZ, LeGeros JP (2008) 16—Hydroxyapatite. In: Kokubo TBT-B and their CA (ed) Woodhead publishing series in biomaterials. Woodhead Publishing, pp 367–394
78. LeGeros RZ, LeGeros JP (2002) Calcium phosphate bioceramics: past, present and future. Key Eng Mater 240–242:3–10. https://doi.org/10.4028/www.scientific.net/KEM.240-242.3
79. LeGeros RZ, Lin S, Rohanizadeh R et al (2003) Biphasic calcium phosphate bioceramics: preparation, properties and applications. J Mater Sci Mater Med 14:201–209
80. LeGeros RZ, Suga S (1980) Crystallographic nature of fluoride in enameloids of fish. Calcif Tissue Int 32:169–174
81. Li L-Y, Cui L-Y, Zeng R-C et al (2018) Advances in functionalized polymer coatings on biodegradable magnesium alloys–A review. Acta Biomater 79:23–36
82. Li Y, Yang H, He Y et al (2020) Properties of UV-irradiated TiO_2, ZrO_2, and TiO_2-ZrO_2 films as pore-sealing layers on micro-arc-oxidized aluminum alloys. J Sol-Gel Sci Technol 93:70–78
83. Ly M, Spinelli S, Hays S, Zhu D (2022) 3D printing of ceramic biomaterials. Eng Regen 3:41–52. https://doi.org/10.1016/j.engreg.2022.01.006
84. Ma H, Feng C, Chang J, Wu C (2018) 3D-printed bioceramic scaffolds: from bone tissue engineering to tumor therapy. Acta Biomater 79:37–59
85. Maier KP, Hoppe-Seyler G, Take H, et al (1973) Enzymatic and metabolic studies on carbohydrate and amino acid metabolism in rat liver during acute uraemia
86. Malmström J, Adolfsson E, Arvidsson A, Thomsen P (2007) Bone response inside free-form fabricated macroporous hydroxyapatite scaffolds with and without an open microporosity. Clin Implant Dent Relat Res 9:79–88
87. McEntire BJ, Bal BS, Rahaman MN et al (2015) Ceramics and ceramic coatings in orthopaedics. J Eur Ceram Soc 35:4327–4369
88. Mi S, Liu Z, Luo C et al (2020) A review on preparing new energy ultrafine powder materials by freeze-drying. Dry Technol 38:1544–1564
89. Mikos AG, Temenoff JS (2000) Formation of highly porous biodegradable scaffolds for tissue engineering
90. Mitthra S, Anuradha B, Prakash V (2020) Recent advances in the material aspects of veneers. Eur J Mol Clin Med 7:2079–2085
91. Montazerian M, Hosseinzadeh F, Migneco C et al (2022) Bioceramic coatings on metallic implants: an overview. Ceram Int
92. Moradi K, Sabbagh Alvani AA (2020) First-principles study on Sr-doped hydroxyapatite as a biocompatible filler for photo-cured dental composites. J Aust Ceram Soc 56:591–598
93. Mozafari M (2014) Bioceramics in the realm of history. Bioceram Dev Appl 4:22–23
94. Munir G, Koller G, Di Silvio L et al (2011) The pathway to intelligent implants: osteoblast response to nano silicon-doped hydroxyapatite patterning. J R Soc Interface 8:678–688

95. Nilawar S, Uddin M, Chatterjee K (2021) Surface engineering of biodegradable implants: emerging trends in bioactive ceramic coatings and mechanical treatments. Mater Adv 2:7820–7841
96. Parithimarkalaignan S, Padmanabhan TV (2013) Osseointegration: an update. J Indian Prosthodont Soc 13:2–6
97. Patel N, Brooks RA, Clarke MT et al (2005) In vivo assessment of hydroxyapatite and silicate-substituted hydroxyapatite granules using an ovine defect model. J Mater Sci Mater Med 16:429–440
98. Pham DT, Gault RS (1998) A comparison of rapid prototyping technologies
99. Piconi C, Sprio S (2021) Oxide bioceramic composites in orthopedics and dentistry. J Compos Sci 5:206
100. Pina S, Reis RL, Oliveira JM (2018) Ceramic biomaterials for tissue engineering. Elsevier Ltd.
101. Piotrowski G, Hench LL, Allen WC, Miller GJ (1975) Mechanical studies of the bone bioglass interfacial bond. J Biomed Mater Res 9:47–61
102. Punj S, Singh J, Singh K (2021) Ceramic biomaterials: Properties, state of the art and future prospectives. Ceram Int 47:28059–28074
103. Qu Y, Liu L (2021) Zirconia materials for dental implants: a literature review. Front Dent Med 2:687983
104. Radin S, Ducheyne P (1996) Effect of serum proteins on solution-induced surface transformations of bioactive ceramics. J Biomed Mater Res 30:273–279
105. Rahaman MN (2017) Ceramic processing and sintering, 2nd edn. CRC Press
106. Rajkumar DRB (2022) Bioceramics in endodontics. Book Rivers
107. Rawlings RD, Wu JP, Boccaccini AR (2006) Glass-ceramics: their production from wastes-A review. J Mater Sci 41:733–761
108. Raza SM, Khurshid Z, Zafar MS et al (2020) Silicon nitride (SiN): an emerging material for dental implant applications. In: Dental implants. Elsevier, pp 287–299
109. Reffitt DM, Ogston N, Jugdaohsingh R et al (2003) Orthosilicic acid stimulates collagen type 1 synthesis and osteoblastic differentiation in human osteoblast-like cells in vitro. Bone 32:127–135
110. Seyhan SA, Alkaya DB, Cesur S et al (2022) Preparation and characterization of pure natural hydroxyapatite derived from seashells for controlled drug delivery. J Aust Ceram Soc 58:1231–1240
111. Shekhawat D, Singh A, Banerjee MK et al (2021) Bioceramic composites for orthopaedic applications: a comprehensive review of mechanical, biological, and microstructural properties. Ceram Int 47:3013–3030
112. Stanley HR, Hench L, Going R et al (1976) The implantation of natural tooth form bioglasses in baboons. A preliminary report. Oral Surg Oral Med Oral Pathol 42:339–356
113. Stanovici J, Le Nail LR, Brennan MA et al (2016) Bone regeneration strategies with bone marrow stromal cells in orthopaedic surgery. Curr Res Transl Med 64:83–90
114. Stavros VG (2014) A bright future for sunscreens. Nat Chem 6:955–956
115. Tanaka Y, Yamashita K (2008) Fabrication processes for bioceramics. In: Bioceramics and their clinical applications. Elsevier, pp 28–52
116. Taton TA (2001) Nanotechnology. Boning up on biology. Nature 412:491–492
117. Tawil PZ, Duggan DJ, Galicia JC (2015) Mineral trioxide aggregate (MTA): its history, composition, and clinical applications. Compend Contin Educ Dent (Jamesburg, NJ 1995) 36:247–252
118. Thavornyutikarn B, Chantarapanich N, Sitthiseripratip K et al (2014) Bone tissue engineering scaffolding: computer-aided scaffolding techniques
119. Timmer MD, Ambrose CG, Mikos AG (2003) In vitro degradation of polymeric networks of poly(propylene fumarate) and the crosslinking macromer poly(propylene fumarate)-diacrylate. Biomaterials 24:571–577
120. Trunec M, Maca K (2014) Advanced ceramic processes. In: Advanced ceramics for dentistry. Elsevier, pp 123–150

121. Turan Y, Kalkandelen C, Palaci Y et al (2022) Synthesis and cytotoxicity analysis of porous β-TCP/starch bioceramics. J Aust Ceram Soc 58:487–494
122. Tut TA, Cesur S, Ilhan E et al (2022) Gentamicin-loaded polyvinyl alcohol/whey protein isolate/hydroxyapatite 3D composite scaffolds with drug delivery capability for bone tissue engineering applications. Eur Polym J 179:111580
123. Udduttula A, Zhang JV, Ren P-G (2019) Bioinert ceramics for biomedical applications
124. Umroh B, Ginting A, Rahman MNA (2020) CO_2 laser machining on alumina ceramic: a review. IOP Conf Ser Mater Sci Eng 1003
125. Ponnudurai VM, Arangannal AS, AarthiJ IR (2020) Portland cement—an effective and cheap alternative to Mta. Eur J Mol Clin Med 7:6502–6508
126. Wang C, Zhou X, Wang M (2004) Influence of sintering temperatures on hardness and Young's modulus of tricalcium phosphate bioceramic by nanoindentation technique. Mater Charact 52:301–307. https://doi.org/10.1016/j.matchar.2004.06.007
127. Wang FR (1990) Experimental study of osteogenic activity of sintered hydroxyapatite–on the relationship of sintering temperature and pore size. Nihon Seikeigeka Gakkai Zasshi 64:847–859
128. Wang K, Zhou C, Hong Y, Zhang X (2012) A review of protein adsorption on bioceramics. Interface Focus 2:259–277
129. Wei ZS, Jun L, Hai XC (2009) Preparation of silver nanopowder by freeze-drying procedure. Dry Technol 27:529–533
130. White JE, Day DE (1994) Rare earth aluminosilicate glasses for in vivo radiation delivery. Key Eng Mater 94–95:181–208. https://doi.org/10.4028/www.scientific.net/KEM.94-95.181
131. Williams DF (1989) A model for biocompatibility and its evaluation. J Biomed Eng 11:185–191
132. Young RA, Elliott JC (1966) Atomic-scale bases for several properties of apatites. Arch Oral Biol 11:699–707
133. Yunoki S, Ikoma T, Tsuchiya A et al (2007) Fabrication and mechanical and tissue ingrowth properties of unidirectionally porous hydroxyapatite/collagen composite. J Biomed Mater Res Part B Appl Biomater 80B:166–173
134. Yves-Christian H, Jan W, Wilhelm M et al (2010) Net shaped high performance oxide ceramic parts by selective laser melting. In: Physics procedia. Elsevier B.V., pp 587–594
135. Zapanta-LeGeros R (1965) Effect of carbonate on the lattice parameters of apatite. Nature 206:403–404
136. Zapanta LeGeros R (1981) Apatites in biological systems. Prog Cryst Growth Charact 4:1–45
137. Zarone F, Leone R, Di MMI et al (2018) No-preparation ceramic veneers: a systematic review. J Osseointegration 10:17–22
138. Zhang P, Li X, Yang J, Xu S (2014) The crystallization and microstructure evolution of lithium disilicate-based glass-ceramic. J Non Cryst Solids 392–393:26–30
139. Zhao J, He Y, Zhang L, Lu K (2016) Preparation of porous TiO_2 powder with mesoporous structure by freeze-drying method. J Alloys Compd 678:36–41
140. Zhekov K, Stefanova VP (2021) Definition and classification of bioceramic endodontic sealers. Folia Med (Plovdiv) 63:901–904
141. Zhou HQ, Xiao HN, Hu XY (2011) Sintering behavior of ceramic green body by thermal analysis techniques. In: Advanced materials research. Trans Tech Publ, pp 226–230
142. Zocca A, Gomes CM, Bernardo E et al (2013) LAS glass-ceramic scaffolds by three-dimensional printing. J Eur Ceram Soc 33:1525–1533

Chapter 7
Drug Delivery Systems for Tissue Engineering

Paul Adrian Tărăbuţă, Ludmila Motelica, Denisa Ficai, Ovidiu Oprea, Anton Ficai, and Ecaterina Andronescu

Abstract Drug delivery systems (DDSs) are developed having in view the need to control the dose (within the therapeutic level) and the time of administration of different biologically active agents being known that higher concentration of the therapeutic agent could be toxic while the lower concentration could be inefficient. These systems can be developed for pure regeneration of the injured bone, skin, nerves, etc. but also for the the treatment of specific diseases such as infections, cancers, osteoporosis, and more. Depending on the applications, various biologically active agents (both natural or synthetic agents) but also supports are used to better address the needs of the patients. As supports, polymers, ceramics and composites were especially considered for highlighting the benefits of these DDSs. Moreover, based on the advances in the last decades, many smart drug delivery systems were developed and additional features were developed such as triggering and targeted delivery, which permits further lowering the dose of the therapeutic agent and ensures the drug delivery to the right organ or tissue, eliminating or at least considerably reducing the systemic side effects. All these features are assuring the premises of "personalized therapy" which are especially important when toxic agents are used (such as chemotherapeutic or antimicrobial agents). The chapter is structured to cover the history of the drug delivery systems (including devices), some information about the smart drug delivery systems as well as some case studies related to enhanced

---

P. A. Tărăbuţă · L. Motelica · A. Ficai (✉) · E. Andronescu
Department of Science and Engineering of Oxide Materials and Nanomaterials, Faculty of Chemical Engineering and Biotechnologies, University Politehnica of Bucharest, 1-7 Gh. Polizu Street, 011061 Bucharest, Romania
e-mail: anton.ficai@upb.ro

P. A. Tărăbuţă · L. Motelica · D. Ficai · O. Oprea · A. Ficai · E. Andronescu
National Centre for Micro and Nanomaterials and National Centre for Food Safety, University Politehnica of Bucharest, 313 Splaiul Independenţei, 060042 Bucharest, Romania

D. Ficai · O. Oprea
Department of Inorganic Chemistry, Physical Chemistry and Electrochemistry, Faculty of Chemical Engineering and Biotechnologies, University Politehnica of Bucharest, 1-7 Gh. Polizu Street, 011061 Bucharest, Romania

O. Oprea · A. Ficai · E. Andronescu
Academy of Romanian Scientists, 3 Ilfov Street, 050045 Bucharest, Romania

O. Gunduz et al. (eds.), *Biomaterials and Tissue Engineering*, Stem Cell Biology and Regenerative Medicine 74, https://doi.org/10.1007/978-3-031-35832-6_7

healing, innovative therapies and the treatment of specific diseases. Based on the major conclusions, some of the perspectives are also presented.

Keywords Drug delivery systems · Biologically active · Agents · Smart drug · Delivery systems · Therapeutic approach

Introduction

Drug delivery systems (DDSs) have been developed to deliver the biologically active agents to the patients being designed for all the administration routes: oral, topical, intravenous, intravaginal, etc. [77]. There are some major advantages for using drug delivery systems, the most important being: lower systemic toxicity, better efficiency, less drug need for the same effect, shorter times of administration, a more constant level of the drug, etc. This can be easily explained considering Fig. 7.1. It can be seen that the classical administration, in multiple doses, can generate important fluctuations which can involve periods of time with sub-therapeutic level but also periods with toxic level while, in the case of the drug delivery systems, the concentration can be maintained at the desired level (within the therapeutical range) for a suitable period of time. Depending on the nature of the biologically active agents, it is very important to keep the concentration within the therapeutical range (in some cases, very toxic active agents, such as cytostatics and antibiotics are used while, in other cases, a sub-therapeutic level is not desired because, for instance, in the case of antibiotics, resistance can be generated) or to design the DDSs to release the active agents for a specific period of time (for instance, antibiotics are usually administered for a period of 5–7 days; a longer release is not desired because of toxicity while a shorter period cannot be efficient and the bacterial strains can survive and even generate resistance). It is important to mention that the ideal drug delivery systems do not exist but important advances have been done in the last time.

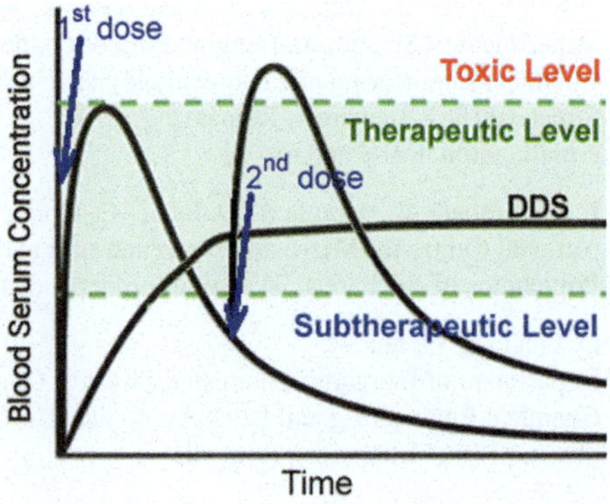

Fig. 7.1 Release profile in classical administration way versus in use of DDSs

Some natural processes such as respiration (both oxygen and carbon dioxide transport), nutrients and degradation products transport, the homeostasis of some species such as Ca^{2+} and PO_4^{3-}, and pH homeostasis can be also assimilated with delivery processes [92]. Considering these natural processes, it is worth mentioning that intrinsic but also extrinsic factors are involved in the control of these processes. The most important challenges in the field are related to: increasing targeted drug accumulation in the desired site, developing release-triggering delivery so, in other words, developing smart features for these drug delivery systems. The biomimetism, a process of optimisation based on mimicking natural solutions, already applied in the natural systems, was also adapted in the development of smart drug delivery systems.

There are two major scopes for the use of the drug delivery systems in tissue engineering: regenerative and therapeutic. There are many diseases which can alter the health of the tissues and in these situations the drug delivery systems are designed to combat these diseases (cancer, infections, etc.) or just to alleviate their effects (such as, for instance, in the case of osteoporosis) or associated effects. Otherwise, drug delivery systems can be used in accidental or congenital defects such as fractures, cuts, burns. In such cases, the main interest is just to assist the body and to assure a fast and safe healing with minimal risks. Many times, there are drug delivery systems that are simultaneously addressed as regenerative and healing. Perhaps, some of the most representative such situations are related to the chitosan-based systems (which are regenerative and also antimicrobial) or silver nanoparticles-based systems which are regenerative but also antimicrobial and even antitumoral [19, 114, 124].

In bone tissue engineering, drug delivery systems are used for several disorders including osteomyelitis, osteoporosis, osteoarthritis, osteosarcoma or bone metastases, but also drug delivery systems can be used for enhancing the healing of fractures (not associated with a disease but induced by a trauma) [16].

Smart Drug Delivery Systems

Considering the increasing needs to develop more efficient therapies, considering the potential of the drug delivery systems to reduce the toxicity and to increase the efficiency, many researches are devoted to the development of smart drug delivery systems. It is important to mention that some drug delivery systems can respond to the needs of the body. There are smart DDSs able to assure personalised release, proportional to the severity of the disease, for instance, the stronger the infection is, the stronger release is assured (if the infection severity is inducing a strong acidity change) and in this way a proper antimicrobial activity is assured (Fig. 7.2).

Triggering factors are extensively used to develop smart DDSs able to answer to the needs of the patients and to assure personalised therapy. These drug delivery systems can be developed considering internal or external factors which can assist the delivery by enhancing or by decreasing the release rate of the active agents but also by assuring a proper, site-specific internalisation. Internal or external triggering

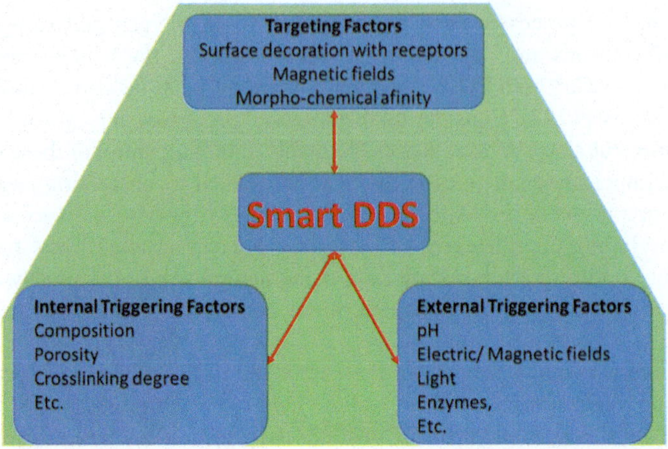

Fig. 7.2 Major features of smart drug delivery systems in terms of triggering and targeting behaviour

factors can be exploited to tune the delivery rate of the active components. The most important internal triggering factors are related to the composition and porosity (additionally influenced by the cross-linking), these factors influencing *the interaction between the* biologically active agent(s) and the support material as well as the diffusion characteristics. External triggering factors can jointly assist the release of the active agents but, they can be activated in certain conditions. For instance, many pH-sensitive DDSs are available, some of them being able to deliver the cargo faster/slower according to the pH (pH changes can occur as a consequence of infections, cancer or, along the digestive tract). There are several materials which can be bio-accumulated by using a magnetic field (such as magnetite-based materials) or can adhere specifically on different surfaces increasing the contact with these surfaces (for instance, there are muco-adhesive polymers which can spend more time in the gastrointestinal tract, as a consequence of their stronger interaction with the mucosa of the stomach/intestines.

Drug Delivery Systems with Enhanced Healing Capacity

Bone is prone to fracture and, when the mass loss is significant, bone grafting materials have to be used to assist the healing. Considering the nature of the bone, mainly based on collagen and hydroxyapatite (HA), many bone grafting materials are developed using these components, alone or in association with other components. To assure fast healing, these grafting materials are often loaded with adequate vitamins, minerals, growth factors or even living progenitor cells [4, 24].

Considering the biological activity of different ions, researchers used them in developing materials with enhanced healing activity. Calcium phosphates [125] and bioactive glasses [121] but even new classes of materials, such as metal organic [142] frameworks (MOFs) are exploited to produce bone grafting materials with enhanced healing (Table 7.1, Fig. 7.3). It is well known that there are some cations which, due to their beneficial role, osteogenic, angiogenic and antimicrobial activity, are extensively studied in the literature. By far, Sr^{2+}, Zn^{2+} and Mg^{2+} are the most used cations for developing doped HA with improved osteogenic activity while Ag^{+} and Zn^{2+} are used for their ability to induce antimicrobial activity. Similarly, bioactive glasses also doped with ions had improved healing capacity (Fig. 7.4). Also, antimicrobial activity was found for similar ions, as a consequence of these ions being released. It is worth mentioning that Zn^{2+}, for instance, is preferred to be used because it can assure both enhanced healing and antimicrobial activity, the latter as a consequence of its release. Synergistic effect can be obtained by combining two or more ions, preventing toxicity and keeping an efficient antimicrobial activity, such as Ag^{+}/Ce^{3+} or Ag^{+}/Ce^{4+} [125]. Certainly, these materials, loaded/doped with ions, can be used alone or in association with other components, including composite materials [140]. For instance, Yu et al. [142] developed MIL-100(Fe) as an iron-based MOF, loaded it with Mg^{2+} and further embedded it into poly(acrylic acid). The as obtained system, due to the release of the Mg^{2+}, assures a slight improvement on the MG-63 osteoblast cells proliferation.

Vitamins are also used to develop bone grafting materials with enhanced healing capacity. Vitamin A, also known as retinol, was proved to be efficient in bone healing as a consequence of enhancing osteoblast cell proliferation and inhibiting osteoclast (bone resorption) activity when loaded in a tricalcium phosphate (TCP) support obtained by 3D printing [130].

Calcitonin is a hormone with a direct role in inhibiting osteoclast activity and, as a consequence, it was also loaded into a Collagen/TCP formulation to develop an injectable bone substitute with improved healing capacity [136].

Bone morphogenetic protein-2 (BMP-2) is one of the most used growth factors being also approved by FDA for use in several orthopaedic diseases, since 2002. This is why many researches involve it in developing bone grafting materials. Wang et al. [132], for instance, loaded BMP-2 into a chitosan/polyethyleneimine (PEI)-modified diatomite composite and proved that these porous composite materials can assure a long-term delivery of BMP-2 and also assure a positive impact on the proliferation and osteogenic differentiation of the bone mesenchymal stem cells, at in vitro level. It is worth mentioning that there are several commercially available drug delivery systems loaded with growth factors devoted for the treatment of bone defects: OP-1 (rhBMP-7 loaded into a collagen (Coll)/carboxymethylcellulose (CMC) support), INFUSE® (rhBMP-2 loaded into a collagen matrix) and AUGMENT® (platelet-derived growth factor—rhPDGF-BB loaded into a collagen matrix) [31, 90], where rh = recombinant human.

Cells were also loaded into different bone grafts, including calcium phosphate-based grafts and these grafts were found to be osteogenic but also angiogenic and thus can be considered promising in the healing of large scale defects [27, 146].

Skin tissue engineering is also an important topic of research. Skin regeneration involves a lot of challenges. Many polymers are used in developing skin grafting

Table 7.1 Enhanced bone grafts based on doped/loaded materials

Bone grafting material	Biological activity and doping/loading agent	Refs.
Doped HA	Osteogenic activity: Li^{+}, Na^{+}, Mg^{2+}, Sr^{2+}, Bi^{3+}, Zn^{2+}, Cu^{2+}, Mn^{2+}, Fe^{2+}, Co^{2+}	[6, 68, 111, 125, 148]
	Antibacterial activity: Mg^{2+}, Ga^{3+}, Bi^{3+}, $Te^{2-/IV+}$, Ag^{+}, Zn^{2+}, Cu^{2+}, Ti^{IV+}, Co^{2+}, Ce^{3+}, La^{3+}, Sm^{3+}, Eu^{3+}	
Bioactive glass	Antibacterial activity: Ag^{+}, Ce^{3+}, Ga^{3+}, Zn^{2+}, Sr^{2+}, Gd^{3+}	[121]
	Osteogenic activity: Sr^{2+}, Zn^{2+}, Cu^{2+}, Mn^{2+}, Mg^{2+}, Sr^{2+}, Co^{2+}	
Mg@MIL-100(Fe)-PAA	Fast healing assured by Mg^{2+} release	[142]
3D-printed TCP@Vitamin A	Osteogenic activity induced by the release of vitamin A	[130]
Coll/CaPs@Calcitonin	Osteogenic activity induced by the release of calcitonin (hormone)	[136]
Chitosan/diatomite@rhBMP-2 Coll/CMC@rhBMP-7 Coll@rhBMP-2 β-TCP-collagen@rhPDGF-BB	Osteogenic activity induced by the release of different growth factors	[31, 90, 132]
BCP@hBMSCs TCP/SCs	High osteogenic and angiogenic activity in vivo	[27, 146]

PAA = Poly(Acrylic Acid). BCP = Biphasic Calcium Phosphate. hBMSCs = human Bone Marrow mesenchymal Stem Cells. SCs = rat Schwann Cells

materials, most of them being suitable for loading with a wide range of biologically active agents: nanoparticles, minerals, vitamins, polyphenols with multiple roles (especially antioxidant, immunomodulatory and regenerative), growth factors and even cells [139].

In chronic diabetic wound management, gene and RNA-based therapies are reported in the literature as being promising solutions for difficult to treat wounds. Unfortunately, because of the short half-life after administration as a subcutaneous injection, most of the trials are focused on developing drug delivery systems [62]. For instance, Shaabani et al. developed two types of nanoparticles (AuNP@chitosan and AuNP@poly(L-Arginine)) loaded with siRNA (small interfering RNA) and found that the two siRNA-loaded nanoparticles have different escape mechanisms because of the different outer shell (chitosan and poly (L-arginine)) and proved that the poly (L-arginine)-based shell is more efficient in transfection and in targeted delivery and thus promotes wound healing [53, 109].

Polyphenols are used in many formulations because of their complex biological activity. For instance, Zhu et al. proposed epigallocatechin-3-gallate (EGCG) in association with small peptide Nap-Phe-Phe-Tyr to be used in wound healing; Nap = α-naphtylacetyl. The self-assembled hydrogel has a major advantage over the pure

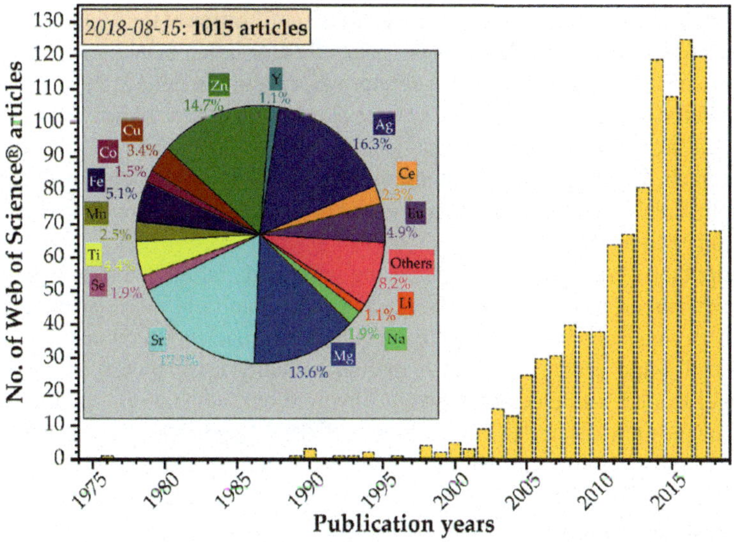

Fig. 7.3 Yearly publications over the 1975–2018 period related to doped hydroxyapatite and their distribution according to the nature of the doping agent (copyright MDPI [125])

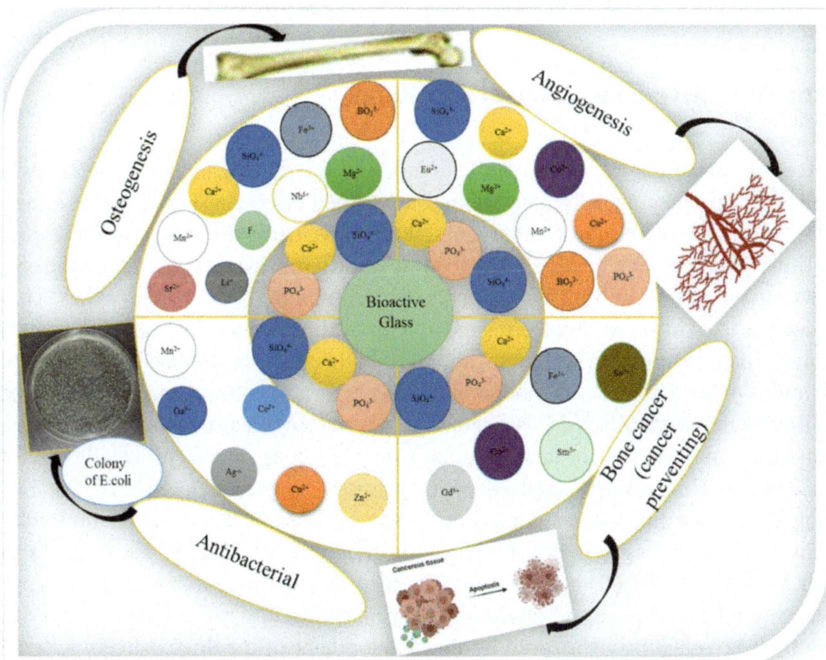

Fig. 7.4 Biological activity induced by the release of specific ions from biological glasses (with the kind permission of Springer [121])

polyphenol because the tetrapeptide assures the stabilization of the EGCG for a period of over 48 h. Due to its multiple roles, especially anti-inflammatory, antioxidant and free radical scavenging, this hydrogel improved healing in mice [151].

Epithelial progenitor cells (EPCs) were also used to promote fast healing. These cells can be loaded alone or in association with some other biologically active agents such as polyphenols. Khojasteh et al. used chitosan/poly(vinyl alcohol)/curcumin/EPCs nanofibers to obtain wound dressings with improved cell adhesion and proliferation [44].

Peripheral nerve grafts are increasingly needed. Drug delivery grafts could assure enhanced healing as a consequence of the multiple mechanisms of action including physical, chemical and therapeutical approaches. Ramburrun et al. [98] developed some grafts based on a thermo-ionically cross-linked gellan-xanthan hydrogel loaded with electrospun poly(3-hydroxybutyric acid-co-3-hydroxyvaleric acid)-magnesium oleate-N-acetyl-cysteine (PHBV-MgOl-NAC) nanofibers which synergistically combine the physical guidance with the release of Mg^{2+} and oleate as well as the sustained release of the anti-inflammatory agent (sodium diclofenac) and nerve growth factor (NGF) over 4 weeks. In vitro and in vivo data show that this complex system can assure an improved healing creating the premises of further investigations at a longer period of time (>6 weeks) and even in long gap peripheral nerve injuries.

Anti-Infective Drug Delivery Systems

Infections are often associated with wounds and this is why many researches were devoted to preventing or treating wound-associated infections [124]. Antimicrobial activity can be achieved in many ways, some of them being associated with the use of antimicrobial grafting materials, the loading of different systems used in tissue engineering with proper antimicrobial agents or by combining the two approaches. These strategies used in tissue engineering are schematically presented in Fig. 7.5. Due to the polycationic nature of the chitosan, antimicrobial activity is induced and, along with its other properties such as biocompatibility, high adsorption capacity, biodegradability, high hydrophilicity. It is often used in tissue engineering alone (especially in soft tissue engineering), or in association with other components, especially calcium phosphates, silicates and bioglasses (in hard tissue engineering) [61, 70, 95, 137]. Doped calcium phosphates (CaPs) but also bioglasses can possess antimicrobial activity if the doping is realized with adequate ions (such as: Ag^{+}, Cu^{2+}, Zn^{2+}) and can be especially used in hard tissue engineering, but also as additional components in polymers and as antimicrobial materials for soft tissue engineering. By far, the highest versatility is assured by the development of the drug delivery systems. Hydrophilic and hydrophobic supports can be combined with antimicrobials to get the desired profile of delivery and thus to get the optimal activity.

Ion-doped bioactive glasses as well as ion-doped HAs and CaPs, in general, are often used in the anti-infective therapies alone or in association with other materials,

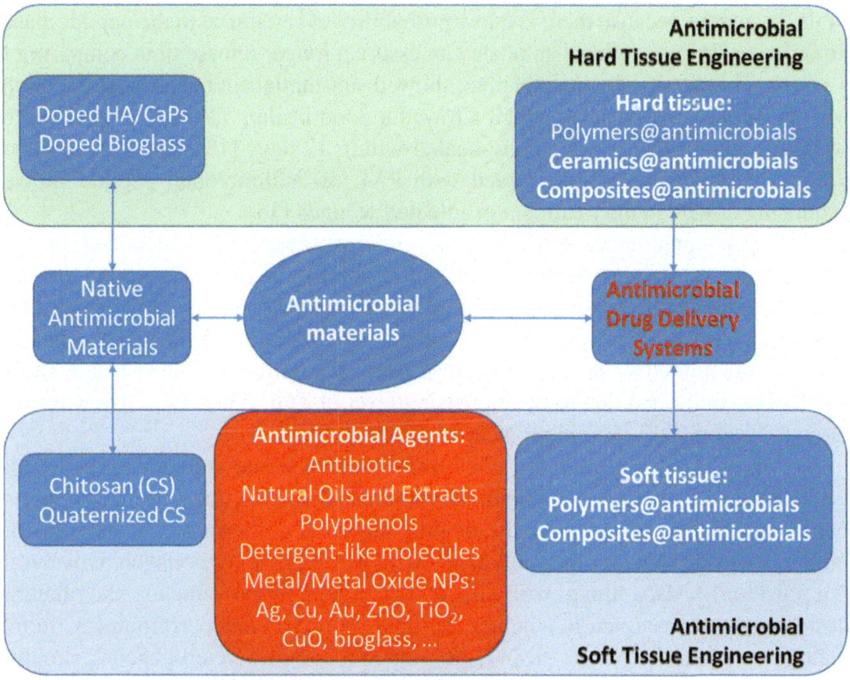

Fig. 7.5 Antimicrobial strategies in tissue engineering

especially embedded into polymeric matrices, even in the infections associated with the skin [33, 54, 110, 121].

Polymers are especially used in soft tissue engineering and by loading them with antimicrobial agents, they become active against the pathogens, being suitable for prevention but also for the treatment of already infected sites, including (multi)resistant bacterial strains. A wide range of antimicrobial agents are available for such applications, including antibiotics, volatile oils and extracts, detergents and detergent-like molecules, complexes, metal and metal oxide nanoparticles [85, 91, 100, 101, 106, 126].

Honey- and bee-derived products are increasingly used in wound healing because of their regenerative but also protective antimicrobial activity as well as in the treatment of the infections [30, 78]. Considering the two major advantages of the honey, wound healing capacity and antimicrobial activity (both depending on the origin of the honey) and the rapid wound healing capacity of calcium alginate (due to a gel forming and dehydration- preventing properties), Mirzaei et al. [78] proposed alginate/honey hydrogel as a topical ointment that protects burned rat wounds against infections and even reduces the hospitalization period.

Methicillin-Resistant Staphylococcus Aureus (MRSA) remains a challenge in the treatment of its associated wound infections, worldwide. Alginate sulphate-based hydrogels, loaded with broad-spectrum antimicrobial peptides confer the premises of the treatment because there is a low probability of resistance to the peptide-based antimicrobials and, alginate sulphate can assure a longer release time comparing to alginate. The results obtained on mice showed substantial antibacterial effect on the MRSA but low cytotoxicity which allowed a good healing capacity over time, the wound infection being completely healed within 12 days [13]. Similarly, hydroxypropyl cellulose-based gel, loaded with PXL150 antimicrobial peptide showed promising effects in the treatment of infected wounds [15].

Complex sanguinarine-loaded gelatin microspheres were embedded into an oxidized dextran-hyaluronic acid hydrogel in order to be used as a wound dressing in the healing of infected wounds. Based on this study, Zhu et al. [152] proved that this wound dressing is efficient even against MRSA and *E. coli* and the healing occurs with limited or no scars due to a better regulation of TGF-β1 and TNF-α (both decreased), and due to the increasing level of TGF-β3; TGF = Transforming Growth Factor; TNF = Tumour Necrosis Factor.

Collagen-based scaffolds are commonly used in tissue engineering especially because it is a major component in human and animal skin, cartilage, nerve, bone, etc. In order to improve the healing capacity, collagen-based materials are often compounded with other polymers (such as cellulose) or collagen hydrolysate (i.e., gelatin). Also, along with the compositional improvements, the material design is also considered to enhance the final properties and performances. In this regard, for instance, Guo et al. [47] developed a complex heterogeneous structure where curcumin (Cur)/gelatin microspheres were embedded in a collagen-cellulose nanocrystalline composite matrix and, as a consequence, an important change of the release profile of curcumin was obtained, the complete release being extended with up to 5 or even 10 days. Due to the presence of curcumin, the antimicrobial activity (diameter of the inhibition zone was 14–17 mm) and the anti-inflammatory activity are suitable, even for infected burns.

A self-healing HyA –Fe^{3+}– EDTA hydrogel can be obtained by combining hyaluronic acid with EDTA and Fe^{3+} in a supramolecular assembly; HyA = hyaluronic acid; EDTA = ethylenediaminetetraacetic acid. The mechanism of action is quite complex: the hyaluronidase excreted by the surrounding bacteria degrades the Fe^{3+} which, according to a Fenton process kills the bacterial cells. The hydrogel degradation and formation are reversible processes and, consequently, the degraded hydrogel is able to self-heal (in the absence of the hyaluronidase). The self-healing takes place in several minutes. The antimicrobial activity is assured by the free Fe^{3+} via the Fenton process (Fig. 7.6) and also, during the bacterial degradation of the hydrogel, along with the Fe^{3+} release, PDGF-BB can be also released and can thus assist the regeneration by promoting angiogenesis and epithelialization [123].

Bioactive glasses as well as their related composites are also used in bone grafting and bone repair. This is why, by loading them with adequate antimicrobial agents, antimicrobial systems can be developed, even for the treatment of some bone-related infections such as osteomyelitis. Mostafa et al. [79] loaded chitosan/bioactive glass composite scaffolds with ciprofloxacin and proved that a scaffold with 5% ciprofloxacin is a biocompatible support which could help in the loco-regional treatment of osteomyelitis. On the other hand, Boulila et al. [17, 18] developed

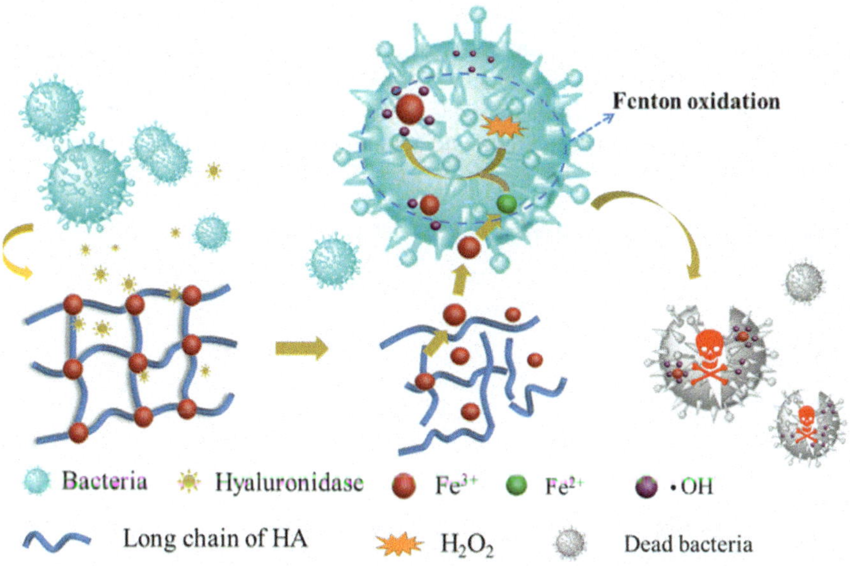

Fig. 7.6 Mechanism of degradation and on-demand release triggered by bacterial activity (with the kind permission of the American Chemical Society [123])

some similar composites, using chitosan or PVA as polymers but loaded with 20% ciprofloxacin and found, in vivo, that the higher antibiotic content leads to a delay in the formation of the apatite layer but also to a pro-oxidative effect.

Ternary polymeric supports based on hyaluronic acid (HyA), poly(ethylene oxide) (PEO) and poly(3-hydroxybutyrate-co-hydroxyvalerate) (PHBV) loaded with curcumin and 3,4-difluorobenzylidene curcumin (CDF) were used to fabricate nanofiber mats and were found to be antimicrobial and, moreover, to assure better cell attachment, faster cell growth and efficient wound healing [94].

Essential oils and natural extracts are often used in developing wound dressings because they possess regenerative and antimicrobial activity, these properties being dependent on the nature of the essential oil, the technology of production, origin, etc. For instance, Saha and Tayalia [103] used clove oil to obtain a wound dressing based on gelatin and chitosan and proved the benefits of these would dressings: enhancement of the healing due to a sustained release of clove oil over a 2-week period and induction of antibacterial and anti-inflammatory activity.

Release-triggering delivery is desired in medical applications because it can assure personalized delivery and treatment for many diseases, depending on the needs of each patient. For instance, PLA/GO/quercetin was found to be an efficient drug delivery system with potent electric triggering capacity due to the presence of GO (graphene oxide) [26]. The electric triggering was tested with an applied electric field of 10 Hz and it was found that the delivery can increase up to 6000 times faster at 0.5% GO and 8640 times faster at 1% GO; when the frequency of the electric field is 50 Hz, the delivery is 750 times faster at 0.5% GO and 864 times faster at

0.1% GO. These results confer the premises of personalized therapy with external control of the delivered drugs. In this case, quercetin was used as a model drug with antimicrobial activity but, many other drugs can be used (Table 7.2).

Calcium phosphates and bioactive glasses are used in orthopaedic and dental applications. They can be improved by inducing antimicrobial activity if proper doping

Table 7.2 Representative drug delivery systems with potential use in the treatment of infected wounds

	Materials	Observations	Refs.
Soft tissue engineering	Alginate/Honey	Antimicrobial wound dressings as well as ointments can be obtained; use of an alginate/honey hydrogel induces a slightly faster healing on a rat model, in vivo (14 days instead of 16)	[78]
	Alginate/CM11 peptide Alginate Sulphate/CM11 Peptide	Alginate sulphate can better tune the release of the antimicrobial peptide (~3 weeks comparing to ~1 week in the case of alginate) and the wound infection is completely healed within 12 days	[13]
	Dex-HyA/(GMs/SA)	Sanguinarine (SA)-loaded gelatin microspheres (GMs) were embedded into a dextran-hyaluronic acid (Dex-HyA) hydrogel and were found to be efficient in the healing of infected wounds associated with *MRSA* and *E. coli*	[152]
	HyA – Fe^{3+} – EDTA HyA – Fe^{3+} – EDTA/PDGF	Self-healing HyA – Fe^{3+} – EDTA hydrogel with on-demand release capacity (of Fe^{3+} and PDGF) was developed and proved to be able to simultaneously treat *E. coli*-and *S. aureus*-associated skin infections and also to assure wound healing	[123]
	Coll/Cell/Curcumin	Natural agents such as polyphenols are efficient antimicrobial agents which can be used in the healing of infected burns	[48]

(continued)

Table 7.2 (continued)

	Materials	Observations	Refs.
	PHBV/PEO/HyA/Cur PHBV/PEO/HyA/CDF	Antimicrobial and regenerative support was developed in a ternary system and loaded with curcumin	[94]
	Gel/CS/Clove Oil	By loading a gelatin/chitosan cryogel with clove oil, a long-term biodegradable and antimicrobial dermal substitute can be obtained	[103]
	PLA/GO/Quercetin	Electric-triggering drug delivery systems based on poly(L-lactic acid), graphene oxide and quercetin were developed for wound dressing	[26]
Hard tissue engineering	Doped CaPs(/antimicrobials)	Calcium phosphates and bioactive glasses, if properly doped, can assure antimicrobial activity as a consequence of the releasing ions, especially: Ag^{+}, Cu^{2+}, Zn^{2+}, Ce^{3+}, Ga^{3+}, Mn^{2+}. Moreover, loading these materials with antimicrobial agents, such as antibiotics, can enhance the antimicrobial activity	[52, 69, 75, 128]
	Doped bioactive glass (/antimicrobials)		[121]
	Mesoporous materials	Mesoporous materials, such as mesoporous silica are suitable for loading and release of antimicrobial drugs	[5, 56, 116, 119]
	Polymer/bioactive glass (/antibiotics)	Many polymers and ceramics are used to develop nanocomposites for antibacterial bone grafts	[12, 18, 66, 79, 147]
	Coll/HA/antibiotics	Synthetic grafts based on collagen and hydroxyapatite (doped with Mg^{2+}) and loaded with antibiotics have loco-regional antimicrobial activity	[80]

(continued)

Table 7.2 (continued)

	Materials	Observations	Refs.
	BCP/biopolymers/ rifampicin	Biphasic calcium phosphate scaffolds coated with PCL or PEU loaded with rifampicin can be also an effective antimicrobial graft with regenerative and preventive role	[86]
	Lyophilized human bone allograft/ antibiotics	Lyophilized human bone allograft was loaded with antibiotics and the antimicrobial and regenerative activity was assessed in vivo, in a rabbit model	[25]
	DEX/ Zn–Mg-MOF74/ PEEK	The surface of the PEEK implant was modified with the dual- metal–organic framework Zn–Mg-MOF74 via a biomimetic way using polydopamine and then further loaded with dexamethasone to develop an antibacterial, angiogenic and osteogenic surface—well suitable for bone grafting	[138]
	RGO/HA/antibiotics	Reduced graphene oxide (RGO)-nHA composite scaffold loaded with vancomycin can be a promising antimicrobial scaffold used in the treatment of infections	[150]

HyA—Hyaluronic Acid; CDF—3,4-difluorobenzylidene curcumin; PEO—poly (ethylene oxide); PHBV—poly (3-hydroxybutyrate-co-hydroxyvalerate); CS-Chitosan; Coll-collagen; HA-hydroxyapatite; BCP- biphasic calcium phosphate; PEEK—Polyetheretherketone; Dex-dexamethasone; Zn–Mg-MOF74—dual-metal–organic framework; CaPs-Calcium Phosphates; PCL—poly(ε-caprolactone); PEU-poly(ester urea); nHA– nanohydroxyapatite

ions are used: Ag^{+}, Cu^{2+}, Zn^{2+}, Ce^{3+}, Ga^{3+}, Mn^{2+}. Some of these ions, according to Taye et al. [121] are also osteogenic and thus recommended for bone grafting and repair. Sun et al. [116], for instance, co-doped mesoporous calcium-silicon nanoparticles (MCSNs) with Ag^{+} and Zn^{2+} and used these materials for the successful treatment of refractory root canal infections. Considering the two compositions ($Ag_{0.5}Zn_3$-MCSNs and $Ag_{0.5}Zn_{10}$-MCSNs), they found out that the second composition can assure an optimal Ag^{+}:Zn^{2+} ionic ratio release of 1:12 and these materials can highlight a strong preventive effect against an *E. faecalis* strain, specific for such infections. Also, coatings based on calcium phosphates and bioactive glasses were developed for the efficient treatment of infections, some of them being just loaded with antibiotics but others being co-loaded with fibroblast growth factor-2 (FGF-2) [42]. Hollow bioactive glasses doped with Cu^{2+} and loaded with danofloxacin were also studied and found efficient in the local treatment of bone infections (including against *S. aureus* biofilms) [63]. Several CaPs doped with Fe^{2+}, Sr^{2+}, Zn^{2+}, Ag^{+} were also loaded with adequate antimicrobial agents such as poly(L-lysine), tobramycin, ciprofloxacin, vancomycin; these doped CaPs proved their antimicrobial activity and their potential use in bone-related infections [21, 32, 52, 69, 75, 128]. Composite materials based on PLA and HA can be also used in the treatment of osteomyelitis if loaded with vancomycin. For this regard, Zhao et al. [147] developed electrospun scaffolds and proved the benefits of these fibrous scaffolds in the treatment of osteomyelitis highlighting the sustained release of vancomycin and also the ability of these supports to promote adhesion and proliferation of osteoblasts.

Lyophilized human bone allografts remain the gold standard in bone grafting but, because of their limited availability, synthetic grafts are needed. Coraca-Huber et al. [25] used such grafts to prove the effect of the loading of these grafts with antibiotics (gentamicin sulphate, 80 mg/ml, vancomycin hydrochloride, 50 mg/ml, clindamycin phosphate, 150 mg/ml, rifampicin, 60 mg/ml and a mixture of rifampicin and vancomycin, 110 mg/ml) over 4 microbial strains *(Methicillin-Sensitive Staphylococcus aureus* ATCC 29,212 (*MSSA*), M*ethicillin-Resistant Staphylococcus aureus* (*MRSA*), *Staphylococcus epidermidis* ATCC 12,228 and *Staphylococcus epidermidis* isolated from clinic). In vitro, it was proved that the simultaneous use of vancomycin and rifampicin potentiates the antimicrobial effect even against the clinically isolated microorganisms. In vivo, it was found that clindamycin, vancomycin and the mixture of vancomycin and rifampicin remain above the MIC90 (minimum inhibitory concentration for 90% of the isolates of a species) up to 3 days post-implantation but gentamicin fell below this value within the first day. The new bone formation was monitored in all the groups, at 120 days post-implantation and it was found that osteosynthesis was 37.84 ± 6.33% (control group), 28.28 ± 7.81% (Vancomycin group), 21.67 ± 8.19% (Clindamycin group), 35.86 ± 8.65% (Gentamicin group) and 26.81 ± 6.92% (Vancomycin + Rifampicin group), relative to the total implant area.

Collagen/Hydroxyapatite composite materials are compositionally the most similar with natural bone and this is why these materials are mostly studied as bone grafting materials [9, 11, 34–38, 40, 57, 84] and as drug delivery systems for bone-related diseases, including infections [3, 7, 39, 43, 55, 59, 76, 80, 87, 97, 102, 133]. Mulazzi et al. [80] developed some composite materials based on collagen and hydroxyapatite (doped with Mg^{2+}) and loaded them with gentamicin (6.25%wt.) and vancomycin (12.5%wt.) as antimicrobial agents—antibiotics. The loading was done by the classical route, by soaking these supports with an appropriate volume of solution to get the desired amount of antibiotics (Fig. 7.7). It was found that the release and loading are dependent on the amount of MgHA, a higher amount of MgHA leading to a higher antibiotic retention, a slower release and possibly efficient use in prevention of infections consequent to implantation. Both drug delivery systems are able to assure a sustained delivery for ~20 days. It is also expected that

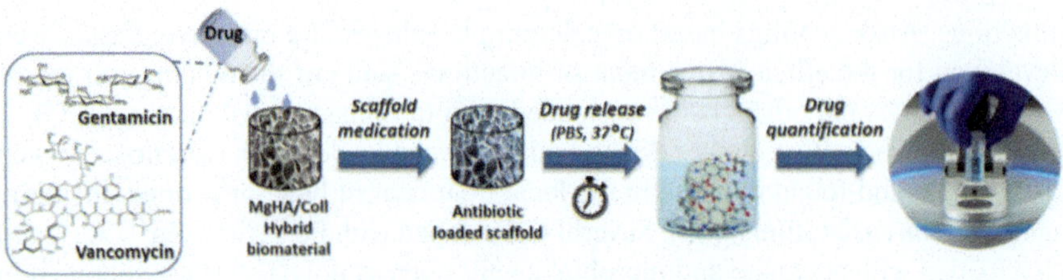

Fig. 7.7 Loading procedure of antibiotics into a porous MgHA/Coll composite material, by soaking (reproduced from [80], with kind permission of MDPI)

these materials can be used in the treatment of osteomyelitis or other infections but especially if the design will involve hybrid particles or beads loaded with antibiotics and then embedded into such composites. Li et al. [71] used strontium-doped HA/alginate as a drug delivery system for releasing vancomycin for bone regeneration. Similar results were obtained by Neto et al. [86] using a biphasic calcium phosphate system coated with a biopolymer and loaded with an antibiotic—rifampicin.

Metal organic frameworks were found to be suitable also in the field of bone regeneration. For instance, Xiao et al. [138] used a dual-metal–organic framework (Zn–Mg-MOF74) coating on the PEEK surface (coating realized via a mussel-inspired polydopamine interlayer) and followed by loading dexamethasone. The strong antimicrobial activity was proved against two common bacterial strains, *E. coli* and *S. Aureus*. In vivo, both antibacterial and angiogenic abilities were proved along with an osteogenic ability which recommends these structures as potential bone grafting materials.

Core – Shell Gold Nanospheres/Mesoporous Silica Nanoparticles loaded with vanillic acid were proposed by Huang et al. [56] for the treatment of orthopaedic infections. The as obtained mesoporous materials are IR-sensitive and, by irradiating them with a NIR 808 nm laser, a temperature increase from 24 to 60 °C can be reached within 12 s. As a result of the temperature increase, the release of the vanillic acid can be tuned, the system being temperature-sensitive (a total cumulative release of 78.95 ± 1.41% at 42 °C versus 42.61 ± 1.71% at 37 °C, after 72 h). In conclusion, the synergic effect of photothermia and vanillic acid release (triggered by NIR radiation) can lead to an excellent antibacterial effect against *S. aureus*.

Akram et al. [5] developed some pH-sensitive mesoporous systems modified with poly(L-glutamic) acid for the release of chlorhexidine for dentin adhesives. The surface preparation of the mesoporous silica involves a silanization step with amino-propyl trimethoxysilane followed by γ-benzyl-L-glutamate polymerisation via a N-carboxy anhydride intermediate able to improve the loading and release behaviour of chlorhexidine. The surface modification is important because it induces a pH sensitivity but also extends the release of the chlorhexidine for several weeks. Embedding these mesoporous systems at 5 or 10% into an experimental resin-based dentin adhesive lead to improvement of the antibacterial activity, resin-dentin bonding integrity and durability.

Carbon-based materials and especially carbon nanotubes (CNTs) and graphene-related materials (GRMs) are extensively used in tissue engineering because of their properties. Moreover, in their related oxides, the properties are strongly correlated with the oxygen content, i.e. oxidation degree and thus, the sorption–desorption of the drugs can be easily tuned. For instance, reduced graphene oxide (RGO)-nHA, a three-dimensional, porous composite scaffold was obtained and loaded with vancomycin to get a long-lasting antibacterial scaffold for the treatment of bone-related infections [150].

Anti-Tumoral Drug Delivery Systems

Cancer is the second leading cause of premature deaths worldwide. In 2020, there were an estimated 19.3 million new cases and almost 10.0 million deaths of cancer worldwide. By 2040, new cancer cases are projected to rise to 28.4 million. Around 1 in 5 men (22.60%) or women (18.55%) develop cancer during their lifetime [117]. Progress in the treatment of cancer is needed for reducing cancer mortality. Historically, the development of the cisplatin represented an important milestone in modern oncotherapy and the survival rate increased consistently [23].

Surgery, radiotherapy, and chemotherapy are the three core strategies in oncotherapy. Besides these three, targeted therapy and immunotherapy constitute two alternative strategies [1].

Drug delivery systems are relevant to chemotherapy, immunotherapy and targeted therapy [29, 41, 46, 60, 129]. Chemotherapy can be defined broadly as therapy with chemical substances [28]. Conventional chemotherapy usually refers to small-molecule drugs [28, 29]. Immunotherapy is broadly defined as the clinical suppression or activation of the immune system for a therapeutic purpose [113]. Immunotherapeutic agents range from molecules to vaccines and cells [113]. Targeted therapy is based on molecules with specific cellular targets and, thereby, interfering with oncogenic processes at cellular level [29, 127]. There are two categories of drugs for targeted therapy: small molecules and monoclonal antibodies [29].

Immunotherapy and targeted therapy are approaches with applicability in precision medicine [29, 73]. Precision medicine, also called personalized medicine, aims to tailor cancer treatment to the specific parameters of a subgroup of patients [73].

Conventional chemotherapy is currently the best option for the treatment of metastases [29, 89]. However, none of the approximately 60 drugs used in conventional chemotherapy target tumour tissues selectively [29]; conventional drugs also target healthy tissues which translates into high toxicity associated with the cancer therapy [50]. In fact, as a consequence of these shortcomings, two approaches are known in the literature: to develop new cytostatics with improved antitumoral activity and lower side effects and to develop new administration ways to better target the tumoral cells/tissues (including smart drug delivery systems with targeted delivery) [81].

Because of the high systemic toxicity, the targeting and the loco-regional delivery are important features able to improve the efficiency by concentrating the active agents in the targeted tissue/organ [8, 22, 74, 145, 149]. Many antitumoral systems are used or researched for cancer treatment, starting from different supports and bioactive agents, systems with triggering and targeting (if any) mechanisms and systems with release mechanisms.

In certain cases, loco-regional drug delivery systems can be developed and used for the treatment of cancer. Such approaches are especially suitable when loco-regional delivery is required: in skin cancer, colon cancer, primary bone cancer treatment, etc. In primary bone cancer therapy, the treatment protocols are still mainly based on surgery, radio- and chemotherapy. That means that along with the surgical intervention of resection of the tumoral bone tissue, a bone defect is created and this defect can be filled with adequate, preferably multifunctional, materials able to prevent remnant tumour cells development and thus to prevent recurrences [7, 8, 74]. These systems, depending on the composition, can assure antitumoral activity by the produced hyperthermia (if magnetite is loaded) or released cytostatics.

Multiple drug delivery systems based on antitumoral drugs are reported in the literature. Hydroxyapatite (HA)-based supports are especially used in localized bone cancer treatment, considering the following issues: HA is the most abundant component of bone, and such supports can manage the release of antitumoral agents for a long period of time. For instance, HA-based ceramic materials obtained by centrifugation can release cisplatin for up to 8–12 weeks when administered in the thigh muscles of the mouse [2, 82]. It is important to mention that based on the results of Nadar et al. [82], cisplatin is mostly accumulated in the tumoral site and only a negligible amount of cisplatin is dispersed in the blood which means that the systemic toxicity is lower. Similar results were obtained with many other drugs, including doxorubicin and methotrexate. Porosity, along with the nature of the drug and support, is an important intrinsic characteristic of the materials which is used to achieve the desired release level and to avoid undesired toxicity. In the case of HA, a porosity of 35–48% seems to be suitable for the optimized release of methotrexate [115].

More complex systems, such as those proposed by Anirudhan et al. [10] were obtained by starting from hydroxyapatite, heparin (Hep), triethylenetetramine (TETA), oleic acid (OA) and folic acid (FA). The as obtained composite hydrogel/micelles were further loaded with cisplatin (CDDP) and curcumin (Cur) (the synergistic/beneficial effect of the CDDP-Cur association being known [108]) and further evaluated, in vitro, against normal and HCT-116 colon cancer cells. Based on these studies it was found that this formulation can be promising in colon cancer treatment, with high efficacy and minimal toxicity.

From a supramolecular irinotecan (IRN)-loaded hydrogel based on alginate and peptide (CDDP/Pept-AlgNP/IRN), antitumoral systems were obtained and implanted subcutaneously in an A549-xenografted mouse. It was found that the releases of the two cytostatics are independent (as presented in Fig. 7.8), the two cytostatics being independently dispersed in the peptide and respectively in the alginate moiety of the hydrogel. Moreover, the combination of the two agents leads to significant reduction of the viability of the tumoral cells, in vitro, and a tumour volume reduction, in vivo.

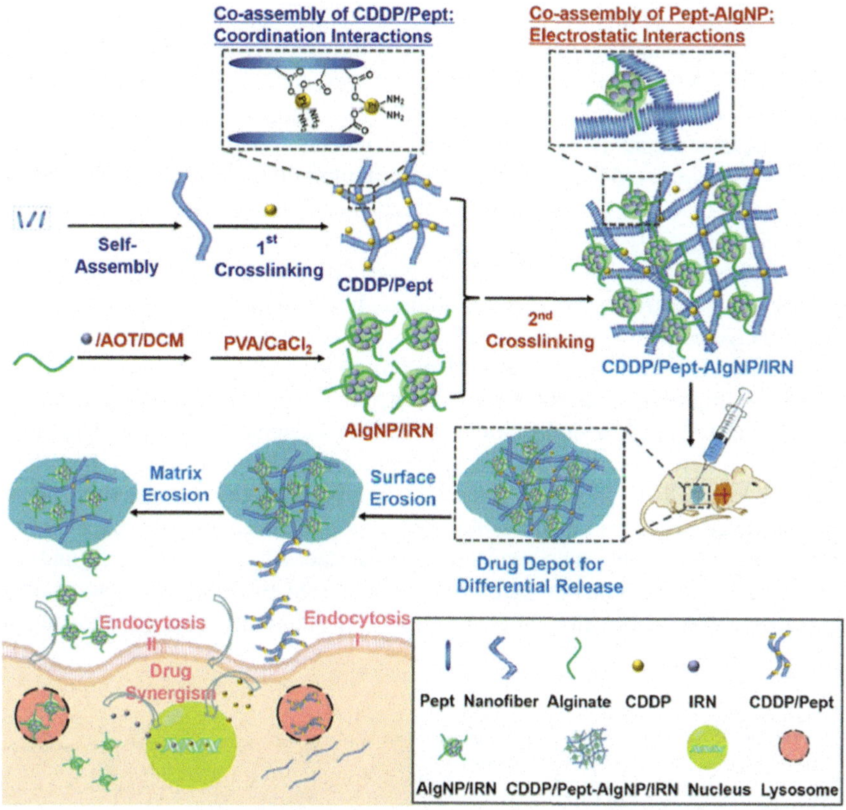

Fig. 7.8 Double-cross-linked CDDP/Pept-AlgNP/IRN nanocomposite hydrogel for differential release of CDDP and IRN (with the kind permission of Elsevier [135])

Chitosan (CS)-based nanocomposites obtained by entrapping magnetite (M), silicon dioxide (S), and/or graphene oxide and further loaded with cisplatin were studied as potential drug delivery systems in cancer treatment [2]. These systems were found to be suitable for both loading and release of cisplatin. Because of the protonation ability of the amino moieties of the chitosan, these systems were found to be pH-sensitive over the pH range of 5.8–7.4. The best cisplatin recoveries were obtained at pH 6.5, meaning 89, 88 and 91% for the samples CS/M/S, CS/M/S/GO and CS/M, respectively. The results were explained based on the DFT (density functional theory) calculations, which reveal the interaction between the CDDP and the composite support as well as the most relevant binding energies.

Graphene oxide is an interesting material increasingly used in drug delivery because of its high versatility. It can be obtained by controlling the oxygen content and thus the hydrophilic: hydrophobic ratio can be tuned. Moreover, due to the hydroxyl and carboxyl moieties, it can be easily modified via chemical or physical modification routes. Nandi et al. [83] used the PEGylation route and the ethylenediamine-modified poly-isobutylene-maleic anhydride (PMA-ED) route, respectively, to modify graphene oxide. GO-PEG and GO-PMA both reacted with 7-ethyl-10-hydroxycamptothecin (SN38) as a topoisomerase I inhibitor and were then loaded with cisplatin as a DNA alkylating/damaging drug. The self-assembled nanoparticles are quickly internalized into HeLa cells—within 6 h, and synergistically act and assure effective combination antitumoral therapy. The self-assembling strategy for obtaining these antitumoral graphene oxide-based nanoparticles is presented in Fig. 7.9.

Micro- and mesoporous materials are also exploited as potential drug delivery systems in cancer treatment. Montmorillonite, for instance, is well known as a pH-sensitive layered double hydroxide, able to be modified and loaded with biologically active molecules which can be released according to a pH-sensitive mechanism, at a slightly alkaline pH, for instance, in colon [58, 65]. Kar et al. [65], for instance, developed organo-modified montmorillonite (OMMT) by using cetyltrimethylammonium bromide—(CTAB) and loaded it with curcumin and methotrexate (Fig. 7.10). The as obtained drug delivery system is able to assure targeted delivery of the two drugs in the tumoral cells. The combination therapy is especially beneficial because curcumin release can enhance the folate receptor on the surface of the tumoral cells and thus mediate the methotrexate into HeLa cells without affecting too much the uptake in normal cells. That means that targeted methotrexate uptake is mediated by the release

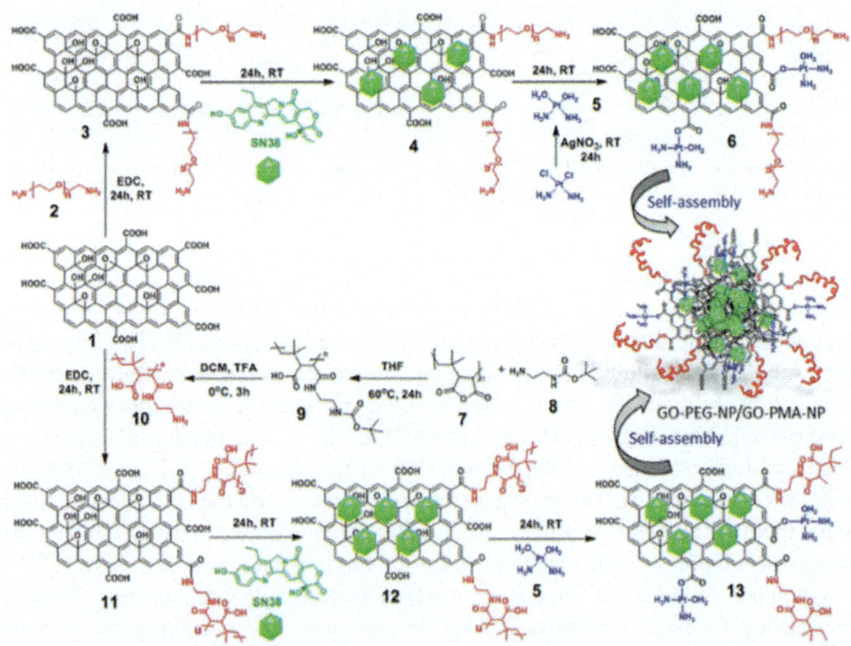

Fig. 7.9 Self-assembling procedure of producing GO-PEG-SN38-CDDP and GO-PMA-SN38-CDDP conjugates (reproduced with the kind permission of the royal society of chemistry [83])

of curcumin. The overall antitumoral activity of the complex system OMMT-MTX-CUR (HeLa cells viability is 6%) is much higher than the antitumoral activity of the simpler systems (HeLa cells viability is 67% for OMMT-MTX – and 63% for OMMT-CUR –) proving the benefit of the combination therapy.

Also, montmorillonite-based drug delivery systems were intended for oral administration and colon delivery of antitumoral agents. In this case, the drug is protected from the harsh conditions of the stomach but, following the transit into the intestines, as a consequence of the neutral or slightly alkaline pH, the release rate of the drug is starting to increase. Considering the normal gastro-intestinal transit, in order to increase the stationary time in the intestine level and to increase the recovery of the loaded drugs, these systems can be embedded into muco-adhesive polymers (such as chitosan or alginate) which allows the targeted and triggering capacity to be tuned adequately [58].

Smart antitumoral systems can be obtained by combining adequate support materials such as Coll/HA composite materials with magnetite and cytostatics and, in this case, a synergy can be obtained. The cytostatic release can be additionally tuned by hyperthermia and, in this case, depending on the patient's needs, enhanced. As a consequence of the higher concentration of the released cytostatic, the antitumoral activity can also be enhanced. This is important because hyperthermia can

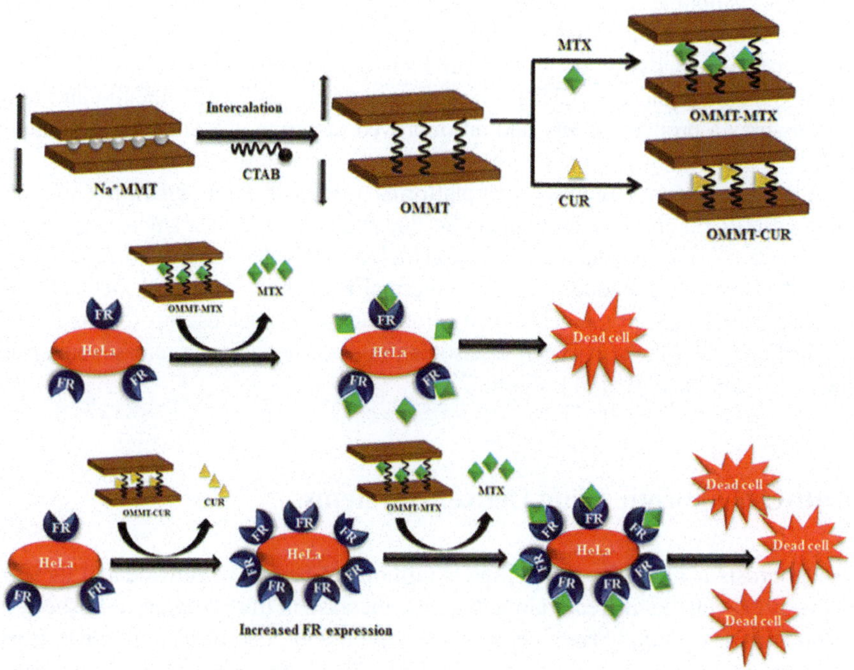

Fig. 7.10 MMT modification by CTAB followed by curcumin and methotrexate intercalation (copyright granted by Elsevier [65])

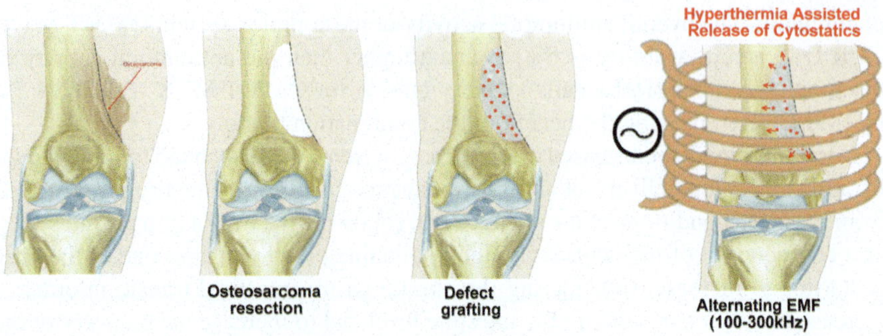

Fig. 7.11 Schematic illustration of a loco-regional use of a multifunctional drug delivery system in a surgical-chemotherapeutical approach

be generated by external irradiation with a proper alternating electromagnetic field (100–300 kHz) and, depending on the evolution of the treatment, the oncologist can decide the dose of the released cytostatic agent by changing the field parameters (potential, current intensity, time of irradiation, characteristics of the coil, etc.) and thus can create the proper conditions for assuring personalized therapy [39, 122] (Fig. 7.11).

Copper sulphide nanoparticle–engineered covalent organic frameworks can be used as efficient drug delivery platforms for doxorubicin, its release being tuneable according to a synergic mechanism blending photothermal therapy (PTT) and chemodynamical therapy (CDT). The presence of bovine serum albumin and folic acid assures a proper stability and an improved targeting efficiency in the tumour [131].

A multifunctional supramolecular platform based on cisplatin and a mitochondria-targeted NIR photosensitizer, named IR780, was proposed as a smart antitumoral system. Its complex mechanisms of action include a targeted mitochondrial dysfunction of cancer cells (induced by photothermal and photodynamical mechanisms) which attenuates the crosstalk between mitochondria and nucleus and down-regulates the DNA repair which, in synergic association with the cisplatin release, enhances the antitumoral activity [144] (Table 7.3).

Anti-Osteoporotic Drug Delivery Systems

Osteoporosis is a disease which affects millions of people worldwide and its incidence is expected to increase along with the increase in life expectancy. In short, in osteoporosis, the equilibrium between resorption and bone formation is perturbed and consequently the bone properties are altered [153]. Based on a recent review, anti-osteoporotic drugs can be administered to the human body by several ways such as: oral, intranasal, injectable, transdermal, or even implantable. Among these, only oral delivery is at clinical study level but, implant-based delivery and injectable delivery,

Table 7.3 Major antitumoral systems, their mechanism of action and targeting/triggering ability

Antitumoral system(s)	Characteristics		References
	Antitumoral mechanism(s)	Targeting/ triggering ability	
Coll/HA/Fe_3O_4/CDDP	Chemotherapy, Hyperthermia	Yes*/Yes	[39, 122]
HA/Hep/TETA/OA/CDDP / Cur	Combinatorial chemotherapy	Yes/No	[10]
Alg/MMT/IRN	Chemotherapy	Yes/Yes	[58]
OMMT-MTX-CUR	Combinatorial chemotherapy	Yes/Yes	[65]
CDDP/Pept-AlgNP/IRN	Combinatorial chemotherapy	No/No	[135]
GO-PEG-SN38-CDDP and GO-PMA-SN38-CDDP	Combinatorial chemotherapy	Yes*/Yes	[83]
CuS@COFs-BSA-FA/DOX	PTT, CDT, combinatorial chemotherapy	Yes/Yes	[131]
IR780@PtNP suprastructures	PTT, chemotherapy, molecular level	Yes/Yes	[144]
CS/Fe_3O_4/SiO_2/GO/CDDP	Chemotherapy, Hyperthermia	Yes/Yes#	[2]

*for loco-regional delivery; #just pH triggering is proved but because of the presence of the magnetite, it is expected that these materials can respond to electromagnetic stimuli, as well

at preclinical level, represent 35 and 39% of the preclinical studies, respectively, which confers hopes in developing implantable devices and injectable solutions to osteoporosis [105].

There are three major classes of anti-osteoporotic drugs: *anti-resorptive agents* such as bisphosphonates, cathepsin K (CTSK) inhibitors, calcitonin, receptor activator of nuclear factor κB (NF-κB) ligand inhibitors, *anabolic agents* and *dual agents* which are able to simultaneously assure anti-resorption and bone formation [143]. Among these classes, there are 11 drugs already approved by the FDA (United States Food and Drug Administration) for the treatment of osteoporosis, as highlighted in Table 7.4. These drugs can be loaded in several supports which could assure a loco-regional, long-term delivery of these agents and, thus, a rebalancing of the disrupted bone resorption/bone formation equilibrium, at least loco-regionally; these supports are presented in Table 7.4

Bisphosphonates are, most likely, the most used anti-osteoporotic drugs which were loaded in most common bone grafting materials, including polymers (chitosan, alginate, collagen, etc.), ceramics (especially calcium phosphates), composites (including Coll/HA, Coll/CS/β-TCP) but also in micro- and mesoporous materials (such as mesoporous bioactive glasses, mesoporous silica or metal–organic frameworks) [96].

Table 7.4 Drug delivery systems with anti-osteoporotic activity

Drug classes/representatives	Support	References
Antiresorptive drugs		
Bisphosphonate: Alendronate[FDA]	Bone cement Layered Double Hydroxide/PC	[153] [14]
Bisphosphonate: Zoledronate[FDA]	Mesoporous Bioactive Glass/PC Zeolitic imidazolate framework-8 (ZIF-8)	[45] [93]
Bisphosphonate: risedronate[FDA]	Ti-6 Al-4 V/zinc titanate	[107]
Bisphosphonate: ibandronate[FDA]	Collagen	[47]
Cathepsin K inhibitor: odanacatib	PLGA/$(Asp)_8$	[143]
Estradiol (hormone replacement therapy)/alendronate	PLGA/Fe_3O_4 NPs	[49]
Calcitonin	HA Mesoporous silica—decorated with a pentapeptide	[67, 72] [141]
Raloxifene[FDA]	Cyclodextrins/chitosan Doped-HA/Alg Composite Beads Pluronic® F68 and Gelucire® GL44	[64, 104, 120, 134]
Bazedoxifene[FDA]	NA*	NA*
Denosumab[FDA]	NA*	NA*
Anabolic drugs		
Teriparatide[FDA]	Liposomes Hyaluronic acid—dissolving microneedles	[99] [112]
Abaloparatide[FDA]	Methacrylated gelatin (GelMA) hydrogel	[88]
Romosozumab[FDA]	NA*	NA*
Dual drug: bone-forming agent with antiresorptive capacity		
Strontium ranelate	HyA/CaPs	[118]

PLGA—poly (D, L-lactide-co-glycolide); PC—poly(ε–caprolactone); [FDA]—FDA approved drug for osteoporosis treatment; NA*—no drug delivery system was found in the literature

Risedronate, a representative of the bisphosphonate class was also absorbed onto the surface of the zinc titanate-coated Ti-6 Al-4 V alloy. According to the results published by Sandomierski et al. [107], the drug is slowly released over a period of 7 days, without a burst-like initial release which is mainly explained by the proper interaction between zinc titanate and risedronate.

Kotak and Devarajan [67] loaded salmon calcitonin into HA nanoparticles and tested the efficiency of this drug delivery system for sublingual administration as a non-invasive administration way. But there are also some systems, based on hydroxyapatite and loaded with a (Asp6) hexapeptide-conjugated salmon calcitonin with targeting capability. In comparison with salmon calcitonin, the prolonged circulation time, threefold higher femur tissue accumulation and the 5.4 times higher adsorption on hydroxyapatite make this conjugate suitable for targeted therapy of osteoporosis [72].

Raloxifene was also loaded in several supports, including polymers and composite beads, the latter being based on a $Mg^{2+}Si^{4+}$-HA powder (46.67%) and on sodium alginate (33.33%) along with two additives, chondroitin sulphate and keratin (20%) [120]. It was found that the cumulative release is strongly dependent on the presence of the two additives, the cumulative release after 12 weeks being at most, 60%, for the sample containing both chondroitin sulphate and keratin (each at 10% content), less than 50% for the beads containing 20% chondroitin sulphate and slightly over 32% for the composite beads containing 20% keratin.

Teriparatide, an analogue of the parathyroid hormone—PTH, which contains the first 34 aminoacids of the 84 aminoacids of the human PTH is one of the most used anabolic agents. It was loaded into liposomes; in this way, it is protected against proteases and the short half-life associated disadvantages can be overpassed. The release occurs according to a Higuchi model and the total release takes ~35 h [99].

A 3D porous structure of a methacrylated gelatin (GelMA) hydrogel was found to be a proper support for the delivery of the abaloparatide analogue of human recombinant parathyroid hormone protein (hrPTHp) [88]. This system can release abaloparatide for over 12 days and can promote bone healing, in vivo, in rats.

Combination therapy was also evaluated, in vivo, on a rat model. For this reason, remote-controllable bone-targeted co-delivery of estradiol and alendronate was assured using a PLGA-based matrix embedding Fe_3O_4 and co-loaded with estradiol and alendronate. Magnetite was used for magnetic targeting and magneto-thermally-triggered drug delivery [49]. Another interesting paper deals with a combination therapy involving raloxifene and alendronate, the two active agents being loaded into the thin mesoporous titanium (IV) oxide films coating over titanium screws [51].

Strontium ranelate is a dual acting drug which not only acts as an antiresorptive agent, but also stimulates bone formation [20]. Its biological activity is a consequence of the presence of Sr^{2+} but, it was also proved that its activity is much higher than that of $SrCl_2$ or of other strontium salts which means that a synergetic effect appears due to the Sr^{2+} and ranelate ions.

Conclusions and Perspectives

Advances in drug delivery systems for tissue engineering make them able to assure fast healing but also to treat different diseases associated with tissues. Relative to the total amount of grafting materials, ~49% are bone grafting materials and ~10% are skin grafts; the subchapter on drug delivery systems with enhanced healing capacity is mainly focused on such applications. Considering this high incidence of bone fractures (many times associated with infections, osteoporosis, cancer, or other diseases) and skin wounds, including burns (many times associated with infections or high risks of infections, diabetes, cancer) many researchers have developed drug delivery systems for the treatment of these diseases. The major advantages of the use of drug delivery systems compared to the other, traditional treatment protocols, are: a more constant level of the active agent (at a proper, therapeutic level), local or targeted delivery which increases the efficiency and reduces the systemic toxicity, and fewer administered drug doses. Several smart DDSs were presented in the literature. They can either respond to the internal, environmental conditions (pH, temperature, enzyme concentration, etc.) or to the external factors (light, magnetic and electric fields, mechanical stress, etc.) and thus can enhance or decrease the release rate. Consequently, the activity can increase or decrease according to the needs of the patient and thus, personalized therapy can be assured. In fact, some of the challenges in the field are related to the development and optimization of these drug delivery systems for personalized therapy using specific stimuli, support materials and active agents.

References

1. Abbas Z, Rehman S (2018) An overview of cancer treatment modalities. In: Neoplasm. IntechOpen, pp 139–157
2. Abdel-Bary AS, Tolan DA, Nassar MY, Taketsugu T, El-Nahas AM (2020) Chitosan, magnetite, silicon dioxide, and graphene oxide nanocomposites: synthesis, characterization, efficiency as cisplatin drug delivery, and DFT calculations. Int J Biol Macromol 154:621–633
3. Absalan F, Sadjadi MS, Farhadyar N, Sadr MH (2022) Bone tissue engineering of HA/COL/GO porous nanocomposites with the ability to release naproxen: synthesis, characterization, and in vitro study. J Inorg Organomet Polym Mater 32(9):3260–3275
4. Ahsan SM, Thomas M, Reddy KK, Sooraparaju SG, Asthana A, Bhatnagar I (2018) Chitosan as biomaterial in drug delivery and tissue engineering. Int J Biol Macromol 110:97–109
5. Akram Z, Daood U, Aati S, Ngo H, Fawzy AS (2021) Formulation of pH-sensitive chlorhexidine-loaded/mesoporous silica nanoparticles modified experimental dentin adhesive. Mater Sci Eng C-Mater Biol Appl 122
6. Ananth, K. P., Sun, J., & Bai, J. (2018). An Innovative Approach to Manganese-Substituted Hydroxyapatite Coating on Zinc Oxide(-)Coated 316L SS for Implant Application. Int J Mol Sci, 19(8).
7. Andronescu E, Ficai A, Georgiana M, Mitran V, Sonmez M, Ficai D, Ion R, Cimpean A (2013) Collagen-hydroxyapatite/cisplatin drug delivery systems for locoregional treatment of bone cancer. Technol Cancer Res Treat 12(4):275–284
8. Andronescu E, Ficai M, Voicu G, Ficai D, Maganu M, Ficai A (2010) Synthesis and characterization of collagen/hydroxyapatite: magnetite composite material for bone cancer treatment. J Mater Sci Mater Med 21(7):2237–2242
9. Andronescu E, Voicu G, Ficai M, Mohora IA, Trusca R, Ficai A (2011) Collagen/hydroxyapatite composite materials with desired ceramic properties. J Electron Microsc 60(3):253–259
10. Anirudhan TS, Suriya R, Anoop SN (2022) Polymeric micelle/nano hydrogel composite matrix as a novel multi-drug carrier. J Mol Struct 1264
11. Ardelean IL, Gudovan D, Ficai D, Ficai A, Andronescu E, Albu-Kaya MG, Neacsu P, Ion RN, Cimpean A, Mitran V (2018) Collagen/hydroxyapatite bone grafts manufactured by homogeneous/heterogeneous 3D printing. Mater Lett 231:179–182
12. Auregan JC, Begue T (2015) Bioactive glass for long bone infection: a systematic review. Injury-Int J Care Injured 46:S3–S7

13. Babavalian H, Latifi AM, Shokrgozar MA, Bonakdar S, Mohammadi S, Moghaddam MM (2015) Analysis of healing effect of alginate sulfate hydrogel dressing containing antimicrobial peptide on wound infection caused by methicillin-resistant staphylococcus aureus. Jundishapur J Microbiol 8 (9)
14. Baniahmad F, Yousefi S, Rabiee M, Shafieil SS, Faghihi S (2021) Alendronate sodium intercalation in layered double hydroxide/poly (epsilon-caprolactone): application in osteoporosis treatment. Iran J Biotechnol 19(1):48–59
15. Bjorn C, Noppa L, Salomonsson EN, Johansson AL, Nilsson E, Mahlapuu M, Hakansson J (2015) Efficacy and safety profile of the novel antimicrobial peptide PXL150 in a mouse model of infected burn wounds. Int J Antimicrob Agents 45(5):519–524
16. Bordone M, Bettencourt A (2023) Management of bone diseases: looking at scaffold-based strategies for drug delivery. Drug Deliv Transl Res 13(1):79–104
17. Boulila S, Oudadesse H, Badraoui R, Lefeuvre B, Mabrouk M, Chaabouni K, Mostafa A, Makni-Ayedi F, Barroug A, Rebai T, Elfeki A, Elfeki H (2017) Antioxidative/oxidative effects and retarding osteoconductivity of ciprofloxacin-loaded porous polyvinyl alcohol/bioactive glass hybrid. Med Biol Eng Compu 55(1):17–32
18. Boulila S, Oudadesse H, Kallel R, Lefeuvre B, Mabrouk M, Chaabouni K, Makni-Ayedi F, Boudawara T, Elfeki A, Elfeki H (2017) In vivo study of hybrid biomaterial scaffold bioactive glass-chitosan after incorporation of ciprofloxacin. Polym Bull 74(10):4153–4173
19. Burdusel AC, Gherasim O, Andronescu E, Grumezescu AM, Ficai A (2022) Inorganic nanoparticles in bone healing applications. Pharmaceutics 14(4)
20. Burlet N, Reginster JY (2006) Strontium ranelate—the first dual acting treatment for postmenopausal osteoporosis. Clin Orthopaedics Related Res 443:55–60
21. Chandra VS, Baskar G, Suganthi RV, Elayaraja K, Joshy MIA, Beaula WS, Mythili R, Venkatraman G, Kalkura SN (2012) Blood compatibility of iron-doped nanosize hydroxyapatite and its drug release. ACS Appl Mater Interfaces 4(3):1200–1210
22. Chau Y, Padera RF, Dang NM, Langer R (2006) Antitumor efficacy of a novel polymer-peptide-drug conjugate in human tumor xenograft models. Int J Cancer 118(6):1519–1526
23. Chen D, Milacic V, Frezza M, Dou QP (2009) Metal complexes, their cellular targets and potential for cancer therapy. Curr Pharm Des 15(7):777–791
24. Chen JX, Ashames A, Buabeid MA, Fahelelbom KM, Ijaz M, Murtaza G (2020) Nanocomposites drug delivery systems for the healing of bone fractures. Int J Pharm 585
25. Coraca-Huber DC, Steixner SJM, Najman S, Stojanovic S, Finze R, Rimashevskiy D, Saginova D, Barbeck M, Schnettler R (2022) Lyophilized human bone allograft as an antibiotic carrier: an in vitro and in vivo study. Antibiotics-Basel 11(7)
26. Croitoru AM, Karacelebi Y, Saatcioglu E, Altan E, Ulag S, Aydogan HK, Sahin A, Motelica L, Oprea O, Tihauan BM, Popescu RC, Savu D, Trusca R, Ficai D, Gunduz O, Ficai A (2021) Electrically triggered drug delivery from novel electrospun poly(lactic acid)/graphene oxide/quercetin fibrous scaffolds for wound dressing applications. Pharmaceutics 13(7)
27. Deng KH, Liu ZX, Dou WXF, Cai Q, Ma W, Wang SH (2022) HA/beta-TCP biphasic calcium phosphate ceramics derived from butterfish bones loaded with bone marrow mesenchymal stem cells promote osteogenesis. Frontiers Mater 9
28. DeVita VT Jr, Chu E (2008) A history of cancer chemotherapy. Cancer Res 68(21):8643–8653
29. Drug Delivery in Oncology From Basic Research to Cancer Therapy (2012) Wiley-VCH
30. Dumitru CD, Neacsu IA, Grumezescu AM, Andronescu E (2022) Bee-derived products: chemical composition and applications in skin tissue engineering. Pharmaceutics 14(4)
31. El Bialy I, Jiskoot W, Nejadnik MR (2017) Formulation, delivery and stability of bone morphogenetic proteins for effective bone regeneration. Pharm Res 34(6):1152–1170
32. Ewald A, Hosel D, Patel S, Grover LM, Barralet JE, Gbureck U (2011) Silver-doped calcium phosphate cements with antimicrobial activity. Acta Biomater 7(11):4064–4070
33. Fan C, Xu Q, Hao RQ, Wang C, Que YM, Chen YX, Yang C, Chang J (2022) Multifunctional wound dressings based on silicate bioactive materials. Biomaterials 287. 1016/j.biomaterials.2022.121652

34. Ficai A, Albu MG, Birsan M, Sonmez M, Ficai D, Trandafir V, Andronescu E (2013) Collagen hydrolysate based collagen/hydroxyapatite composite materials. J Mol Struct 1037:154–159
35. Ficai A, Andronescu E, Ficai D, Sonmez M, Albu MG, Voicu G (2012) Mimicking the morphology of long bone. Cent Eur J Chem 10(6):1949–1953
36. Ficai A, Andronescu E, Trandafir V, Ghitulica C, Voicu G (2010) Collagen/hydroxyapatite composite obtained by electric field orientation. Mater Lett 64(4):541–544
37. Ficai A, Andronescu E, Voicu G, Ghitulica C, Vasile BS, Ficai D, Trandafir V (2010) Self-assembled collagen/hydroxyapatite composite materials. Chem Eng J 160(2):794–800
38. Ficai A, Andronescu E, Voicu G, Manzu D, Ficai M (2009) Layer by layer deposition of hydroxyapatite onto the collagen matrix. Mater Sci Eng C-Mater Biol Appl 29(7):2217–2220
39. Ficai D, Sonmez M, Albu MG, Mihaiescu DE, Ficai A, Bleotu C (2015) Antitumoral materials with regenerative function obtained using a layer-by-layer technique. Drug Des Dev Ther 9:1269–1279
40. Ficai M, Andronescu E, Ficai D, Voicu G, Ficai A (2010) Synthesis and characterization of COLL-PVA/HA hybrid materials with stratified morphology. Colloids Surf B-Biointerfaces 81(2):614–619
41. Foss FM (2000) DAB(389)IL-2 (denileukin diftitox, ONTAK): a new fusion protein technology. Clin Lymphoma 1:S27–S31
42. Gherasim O, Grumezescu AM, Grumezescu V, Negut I, Dumitrescu MF, Stan MS, Nica IC, Holban AM, Socol G, Andronescu E (2021) Bioactive coatings based on hydroxyapatite, kanamycin, and growth factor for biofilm modulation [Article]. Antibiotics 10(2):1–19, Article 160
43. Ghica MV, Ficai A, Marin S, Marin M, Ene AM, Patrascu JM (2015) Collagen/bioactive glass ceramic/doxycycline composites for bone defects. Revista Romana De Materiale-Romanian J Mater 45(4):307–314
44. Golchin A, Hosseinzadeh S, Staji M, Soleimani M, Ardeshirylajimi A, Khojasteh A (2019) Biological behavior of the curcumin incorporated chitosan/poly(vinyl alcohol) nanofibers for biomedical applications. J Cell Biochem 120(9):15410–15421
45. Gomez-Cerezo N, Casarrubios L, Saiz-Pardo M, Ortega L, de Pablo D, Diaz-Guemes I, Fernandez-Tome B, Enciso S, Sanchez-Margallo FM, Portoles MT, Arcos D, Vallet-Regi M (2019) Mesoporous bioactive glass/epsilon-polycaprolactone scaffolds promote bone regeneration in osteoporotic sheep. Acta Biomater 90:393–402
46. Green MR, Manikhas GM, Orlov S, Afanasyev B, Makhson AM, Bhar P, Hawkins MJ (2006) Abraxane, a novel Cremophor-free, albumin-bound particle form of paclitaxel for the treatment of advanced non-small-cell lung cancer. Ann Oncol 17(8):1263–1268
47. Guo JL, Zhang Q, Li J, Llu YS, Hou ZY, Chen W, Jin L, Tian Y, Ju LL, Liu B, Dong TH, Zhang F, Zhang YZ (2017) Local application of an ibandronate/collagen sponge improves femoral fracture healing in ovariectomized rats. Plos One 12(11)
48. Guo R, Lan Y, Xue W, Cheng BA, Zhang YM, Wang CY, Ramakrishna S (2017) Collagen-cellulose nanocrystal scaffolds containing curcumin-loaded microspheres on infected full-thickness burns repair. J Tissue Eng Regen Med 11(12):3544–3555
49. Guo YY, Liu YW, Shi C, Wu TT, Cui YZ, Wang SY, Liu P, Feng XB, He Y, Fu DH (2021) Remote-controllable bone-targeted delivery of estradiol for the treatment of ovariectomy-induced osteoporosis in rats. J Nanobiotechnology 19(1)
50. Halazun JF, Wagner HR, Gaeta JF, Sinks LF (1974) Daunorubicin cardiac toxicity in children with acute lymphocytic leukemia. Cancer 33(2):545–554
51. Harmankaya N, Karlsson J, Palmquist A, Halvarsson M, Igawa K, Andersson M, Tengvall P (2013) Raloxifene and alendronate containing thin mesoporous titanium oxide films improve implant fixation to bone. Acta Biomater 9(6):7064–7073
52. He WC, Wu Z, Wu YM, Cai YY, Cui Z, Yu B, Hong YL (2021) Construction of antimicrobial material-loaded porous tricalcium phosphate beads for treatment of bone infections. ACS Appl Bio Mater 4(8):6280–6293
53. Herce HD, Garcia AE, Litt J, Kane RS, Martin P, Enrique N, Rebolledo A, Milesi V (2009) Arginine-rich peptides destabilize the plasma membrane, consistent with a pore formation translocation mechanism of cell-penetrating peptides. Biophys J 97(7):1917–1925

54. Homaeigohar S, Li M, Boccaccini AR (2022) Bioactive glass-based fibrous wound dressings. Burns Trauma 10
55. Hong MH, Lee JH, Jung HS, Shin H, Shin H (2022) Biomineralization of bone tissue: calcium phosphate-based inorganics in collagen fibrillar organic matrices. Biomater Res 26(1)
56. Huang Y, Chen JR, Lin J, Lin JH, Chen XW (2021) Preparation of vanillic acid-loaded core-shell gold nanospheres/mesoporous silica nanoparticles for the treatment of orthopedic infection. ACS Omega 6(4):2899–2905
57. Ilie A, Andronescu E, Ficai D, Voicu G, Ficai M, Maganu M, Ficai A (2011) New approaches in layer by layer synthesis of collagen/hydroxyapatite composite materials. Cent Eur J Chem 9(2):283–289
58. Iliescu RI, Andronescu E, Ghitulica CD, Voicu G, Ficai A, Hoteteu M (2014) Montmorillonite-alginate nanocomposite as a drug delivery system incorporation and in vitro release of irinotecan. Int J Pharm 463(2):184–192
59. Ionescu O, Ciocilteu MV, Manda CV, Neacsu IA, Ficai A, Amzoiu E, Stiolica AT, Croitoru O, Neamtu J (2019) Bone—graft delivery systems of type PLGA-gentamicin and collagen—hydroxyapatite—gentamicine. Materiale Plastice 56(3):534–537
60. Jain K (2010) Advances in the field of nanooncology. BMC Med 83(8):1–11
61. Jebahi S, Saoudi M, Farhat L, Oudadesse H, Rebai T, Kabir A, El Feki A, Keskes H (2015) Effect of novel curcumin-encapsulated chitosan-bioglass drug on bone and skin repair after gamma radiation: experimental study on a Wistar rat model. Cell Biochem Funct 33(3):150–159
62. Jiang T, Li QY, Qiu JM, Chen J, Du S, Xu X, Wu ZH, Yang XF, Chen ZB, Chen TK (2022) Nanobiotechnology: applications in chronic wound healing. Int J Nanomed 17:3125–3145
63. Jimenez-Holguin J, Sanchez-Salcedo S, Cicuendez M, Vallet-Regi M, Salinas AJ (2022) Cu-doped hollow bioactive glass nanoparticles for bone infection treatment. Pharmaceutics 14(4)
64. Kanade R, Boche M, Pokharkar V (2018) Self-assembling raloxifene loaded mixed micelles: formulation optimization, in vitro cytotoxicity and in vivo pharmacokinetics. AAPS PharmSciTech 19(3):1105–1115
65. Kar S, Kundu B, Reis RL, Sarkar R, Nandy P, Basu R, Das S (2019) Curcumin ameliorates the targeted delivery of methotrexate intercalated montmorillonite clay to cancer cells. Eur J Pharm Sci 135:91–102
66. Kargozar S, Montazerian M, Harnzehlou S, Kim HW, Baino F (2018) Mesoporous bioactive glasses: promising platforms for antibacterial strategies. Acta Biomater 81:1–19
67. Kotak DJ, Devarajan PV (2020) Bone targeted delivery of salmon calcitonin hydroxyapatite nanoparticles for sublingual osteoporosis therapy (SLOT). Nanomed-Nanotechnol Biol Med 24
68. Kulanthaivel S, Roy B, Agarwal T, Giri S, Pramanik K, Pal K, Ray SS, Maiti TK, Banerjee I (2016) Cobalt doped proangiogenic hydroxyapatite for bone tissue engineering application. Mater Sci Eng C Mater Biol Appl 58:648–658
69. Kumar GS, Govindan R, Girija EK (2014) In situ synthesis, characterization and in vitro studies of ciprofloxacin loaded hydroxyapatite nanoparticles for the treatment of osteomyelitis. J Mater Chem B 2(31):5052–5060
70. Lee EJ, Huh BK, Kim SN, Lee JY, Park CG, Mikos AG, Bin Choy Y (2017) Application of materials as medical devices with localized drug delivery capabilities for enhanced wound repair. Prog Mater Sci 89:392–410
71. Li HB, Jiang F, Ye S, Wu YY, Zhu KP, Wang DP (2016) Bioactive apatite incorporated alginate microspheres with sustained drug-delivery for bone regeneration application. Mater Sci Eng C-Mater Biol Appl 62:779–786
72. Liu YP, Yu P, Peng X, Huang Q, Ding MM, Chen YT, Jin RT, Xie J, Zhao CS, Li JS (2019) Hexapeptide-conjugated calcitonin for targeted therapy of osteoporosis. J Control Release 304:39–50
73. Low SK, Nakamura Y (2019) The road map of cancer precision medicine with the innovation of advanced cancer detection technology and personalized immunotherapy. Jpn J Clin Oncol 49(7):596–603

74. Marques C, Ferreira JMF, Andronescu E, Ficai D, Sonmez M, Ficai A (2014) Multifunctional materials for bone cancer treatment. Int J Nanomed 9:2713–2725
75. Marques CF, Lemos A, Vieira SI, Silva OABDE, Bettencourt A, Ferreira JMF (2016) Antibiotic-loaded Sr-doped porous calcium phosphate granules as multifunctional bone grafts. Ceram Int 42(2):2706–2716
76. Martins VCA, Goissis G (2000) Nonstoichiometric hydroxyapatite-anionic collagen composite as support for the double sustained release of gentamicin and norfloxacin/ciprofloxacin. Artif Organs 24(3):224–230
77. Mehtani D, Seth A, Sharma P, Maheshwari N, Kapoor D, Shrivastava SK, Tekade RK (2019) Biomaterials for sustained and controlled delivery of small drug molecules. In: Tekade RK (ed) Biomaterials and bionanotechnology. Academic Press, pp 90–153
78. Mirzaei B, Etemadian S, Goli HR, Bahonar S, Gholami SA, Karami P, Farhadi M, Tavakoli R (2018) Construction and analysis of alginate-based honey hydrogel as an ointment to heal of rat burn wound related infections. Int J Burns Trauma 8(4):88–97
79. Mostafa AA, El-Sayed MMH, Mahmoud AA, Gamal-Eldeen AM (2017) Bioactive/natural polymeric scaffolds loaded with ciprofloxacin for treatment of osteomyelitis. AAPS PharmSciTech 18(4):1056–1069
80. Mulazzi M, Campodoni E, Bassi G, Montesi M, Panseri S, Bonvicini F, Gentilomi GA, Tampieri A, Sandri M (2021) Medicated hydroxyapatite/collagen hybrid scaffolds for bone regeneration and local antimicrobial therapy to prevent bone infections. Pharmaceutics 13(7)
81. Munir MU, Salman S, Javed I, Bukhari SNA, Ahmad N, Shad NA, Aziz F (2021) Nano-hydroxyapatite as a delivery system: overview and advancements. Artificial Cells Nanomed Biotechnol 49(1):717–727
82. Nadar RA, Asokan N, Degli Esposti L, Curci A, Barbanente A, Schlatt L, Karst U, Iafisco M, Margiotta N, Brand M, van den Beucken JJJP, Bornhauser M, Leeuwenburg SCG (2020) Preclinical evaluation of platinum-loaded hydroxyapatite nanoparticles in an embryonic zebrafish xenograft model. Nanoscale 12(25):13582–13594
83. Nandi A, Ghosh C, Basu S (2019) Polymer conjugated graphene-oxide nanoparticles impair nuclear DNA and topoisomerase I in cancer. Nanoscale Adv 1(12):4965–4971
84. Neacsu IA, Arsenie LV, Trusca R, Ardelean IL, Mihailescu N, Mihailescu IN, Ristoscu C, Bleotu C, Ficai A, Andronescu E (2019) Biomimetic collagen/Zn2+-substituted calcium phosphate composite coatings on titanium substrates as prospective bioactive layer for implants: a comparative study spin coating vs. MAPLE. Nanomaterials 9(5)
85. Nedelcu IA, Ficai A, Sonmez M, Ficai D, Oprea O, Andronescu E (2014) Silver based materials for biomedical applications. Curr Org Chem 18(2):173–184
86. Neto AS, Pereira P, Fonseca AC, Dias C, Almeida MC, Barros I, Miranda CO, de Almeida LP, Morais PV, Coelho JFJ, Ferreira JMF (2021) Highly porous composite scaffolds endowed with antibacterial activity for multifunctional grafts in bone repair. Polymers 13(24)
87. Ni T, Zhu YM, Hao L, Chen Y, Cheng T (2022) Preparation of photothermal-sensitive PDGF@ZIF-8-PDA@COL/PLGA-TCP composite scaffolds for bone defect repair. Mater Des 217
88. Ning ZY, Tan BY, Chen B, Lau DSA, Wong TM, Sun TH, Peng SL, Li ZY, Lu WW (2019) Precisely controlled delivery of abaloparatide through injectable hydrogel to promote bone regeneration. Macromol Biosci 19(6)
89. O'Shaughnessy J (2005) Extending survival with chemotherapy in metastatic breast cancer. Oncologist 10:20–29
90. Ogay V, Mun EA, Kudaibergen G, Baidarbekov M, Kassymbek K, Zharkinbekov Z, Saparov A (2020) Progress and prospects of polymer-based drug delivery systems for bone tissue regeneration. Polymers 12(12)
91. Oprea O, Andronescu E, Ficai D, Ficai A, Oktar FN, Yetmez M (2014) ZnO applications and challenges. Curr Org Chem 18(2):192–203
92. Ozbek M, Domack U, Barnikol WKR (1999) A model for evaluation of artificial oxygen carriers regarding circulation, respiration and metabolism in anesthetized spontaneously breathing guinea pigs. Oxygen Transport Tissue Xxi 471:17–26

93. Pan YX, Wang JH, Jiang ZC, Guo Q, Zhang Z, Li JY, Hu YH, Wang L (2022) Zoledronate combined metal-organic frameworks for bone-targeting and drugs deliveries. Sci Rep 12(1)
94. Patel PR, Singam A, Iyer AK, Gundloori RVN (2022) Bioinspired hyaluronic acid based nanofibers immobilized with 3, 4-difluorobenzylidene curcumin for treating bacterial infections. J Drug Delivery Sci Technol 74
95. Pishbin F, Mourino V, Flor S, Kreppel S, Salih V, Ryan MP, Boccaccini AR (2014) Electrophoretic deposition of gentamicin-loaded bioactive glass/chitosan composite coatings for orthopaedic implants. ACS Appl Mater Interfaces 6(11):8796–8806
96. Poorirani S, Taheri SL, Mostafavi SA (2022) Scaffolds: a biomaterial engineering in targeted drug delivery for osteoporosis. Osteoporosis Int
97. Pupo YM, Leite LMB, Senegaglia AC, Antunes L, Nadal JM, de Lara EL, Saito RE, Antunes SRM, Lacerda WF, Farago PV (2021) Effect of hydroxyapatite microspheres, amoxicillin-hydroxyapatite and collagen-hydroxyapatite composites on human dental pulp-derived mesenchymal stem cells. Materials 14(24)
98. Ramburrun P, Kumar P, Ndobe E, Choonara YE (2021) Gellan-Xanthan hydrogel conduits with intraluminal electrospun nanofibers as physical, chemical and therapeutic cues for peripheral nerve repair. Int J Mol Sci 22(21)
99. Rana D, Salave S, Jain S, Shah R, Benival D (2022) Systematic development and optimization of teriparatide-loaded nanoliposomes employing quality by design approach for osteoporosis. J Pharm Innovation
100. Ribeiro DML, Carvalho AR, de Macedo GHRV, Chagas VL, Silva LD, Cutrim BD, Santos DM, Soares BLL, Zagmignan A, de Miranda RDM, de Albuquerque PBS, da Silva LCN (2020) Polysaccharide-based formulations for healing of skin-related wound infections: lessons from animal models and clinical trials. Biomolecules 10(1)
101. Rodriguez I, Fernandez-Vega L, Maser-Figueroa AN, Sang BL, Gonzalez-Pagan P, Tinoco AD (2022) Exploring titanium(IV) complexes as potential antimicrobial compounds. Antibiotics-Basel 11(2)
102. Rusu LC, Nedelcu IA, Albu MG, Sonmez M, voicu g, radulescu m, ficai d, ficai a, negrutiu ml, sinescu c (2015) tetracycline loaded collagen/hydroxyapatite composite materials for biomedical applications. J Nanomater 2015
103. Saha R, Tayalia P (2022) Clove oil-incorporated antibacterial gelatin-chitosan cryogels for tissue engineering: an in vitro study. ACS Biomater Sci Eng 8(8):3557–3567
104. Saini D, Fazil M, Ali MM, Baboota S, Ameeduzzafar AJ (2015) Formulation, development and optimization of raloxifene-loaded chitosan nanoparticles for treatment of osteoporosis. Drug Delivery 22(6):823–836
105. Salamanna F, Gambardella A, Contartese D, Visani A, Fini M (2021) Nano-based biomaterials as drug delivery systems against osteoporosis: a systematic review of preclinical and clinical evidence. Nanomaterials 11(2)
106. Salatin S, Bazmani A, Shahi S, Naghili B, Memar MY, Dizaj SM (2022) Antimicrobial benefits of flavonoids and their nanoformulations. Curr Pharm Des 28(17):1419–1432
107. Sandomierski M, Jakubowski M, Ratajczak M, Voelkel A (2022) Drug distribution evaluation using FT-IR imaging on the surface of a titanium alloy coated with zinc titanate with potential application in the release of drugs for osteoporosis. Spectrochimica Acta Part a-Mol Biomol Spectroscopy 281
108. Satar NA, Ismail MN, Yahaya BH (2021) Synergistic roles of curcumin in sensitising the cisplatin effect on a cancer stem cell-like population derived from non-small cell lung cancer cell lines. Molecules 26(4)
109. Shaabani E, Sharifiaghdam M, Lammens J, De Keersmaecker H, Vervaet C, De Beer T, Motevaseli E, Ghahremani MH, Mansouri P, Smedt SD, Raemdonck K, Faridi-Majidi R, Braeckmans K, Fraire JC (2021) Increasing angiogenesis factors in hypoxic diabetic wound conditions by siRNA delivery: additive effect of LbL-gold nanocarriers and desloratadine-induced lysosomal escape. Int J Mol Sci 22(17)
110. Sharaf SS, El-Shafei AM, Refaie R, Gibriel AA, Abdel-Sattar R (2022) Antibacterial and wound healing properties of cellulose acetate electrospun nanofibers loaded with bioactive glass nanoparticles; in-vivo study. Cellulose 29(8):4565–4577

111. Sheikh L, Sinha S, Singhababu YN, Verma V, Tripathy S, Nayar S (2018) Traversing the profile of biomimetically nanoengineered iron substituted hydroxyapatite: synthesis, characterization, property evaluation, and drug release modeling. RSC Adv 8(35):19389–19401
112. Sim J, Kang G, Yang H, Jang M, Kim Y, Ahn H, Kim M, Jung H (2022) Development of clinical weekly-dose teriparatide acetate encapsulated dissolving microneedle patch for efficient treatment of osteoporosis. Polymers 14(19)
113. Song W, Musetti SN, Huang L (2017) Nanomaterials for cancer immunotherapy. Biomaterials 148:16–30
114. Spoiala A, Ilie CI, Ficai D, Ficai A, Andronescu E (2021) Chitosan-based nanocomposite polymeric membranes for water purification-a review. Materials 14(9)
115. Sun HN, Liu SS, Zeng XF, Meng XG, Zhao LN, Wan YZ, Zuo GF (2017) Morphology effect of nano-hydroxyapatite as a drug carrier of methotrexate. J Mater Sci-Mater Med 28(10)
116. Sun Q, Duan MT, Fan W, Fan B (2021) Ca-Si mesoporous nanoparticles with the optimal Ag-Zn ratio inhibit the Enterococcus faecalis infection of teeth through dentinal tubule infiltration: an in vitro and in vivo study. J Mater Chem B 9(9)
117. Sung H, Ferlay J, Siegel RL, Laversanne M, Soerjomataram I, Jemal A, Bray F (2021) Global cancer statistics 2020: GLOBOCAN estimates of incidence and mortality worldwide for 36 cancers in 185 countries. CA Cancer J Clin 71(3):209–249
118. Svarca A, Grava A, Dubnika A, Ramata-Stunda A, Narnickis R, Aunina K, Rieksta E, Boroduskis M, Jurgelane I, Locs J, Loca D (2022) Calcium phosphate/hyaluronic acid composite hydrogels for local antiosteoporotic drug delivery. Frontiers Bioeng Biotechnol 10
119. Szewczyk A, Skwira A, Konopacka A, Sadej R, Prokopowicz M (2021) Mesoporous silica-bioglass composite pellets as bone drug delivery system with mineralization potential. Int J Mol Sci 22(9)
120. Szurkowska K, Kazimierczak P, Kolmas J (2021) Mg,Si-Co-substituted hydroxyapatite/alginate composite beads loaded with raloxifene for potential use in bone tissue regeneration. Int J Mol Sci 22(6)
121. Taye MB (2022) Biomedical applications of ion-doped bioactive glass: a review. Appl Nanosci 12(12):3797–3812
122. Teodor ED, Gatea F, Ficai A, Radu GL (2018) Functionalized magnetic nanostructures for anticancer therapy. Curr Drug Targets 19(3):239–247
123. Tian R, Qin XY, Yuan PY, Lei K, Wang L, Bai YK, Liu SY, Chen X (2018) Fabrication of self-healing hydrogels with on-demand antimicrobial activity and sustained biomolecule release for infected skin regeneration. ACS Appl Mater Interfaces 10(20):17018–17027
124. Tiplea RE, Lemnaru GM, Trusca RD, Holban A, Kaya MGA, Dragu LD, Ficai D, Ficai A, Bleotu C (2021) Antimicrobial films based on chitosan, collagen, and ZnO for skin tissue regeneration. Biointerface Res Appl Chem 11(4):11985–11995
125. Tite T, Popa AC, Balescu LM, Bogdan IM, Pasuk I, Ferreira JMF, Stan GE (2018) Cationic substitutions in hydroxyapatite: current status of the derived biofunctional effects and their in vitro interrogation methods. Materials 11(11)
126. Torky HA, Khaliel SA, Sedeek EK, Tawfik RG, Bkheet AAE, Ebied SK, Amin HS, Zahran SI, Emara HA, Nofal AM, Elghazaly EM (2022) Silver nanoparticle effect on Salmonella enterica isolated from northern west egypt food, poultry, and calves. Appl Microbiol Biotechnol 106(17):5701–5713
127. Tsimberidou AM (2015) Targeted therapy in cancer. Cancer Chemother Pharmacol 76(6):1113–1132
128. Venkatasubbu GD, Ramasamy S, Ramakrishnan V, Kumar J (2011) Nanocrystalline hydroxyapatite and zinc-doped hydroxyapatite as carrier material for controlled delivery of ciprofloxacin. 3 Biotech 1(3):173–186
129. Vose JM (2004) Bexxar®: novel radioimmunotherapy for the treatment of low-grade and transformed low-grade non-Hodgkin's lymphoma. Oncologist 9(2):160–172
130. Vu AA, Kushram P, Bose S (2022) Effects of vitamin A (Retinol) release from calcium phosphate matrices and porous 3D printed scaffolds on bone cell proliferation and maturation. ACS Appl Bio Mater 5(3):1120–1129

131. Wang SL, Pang Y, Hu SY, Lv J, Lin YL, Li M (2023) Copper sulfide engineered covalent organic frameworks for pH-responsive chemo/photothermal/chemodynamic synergistic therapy against cancer. Chem Eng J 451. https://doi.org/10.1016/j.cej.2022.138864
132. Wang XY, Li YF, Ren WJ, Hou RX, Liu HF, Li R, Du SJ, Wang L, Liu JY (2021) PEI-modified diatomite/chitosan composites as bone tissue engineering scaffold for sustained release of BMP-2. J Biomater Sci-Polym Edn 32(10):1337–1355
133. Wang Z, Hou S, Yao SY, Shang YL, Deng S, Peng Y, Zhou WW, Lv XF, Ren B, Peng C, Yang J, Huang ZH (2022) Osteogenesis of aspirin microsphere-loaded tilapia collagen/hydroxyapatite biomimetic scaffolds. J Mater Sci 57(25):11882–11898
134. Wang ZH, Li Y (2018) Raloxifene/SBE-beta-CD inclusion complexes formulated into nanoparticles with chitosan to overcome the absorption barrier for bioavailability enhancement. Pharmaceutics 10(3)
135. Wu C, Liu J, Zhai ZR, Yang LQ, Tang X, Zhao LZ, Xu KM, Zhong WY (2020) Double-crosslinked nanocomposite hydrogels for temporal control of drug dosing in combination therapy. Acta Biomater 106:278–288
136. Wu K, Chen YC, Lin SM, Chang CH (2021) In vitro and in vivo effectiveness of a novel injectable calcitonin-loaded collagen/ceramic bone substitute. J Biomater Appl 35(10):1355–1365
137. Xia YD, Wang DX, Liu D, Su JY, Jin Y, Wang D, Han BB, Jiang ZP, Liu B (2022) Applications of chitosan and its derivatives in skin and soft tissue diseases. Frontiers Bioeng Biotechnol 10
138. Xiao TH, Fan L, Liu RT, Huang XW, Wang SH, Xiao LG, Pang YY, Li D, Liu J, Min YG (2021) Fabrication of dexamethasone-loaded dual-metal-organic frameworks on polyetheretherketone implants with bacteriostasis and angiogenesis properties for promoting bone regeneration. ACS Appl Mater Interfaces 13(43):50836–50850
139. Xu J, Zheng SS, Hu XY, Li LY, Li WF, Parungao R, Wang YW, Nie Y, Liu TQ, Song KD (2020) Advances in the research of bioinks based on natural collagen, polysaccharide and their derivatives for skin 3D bioprinting. Polymers 12(6)
140. Youness RA, El-deen DMT, Taha MA (2022) A review on calcium silicate ceramics: properties, limitations, and solutions for their use in biomedical applications. Silicon
141. Yu P, Chen Y, Wang Y, Liu Y, Zhang P, Guo Q, Li S, Xiao H, Xie J, Tan H, Li J (2019) Pentapeptide-decorated silica nanoparticles loading salmon calcitonin for in vivo osteoporosis treatment with sustained hypocalcemic effect. Mater Today Chem 14
142. Yu YS, Hsu CH, Cheng PH, Wu KCW, Liu CH (2022) Poly(acrylic acid)-grafted metal-organic framework carrying Mg ions for bone repair. Mater Chem Phys 292
143. Zeng YY, Shen YD, Wu SY, Cai L, Wang Z, Cai KX, Shen JT, Yie KHR, Zhang HL, Xu LH, Liu JS (2022) Bone-targeting PLGA derived lipid drug delivery system ameliorates bone loss in osteoporotic ovariectomized rats. Mater Des 221
144. Zhang W, Du XF, Ben L, Li CR, Long J, Zhao MX, Yao ZY, Liang XJ, Lai YX (2022) Engineering supramolecular nanomedicine for targeted near infrared-triggered mitochondrial dysfunction to potentiate cisplatin for efficient chemophototherapy. ACS Nano 16(1):1421–1435
145. Zhang XX, Chen HE, Zhang Y, Huang QJ, Feng JJ, Xing HY, Fu XG, Yan XF, Zhang YY, Xu Q, Liang JM (2022) HA-DOPE-modified honokiol-loaded liposomes targeted therapy for osteosarcoma. Int J Nanomed 17:5137–5151
146. Zhang XX, Jiang XR, Jiang S, Cai XY, Yu SJ, Pei GX (2021) Schwann cells promote prevascularization and osteogenesis of tissue-engineered bone via bone marrow mesenchymal stem cell-derived endothelial cells. Stem Cell Res Therapy 12(1)
147. Zhao XY, Han Y, Zhu TT, Feng NB, Sun YF, Song ZM, Li SQ, Liu JG, Ding JX (2019) Electrospun polylactide-nano-hydroxyapatite-vancomycin composite scaffolds for advanced osteomyelitis therapy. J Biomed Nanotechnol 15(6):1213–1222
148. Zhong CL, Qin BY, Xie XY, Bai Y (2013) antioxidant and antimicrobial activity of tellurium dioxide nanoparticles sols. J Nano Res 25:8–15
149. Zhong J, Wen WY, Wang JJ, Zhang MY, Jia YJ, Ma XW, Su YX, Wang YJ, Lan XM (2022) Bone-targeted dual functional lipid-coated drug delivery system for osteosarcoma therapy. Pharm Res

150. Zhou GD, Lin JX (2020) Graphene-based nanomaterials combined with early exercise rehabilitation training in the treatment of patients with infectious bone defects. Int J Nanotechnol 17(2–6):291–307
151. Zhu J, Hou R, Liu M, Wang L, Chen W, Sun Y, Wei W, Ye S (2022) A novel wound dressing based on epigallocatechin-3-gallate self-assemble hydrogels promotes effects on wound healing. Mater Today Sustain 18
152. Zhu QY, Jiang M, Liu Q, Yan SN, Feng LB, Lan Y, Shan GQ, Xue W, Guo R (2018) Enhanced healing activity of burn wound infection by a dextran-HA hydrogel enriched with sanguinarine. Biomater Sci 6(9):2472–2486
153. Zogakis PN, Teles AR, Zafeiris EP, Zafeiris CP (2022) Bisphosphonate-loaded bone cement: background, clinical indications and future perspectives. J Musculoskelet Neuronal Interact 22(4):587–595

Chapter 8
Cell-Materials Interaction

Jennifer O. Buitrago, Begoña M. Bosch, and Román A. Pérez

Abstract Cells are able to respond to different physicochemical stimuli due to a complex molecular system that governs that response. The cell response to these stimuli can be translated into essential processes like cell survival, cell proliferation, cell differentiation, or protein synthesis. Cells need to respond to signals from their interactions, whether those interactions are physical with neighbouring cells, with the extracellular matrix (ECM) or chemical with the environment. These cell responses are of important relevance in the field of biomaterials for tissue engineering applications. Biomaterials, either from a natural or synthetic origin, have been designed for the replacement of injured tissues, to restore the tissue functionality, or for the release of therapeutical cues or cells. Biocompatibility is crucial for a correct implantation and functionality restoration, as biomaterials, in permanent contact with tissues or body fluids, must elicit an appropriate host response. Several factors, such as biophysicochemical properties, time of implantation, and material degradation, to name a few, will determine whether biomaterials succeed or fail to overcome host response. In this chapter, we will go over the fundamental concepts of cell interaction with ECM and other cells, as well as how these cells interact similarly when they come into contact with the surface of biomaterials. Furthermore, we will define the various cellular functions that are dependent on these interactions, as well as the host response to biomaterial implantation.

Keywords Cell adhesion · Cell–cell interaction · Cell-material interaction · Integrins · Cadherins · Extracellular matrix · ECM · Cell migration · Cell spreading · Cell signaling · Host response

J. O. Buitrago · B. M. Bosch (✉) · R. A. Pérez
Bioengineering Institute of Technology (BIT), Universitat Internacional de Catalunya (UIC), c/ Josep Trueta S/N, Sant Cugat del Vallès, 08195 Barcelona, Spain
e-mail: bbosch@uic.es

Deparment of Basic Sciences, Faculty of Medicine and Health Sciences, Universitat Internacional de Catalunya (UIC), c/ Josep Trueta S/N, Sant Cugat del Vallès, 08195 Barcelona, Spain

O. Gunduz et al. (eds.), *Biomaterials and Tissue Engineering*, Stem Cell Biology and Regenerative Medicine 74, https://doi.org/10.1007/978-3-031-35832-6_8

Introduction to Cell Interaction

Cells use signals from their extracellular and intracellular environments to control basic processes like growth, proliferation, and survival. These cellular decisions are heavily influenced by nutrient availability and cellular metabolism. Understanding the mechanism of action underlying these processes could aid in the restoration of cellular homeostasis in a variety of pathologies. These interactions take place directly between the cell surface and the extracellular matrix (ECM) or between cells. Cells can communicate with one another in response to changes in their microenvironment through these interactions. Cell survival depends on the ability to send and receive signals.

In multicellular organisms, the cells that form the tissues are in direct contact with each other by intercellular connections visible only through electron microscope images. Cells must also be anchored to the ECM, which consists of a 3D network of proteins, glycosaminoglycans, proteoglycans and glycoproteins in which they are embedded. However, not always cell communications are done by direct contact through cellular unions, signals can be also sent through soluble molecules to the target cells that can be far from the sending cell. Stimulus of cells are of various kinds and not only hydrophilic or liposoluble molecules can serve as signaling ligands, but physical stimuli can also activate signaling pathways and cells can respond to them. Understanding how these ligands or stimuli regulate cell function and state is crucial for the overall tissue homeostasis. Disruption of this homeostasis as a consequence of different events, such as uncontrolled cell cycle, changes in cell signaling, inflammatory processes or tissue trauma, can result in tissular dysfunction. Although in many occasions, the body can regulate cellular processes to restore tissue functions, in some others, biomaterials and medical devices are used to fulfil the needs of such processes.

In the last few decades, materials scientists have focused their research in the finding of new biomaterials for Tissue Engineering and Regenerative Medicine applications. During the design of biomaterials not only is important to study the raw material they are composed of (metals, ceramics or polymers) or the featuring physical properties, but also any design should include a study about the interaction between cells and such biomaterials. On most occasions, biomaterials are in physical contact with cells or can deliver soluble molecules, like drugs, proteins, or nucleic acids. Thus, understanding and controlling cell to material interaction is, if not, one of the most crucial events for the successful material integration into tissue and function restoration. Researchers evaluate in vitro this cell interaction under a controlled environment. However, in vitro assays cannot reproduce all the physiological events that take place in the body and in vivo assays are usually needed to comprehend the whole process of tissue regeneration. Inflammation after implantation of a material is a common event that has to be controlled. This control will depend on the physical and biochemical properties of the materials and the immune system should respond with minimal inflammation as a response to foreign body reaction, and so proceed with the tissue regeneration process. On the other hand, if the reaction to the material could not be controlled, macrophages would try to eliminate the material or at least, minimize hazardous potential effects on the surrounding healthy tissue by encapsulating the materials in a fibrotic capsule.

Learning Outputs

- To know the cellular unions involved in cell interactions with other cells or ECM
- To know the pathways in which tissue cells are able to communicate with other cells and/or with the outside
- To know cellular mechanisms that are involved in the response of external stimuli (cell–cell or cell-ECM / biomaterial)
- To know the host response processes to implanted biomaterials

Types of Receptor-Ligand Interaction (Cell-ECM/Cell–cell)

Tissues are made up of a network of cells with similar structures and functions that connect to one another and to the ECM. The main body tissues are epithelial, muscular, connective, and nervous tissues, each with a distinct function due to cellular organisation. Blood tissue, on the other hand, is made up of free cells in suspension and a liquid ECM. Cellular organization occurs as a result of the various contacts that cells make with other cells and/or the extracellular matrix. To fully understand these types of unions, epithelial cells are useful in studying the various contacts. The main functions of epithelial tissues are to serve as selective barriers to external agents such as viruses, bacteria, or toxic substances; to have biochemical functions such as hormone release, milk, or tears; to absorb nutrients and water; or to receive physical signals such as light or sound. This would not be possible unless epithelium was organised strictly and hierarchically. Simple epithelium (e.g., the gut) has only one unique type of cell, whereas stratified epithelium (e.g., the skin) has multiple layers of cells. Cells in the epithelium do not all have the same shape and function, as columnar, cuboidal, or squamous shapes can be found. However, epithelial cells are typically polarized, which means they have an apical part that is exposed to the outside, such as air or other fluids, and a basal part that is in direct contact with the basal lamina of the ECM. Not only are they located differently, but their functions will differ as well because polarization defines the internal organization of their organelles. This polarization is the result of various cell unions, which consist of proteinic structures that physically bind two surfaces (cell-to-cell or cell-to-ECM). These junctions aid in cell communication and structural support, as well as acting as barriers [4, 11]

Types of Cell Unions, Transmembrane Proteins and Ligands

The composition and function of cell-to-cell and cell-to-ECM unions differ (Fig. 8.1). They are typically composed of a transmembrane protein that connects internal cytoskeleton components via the inner domain, while the external domain anchors to other similar proteins from neighbouring cells. **Tight junctions** (also known as zonula occludens) are multiprotein complexes formed at the apical part of cells by the transmembrane proteins claudins and occcludins. Tight junctions, in particular, seal the intercellular space within cells, preventing free flux of molecules from the apical to the basal part of the cell and vice versa, requiring selective molecule transport mediated by carrier proteins at the apical or basal part. **Cadherins**, which play a role in cell-to-cell junctions, are calcium dependent transmembrane proteins that bind homophylically to another cadherin from a neighbouring cell via the external domain and to cytoskeleton proteins via the inner domain. Because the affinity of these cadherins for cytoskeleton proteins varies, classical cadherins bind to actin filaments in **adherens junctions** and nonclassical cadherins, such as desmoglein and desmocollin, bind to intermediate filaments in **desmosome unions**. Cell-to-ECM junctions, on the other hand, are mediated by **integrins**, another type of transmembrane protein that binds ECM components via the external domain and actin filaments or intermediate filaments via the inner domain in **cell–matrix anchoring junctions** (also known as focal adhesions) or **hemidesmosomes**, respectively. Hemidesmosomes, however, are found in keratinocytes of the skin epidermis, while focal adhesions are involved in cell traction, cell adhesion or ECM reorganization. While the function of adherens junctions and desmosomes is to bind neighbouring cells, desmosomes are a very resistant union due to anchoring with intermediate filaments and giving strength to epithelia, preventing cell separation as a result of mechanical or pressing forces. Similarly, cells are firmly attached to the basal lamina via hemidesmosomes, which maintains cell-to-ECM union under high mechanical stress. Lastly, **gap unions** formed by connexins and innexins proteins form a canal that allows communication between the cytoplasms of both adjacent cells as well as diffusion of small molecules such as ions or hydrophilic molecules with low molecular weight (MW 1000 Da) [4, 11].

It is important to note that transient unions, such as those found in blood vessels, are required in some cases. These unions are important in blood vessels because leukocytes must bind to endothelial cells in blood vessel lumens before migrating to tissues. **Selectins** are the proteins involved in these unions, and they, like cadherins and integrins, require calcium ion to form the union; however, instead of other selectins, they use heterophilic binding, binding to specific carbohydrate groups of **mucin** proteins at the membrane of the target cells [13].

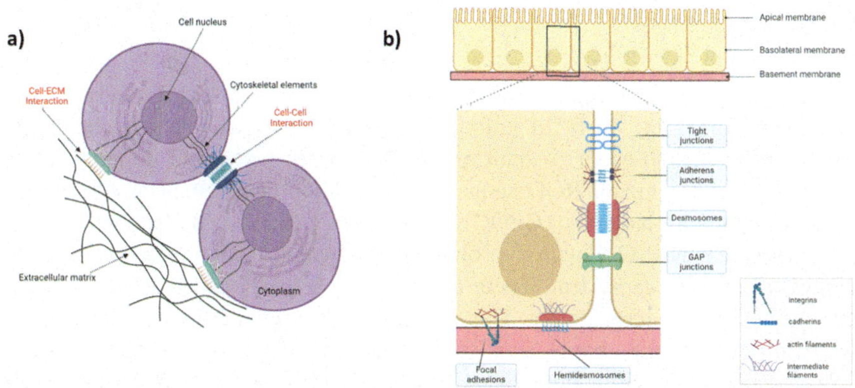

Fig. 8.1. **a** Cell-to-cell and cell-to-ECM unions. **b** Types of unions within epithelial cells and to the ECM

Cadherins

Cadherins are cellular adhesion molecules (also known as CAM) that mediate adherens junctions and desmosome unions in cell-to-cell contact (Fig. 8.2a). There are various types of cadherins that belong to the cadherins superfamily, and while they differ depending on the cellular type in which they are found, they all rely on the presence of calcium to form a union. Their unions are usually homophilic and symmetric, since they are formed by two cadherins of the same subtype at the external domain and bind the same ligand, e.g. actin filaments, at both internal domains for each cadherin [11]. When the concentration of calcium exceeds 1 mM, calcium ions bind to the N-terminus of the external domain, causing the structure to become rigid, not flexible, and vulnerable to union with the other cadherin under the same conditions. When calcium concentrations are less than 0.05 mM, calcium separates and the structure bends, becomes flexible, and is not exposed, preventing union with other cadherins [4].

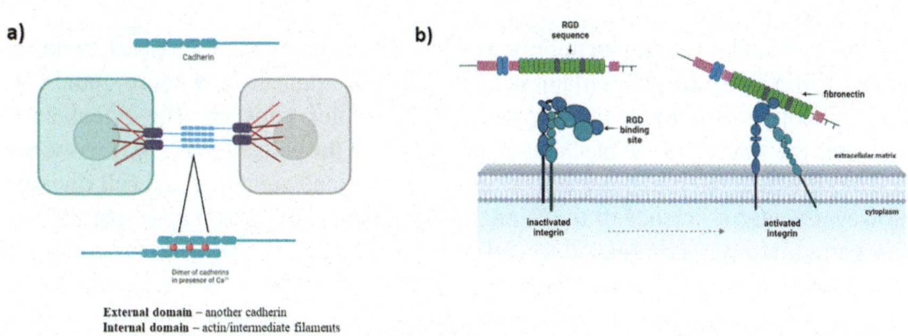

Fig. 8.2. **a** Cadherins are the membrane receptors in cell-to-cell unions and **b** integrins are the membrane receptors in cell-to-ECM unions

Integrins

Integrins are a type of cellular adhesion molecule found in focal adhesions and hemidesmosomes in cell-ECM unions (Fig. 8.2b). Integrins are transmembrane heterodimers composed of α-chain and β-chain, with both head domains at the external part recognizing RGD sequences of ECM proteins such as fibronectin, vitronectin, and fibrinogen, or GFOGER sequences (where single letter amino acid nomenclature is used, O = hydroxyproline) in collagen [40]. This union, like cadherin unions, is calcium dependent. There are approximately 24 and 8 of the subtypes of α and β chains, respectively. This means that one unique β chain can interact with a different α or β chains, which combination is unique for each ECM ligand. For example, the combinations of α1β1 or α2β1 bind to collagen, but α6β1 and α5β1 recognize laminin and fibronectin, respectively [22]. Integrins, like cadherins, require a conformational change from inactive to active. In an inactive state, both chains are closed to each other and both heads are bent, hiding the ECM binding sites. In an active state, integrins separate from each other and unfold, revealing an external extended conformation with high affinity for ECM ligands and talin, a protein regulator, at the cytosolic domain [4].

Integrins play a role in cellular signaling by acting as a link between the ECM and the cytoskeleton. As integrins are physically connected to the cytoskeleton and to the ECM, they can act as mechanical sensors generating signals that affect cell physiology that involves signaling mechanisms. Integrins are bidirectional signaling receptors that participate in both inside-out and outside-in signaling. This is due to integrin activation from inside because of the union talin to β chain resulting in inside-out signaling; or integrin activation from outside as a result of the interaction with components of the ECM resulting in outside-in signaling. Integrins participate in signaling pathways for cell proliferation, growth, survival, adhesion, motility and cell migration [13, 22].

Mechanisms of the Cytoskeleton

The cell cytoskeleton plays an important role in several cellular processes. Composed by the fibrillar proteins actin filaments, intermediate filaments and microtubules, the cytoskeleton constitutes a very dynamic and complex network. The cytoskeleton regulates the shape and the mechanical properties of the cell, but it is also involved in cell motility, the internal organization of organelles, cell signaling and cell division. However, material science or tissue engineering fields have based their interest on a few particular processes related to cell-material interaction that determine material success in restoring truncated tissue functions. Knowing the key processes involved in these interactions will allow us to design biomaterials able to meet the needs of cells with a high success rate. In vitro studies have elucidated the steps that cells follow during cell-surface (ECM or materials) interactions which are in order of occurrence: adhesion, spreading, polarization and migration.

Cell Adhesion to ECM and Cell Spreading

As above mentioned, integrin adhesion receptors regulate cell-ECM interactions. Integrins are present in hemidesmosomes when are linked internally to intermediate filaments, while are a part of the focal adhesion complexes when linked to actin filaments. These integrin-mediated adhesions must be highly regulated to control cell and tissue homeostasis, as they provide the anchorage to the ECM regulating cell signaling that control cell survivability or cell migration processes for example [9]. Although integrins mediate these junctions, some adaptor proteins are required in the cytosolic domain of integrins to regulate the union with the cytoskeletal proteins. Plectin and BP230 proteins assist in the union of integrin and keratin filaments on hemidesmosomes. Focal adhesions, on the other hand, are multiprotein complexes composed of integrins that bind directly or indirectly to other adapter proteins such as talin, paxillin, vinculin, and kindlin (Fig. 8.3) [4]. Integrins, once activated, promote the clustering of more activated integrins to form a dense plaque and thus forming the adhesion complex. This complex is responsible for the regulation of other kinases protein, tyrosine kinases Src and focal adhesion kinase (FAK), which regulate the activation of signaling pathways that control actin filament association, and so adhesion, proliferation, survival, or migration may be promoted. Moreover, focal adhesion complexes also regulate the small Rho GTPases (RhoA, Rac1, Cdc42) and Rho-associated protein kinase (ROCK), which regulate the cytoskeletal reorganization or remodelling and also participate in cell spreading and migration [9, 12].

In vitro experiments on cell adhesion have identified three stages (Fig. 8.4): (1) **Sedimentation**: initial attachment in which the cell interacts with the substrate through electrostatic interaction and guided by gravity, and there is a minimal area in contact with the surface, starting cell reshaping; (2) **Cell attachment**: the cell flattens by integrin-mediated adhesion and there is more surface in contact with the substrate; and (3) **Cell spreading and stable adhesion**: cell is fully spread arriving to the maximal spread area, and integrins form a stronger bond via focal adhesion complexes. However, cellular behavior depends on the various chemical or mechanical characteristics of ECM, and factors like elasticity, topography, or porosity can affect how the cell reacts to any change on them [1, 15].

Cell Polarization and Migration

With the exception of sperm, the majority of the cells move by crawling through tissues rather than swimming. Cellular migration, the term used to describe this movement that resembles crawling, depends on the actin cytoskeleton. The cellular

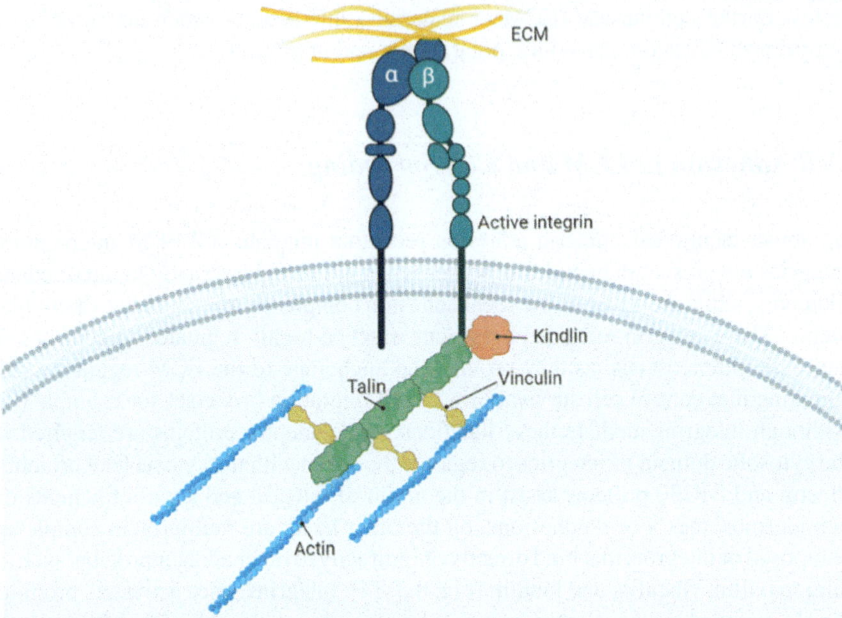

Fig. 8.3 Adhesions complex in focal adhesions

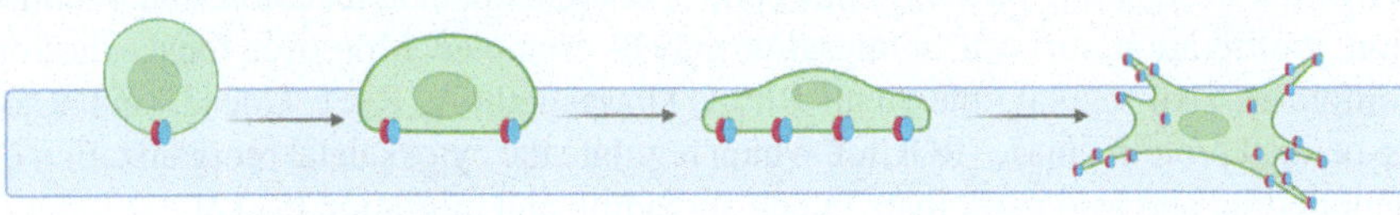

Fig. 8.4 Cell spreading process

cortex contains the actin cytoskeleton, which is involved in migration as well as cellular shaping, along with other proteins. All migrating cells must exhibit polarity, with one part in the front and one in the back. Cells would move in any direction if they were not polarized [23]. Actin and tubuline filaments with polarized front and back ends are required for this polarity. Due to the presence of additional proteins like Arp2/3 that can crosslink such filaments, these filaments may take on 3D structures. Different structures, including lamellipodia (actin filament in lamin or dendritic network) and filopodia (actin filament in tight parallel bundles), can be seen at the front of a migrating cell. They develop as a result of the polimerization of actin filaments, which pushes the front membrane. Integrins are important in migration because they help filopodia and lamellipodia attach to the substrate at the front of the cell, forming a powerful adhesion complex [16]. At some point, the back of the cell disassembles these adhesion complexes, causing the cell to separate from the substrate [41]. Cellular migration takes place in 3 main steps as seen in Fig. 8.5:

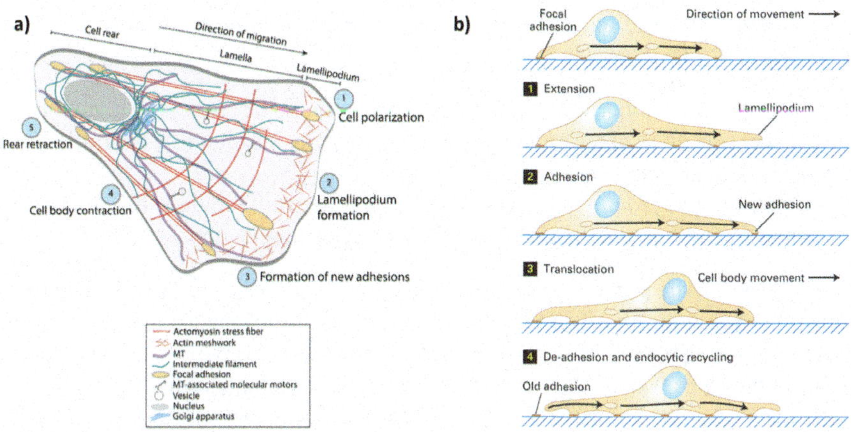

Fig. 8.5 Cell migration process

- **Extension**: actin polymerization at the cell leading edge pushes the membrane and protudes, forming filopodia and lamellipodia
- **Adhesion**: both filopodia and lamellipodia attach to the surface through focal adhesions formed by integrins, which are followed by the rearward movement of actin filaments
- **Traction**: by interacting with miosin II, focal adhesions exert a force that causes actin filaments to contract at the back of the cell. The cell is propelled forward by the pushing force and the dissolution of focal adhesions at the back of the cell.

Migration regulation is mediated by Rho GTPases (Cdc42, Rac1 and RhoA) as they control cell polarity. Each of these proteins regulates one of the 3 steps that occur during migration. Actin filaments polymerize and come together to form filopodia at the leading edge of the cell as a result of Cdc42 activation at the front of the cell. Lamellipodia are produced as a result of the simultaneous activation of Rac1 by Cdc42 and Arp2/3, which is activated by Rac1. However, Rac1 at the rear of the cell activates RhoA, which, by activating miosin II, regulates the formation of stress fibres and the contraction of actin filaments [9, 12].

Cell Adhesion to Biomaterials

Biomaterials for tissue engineering applications support cells by acting as an artificial extracellular matrix and indirectly, sending signals to cells. The basis for designing materials is to understand the cellular mechanisms involved in the interaction of cells with the surface of the materials. Cells are controlled by soluble factors, cell–cell interactions and adhesion to the extracellular matrix. In materials science, there are two ways to modulate the cell-material interaction by designing different types of materials [21] Fig. 8.6:

1. Materials that do not allow cell adhesion or protein absorption as a measure to prevent activation of the immune system, thrombosis, coagulation or proliferation of microorganisms, etc.
2. Materials that promote cell adhesion, migration, proliferation, differentiation, viability and functionality for tissue regeneration applications.

Focusing on the second type of materials, as mentioned in the previous subsections, the first thing that has to happen for the cell to respond is adhesion to the surface of the material. The molecular mechanisms on the part of the cell should occur in the same way as they would do with the extracellular matrix under normal conditions (sedimentation, cell adhesion, cell spreading and migration) [1]. Since

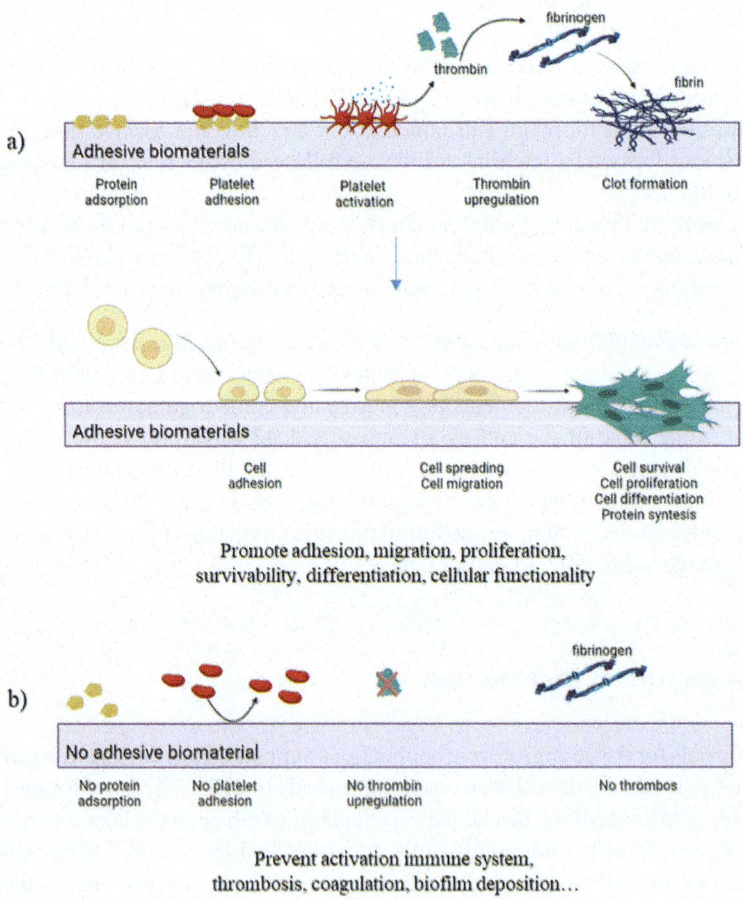

Fig. 8.6 Types of biomaterials

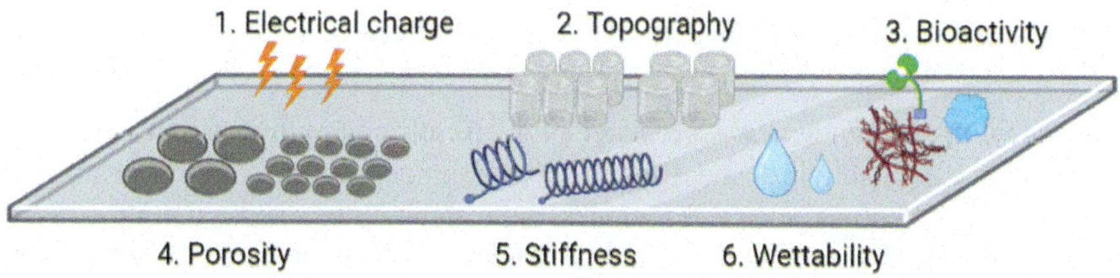

Fig. 8.7 Cell-material interactions can be modulated through different properties of materials

most materials do not have the same physicochemical properties as the extracellular matrix, sometimes this initial adhesion cannot take place and therefore the survival signals to the cell cannot be guaranteed. However, the study of materials has shown that there are different ways of modulating this interaction to make it as similar as possible to the interaction with the extracellular matrix. Some of these techniques involve controlling the physical, chemical and biological properties of the material. Among these, the most studied are topography, hydrophobicity/hydrophilicity, surface electrical charge, porosity, mechanical properties (stiffness) and bioactivity (Fig. 8.7).

- *Topography*: topography controls cell morphology and alignment and can control adhesion points by creating patterns [24].
- *Porosity*: related to topography, porous surfaces tend to favor adhesion and cell spreading. Also note the different levels of porosity such as macroporosity, microporosity and nanoporosity, and the degree of porosity, whether low or high [5, 6].
- *Hydrophobicity/Hydrophilicity*: the ability of a liquid to spread on a material due to intermolecular interactions between water and the surface of the material by studying the contact angle on the material. A small contact angle indicates that the material is hydrophilic and would favor cell adhesion [36].
- *Stiffness*: the elasticity of the materials modulates the morphology of cells through the tensile force generated between the cell and the material. This change in morphology can activate differentiation pathways, for example, simply by changing the degree of elasticity [29, 37].
- *Electrical charge*: the electrical charge of the cell surface is negative, so to favor adhesion by electrostatic forces, materials would be created with a positive charge on their surface, thus obtaining attractive (±) rather than repulsive (−/−) forces [38].
- *Bioactivity*: the presence of peptide motifs, such as RGD peptides on the surface of the materials, with high affinity for integrins would favor adhesion. By involving integrin binding, the binding will be of a stronger character than the other bonds such as those established by electrostatic forces [30].

Basics of Cell Signaling and Receptor Types

Cell signaling is a fundamental process by which cells communicate with each other to coordinate their activities and respond to environmental stimuli. Cells use a variety of signaling molecules, such as hormones, neurotransmitters, and growth factors, to transmit signals from one cell to another. The process of cell signaling begins with the production and release of signaling molecules from one cell. These molecules can be released into the extracellular space, where they can bind to specific receptors on the target cells. When a signaling molecule binds to its receptor, it triggers a series of intracellular signaling events that can result in changes in gene expression, protein activity, or other cellular responses. Cell signaling is essential for many biological processes, including development, growth and metabolism. Disruptions in cell signaling can contribute to many diseases, including cancer, diabetes, and neurological disorders. Therefore, understanding the basics of cell signaling is critical for the development of effective treatments for these diseases.

Cell Signaling

Cell signaling consists in the transformation of extracellular signals into a cellular response. This process is known as signal transduction and has three main stages: signal reception, signal transduction and cellular response. This process can have intermediate steps or signal transduction pathways or systems, depending on intracellular signaling proteins.

Signaling molecules or ligands are responsible for the first stage of signal transduction and can act locally or at distance. There are four types of cellular signaling based on how the signaling molecules arrive at the target cell: endocrine, paracrine, autocrine and contact-dependent (Fig. 8.8). The distant signaling is the **endocrine**, which is based on the secretion of hormones from an endocrine gland into the bloodstream and its arrival to the target cell that is usually distant. **Paracrine** signaling occurs when a secretory cell releases molecules such as neurotransmitters or cytokines that will act in an adjacent target cell. **Autocrine** signaling takes place when the secretory cell which releases the molecule and the target cell is the same. Some of these molecules are cytokines and eicosanoids like prostaglandins, thromboxanes, leukotrienes and lipoxins. Lastly, **contact-dependent** signaling occurs when there is direct contact between the secretory and the target cell. It is a signaling which requires plasma-membrane-attached proteins and permits the interaction of molecules as ions and surface proteins between two adjacent cells. This type of signaling is present in gap junctions between adjacent cells and in immune cells.

The first stage of signal transduction is the reception of the ligands or molecules that can take place at the cell surface or in the cytoplasm, depending on the signal molecule solubility. Hydrophilic molecules (such as small molecules, peptides or proteins) are not able to diffuse through the cell membrane and, in consequence,

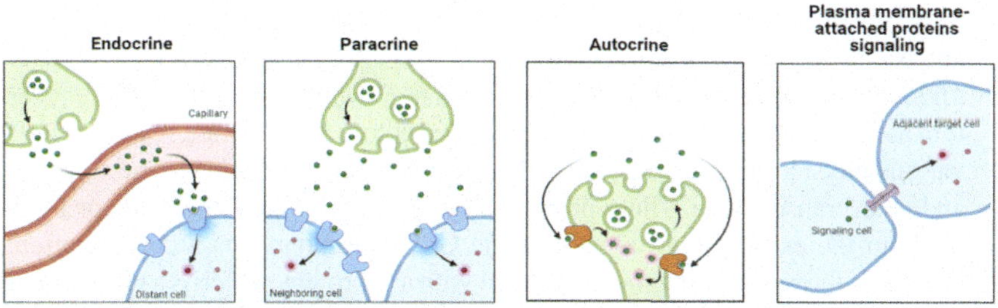

Fig. 8.8 Types of cell signaling

they bind to receptors in the plasma membrane. In contrast, hydrophobic molecules (such as steroids, retinoids or thyroxine) can pass through the plasma membrane and they bind to cytosolic receptors [4].

Receptor Types

In cell signaling, receptors are proteins located on the surface of the cell membrane or inside the cell that detect and respond to specific molecules called ligands. Regarding the receptors on the cell surface or transmembrane receptors there are several types (Table 8.1) [4, 10, 11]:

- *Ionotropic Receptor or Ion Channel Receptors*: These receptors are also known as ligand-gated ion channels. They are composed of a pore-forming transmembrane protein that allows ions to cross through the membrane and permits the binding of a ligand to the extracellular domain of the receptor. Examples of ion channel

Table 8.1 Summary of receptor types, localization of receptor and examples

Receptor type	Localization	Examples	References
Ion channel receptor	Cell membrane	Nicotinic acetylcholine, purine, glutamate-gated…	[4, 11, 31]
G Protein-Coupled receptor	Cell membrane	Opioid receptors (adrenergic), histamine, serotonin…	[8, 11]
Catalytic receptors	Cell membrane	Insulin receptor, EGFR, VEGF receptor, FGF receptor…	[4]
Nuclear receptors	Cytoplasmic or nuclear	Estrogen, thyroid hormone, androgen, retinoic acid…	[8, 10, 32]

Abbreviation: *EGFR* Epidermal Growth Factor Receptor, *FGF* Fibroblast Growth Factor, Vascular Endothelial Growth Factor receptors include the nicotinic acetylcholine receptor which exchanges ions of K^+ and Na^+ to generate an action potential.

- *Metabotropic Receptors or G Protein-Coupled Receptors (GPCRs)*: These receptors are the most abundant of cell surface receptors and they are responsible for various physiological processes, such as sensory perception, hormone signaling, and neurotransmission. These processes begin when a signal molecule or ligand binds to the extracellular domain of the GPCR, it stimulates a conformational change of the receptor in order to interact with a G protein, which is a type of intracellular signaling molecule that can activate or inhibit a variety of downstream effector proteins. Depending on the type of G protein and the effector protein involved, the downstream effects can be quite diverse, including the activation or inhibition of enzymes, and the opening or closing of ion channels. One of the most well-known examples is the opioid receptors, which includes β-adrenergic receptors and muscarinic acetylcholine receptors.
- *Catalytic or Enzyme-Linked Receptors*: These receptors are composed of a transmembrane protein that has both an extracellular ligand-binding domain and an intracellular enzyme domain. This pathway starts when a ligand binds to the extracellular domain, and in consequence, the enzyme domain becomes activated and catalyzes a biochemical reaction within the cell. There are several subtypes of catalytic receptors, being the most studied the receptor tyrosine kinases (RTKs) and receptor guanylate cyclases (RGCs), each with a distinct enzymatic activity and signaling mechanism. Examples of enzyme-linked receptors include insulin receptor and epidermal growth factor receptor (EGFR).

Regarding the receptors within the cell, these include cytoplasmic or nuclear receptors and are generally involved in gene regulation. They are activated by hydrophobic ligands that diffuse through the cell membrane and bind to the receptor in the cytoplasm or nucleus. Once activated, they can bind to specific regulatory regions of the DNA in order to regulate the expression of genes. Examples of nuclear receptors include the estrogen receptor and the thyroid hormone receptor [8].

In summary, the types of receptors in cell signaling include ion channel receptors, G protein-coupled receptors, enzyme-linked receptors, and cytoplasmic or nuclear receptors. Each type of receptor has a unique structure and function, but they all play important roles in mediating cellular responses to extracellular signals.

Examples of Signaling Pathways

Signaling pathways are the diverse intracellular signaling events that occur in response to extracellular signals. Main signaling pathways include Ras/MAPK, JAK-STAT, TGF-β/Smads, Rho-kinase, Wnt signaling pathways controlled by ubiquitination and protein degradation, and Notch/Delta signaling pathways controlled by protein breakage.

- *Mitogen-activated protein kinase (MAPK) pathway* is involved in cell proliferation, gene expression, differentiation, mitosis, and apoptosis, among others. It is activated by a variety of extracellular signals, including almost all tyrosine kinase and cytokine receptors, and involves a cascade of protein kinases that ultimately lead to the activation of ERKs, which can then translocate to the nucleus and modulate gene expression. Due to the importance of ERK, this pathways is also known as the extracellular signal-regulated kinase (ERK) pathway [17].
- *JAK/STAT pathway* plays key roles in cell division, differentiation, regulation of immune response, cell death and tumor formation. It is activated by cytokines, such as interferons and interleukins, which bind to their receptors and activate janus associated kinases (JAKs). JAKs then phosphorylate and activate STATs, which translocate to the nucleus and modulate gene expression [39].
- *TGF-β/Smads pathway* has an effect in cell proliferation, differentiation, extracellular matrix production and cell death. It is activated by the binding of TGF-β to its receptor on the cell membrane. Upon activation, the receptor phosphorylates and activates the Smad family of transcription factors. The activated Smads then translocate to the nucleus and modulate gene expression [2].
- *Rho-kinase pathway* is involved in the regulation of many cellular functions including cytoskeletal arrangement, cell migration and adhesion, among others. Rho activates Rho-kinase, which can then phosphorylate and activate downstream effectors, including myosin light chain and LIM kinase [20, 25].
- *Wnt signaling pathway* plays a key role in embryonic development, osteogenesis, and the proliferation of stem cells. It is activated by the binding of Wnt ligands to their receptors, which leads to the activation of a cascade of proteins, including β-catenin, that regulate gene expression and modulate cellular processes [33].
- *Notch/Delta signaling pathway* is crucial in the development of most tissues in embryogenesis, the most studied is neuronal development (proliferation, differentiation and apoptosis). It is activated by the binding of Notch ligands to their receptors, which leads to the cleavage of the receptor and the release of the intracellular domain of Notch. The intracellular domain of Notch then translocates to the nucleus and modulates gene expression. This process is dependent on the direct contact between the signaling cell (Delta protein) and the receptor cell (Notch receptor) [34].

In summary, these pathways are the most well-known pathways but they are just a few examples of the great variety of pathways that exist.

Cellular Functions Dependent on Cell-Environment Interaction

Cell-environment interactions are crucial for many cellular functions, as they allow cells to respond and adapt to their surrounding. Therefore, it is important to understand the mechanisms of how molecules interact with cellular receptors which provoke a cascade of intracellular steps that results in gene expression alteration. These changes in gene expression have an effect on the four main processes: cell survival, proliferation, differentiation and protein synthesis (Fig. 8.9) [13].

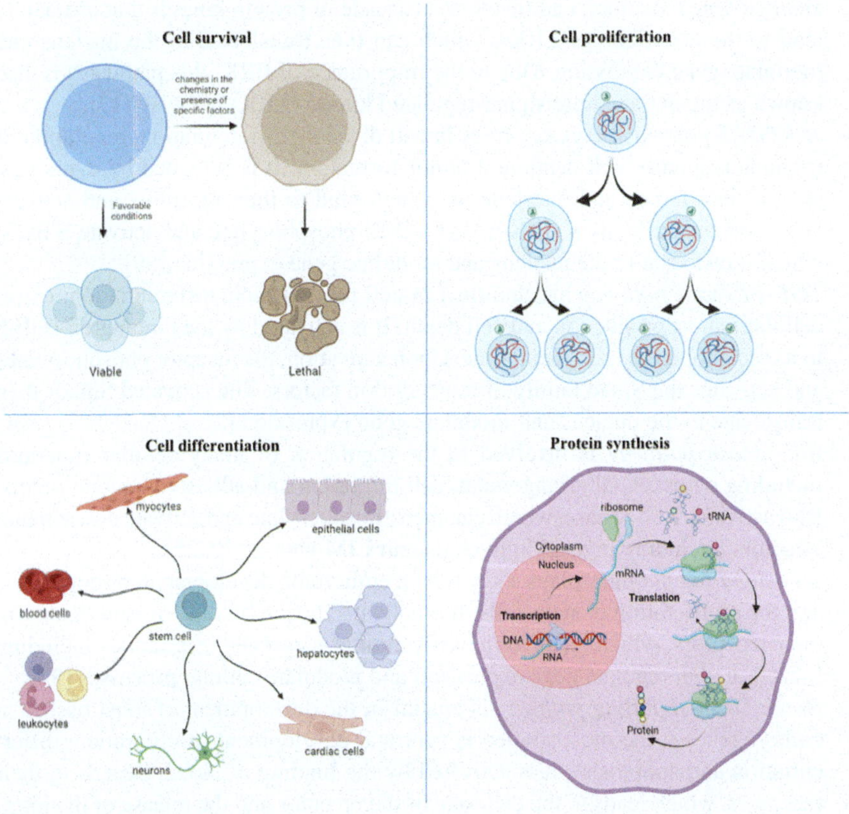

Fig. 8.9 Cell processes

Cell Survival

The extracellular environment has a strong effect on the viability of cells, resulting in cell death in some cases. Cell survival can be defined as the ability of cells to perform specific functions depending on the extracellular stimulus such as metabolism, growth, reproduction, some form of responsiveness, and adaptability. In particular cases, some factors, ions or some conditions such as pH can lead to the induction of cell death. For example, acidic conditions lead to necrotic death, and high concentration of cytosolic calcium ions (Ca^{2+}) results in apoptosis, a type of programmed cell death [10, 42].

Cell Proliferation

Cell proliferation is the process of increasing the number of cells because of controlled cell growth and division. This process is regulated by different cell cycle mechanisms that are affected by the extracellular environment. Based on the cell proliferation ability, there are cells labile, permanent or stable, which are continuously dividing, non-dividing or an intermediate dividing state respectively. Many growth factors and cytokines produce an enhanced effect of cell proliferation, known as mitogens, and other inhibit cell proliferation, known as growth inhibitors. For example, epidermal growth factor (EGF) promotes cell proliferation in many cell types, while transforming growth factor β (TGF-β) inhibits cell cycle progression [13, 27].

Cell Differentiation

Cell differentiation is the process by which progenitor cells or stem cells change their functional or phenotypical type, normally into a more specialized state. Cells rely on extracellular cues to differentiate into specific cell types and to organize into tissues and organs. During development, stem cells interact with neighbouring cells and the extracellular matrix, responding to cues that dictate their fate, position, and function. In fact, pluripotent stem cells (PSCs) have the ability to differentiate into all three germ layers: endoderm, mesoderm, and ectoderm depending on the growth factor. For example, the differentiation of PSCs into endoderm requires the activation of the Nodal signaling pathway, which is stimulated by growth factors such as Activin A and Nodal. In addition, other factors such as FGF2 and Wnt3a also promote endoderm differentiation [7, 14].

Protein Synthesis

Protein synthesis is the process of creating protein molecules. This involves transcription of a sequence of messenger RNA from a gene and occurs at the nucleus. Followed by the translation of this RNA into a sequence of amino acids to form a protein, taking place at the cytosol. This process is affected by various factors such as pH, oxygen levels, growth factors or mechanical cues. For example, it has been shown that cancer cells grown on stiff substrates tend to exhibit increased protein synthesis, migration, and invasion compared to cells grown on soft substrates [18].

Immunological Response to Implanted Biomaterials

When a biomaterial is implanted in the body, the immune system will recognize the material as foreign and will initiate an immune response. The immune response to implanted biomaterials can be divided into five steps: the acute inflammation, the macrophage response, the foreign body reaction, the formation of fibrotic capsule and the chronification [3, 19, 26, 35].

Acute Inflammation

The acute inflammation is characterized by the activation of the innate immune system, which is the body's first line of defense against strange molecules or pathogens. The innate immune system recognizes and responds to biomaterials by releasing inflammatory cytokines, such as interleukin-1 (IL-1), interleukin-6 (IL-6), and tumor necrosis factor-alpha (TNF-α), which recruit immune cells to the site of implantation. Immune cells, including neutrophils and macrophages, phagocytose the biomaterial and try to eliminate it. The acute phase is typically short-lived, lasting only days or a few weeks [28].

Macrophage Response

Once at the site of the injury, macrophages are recruited to the site of the injury through chemotaxis, a process in which they are attracted to the site by chemical signals, such as cytokines and chemokines. These signals are released by other immune cells, such as neutrophils, and damaged tissue cells. Macrophages phagocytose and attempt to eliminate the foreign body through various mechanisms, including degradation by lysosomal enzymes and reactive oxygen species (ROS).

In some cases, macrophages that have eliminated a foreign body may undergo apoptosis (programmed cell death). This can result in the release of pro-inflammatory cytokines, such as IL-1β, IL-6, and TNF-α, which can further recruit other immune cells to the site of the injury. If macrophages can eliminate the biomaterial, the reaction is finished [19, 28].

Foreign Body Reaction

If the foreign body cannot be eliminated by the macrophages, a longer foreign body reaction can occur. This involves the formation of a fibrous capsule around the foreign body, which is composed of extracellular matrix proteins, such as collagen. The fibrous capsule can isolate the foreign body from the surrounding tissue and limit its function [3].

Formation of Fibrotic Capsule

Over time, the fibrous capsule can thicken and become more fibrotic, which is referred to as fibrotic capsule formation. This process is mediated by myofibroblasts, which are activated fibroblasts that can contract and produce extracellular matrix proteins. Fibrotic capsule formation can lead to tissue compression, necrosis, and implant failure [3].

Chronification of the Inflammation

If the biomaterial is not removed during the initial phase, the immune response enters the chronic phase. The chronic phase is characterized by the activation of the adaptive immune system, which involves the activation of T and B lymphocytes. The T lymphocytes recognize and respond to specific antigens on the surface of the biomaterial, while the B lymphocytes produce antibodies against these antigens. This immune response can result in chronic inflammation and fibrous encapsulation of the biomaterial, which can limit its functionality and cause complications, such as implant failure or tissue damage [3, 4].

References

1. Khalili AA, Ahmad MR (2015) Int J Mol Sci 16:18149–18184
2. Hata A, Chen Y-G (2016) Cold Spring Harb Perspect Biol 8:a022061

3. Carnicer-Lombarte A, Chen ST, Malliaras GG, Barone DG (2021) Front Bioeng Biotechnol 9
4. Alberts B, Johnson A, Lewis J, Morgan D, Raff M, Roberts K, Walter P (2017) Mol Biol Cell
5. Murphy CM, O'Brien FJ (2010) Cell Adh Migr 4:377
6. Murphy CM, Haugh MG, O'Brien FJ (2010) Biomaterials 31:461–466
7. Payne C, King J, Hay D (2011) 207–216
8. Mangelsdorf DJ, Thummel C, Beato M, Herrlich P, Schütz G, Umesono K, Blumberg B, Kastner P, Mark M, Chambon P, Evans RM (1995) Cell 83:835–839
9. Cox EA, Sastry SK, Huttenlocher A (2001) Mol Biol Cell 12:265
10. Miller EJ, Lappin SL (2022) Physiol Cellular Receptor
11. Lodish H, Berk A, Kaiser CA, Krieger M, Scott MP, Bretscher A, Ploegh H, Matsudaira P (2016) Molecular cell biology, 8th edn. Macmillan Learning
12. Guan G, Cannon RD, Coates DE, Mei L (2023) Genes (Basel) 14:272
13. Temenoff JS, Mikos AG (2008) Biomaterials: the intersection of biology and materials science. Pearson/Prentice Hall
14. Ameri J, Ståhlberg A, Pedersen J, Johansson JK, Johannesson MM, Artner I, Semb H (2010) Stem Cells 28:45–56
15. van Tam JK, Uto K, Ebara M, Pagliari S, Forte G, Aoyagi T (2012) Sci Technol Adv Mater 13
16. Lock JG, Wehrle-Haller B, Strömblad S (2008) Semin Cancer Biol 18:65–76
17. McCain J (2013) P T 38:96–108
18. Hoerter JE, Ellis SR (2022) Biochemistry, protein synthesis
19. Anderson JM (2001) Annu Rev Mater Res 31:81–110
20. Katoh K (2017) PeerJ 5:e4063
21. Bačáková L, Filová E, Rypáček F, Švorčík V, Starý V (2004) Physiol Res 53:35–45
22. Barczyk M, Carracedo S, Gullberg D (2009) Cell Tissue Res 339:1:269–280
23. Glotzer M, Hyman AA (1995) Curr Biol 5:1102–1105
24. Ventre M, Natale CF, Rianna C, Netti PA (2014) J R Soc Interface 11
25. Amano M, Nakayama M, Kaibuchi K (2010) Cytoskeleton 67:545–554
26. Major MR, Wong VW, Nelson ER, Longaker MT, Gurtner GC (2015) Plast Reconstr Surg 135:1489–1498
27. Yang N, Ray SD, Krafts K (2014) Encyclopedia of toxicology, Elsevier, pp 761–765
28. Lacy P, Stow JL (2011) Blood 118:9–18
29. Janmey PA, Fletcher DA, Reinhart-King CA (2020) Physiol Rev 100:695
30. Fraioli R, Rechenmacher F, Neubauer S, Manero JM, Gil J, Kessler H, Mas-Moruno C (2015) Colloids Surf B Biointerfaces 128:191–200
31. Rao R, Shah S, Bhattacharya D, Toukam DK, Cáceres R, Pomeranz Krummel DA, Sengupta S (2022) Front Physiol 13:332
32. Sever R, Glass CK (2013) Cold Spring Harb Perspect Biol 5
33. Nusse R (2012) Cold Spring Harb Perspect Biol 4:a011163–a011163
34. Kopan R (2012) Cold Spring Harb Perspect Biol 4:a011213–a011213
35. Trindade R, Albrektsson T, Tengvall P, Wennerberg A (2016) Clin Implant Dent Relat Res 18:192–203
36. S.M. Oliveira, N.M. Alves, J.F. Mano, https://doi.org/10.1080/01694243.2012.697776 28 (2014)
37. Tee SY, Fu J, Chen CS, Janmey PA (2011) Biophys J 100:L25
38. Taghian T, Narmoneva DA, Kogan AB (2015) J R Soc Interface 12
39. Vainchenker W, Constantinescu SN (2013) Oncogene 32:2601–2613
40. Wagner WR, Sakiyama-Elbert SE, Zhang G, Yaszemski MJ (2020) Biomaterials science: an introduction to materials in medicine, 1–1586
41. Trepat X, Chen Z, Jacobson K (2012) Compr Physiol 2:2369–2392
42. Zhang Z-X, Gan I, Pavlosky A, Huang X, Fuhrmann B, Jevnikar AM (2017) J Immunol Res 2017:1–10

Chapter 9
Bioreactors for Tissue Engineering

Busra Ahata, Tugce Kan, Beyza Serefoglu Gun, Yigit Tanyeri, Busra Oktay, Aysel Oktay, and Rabia Cakir Koc

Abstract Bioreactors have been widely used in various fields of biological production for many years. Their ability to provide a tightly controlled environment during the process and to allow for monitoring and intervention to the process parameters make them quite favorable to use in biological production lines. Also, bioreactors are widely employed in tissue engineering applications. Ideally, a tissue engineering bioreactor should have the capability to effectively regulate various environmental factors, such as pH, oxygen levels, temperature, nutrient transportation and waste elimination. Additionally, it should facilitate sterile operations, such as sampling and feeding, as well as automated procedures. The general approach for these applications include immobilization of suitable cells within porous, biodegradable and biocompatible scaffolds. These scaffolds serve as frameworks for tissue formation and the cell/scaffold constructs are cultured within a bioreactor, which creates a dynamic in vitro setting conducive to tissue growth. As the technology for these systems and required conditions continue to become more complex, these bioreactor designs will also evolve with time to help treat patients with diseases related to tissue damage. There are specific designs for various kinds of bioreactors (spinner flasks, rotating wall vessel bioreactors, perfusion systems, pulsatile systems, strain systems, hollow fiber systems, wave bioreactors, microfluidic bioreactors, compression and hydrostatic systems) in the market which allows better outcomes for certain applications

B. Ahata · T. Kan · B. Serefoglu Gun · Y. Tanyeri · R. Cakir Koc (✉)
Health Institutes of Turkiye (TUSEB), Istanbul, Türkiye
e-mail: rabiakoc@yildiz.edu.tr

B. Ahata
Department of Medical Biotechnology, Institute of Health Sciences, Acıbadem Mehmet Ali Aydınlar University, Istanbul, Türkiye

T. Kan · B. Serefoglu Gun · B. Oktay · A. Oktay · R. Cakir Koc
Department of Bioengineering, Faculty of Chemical and Metallurgical Engineering, Yildiz Technical University, Istanbul 34220, Türkiye

Y. Tanyeri
Department of Bioengineering, Hacettepe University, Ankara, Türkiye

A. Oktay
Department of Molecular Biotechnology, Turkish-German University, Istanbul 34820, Türkiye

O. Gunduz et al. (eds.), *Biomaterials and Tissue Engineering*, Stem Cell Biology and Regenerative Medicine 74, https://doi.org/10.1007/978-3-031-35832-6_9

such as cardiovascular tissue engineering, bladder tissue engineering, neural tissue engineering, cornea tissue engineering, kidney tissue engineering, musculoskeletal tissue engineering, lung tissue engineering and gastrointestinal tissue engineering. All of these different systems and their special applications for tissue engineering studies are explained in this chapter with their specific advantages and disadvantages which make them favorable with the physicochemical environment they provide. When current developments are examined and evaluated, it is seen that bioreactors will have enhanced designs that will help them better mimic the physiological pathways of cells, tissues and their interaction with the surroundings to have better solutions for whole organ, bone, and regenerative tissue engineering applications in the future.

Keywords Bioreactor · Scaffold · Tissue formation · Tissue engineering

Introduction

The term "bioreactor" is simply referred to a "device" or "system" designed to sustain biologically active conditions, which are necessary for cultivating an organism or conducting a biological reaction. Dating back to 1970's [149] bioreactors are being used for various fields of production including pharmaceuticals, fermentation products, and biotechnological products. Thanks to the ability of providing a tightly controlled environment and to be able to monitor the process closely as well as other design parameters for ensuring a reliable cell growth, they are still being widely used [58]. Various designs of disposable (single use e.g. spinner flasks) and reusable bioreactors (e.g. stainless-steel bioreactors) are present in the market with their distinct advantages and disadvantages for different biological productions process needs. When selecting the type of bioreactor that is going to be used for the production of a biological product, the selection between these bioreactors and their sub-choices (material, size etc.) are substantial to ensure a robust, efficient and reliable process [58]. When it comes to tissue engineering applications, just like any other process, bioreactors are designed to meet the needs of the cells to be cultured in them. Immobilization of cells within porous, biodegradable and biocompatible scaffolds allow them to be cultured within a bioreactor which enables tissue growth in a dynamic in vitro setting [23]. They can be divided into different categorized according to their detailed characteristics, such as the type of flow they have inside the chamber (laminar or turbulent) specific to the characteristic physiology of the tissues, or the type of pseudo-physiological environment they provide with a rotating or non-rotating designs [97].

In general terms, bioreactors are used in tissue engineering in order to mimic the natural physiological environment of cells to provide biochemical and physical regulatory signals, direct cells to differentiate and provide a suitable platform for the development of new tissues by stimulating extracellular matrix production [138]. These approaches enable the conditions for cells to be stimulated which enables the production of the extracellular matrix (ECM) that is essential for vessel branching and vascular network formation [130].

Every system has their own unique advantages and disadvantages which qualify it feasible for specific applications and certain types of tissues. For instance, while stirrer bioreactors have an advantages in bone-tissue engineering applications by providing up-regulation of several growth factors, they reduce the diffusional gradients between the scaffolds, exposing cells to shear stress due to fluid convection [11, 144]. On the other hand, since shear stress is also beneficial for collagen and glycosaminoglycan (GAG) growth, such bioreactors may be useful for cartilage regeneration studies in cartilage-tissue engineering [12].

The simplicity or complexity of the system tends to change regarding their ultimate production goals. Factors such as the necessity for gas exchange, temperature control, how the access around the environment are going to be are some of the main examples that specify the complexity of system. Even though they differ for various applications, almost all of them have similar design parameters (maintaining the concentration of nutrients and gasses, providing mass transport, sustaining critical parameters such as pH and temperature etc.) that allow their usage for almost any biological production [136]. In addition to these basic factors, bioreactors have evolved from these designs with additional features to have different mechanical and biological components necessary to resemble the physiological conditions of the tissue's environment [121]. Advanced studies on this matter and providing new production solutions will make a great impact on improving the health of people with similar conditions.

In this chapter, some of the most common types of bioreactors ranging from basic designs to complex systems will be evaluated in terms of their advantages and disadvantages along with their critical design parameters, applications and future perspectives in tissue engineering. Further studies on this area and the provision of new production solutions will enhance the use of tissue engineering studies in "precision" or "personalized" medicine applications.

Types of Bioreactors Used in Tissue Engineering

The construction of completely functional three-dimensional (3D) artificial tissues and organs utilizing a biomaterials, cells, and signaling molecules is the ultimate objective of tissue engineering research. A dynamic culture that combines convection, perfusion, and diffusion is required to maintain 3D, clinically relevant sizes of tissue-engineered constructs (TECs) [121].

By stimulating cells with biochemical and physical regulatory signals, tissue engineering bioreactors can help the in vitro formation of new tissue by promoting cell proliferation, differentiation, and/or ECM production before in vivo implantation [48, 174]. Successful tissue engineering applications require the ability to maintain high cell populations over extended periods of culture without losing origin

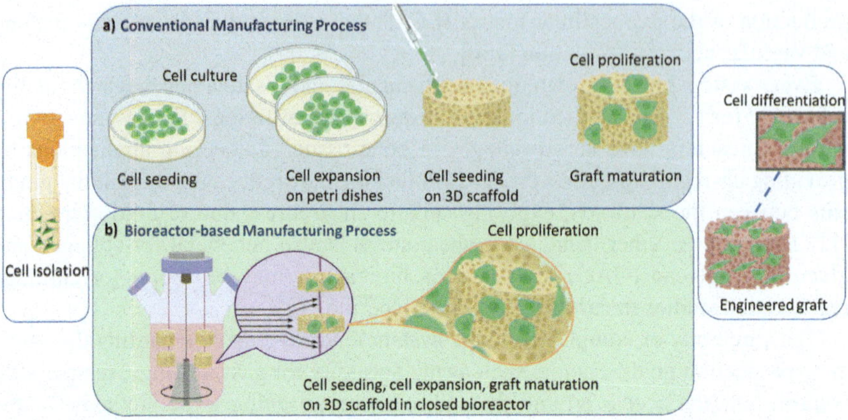

Fig. 9.1 Schematic representation of tissue formation approaches **a** conventional cell seeding **b** bioreactor-based. Created with Biorender (https://app.biorender.com/)

features, which is almost unattainable in static culture methods. Additionally, bioreactors offer a way to validate the functionality of the tissue engineering constructs before implantation under physiologically relevant loading conditions, as well as their preconditioning and maturement for some specific tissue engineering applications that require to tolerate significant mechanical loading immediately after implantation, such as heart valves and vascular tissues [47]. Bioreactors are also essential for tissue engineering applications on a realistic and larger scale because they enable aseptic procedures such as mounting, feeding, and sampling, while allowing control of environmental factors like nutrient provision, pH, pressure, oxygen level, waste removal, and temperature.

A variety of tissue engineering researches use conventional approaches (static seeding), which produce constructions with a thin tissue-like layer at the scaffold's base as a result of the cells' gravitational settling, as shown in Fig. 9.1. On the other hand, convective mixing (using spinner flasks) and convective flow (using flow perfusion) can enhance initial cell seeding and homogeneity, which in turn improves the tissue architecture [47]. Different tissue engineering applications employ a variety of bioreactors, and this chapter covers a range of bioreactors, such as spinner flasks, rotating walls, compression, perfusion, and microfluidic bioreactors (Fig. 9.2).

Spinner Flasks

The most straightforward bioreactor systems are spinner flasks which are cylindrical culture systems that have two arms that can be used to excise the stumps, and a magnetic rod attached to an impeller to circulate media and other culture components in the culture in a dynamic flow. In general, tissue engineering cultures in spinner

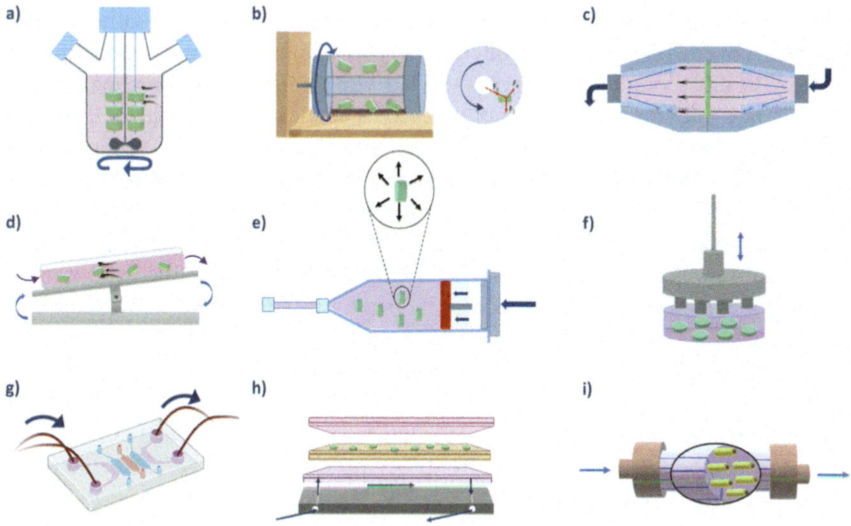

Fig. 9.2 A general scheme of bioreactor types; **a** spinner-flask bioreactor, **b** rotating-wall bioreactor, **c** perfusion bioreactor, **d** wave bioreactor, **e** hydrostatic pressure bioreactor, **f** compression bioreactor, **g** microfluidic bioreactor, **h** perfusion-parallel plate bioreactor and **i** perfusion hollow fiber bioreactor. Created with Biorender (https://app.biorender.com/)

flasks can be done in two ways; the cell/scaffold structures can be remained stationary in the flask or can be left to float freely in the spinner flask with the fluid flow. In the second option, the cells are exposed to less shear force and are less stressed [154]. These shear forces generally depend on the impeller's diameter size and rotational velocity. Therefore, in order to ensure the optimum cell growth, lower rotation rates are advised and implemented. The cell/scaffold constructs are mostly connected to the top of the flask with a needle, or the constructs are seeded onto microcarriers and immersed in the culture medium.

In spinner flasks, the uppermost section of the flask carries out gas exchange and medium oxygenation, while mixing tools like magnetic stirrers provide a well-mixed culture environment and increase the effectiveness of nutrition delivery and cell seeding in the scaffold. Standard stirring rates between 30 and 50 rpm promote dynamic medium mixing while minimizing harm to cells grown on the scaffolds [103]. The flow over the surface of the scaffolds creates vortices in the superficial pores of the scaffolds. Vortices are turbulence-related instabilities, and they may enhance fluid transport to the center of the scaffold, which increases cell viability and proliferation [58].

Spinner-flask technology was first used to allow biomass growth in cultures and has been used in tissue engineering studies for over 18 years. All culture parameters are well defined since it is a long-studied technology. The interactions between fluid dynamics and scaffold structure have become predictable using computational tools [49, 176]. Thus, spinner flasks are frequently encountered in applications involving the expansion and engineering of various embryonic stem cell populations and the maturation of cells derived from adult tissues [150, 175]. Cartilage tissue engineering is one of the most common tissue types studied with spinner flasks. Since the cartilage tissue is not vascularized, basic bioreactor concepts like spinner flasks have high efficiency in cartilage regeneration [58]. Adipose Stem Cells (ASC) chondrogenic differentiation in spinner flasks has also been demonstrated, and this is linked to the creation of a spheroid culture that permits cell–cell contact. To create trachea transplants, rabbit mesenchymal stem cells (MSCs) were grown for four weeks in spinner flasks and macroaggregated on a PLGA (polylactic-glycolic acid) scaffold was demonstrated [89, 177]. In studies on bone tissue engineering, it was observed that the dynamic mixing environment of spinner flasks increased cell formation and osteogenic differentiation on the scaffold more than static culture environments [144].

The most important disadvantage of spinner flasks are the limited size of target tissue that can be cultured on the scaffold. Because the mass transfer (all the media and micronutrients required for cell culture) within the flask is not enough to provide homogeneous cell distribution along the scaffolds. Insufficient transfer causes uneven distribution and growth of cells on the scaffold. Increased mixing speed causes more shear stress on the peripherally located cells on the scaffold. Over time of culture, these cells seeded into the scaffolds cause more ECM and mineral deposition around the scaffolds increased amount of ECM and deposition of minerals around the scaffolds, resulting in a sharp apparent nutrient gradient and waste accumulation in the center of the scaffold, leading to cellular necrosis [41]. Also, inhomogeneous spread of the ECM around the scaffold severely affects the mechanical integrity of these structures. Rotating-wall vessel (RWV) bioreactors with a dynamic culture and low shear stress can be alternatives to spinner flasks, as discussed below.

Rotating Wall Vessel

Shear stress plays a crucial role in regulating the mechanical factors of tissue structures, nevertheless high shear stress leads to the formation of unwanted particules around the tissues. Hence, the need for bioreactors with low shear stress has arisen. The RVW bioreactor is the most common bioreactor in tissue engineering studies and reveals the benefit of low shear stress. It was first established by the National Aeronautics and Space Administration (NASA) to test simulated microgravity conditions, allowing cells to be grown in a microgravity environment [137].

The RWV consists of two concentric cylinders with a rotating outer cylinder and a fixed interior cylinder containing an oxygen-permeable membrane for gas exchange. The bioreactor is connected to a motorized drive system that enables it to rotate the system around the cylinder's axis at a slow, constant speed. The space between the two cylinders contains media and 3D cell/scaffold constructs. These constructs remain close to a "free-fall" state in a microgravity environment where drag forces, centrifugal strength, and net gravitational forces are balanced and subjected to a dynamic laminar flow, as shown in Fig. 9.3 [162]. Because of this design, nutrient transmission increases, while shear stresses and turbulence are reduced. However, the low diffusion rate in the interior of the scaffold causes the inhomogeneous distribution of cells [19, 118]. Today, there are a variety of systems of the RWV, including Slow Turning Lateral Rotation Vessels (STLV), High Aspect Ratio Vessel (HARV), and Rotating Wall Perfused Vessel Systems (RWPV). Currently commercially available, STLVs are configured as an annular space between two concentric cylinders with an interior silicon gas exchange membrane, allowing greater control over culture parameters such as dissolved oxygen, pH, and temperature. On the other hand, although HARVs are similar to STLVs in general principle, they are more advanced in terms of gas exchange and culture rotation speed parameters [41]. RWPVs are designed to improve cell surface diffusion and mass transfer by convective flow under microgravity conditions [168].

Over the past century, RWV bioreactors are used in the culture of the retinal cell line to produce 3D-retina-like structures [39], temporomandibular joint disc [36], cartilage, and cardiac tissue engineering studies. It has been demonstrated that the

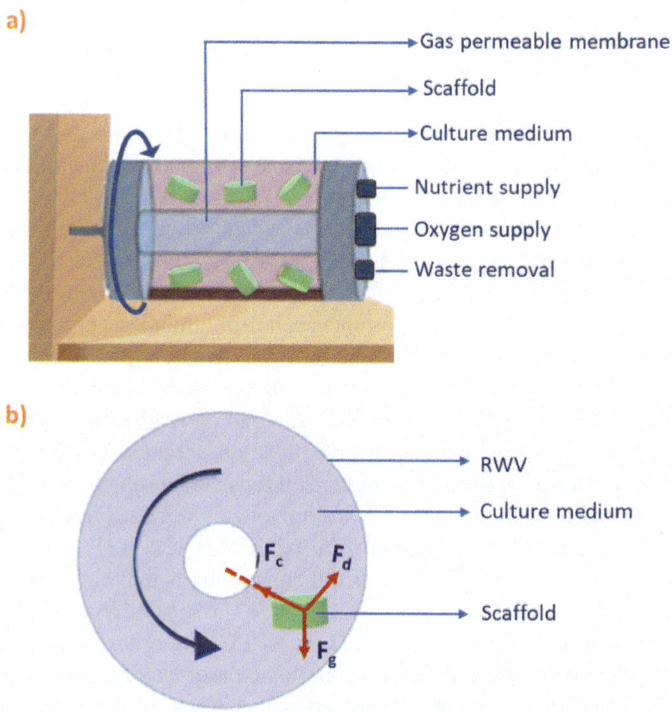

Fig. 9.3 **a** Schematic representation of a RWV bioreactor and **b** the centrifugal force (F_c), the gravity force (F_g) and the drag force (F_d) are balanced by the rotational motion of the bioreactor to keep the scaffold suspended. Created with Biorender (https://app.biorender.com/)

constructs produced by their culturing in RWV are structurally and functionally better than those produced using static or spinner flasks [23]. RWVs are also used in bone tissue engineering studies. Increased contact between cells and media enhances the proliferation and differentiation of osteoblastic cells due to better-controlled oxygen supply and less turbulence [111]. RWV-cultivated cells showed up regulations in markers indicating osteoclastic phenotypes compared to cells cultured in a traditional stationary culture, indicating that this environment can lead to higher bone resorption.

However, there are certain drawbacks associated with the use of RWVs. Cell proliferation is restricted to small scaffolds because sufficient transport of nutrients cannot be assured into the inner part of scaffolds [58]. In some studies, scaffolds have been reported to collide with the walls of the rotating vessel chamber during culture, causing cell damage and impairing the attachment of cells and matrix deposition on the scaffolds. When newly developed scaffolds with a density lower than water (like PLGA) are used, higher levels of ALP and calcium are observed in osteoblasts compared to static culture [24]. Although these findings are encouraging, the lack of effectiveness shown in RWV systems has led academics to look at other dynamic culture systems.

Perfusion Systems

A perfusion system generally includes a medium reservoir, a pump, and a conjunction system with columns, chambers, or cartridges that engage the cell/scaffold constructions, which can be used to improve cell growth [175]. In order to allow the medium to flow through the pores of the scaffold instead of around it, scaffolds are fitted tightly to the bioreactor cartridges. Through improved nutrition delivery to the interior of the scaffold and providing mechanical stimulation from liquid shear, medium flow through the scaffold porosity promotes cell differentiation (Fig. 9.4) [102].

Because perfusion bioreactors use a pressure gradient, they provide more even cell distribution and tissue-specific protein expression compared to constructs stimulated in a spin flask bioreactor. Since the mechanical loading regime most nearly mimics the condition that occurs in vivo, perfusion bioreactors are most widely employed for bone tissue engineering applications. Other cell types, including as MSCs, chondrocytes, keratinocytes, hepatocytes, and cardiomyocytes were successfully cultivated in perfusion systems for the creation of a TECs [52, 122, 131, 140]. Also, they enable the development of cell and scaffold constructions that have been computationally planned and printed into whole tissues [43]. Decellularization, developing technology to remove cells from original tissues' ECM in order to create 3D organ/tissue scaffolds for TE, is another use for perfusion bioreactors. Several tissues and organs have been effectively decellularized with the use of perfusion bioreactors [160].

With the perfusion system that provides continuous media flow, harmful metabolites can be eliminated, and mass transfer is enhanced growth factors and nutrients are continuously supplied. There are different configuration types of these bioreactors;

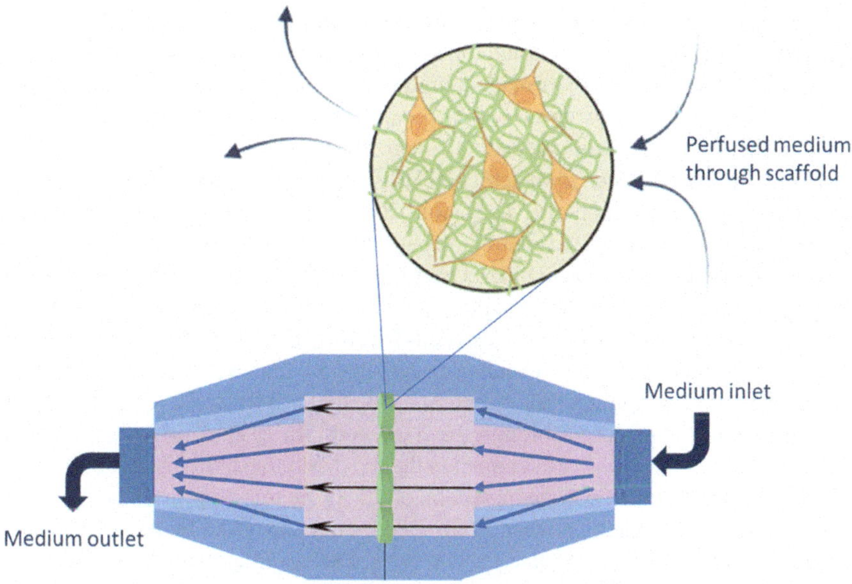

Fig. 9.4 General schematic representation of a perfusion bioreactor. Created with Biorender (https://app.biorender.com/)

like parallel-plate bioreactors, hollow-fiber bioreactors (HFBs), fixed (packed)-bed bioreactors, and fluidized-bed bioreactors [111].

Parallel plate bioreactors are composed of polystyrene plates and each plate has two compartments separated by a gas-permeable/liquid-impermeable membrane. The bottom compartment from those compartments contains the cells adhered as a monolayer to the surface, and is filled with the culture medium, whereas the upper compartment is filled with a mixture of gases. Cell mechanotransduction research with Parallel-Plate bioreactors usually makes use of well-defined shear stress resulting from the laminar flow that simulates intracellular environments [16, 170]. In the early years of tissue engineering, although a few studies were using this type of bioreactors, such as skin tissue and bone tissue studies, it is not widely used today.

HFBs offer high surface area-to-volume ratios (100–200 cm^2/L) and give cells a 3D environment for cellular attachment and proliferation with a low level of shear stress [120]. The system is composed of hair-like hollow fibers made of cellulosic, polysulfone, polypropylene, or polyethylene materials within a tubular cartridge that has inlet and outlet ports for flow around or inside the fibers. The cells are cultured on the interior or exterior surfaces of the hollow fibers, which are semipermeable tubular membranes with pores that range in size from 10 kDa to 0.3 m [172]. These pores of hollow fiber membranes also prevent the passage of unwanted molecules. Hollow fiber bioreactors are used for tissue engineering studies of tubular-shaped tissues such as blood vessels, intestines, and urinary organs, as they mimic the natural capillary system, and facilitate the generation of heterogeneous tissues with these bioreactors. Apart from that, they are well-suited for cultivating cells with high metabolic activity, such as hepatocytes, as they provide a high mass transfer rate [65, 77]. Apart from those a HFB would be less suitable if the research plan involves harvesting cells, due to the difficulty of harvesting cells that adhere to fibers.

Fixed or packed-bed bioreactors (PBRs) are small and compact systems contributing high productivity. They consist of a tank that contains the culture medium and an immobilized matrix of particles compactly packed enclosed in a column. Particles in the immobilized matrix can be composed of porous ceramic beads, macroporous microcarriers, porous glass beads, polyester discs, glass fibers, hydrogels, and alginate beads and allow cells to grow within or on them, while the tank provides oxygen and nutrients to cells through the bed. In these bioreactor configurations, both units are coupled via a circulation loop through which an oxygen-enriched culture medium is perfused through the fixed-bed [38, 111]. Thus, while the cells (adherent or non-adherent) are retained in the fixed-bed, fresh medium is supplied to the cells by perfusion, and toxic metabolic products are removed from the cells. The main and only difference between Fixed-Bed bioreactors and Fluidized-Bed bioreactors is that the particles in which cell growth is achieved are either packed (fixed) or floating (fluidized). In a fluidized bed bioreactor, culture media is continuously pumped upwards to a group of particles, cells, or immobilized cells, causing them to be suspended and behave as though they were fluid. Due to its uniform bed expansion behavior, superior mass transfer qualities, low shear stress, and straightforward scale-up, FBB have grown in popularity in biotechnology operations [33].

Although perfusion systems manage the problems related to other bioreactor systems and static culture, and using these bioreactors and their derivatives seems promising in clinical scenarios, these types of bioreactors have some limitations. Because of improper connections within the system, these systems are vulnerable to contamination and leakage. Also, significant optimization of process parameters and scaffold designs are required for maximizing the yield of the culture process.

Pulsatile Perfusion Bioreactors

Pulsatile perfusion bioreactors have been developed to mimic the pulsatile physical forces and in vitro cardiovascular conditions that vascular cells are exposed to during vasculogenesis. The system, first developed by Niklason et al., to mimic in vitro cardiovascular conditions, provides intraluminal pulsatile flow to four reactors, each of which contains one construct. Thanks to the pump in the perfusion pulsatile system, it can be operated at pulse rates at defined beats/minute intervals by applying pressure at variable stroke volumes [110].

Although the general operation of pulsatile bioreactors is essentially the same, minor modifications in reactor design might result in significantly varied hemodynamic conditions and, thus, varied outcomes for preconditioned cell-seeded heart valves. In the literature, there are several studies conducted by investigators to engineer pulsatile bioreactors to condition intact tissue-engineered heart valves. The first compact heart valve bioreactor, created by Hoerstrup et al., is composed of an air chamber, a media chamber, and a perfusion chamber that holds the heart valve. A reciprocating pneumatic diaphragm located between the air and media chambers in the reactor provides pulsatile flow [62]. On the other hand, Weston & Yoganathan engineered a tubular pulsatile flow bioreactor to evaluate compartmentalized leaflets that were sutured into a tubular structure. A heat exchanger and a gas infusion filter were employed in this bioreactor to enable the system to operate physiologically outside of the incubator. Pumps were utilized to create a constant and pulsatile flow throughout the closed loop system [167]. The bioreactor designed by Lichtenberg et al. includes a pulsatile pump, heart valve reservoir, media reservoir, and oxygenation/compliance chamber. This closed loop system created allows direct control of the flow rate thanks to a pulsatile pump, while monitoring the conditions inside the bioreactor with flow, pressure, and temperature transducers [86].

By virtue of perfusion bioreactors' ability to simulate the physiological and chemical conditions of living tissue, it has been possible to research in vitro cellular responses and develop better, and more effective tissues. Tissue-engineered heart valves still need to be thoroughly studied, and several significant problems need to be overcome before they can be used in clinical practices. There is no established conditioning technique for pulsatile perfusion bioreactors because there are so few clinical studies on the disease. Future research should pinpoint the conditions of these bioreactors that will promote clinical success.

Rocker Platforms—Wave Bioreactors

Wave bioreactors are disposable single-use bioreactors that typically consist of a transparent flexible polymer bag. Wave-induced agitation is achieved by placing the bioreactor bag on a rocking platform. For this reason, they are also referred to in the literature as "Rocker platforms" (Fig. 9.5).

The rocking platform can be either an open system that can be kept in an incubator or a closed system with a controlled environment. To accomplish the essential gas transfer through the headspace of the bag and culture homogenization, depending on the needs of the cultured constructs, the geometry, filling level, rocking angle, and rocking velocity of the wave bioreactor, as well as the viscosity of the medium, must be adjusted [20]. Due of the minimal shear stress generated by rocking without mechanical mixing, mass transfer is increased. Additionally, the technology is appropriate for sensitive cells like stem cells since it offers bubble-free aeration. These systems have a reduced risk of contamination since they use disposable bioreactor bags. As with all closed systems, it allows monitoring and control of temperature, pH and DO [8].

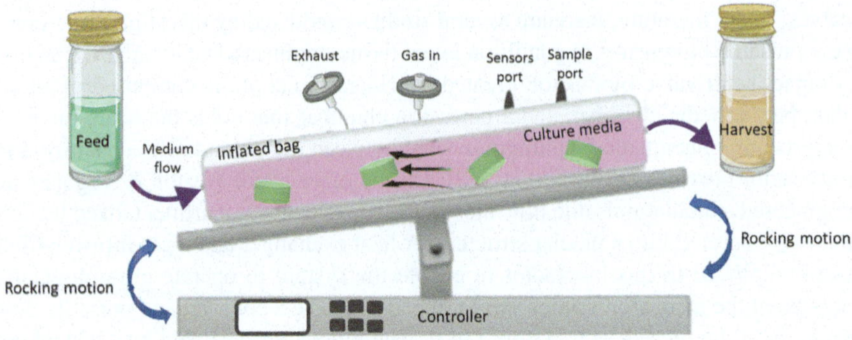

Fig. 9.5 Schematic drawing of a wave bioreactor. Created with Biorender (https://app.biorender.com/)

The initial version of the wave bioreactor was originally released on the market as a disposable replacement for stainless steel bioreactors in the 1990s. The original technology which was created by Singh consisted of a disposable plastic bag positioned on a motorized platform that was controlled outside the incubator and contained culture media and cell inoculum [145]. Without affecting the fluid shear or gas bubbles, the platform performs a rocking motion enabling good mixing and gas transmission. Unified sensor technology and control software not found in the early versions have been added to new designs to enhance automation, reliability, and repeatability [146].

Wave bioreactor can perform in batch, fed-batch, repeated fed-batch, and continuous perfusion modes, and it has auxiliary ports for connecting culture media bags for perfusion. When the bioreactor is performed continuously, the harvest bag and feed bag are integrated to allow continuous supply of fresh medium and removal of the waste medium [81]. A fed-batch system is the most favored wave bioreactor process since its ability ease of use and for eliminating the possibility of substrate/product inhibition seen in batch systems [40]. It has been shown in studies in the literature that scale-up and system automation are facilitated by using the batch feed process, with volumes up to 500 L [156]. It has been demonstrated in propagation cultures of mammalian cells, such as neutrophils from hematopoietic stem cells [156], embryonic feline lung fibroblasts [67], and T cells [57], that waves produced by shaking suspend cells/aggregates, thereby increasing mass transfer [96, 153].

Except for cell proliferation, where static bags are widely common, wave bioreactors are practically insignificant for functional tissue production. Based on the literature, there was only one official report of 3D tissue production in a disposable wave bioreactor. In the study conducted by Halberstadt et al., the production of human dermal replacement was achieved in a system consisting of 16 wave-bag reactors operating in perfusion mode, a 16-channel peristaltic pump, reservoir bag and waste bag. Each bag consists of a biodegradable free-floating 3D mesh scaffold to provide the necessary template for cell growth and skin tissue development. Tissues obtained after 22 days of culture with this system were histologically comparable to tissues produced in both continuously perfused and fed static culture bags of decorin, collagen type I, and fibronectin deposition [56]. Since the outcomes were promising, the perfusion bioreactor was adapted to produce obtainable through commercial channels tissue-engineered molting Dermagraft® in disposable bags [126]. Wave bioreactors do not require being sterilized, and scaling them up is relatively simple, but each research needs to optimize the rotating speed, angle, and bag fill level.

Microfluidic Bioreactors

Microfluidic bioreactors, also known as perfusion microbioreactors, biochips, or cell chips are a miniaturized version of conventional bioreactors with at least one perfused channel with a size in the micrometer range (Fig. 9.6). Similar to macroscale bioreactor systems, microfluidic bioreactor systems integrate monitoring and control components. They were developed to address a number of issues that were present in conventional systems, including the high consumption of growth medium and components like growth factors, limited compliance with high-throughput screening, challenges in controlling parameters and the microenvironment, elevated manufacturing expenses, difficulties with live-cell analysis and imaging, and the inadequate supply tissues with oxygen and nutrients [117]. Through channels, microliter quantities of fluid can be delivered to cells to properly evaluate the impacts of various doses of growth factors or pharmacological drugs.

The requirement to cultivate cells under shear stress has led to the development of microfluidic bioreactors. Unless specific features like actuators or surface modifications are included, the flow regimes within the microfluidic system are always laminar due to the small geometry of the channel [84, 178]. In addition to ensuring a constant flow of nutrients and the elimination of waste materials, the laminar flow regime also applies precise mechanical stress to the cells grown inside the channels. The

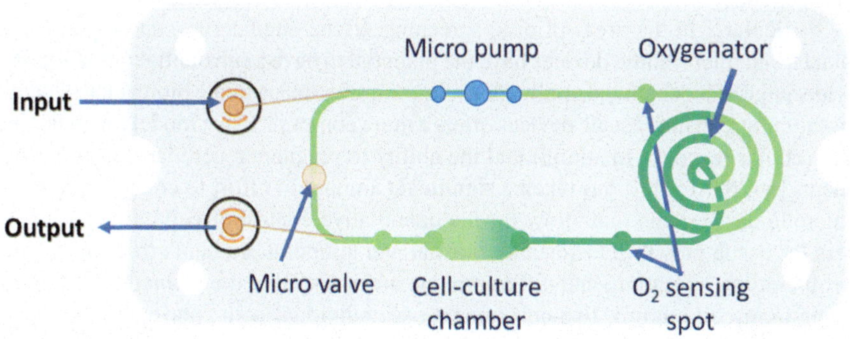

Fig. 9.6 A scheme of a microfluidic bioreactor. Created with Biorender (https://app.biorender.com/)

shear stress sensitivity of stem cells makes microfluidics an advantageous method for exploring stem cell differentiation caused by mechanical stimulation [44].

Microfluidic systems are fundamentally insufficient for growing cells or tissue-engineered products where large cell populations or complex structures are required due to their miniature geometry. The small size, however, has a number of benefits, including quick reaction times and minimum reagent use. Microfluidic technology also lends itself to automation and sensor integration. These characteristics make microfluidics an ideal technology for developing testing devices for toxicity and drug screening as well as for fundamental research [44]. Improvements to these systems have made it feasible to utilize multicellular aggregates, microspheres, and cell encapsulation in high-density 3D cell culture to more accurately mimic the interactions between native tissue cells than is possible in 2D culture. However, it is challenging to investigate how pharmacological substances affect the complex processes of tissues like the heart or lung in microfluidic bioreactors. As a result, most recent lab-on-a-chip bioreactor designs combine physiological factors like airflow and mechanical stimulation that simulate respiration or integrated vascularization and direct blood flow with contractile heart cells [9, 180].

Microfluidic bioreactors have several applications in a broad range of fields, including biochemical analysis, drug development, environmental monitoring, DNA and protein separation, and analysis. Aside from these, it is employed in subsidiary branches of cell biology including adhesion, spreading, proliferation, and differentiation, as well as in monitoring toxicity, counting, and sorting cells, and signaling pathways [2, 17, 78, 88, 141]. With such a diverse set of applications, it is possible to examine single cells, cell populations, tissues, and even complete organs like the skin-on-a-chip, vasculature-on-a-chip, bone-on-a-chip, brain-on-a-chip, kidney-on-a-chip, muscle-on-a-chip, heart-on-a-chip, lung-on-a-chip, liver-on-a-chip, gut-on-a-chip, multiorgan-on-a-chip, or tumor-on-a-chip in vitro [6, 53, 59, 74, 76, 106, 112, 143, 182]. Accordingly, the use of microfluidic bioreactors for cell studies is spreading quickly, with novel designs and microenvironments continually arising as a result of the incorporation of various materials, processing methods, and functional components.

Particularly in the areas of drug screening, tissue engineering, and organ transplantation, microfluidic devices have the potential to have a substantial influence on a wide range of biochemical applications. In contrast to macroscale bioreactor systems, tissue culture in microscale devices offers a more comprehensive model for analyzing the cellular response to stimuli and the ability to regenerate cellular microenvironments [64, 68, 166]. It has taken a significant amount of effort to create appropriate microfluidic systems that allow for the quantitative control of cell culture parameters for tissue growth. For instance, spatial and temporal gradients that control cell proliferation, migration, and differentiation are essential to the formation of tissues. Long-term cell culture, live-cell imaging of individual cells, and cell tracking to ascertain destiny are all possible with microfluidic technology [128]. Systems based on smart phone designed for internal environment and hybrid materials monitorization that enable point-to-point cell manipulation inside the bioreactor are some of the most recent developments in lab-on-a-chip technology [27]. The development of increasingly sophisticated human-on-a-chip systems that will investigate the effects of pharmaceuticals on various organ systems both directly and indirectly is predicted by current developments.

Strain Bioreactors, Compression Bioreactors and Hydrostatic Pressure Bioreactors

Strain bioreactors have been designed to directly apply mechanical stress to various mechanically responsive tissue cultures, such as bone, ligament, tendon, cartilage, and cardiovascular tissues [34, 50, 134]. Through a strain bioreactor, a direct mechanical strain may be imposed in the ways of stretching, compression, and bending. In strain bioreactors, the clamps attached to the scaffold are often employed to transfer the tensile force for 3D constructions, and linear actuators with digital control are used to manage tension [107]. To minimize structural damage during loading, the design of clamps must be altered based on the intended use. Various types of clamps have been developed to optimize scaffold assembly because it is pivotal that they do not cause cracking or tearing in the scaffolding they are mounted on. For example, spiral grips and attachment hooks are used for thicker structures, while grip pins and standard clamps are used for thinner structures [18, 99]. Alignment of cells at a 90-degree angle to the stretch direction is known to be induced by cyclic stretching and may produce homology to the target native tissue [134, 159]. According to several studies, the degree of alignment relies on the waveform of the stretch, frequency, and magnitude. In order to better simulate the physiological circumstances in tissues like the peritoneum, skin, and aortic valves, stretching can also be biaxial or equiaxial [71, 83].

In engineering studies of tissues exposed to compression in the natural environments like cartilage in the knee joints and bones, a compression bioreactor is adopted, which can provide both static and dynamic loading. Only the manner by which the force is applied to the structure differentiates these bioreactors apart from strain bioreactors. A standard basic compression bioreactor is made up of a motor that can apply linear motion and a control system that allows the operator to choose between various magnitudes and frequencies [63].

The bioreactor offers a regulated environment to establish a compression load bioreactor, it is essential to identify the compression type (dynamic or static compression) and determine its strain amplitude, frequency, and duration, in order to build a compression load bioreactor. Such bioreactors may be designed to offer both dynamic and static loading, allowing them to be adjusted for various application types [121]. Dynamic loading, which simulates more physiological loading, showed improved results than many other stimuli, despite the fact that static loading, which only allows for limited mass transport, has a negative impact on cartilage growth. Further studies have demonstrated the stimulatory effects of compressive strain on the scaffold elastic modulus, sulfated glycosaminoglycan (GAG), and hydroxyproline concentration in cartilage tissue engineering [31, 42, 101].

Another method of providing mechanical stimulation to structures made of tissue engineering is the use of hydrostatic pressure bioreactors. By covering a monolayer of cells grown in a petri dish with culture media and putting them in a pressure chamber where a gas phase works on both sides of the dish, hydrostatic pressure may be transmitted to the cells. Scaffolds are often statically cultured in cartilage tissue engineering research before being moved to a hydrostatic chamber for loading for a prescribed time frame [25, 61, 181]. The essential components of a hydrostatic pressure bioreactor are a chamber with the capacity to resist the applied pressures, pumps or pistons to apply that pressure, filters for ventilation, and non-return valves. As an example, an actuator-controlled piston may be used to pressurize a pressure chamber that is filled with media. The plunger can pressurize through an impermeable membrane to maintain sterility while preventing direct contact between the plunger and the culture media. A water-filled pressure chamber that uses a variable back pressure valve and an actuator to pressurize a media-filled chamber via an impermeable film is one variation of this idea [127, 165]. The hydrostatic pressure application's ideal magnitude, frequency, and period have not yet been determined. Dynamic hydrostatic pressure was described as preferable to static hydrostatic pressure in terms of how effectively chondrocytes proliferated in a monolayer [181].

To determine the most effective magnitude, frequency, and duration of applying strain, compression, or hydrostatic pressure using bioreactors, a case-specific approach is necessary. This approach must consider factors such as scaffold type and shape, as well as changes in cell number, porosity, and elastic moduli resulting from deposited ECM during the culture period. While it is possible to design and construct bioreactors that can apply various types and magnitudes of strain, compression, or hydrostatic pressure, a case-specific approach is needed (Fig. 9.7).

Combined Systems

In contrast to the basic loading circumstances caused by the different kinds of bioreactors discussed in this chapter, physiological loading conditions in the body are significantly more complicated [8]. As given in Table 9.1, the applications can be varied according to advantages and disadvantages of bioreactors. Combined systems are used to overcome their disadvantages or increase their advantages. Combinations of several bioreactor types can be employed to simulate the in vivo environment in vitro more effectively, fulfill the loading requirements for tissue-specific applications, and more accurately model the original tissue microenvironment. Stretching, compression, or perfusion cycling on HP bioreactors is the most popular use for combination bioreactors. Nutrient exchange is made possible in these various bioreactors designed for engineering certain tissues by perfusion, while stimulation is made possible through various mechanical stimuli [13, 165].

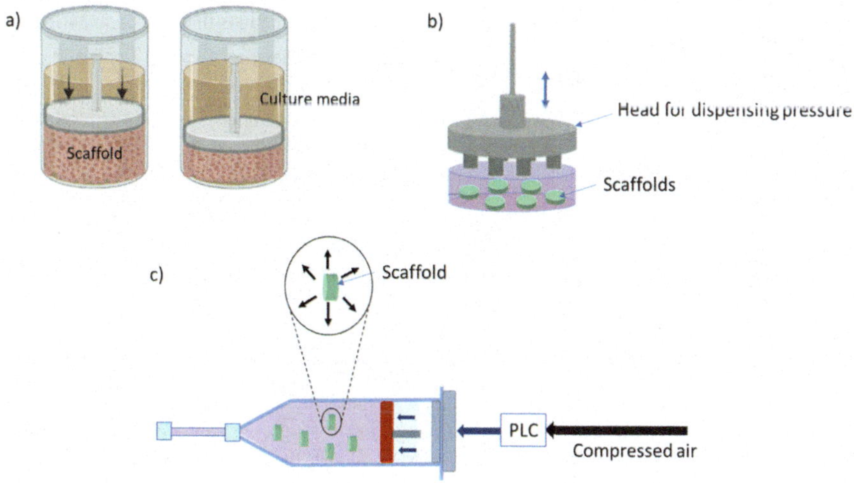

Fig. 9.7 **a** Strain bioreactor **b** compression bioreactor and **c** hydrostatic pressure bioreactor. Created with Biorender (https://app.biorender.com/)

A perfusion-loop tension and vibration bioreactor made for vocal cord tissue engineering that may simulate airflow-induced stimulation is one of the most well-known instances of a combination bioreactor. The bioreactor, which consisted of two synthetic vocal fold replicas in a silicon body, used airflow-induced self-oscillations, which have been demonstrated to create mechanical loading and contact forces that replicate human phonation. It was determined at the end of the study that the phonomimetic bioreactor supports ECM protein production and cell survival as projected [82, 157]. In research by Dermenoudis and Missirlis to simulate blood vessels, four mechanical stimuli have been established and developed: (1) normal and (2) blood pressure-related environmental stress, (3) shear stress from blood flow, and (4) individually controlled rotation-induced gravitational field. Rotation was found to be the most complicated stimulus in the study,when used alone, it causes the cells' polarity axes to shift frequently, and when coupled with other stimuli, it prevents elongation without changing the orientation profile [35].

In a study, scaffold-free cartilage constructions produced by porcine chondrocytes were cultivated under static and compression conditions to investigate the effects of perfusion and cyclic compression. GAG content was discovered to be considerably higher in the mechanically loaded group than in the statically loaded group and native tissue at the conclusion of the research [161].

Although combined systems provide a superior degree of tissue, size, and scaffold specific in vivo stimulation, they also add complexity and offer less control over testing parameters. Biological reactions to combined loading are typically challenging to predict and certainly do not total to the sum of the individual effects. There are multiple interactions among various cellular components, making it more

Table 9.1 Bioreactor types, advantages, disadvantages and applications in tissue engineering

Type of bioreactor	Type of stimulation	Advantages	Disadvantages	Applications
Spinner Flask	Shear stress	Exposure of the scaffold to shear results in enhanced cell proliferation rate, matrix deposition, and expression of proteins specific to the phenotype	Inhomogeneous distribution of the ECM Limited size and mass transfer	Musculoskeletal tissue engineering [1] Cartilage tissue engineering [58]
Strain	Compression Tension Bending Torsion Pressure	Ability to simulate physiological loading conditions Enhanced cell growth rate, matrix maturation, and expression of a variety of phenotype specific proteins	Risk of construct damage caused by the mounting of scaffolds and the application of direct strains Limited mass transfer	Cardiovascular tissue engineering [4] Musculoskeletal tissue engineering [34, 50, 134]
Rotating wall vessel	Low shear stress, reduced gravity conditions	Protection of cells from exorbitant shear stress and turbulence Simulation of microgravity	Time-consuming for optimization of the culture conditions Cells damage caused from scaffold colliding to the bioreactor wall Limited mass transfer Limited cell proliferation	Musculoskeletal tissue engineering [1] Cardiovascular tissue engineering [4] Ocular tissue engineering [39]

(continued)

difficult to optimize the right timing, quantity, and frequency of the parameters as the number of stimulation factors rises.

Future Perspectives on Tissue Engineering Bioreactors

Since the bioreactors have the potential to increase process efficiency, particularly for the clinical application of tissue engineering constructions, they are quickly becoming an essential component of tissue engineering research. Improved mass

Table 9.1 (continued)

Type of bioreactor	Type of stimulation	Advantages	Disadvantages	Applications
Perfusion	Shear stress	Better cell distribution and tissue-specific protein expression Can be automated Mimicking in vivo physiological environment of the tissue Limited turbulence	Appropriate only for scaffolds that are both mechanically strong and highly porous Vulnerable to contamination and leakage Optimization of flow rates is vital High rates of flow induced shear can cause cell and membrane disruption	Musculoskeletal tissue engineering [1] Cardiovascular tissue engineering [4] Kidney tissue engineering [147] Lung tissue engineering [116]
Perfusion-hollow fiber	Low shear stress	Limited contamination Increased surface to volume ratio	Not suitable for cell imaging Difficult to harvest cells expensive commercial HFBs	Urinary tissue engineering [77]
Perfusion-parallel-plate bioreactor	Shear stress	Well-defined shear stress Simulations of intracellular environments Easy to manufacture Inexpensive	Difficult to employ for 3D constructs	Skin tissue engineering [170]
Pulsatile perfusion bioreactor	Pulsatile physical forces Low shear stress	Simulation the physiological and chemical conditions of living tissue	Requirement for maintaining medium reservoir's temperature a little higher than desired temperature for the valve chamber due to heat loss	GI system tissue engineering [70] Cardiovascular tissue engineering [4] Kidney tissue engineering [119]

(continued)

Table 9.1 (continued)

Type of bioreactor	Type of stimulation	Advantages	Disadvantages	Applications
Microfluidic	Shear stress Tension Compression Pressure	Completely controlled mechanical stimuli Reducing the use of culture components Enabling high throughput screening and lowering production costs	Diffuculties for obtaining Organ-size products Requirement for adjustment for each study due to the decreased scale	GI system tissue engineering [173] Lung tissue engineering [80] Organ-on-chip [182]
Wave	Shear stress	Low shear stress Suitability for sensitive cells due to bubble-free system Easy to scale-up	The requirement to optimize the rocking rate, velocity, angle, and bag filling level for each study Limited mass transport High cost	Dermal tissue engineering [56]
Combined	Shear stress Compression Tension Bending Torsion Pressure Electromagnetic	Ability to apply different kinds of stimuli simultaneously	Requires a higher level of expertise Difficult optimization due to the increased number of parameters	

transfer, tightly regulated culture conditions, physiologically suitable stimuli, continuous medium supply, reduction of process steps, automated sampling for quality control, and standardization may all be offered by tissue engineering bioreactors [8].

Since poor cell viability caused by a lack of vascularization has been the rate-limiting factor in the efficient implementation of tissue engineering constructions in clinics, improved mass transfer is by far the main goal of employing tissue engineering bioreactors. With the help of porous scaffolds and precise perfusion, bioreactors provide appropriate oxygen, nutritional, and biosignal availability to the interior of tissue engineering constructions while supporting the development of the tissue in bigger dimensions than statically diffusible 100–200 m layers. When designing a new generation of modern bioreactors to simulate the physiological tissue microenvironment involving biochemical, biophysical, mechanical, and electromechanical parameters more accurately, the following factors should be taken into account.

- A sufficient environment for in vivo vascularization should be supplied after implantation, as in vitro vascularization of tissues is a priority in tissue engineering.

- The inflammatory environment must be considered as a crucial element of the mammalian host tissue response for a biomimetic approach.
- Continuous monitoring of the bioreactor environment and tissue growth using advanced imaging and sensing methods to track cell fate and tissue development in the intricate 3D environment [20].

The optimum automation of bioreactor control requires real-time and non-destructive evaluation of tissue and organ regeneration. To report inputs of environmental signals (such as mechanical actuation, oxygenation, or transfer of biological components), imaging and sensor data can be employed in a feedback loop. The advent of biotechnology and nanotechnology has altered how we perceive medicine and what we expect from healthcare systems. Future medical procedures, including tissue engineering, must be customized to each patient's needs in order to practice "precision" or "personalized" medicine, which is becoming more and more significant. Personalization of tissue engineering constructions would need bioreactors created specifically to meet the requirements of the individual patient. The use of patient-specific culture medium with specified loading conditions, patient-derived cells, and scaffolds that have been (bio)printed in the size and form of the desired defect is nonfiction.

Critical Parameters in Bioreactor Design

Factors to be ensured during three-dimensional tissue fabrication in bioreactors are maintaining a uniform cell concentration when seeding cells on the scaffold, control of microenvironmental parameters (temperature, pH, pressure, DO, metabolites, shear stress, and agitation), and aseptic parameters. In order to deliver nutrients and oxygen to cells and regulate the elimination of metabolic waste from the environment, a bioreactor offers a biomechanical and biochemical environment [8].

One of the most critical parameters during cell seeding on scaffolds in bioreactors is mass transfer. During long-term culture, cell viability needs to be maintained in the interior of the construct after cells are seeded on porous scaffolds. In order to maintain cell viability, nutrients, oxygen, and regulatory molecules must be transferred efficiently from the culture medium to tissue surfaces and inner cells of the tissue structure is required. At the same time, CO_2 and metabolites from the tissue's cells must be transferred to the bulk medium [87] (Fig. 9.8).

Mass transfer between a moving fluid and a surface is called convective mass transfer. In a bioreactor, the external mass transfer rate depends on hydrodynamic conditions. In a system with concentration difference, mass transfer occurs either by molecular (diffusion) or by convection. The internal mass transfer rate depends on the tissue structure, size and porosity of the scaffold, the diffusion rate of the cell and molecules from the biomaterial, and utilizing of both convection and diffusion mechanisms. According to the design of the scaffold, an efficient mass transfer can

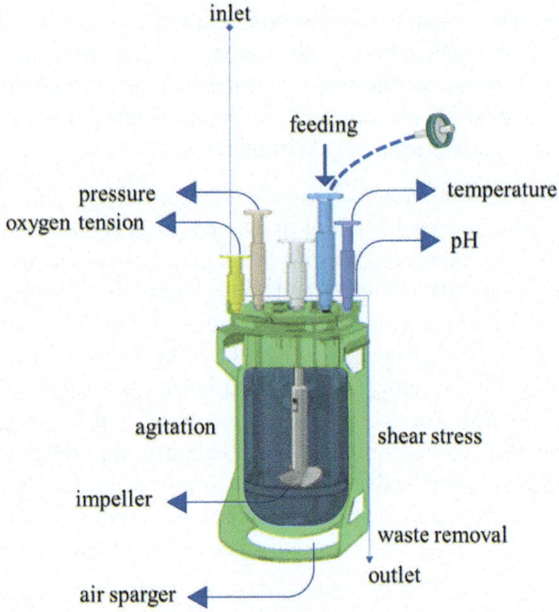

Fig. 9.8 Design parameters of bioreactors. Created with Biorender (https://app.biorender.com/)

be achieved by determining the flow direction and flow rate of the mass transfer [28, 87].

Every organism has an optimum temperature value at which it performs its metabolic activities. The optimal physiological temperature at which most mammalian cells continue their vital functions is 37 °C. At a temperature above 38 °C, it has a rapidly lethal effect on cell viability. At lower temperatures, cell metabolism slows down. In this case, the temperature of the culture medium in the bioreactor must be kept uniformly constant. The bioreactor's water jacket, temperature control unit, and temperature sensor work together to regulate this. The jacket that surrounds the bioreactor tank is a water-containing system. The jacket in contact with the bioreactor ensures that the temperature of the culture medium is balanced and remains constant. The temperature control unit receives a signal from the temperature sensor, which subsequently adjusts the temperature based on the culture medium's actual temperature. Water or any other heat transfer fluid circulating in the jacket around the bioreactor tank is heated or cooled by the temperature control device. As a result, the bioreactor's culture medium temperature is balanced by comparing it to the jacket's temperature [10, 28].

Hydrogen ion concentration (pH) significantly affects the metabolic activities and growth of organisms. Every organism has an optimum pH range in which it shows maximum activity. The optimal physiological pH required for mammalian cells to continue their vital functions is in the range of 7.0–7.4. As a result of cellular metabolisms converting glucose to lactate, the pH value of the culture medium decreases due to the production of CO_2 and water. For this reason, the culture medium becomes more acidic when no treatment is done during the culture. A bicarbonate buffer is used to keep the pH value of the culture medium at an optimal value. The pH is balanced by changing the amount of bicarbonate used in the culture. In addition, CO_2 can be added to the culture medium with a sparger to decrease the pH value, or air can be added to the sparger to increase the pH value. When the pH value of the culture medium in the bioreactor needs to be increased, a basic solution such as NaOH or Na_2CO_3 can be added. As a result of adding air, CO_2 or basic solution to the bioreactor, the pH probe in the bioreactor measures the pH value of the environment and is automatically brought to the optimal value by the controller [28, 87].

Oxygen, which is one of the components that must be present in the culture medium, is rapidly consumed by the respiration of the cell. Oxygen, which has low water solubility, must be continuously supplied to the culture medium. In order to avoid oxygen limitation in the culture medium, it is aimed that the OTR be greater than the oxygen utilization rate. For this reason, air, air and O_2 mixture or pure O_2 can be added to the bioreactor by air sparger to ensure the continuation of the culture medium. Continuous gas entry into the bioreactor is provided by the sparging air, which is usually located under the impeller. With the sparging air under the impeller, the circulation of the gas given to the bioreactor is ensured. Another important parameter in the culture medium is the DO level. Most mammalian cell cultures are able to continue their metabolic activity at DO of around 20–50% of the saturation with air. The DO level in the environment is detected by a sensor. The addition of air/O_2 to the bioreactor is managed by the controller according to the difference between the value measured by the DO sensor immersed in the culture and the desired value. When there is a higher level of DO in the medium than the desired value, nitrogen can be added to the bioreactor via the sparger to remove the oxygen from the culture medium. In another widely used method, cells are allowed to consume oxygen up to a certain point. As a result of O_2 consumption, air, O_2 or air/O_2 mixture is given with the sparger to raise the DO level again below the determined value. Allowing the cells to consume oxygen until the target value is reached is an alternative and more typical common approach. When a setpoint is not reached by the process value, a mixture of air and O_2 is added to raise the process value back to the setpoint for DO [105, 123, 135].

Oxygen transfer in the culture medium is important because of the poor solubility of oxygen in the culture medium. A balance must be maintained between the oxygen supplied to the cells and the oxygen consumed by the cells. Therefore, another critical parameter during the design of the bioreactor is the oxygen tension setting. In tissue engineering applications, the oxygen requirement of the cells in the culture medium varies according to the phases in the growth curve. During the initial expansion phase, the overall oxygen demand increases as the cell density increases with time. In the next process, the cells go from the state of reproduction to the state of differentiation, and in the case of differentiation, the oxygen requirement of the cells that use less oxygen decreases [126].

The culture in a reactor can be aerated by aeration, direct scattering, indirect and/or membrane aeration (diffusion), medium perfusion, this helps to increase the atmospheric pressure and the partial pressure of oxygen. In a bioreactor, DO can be transported by global mass transfer, internal mass transfer, or external mass transfer.

The rate at which oxygen is given to the environment at the gas–liquid interface is constrained by how soluble oxygen is in water. The flow area of the vessel and the net consumption or production rate determine the oxygen concentration distributions in the culture medium for global mass transfer. The equilibrium between the oxygen provided to the environment, known as the OTR, and the oxygen consumed by cells, known as the oxygen absorption rate, determines the oxygen concentration in the environment. The liquid phase mass transfer coefficient (kLa) and productivity are significantly impacted by the OTR ratio [93, 105].

It is acknowledged that mechanical stimulation—such as pressure, tension, hydrodynamic pressure, and fluid flow—is crucial for the maturation of organs in the bioreactor. Mechanical interactions between the culture medium and the scaffold during tissue growth determine whether cells form cell clumps or disperse on the scaffold. Determining optimal physical parameters is complex due to the diversity of cell types, scaffolds, forces, regimens applied, and culture medium available. An impeller system is used to ensure a homogeneous distribution of the culture medium in the bioreactor and for air circulation [105]. The agitation system basically consists of a rotor, a drive mechanism (magnetic or direct) and a motor. The bioreactor can be powered to achieve effective mixing and a uniform distribution of temperature, DO, and pH in the culture medium. The spreader's design, the impeller's type, size, and placement, as well as the influence of shear stress from hydrodynamics and aeration, define the process's possible effects on cells and the process. Determining the pressure inside the bioreactor is also an important parameter. The pressure is measured with a sensor connected to the bioreactor. A gas insertion lockout strategy is put in place if the pressure rises due to clamp-on or clogged vent filters as a result of excessive foaming, for example, due to mishandling of the bioreactor. As a result of excessive foaming due to misuse of the bioreactor, the ventilation filters are clogged, and therefore, when pressure increases, the A gas addition interlock strategy is applied. All these critical parameters are determined and controlled in bioreactor design. A signal from each probe is evaluated and accordingly the system is regulated at the desired level [8].

Shear stress has an effective effect on tissue function and viability. There are different values for the maximum sustained shear stress for each cell type. The high shear stress generated on the scaffold surface by a fluid flow can strip the attached cells, in which case tissue growth can be significantly slowed down compared to static cultures. Simply put, the fluid flow affects the shear stress, the orientation and function of the cells. For example, it has been observed that shear stress affects endothelial cell proliferation and directs them downstream (Fig. 9.9) [69, 123].

Although the operational process of sterilization varies little depending on the organism, it must be carefully adjusted to the bioreactors' geometrical design and material composition. With the determination of aseptic parameters, the sterilization procedure of the bioreactor is applied [135].

Tissue engineering bioreactors are mostly laboratory-scale (lab-scale) bioreactors that involve tissue production and tissue modeling. Experiments require multiple samples which are conducted in T-flasks or spinner flasks in incubators to observe cell growth and to perform substrate or product assays. Many tissue cultures are performed in lab-scale as experiments require relatively low numbers of cells. In this scale, T-flasks with a surface area ranging from 25 to 225 cm^2 are used. The maximum number that flasks with a surface area of 225 cm^2 can reach is 1×10^7 per ml. The scale-up process is accomplished by gradually increasing the culture from the lab-scale to the industrial scale. This method gains functionality by adapting directly to the environment in which the cells are transferred and proliferating [21]. The scale-up is divided into two categories: scale-up in suspended cultures and scale-up in monolayer culture. In monolayer culture, cells proliferate by attaching to the flask surface. Therefore, it is necessary to increase the surface area and medium volume to scale-up monolayer cultures. Monolayer cells are more difficult to scale than suspended cells. Transfer of cells to the new medium cannot be accomplished as a simple fluid transfer, as the cells must be released from the substrate mostly using an enzyme. This consists of a highly variable and contamination-prone process that is labor-intensive in large-scale operations [51]. The advantages of monolayer cultures are ease of medium exchange, washing and cell perfusion, high production of pharmaceutically important components such as hormones, vaccines, insulin and interferon, and repeated use with different cells and mediums with the same experimental setup and equipment. The disadvantages are that it is tiring and costly, requires a lot of free space, cannot effectively monitor cell growth, and is difficult to measure important process parameters. The scale-up of monolayer cells can be performed in roller bottle culture, roux bottle culture, multi-surface culture, microcarrier culture, fixed-bed reactors, fluidized-bed reactors, and hollow-fiber reactors [125]. The scale-up is much simpler and more controllable for cells growing in suspension, as stirred vessels show similar design properties at all scales. Scaling-up the suspension culture is accomplished primarily by increasing the culture volume. Spinner flasks (100–1000 ml) and bench-top bioreactors (1–50 L) are used in lab-scale for the development of suspended cultures. After the reproducibility and repeatability of this bioprocess is possible and the process parameters are optimized, a pilot scale process (50–10,000 L) is designed to maintain optimum operating conditions. After lab-scale and pilot scale studies are successful, the plant scale (> 10,000 L) is designed for co mmercial and large-scale production (Fig. 9.9) [92]. By scale-up, it is aimed to successfully transfer the optimum conditions obtained in small-scale bioreactors to large-scale bioreactors. Scale-up studies are very critical and indispensable in order to create suitable parameters and conditions to change the scale without harming the kinetic behavior and growth performance of cells. However, the kinetic behavior of cells is significantly affected by local environmental conditions such as temperature, pH, DO, and nutrient concentration. Therefore, small-scale studies may tend to overestimate the process performance at larger scales if inconsistencies in scale-up are not resolved. For this purpose, environmental conditions and parameters should be kept under control and constantly monitored. This is done taking into account physical, biochemical and bioprocess factors. Physical factors include mixing parameters, heat and mass transfer, power consumption, DO, temperature, pH and shear stress. Biochemical factors are mainly media components and their physicochemical properties and concentrations in the

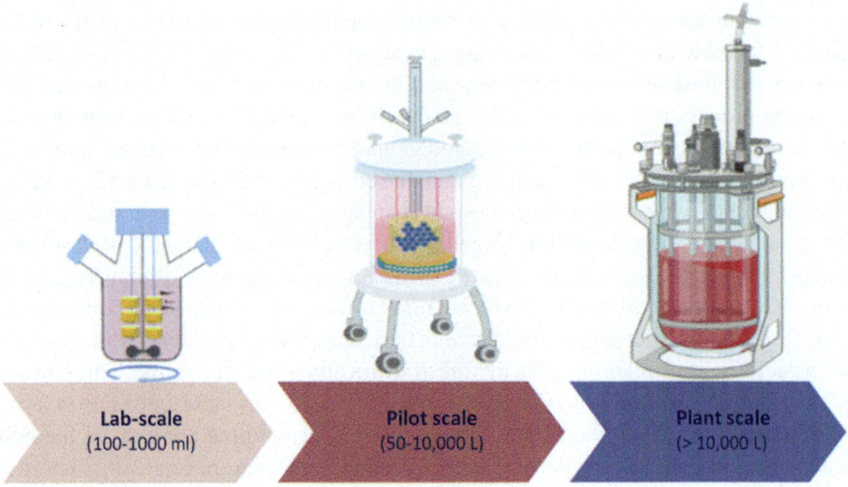

Fig. 9.9 Scale-up of bioreactors from lab-scale to plant scale. Created with Biorender (https://app.biorender.com/)

bioreactor. In addition, bioprocess factors such as pre-culture conditions, sterilization quality, and inoculation rate also determine how the scale will be administered successfully [138].

The conventional method used to scale up the bioreactor includes determining the geometry of the bioreactor, the stirrer speed, and the aeration rate of the large-scale bioreactor, taking into account the experimental results of the lab-scale bioreactor. The most widely used method of scale-up is to determine the dimensions of the large-scale reactor while maintaining the geometric similarity of the bioreactors. After the volume of the large-scale bioreactor has been determined, its geometric properties such as tank height, tank diameter, and agitator size are estimated using certain predetermined and accepted ratios and calculations. Typical bioreactors are cylindrical and designed to have a height-to-diameter ratio of 2/1–3/1. This ratio can be used as one of the simplest scale-up strategies. However, this ratio may not be so simple when applied to reality. Enlarging the bioreactor diameter by 5 times and keeping the height-to-diameter ratio constant will increase the reactor volume by 125 times, which undoubtedly makes the production of 3D textures on a larger scale quite different. Empirical correlations are needed to determine impeller speed and aeration rate and to keep the parameters related to the change in scale constant. Evaluation of the impeller speed is accomplished by keeping agitation power input per unit volume, volumetric oxygen mass kLa, or impeller tip speed constant. The aeration rate can also be determined by parameters such as equal superficial gas velocity, specific gas flow rate or gas flow number. In scale-up, one or more process parameters are kept constant by the engineers, estimations on other parameters are made accordingly and strategies are created in this direction [51, 92].

In summary, scale-up techniques in bioreactors bring some problems that need to be overcome. These problems are parameters that need to be optimized such as operating time, reactor capacity, oxygen, pH, temperature, gas exchange, mass transfer, continuous monitoring, product recovery, control of secondary processes, depletion of nutrients and oxygen, formation of toxic metabolites and production efficiency. The potential applications of bioreactors in tissue engineering can be better achieved when working with reproducible and repeatable systems with high degree by optimizing scale-up parameters and conditions.

Application of Bioreactors

Tissue engineering of all 3D tissues requires homogeneous cell distribution to develop homogeneous tissue [108]. With bioreactors, the biomechanical and biochemical environment that is effective in cell and tissue growth can be provided in a controlled manner. Therefore, functional cells and tissues can be grown suitable for transplantation using bioreactor technology (Fig. 9.10). These systems' major goals are to maintain ideal gas and nutrient concentrations in the culture medium, ensure homogeneity of cell distributions on 3D scaffolds, and expose the developing tissue to the similar physical stimuli. In vitro bioreactor systems based on controlled management of cell culture parameters ensure high reliability and reproducibility of experiments. Additionally, unlike bioreactors, static cultures on plates or flasks do not offer a flexible environment for studying cell-scaffold interactions under various pressure settings [139].

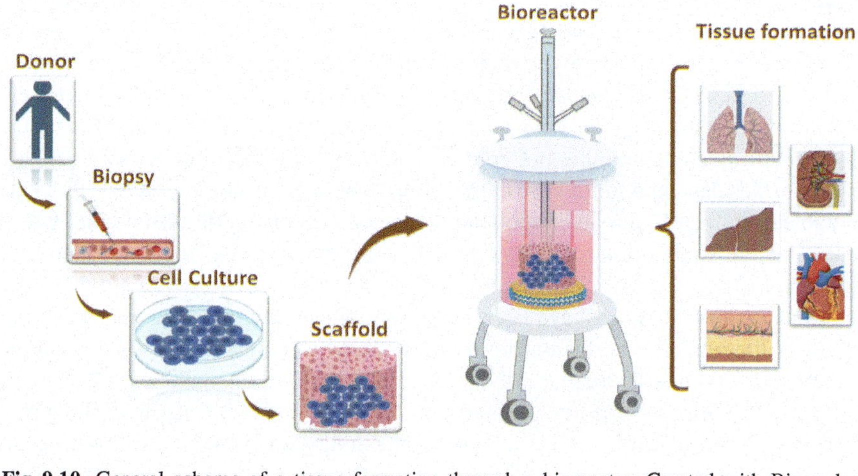

Fig. 9.10 General scheme of a tissue formation through a bioreactor. Created with Biorender. (https://app.biorender.com/)

To overcome the difficulties with large-scale cell production, big and uniform cell development in bulky tissue, high-efficient nutrient supply, important environmental stimulation delivered to cells, as well as metabolite removal, tissue engineering uses bioreactors [179]. The purpose of using a bioreactor can be listed as to promote cell proliferation, development, and placement within the scaffold to promote maturation and in vitro simulation of physiological or pathophysiological dynamic conditions [94].

Cell biology and tissue engineering studies require multiple isolations to maintain the cell source, cell expansion, cell viability, and long-term phenotypic stability. The standardization and quality control of biomanufacturing in this context is the next step that can make bioengineering a regular application in clinics [29].

As a result, bioreactor culture is required in several fields of tissue engineering. Bioreactors have been used in many different applications including whole organ, bone, skin tissue engineering [97]. In this section, use of bioreactors in tissue engineering of the gastrointestinal (GI) system, musculoskeletal, neural, cardiovascular, bladder, uterine, cornea, kidney, lung tissue will be examined.

Bioreactors for Gastrointestinal System Tissue Engineering

The GI tract is a complex system that involves the integration of different cell elements, immune, absorption, secretory, and motility signals. Intestinal motility, which is the process of coordinated contraction and relaxation of the smooth muscle in the GI tract, is a crucial aspect of intestinal physiology. This process, also known as peristalsis, occurs in various patterns of contraction and relaxation. Disruptions in peristalsis cause various GI diseases and disorders. Cell–cell interactions and GI disorders were understandable using 3D bioengineered models [29]. The cultivation of intestinal cells and tissues in dynamic bioreactor systems using 3D techniques is used to develop alternative treatments for intestinal diseases and to represent intestinal microenvironments in vivo. In a 2018 study by Zhou et al., a multifunctional bioreactor system containing pre-epithelized 3D silk scaffolds in a dynamic culture medium was designed for in vitro engineering of human intestinal tissues [184]. One notable example of bioreactor application is the development of gut-on-chip, which can simulate some human physiological features in a precise and controlled way [66, 95, 173].

Considerable attempts have been made to replicate the dynamic microenvironment of the gut. The perfusion bioreactor enabled to cultivate intestinal organoid units on biodegradable tubular polymer scaffolds that are compatible with live cell attachment. This method has been successful in maintaining organoids for up to two days, with potential implications for the long-term cultivation and bioengineering of intestinal cells [75]. Another bioreactor model, an electro-reactive elastomeric membrane utilized for in vitro modeling, imitate the mechanical patterns of intestinal tissue's contraction and relaxation cycles [22]. Pulsatile perfusion bioreactors have been employed to enhance the production of smooth muscle cells (SMCs) and collagen in 3D PCL scaffolds which were exposed to pulsatile stretching and shear stress for up to eight weeks [70]. Bioreactors have also been employed to induce differentiation of adipose-derived stem cells into SMC of decellularized scaffolds that increase SMC phenotype expression and examine their contractile phenotype on collagen gel coating.

Bioreactors for Musculoskeletal Tissue Engineering

Musculoskeletal tissue, comprising cartilage, bone, skeletal muscle, ligament and tendons, may experience dysfunction as a result of various factors such as trauma and natural injury. Tissue engineering ensures a practical solution to the limited availability of natural implants and the inadequacy of current treatment methods for musculoskeletal tissue defects. Engineered constructs must be kept under sterile conditions and suitable stimuli that mimic the natural tissue with biochemical and biomechanical settings. Several types and configurations of bioreactors have been developed for the renewal of musculoskeletal tissue, such as spinner flasks, RWV, flow perfusion systems, and mechanical loading devices. Bioreactor designs that utilize dynamic flow (for cartilage and bone tissue) and mechanical cyclic stretching (for tendons, ligaments, and bone) are among the most commonly preferred for orthopedic tissue. There are several bioreactors available for culturing musculoskeletal tissue [1]. In addition to providing an in vitro environment that simulates in vivo conditions for the tissue growth bioreactors are important in tissue engineering also by enabling systematic investigations of the living tissues responses to a wide variety of mechanical and biochemical signals [5].

Numerous bioreactors are designed to apply mechanical stimulation along a specific direction, which enables the growth of oriented 3D muscle bundles capable of contraction. This approach results in the production of in vitro constructs with aligned muscle fibers that mimic the anatomical structure of skeletal muscles. However, the development of bioreactors that provide biaxial or radial stimulation is less common in the literature. While these types of bioreactors are less prevalent, they could offer unique advantages for the engineering of complex tissues and organs that require multidirectional mechanical stimulation. Further research and development of such bioreactors may lead to new approaches for the engineering of tissues with intricate anatomical features and functions. In the study by [158] a bioreactor capable of mechanical stimulation of porcine derived diaphragmatic scaffolds in a radial manner was designed to promote alignment of cell and muscular fiber development in clinically relevant diaphragmatic constructs [158].

For example, the RWV bioreactor has been tested to improve transportation of nutrient and promote tissue growth and differentiation in cartilage tissue engineering applications. The RWV bioreactor ensures a suitable hydrodynamic environment for cartilage tissue growth and phenotype differentiation [133]. Flow perfusion bioreactors are a viable culturing technique for bone structures. Interstitial flow plays an important role in bone homeostasis. Bone grafts designed for accomplished cell stimulation, nutrient transport, and bone regeneration must be adequately perfused.

Bioreactors for Neural Tissue Engineering

Central nervous system (CNS) and peripheral nervous system (PNS) have limited self-renewal capacity in mammals, and disease and injury go about with persistent lack of functionality. Thus, CNS and the PNS renewal and renovation have significant challenges in tissue engineering [15]. The regeneration of damaged tissue is blocked because of the formation of astrocytes and scar of glia in the CNS. Mainly, research has focused on preventing further damage and stabilizing the affected area. But there are also studies focusing on repair processes to improve healing of loss functions related with CNS damage [142]. Recent neural tissue engineering strategies are developing for CNS and PNS tissue regeneration as potential treatments. Nervous tissue requires strict culture conditions and is even more difficult to induce differentiation and integration [152]. One of the most recent approaches often used to culture NSCs in vitro is the use of bioreactors in which biochemical or biological processes are tightly controlled and closely monitored [132]. In the study by Sun et al. [152], an approach using a closed-loop conduit bioreactor was used to introduce and culture Schwann cells on microfibers of longitudinally aligned viscose rayon and polystyrene model materials [152].

Bioreactors for Cardiovascular Tissue Engineering

Cardiac tissue engineering holds great promising approach for heart regeneration and modeling the pathophysiology of the human heart. Bioreactors are an essential tool in vascular tissue engineering and regulate physical and chemical parameters [114]. Bioreactors have been utilized for the amplification and differentiation of progenitor cells into the cardiomyocyte lineage [100]. Significant efforts have been made to develop functional and biomimetic cardiac structures. A number of bioengineered heart valve structures have shown encouraging results reaching clinical trials. In addition, small myocardial grafts have been well engineered using 3D bioreactors that provided precise control of specific stimulation parameters [55, 109].

Recently, a number of bioreactor systems have been developed in cardiovascular tissue engineering that mimic mechanical and chemical stimuli in vitro. The designs of these bioreactors are primarily concerned with tissue engineering of heart valves and blood vessels.

Different types of bioreactors have been used to develop supereminence heart valve tissue constructs. For example, dynamic and hydrodynamic, rotating, pulsatile, perfusion, and controlled cyclic stretching are the frequently used bioreactors [4]. A conventional vascular bioreactor typically consists of four main components: a cultivation room, an electric pump, a media reservoir, and a temperature controller. The equipment has been developed in vitro to mimic maintenance of blood flow balance. In 2018, Wolf et al. developed a compact, portable and versatile bioreactor system that enables cost-effective large scale and centralized production of autologous tissue-engineered vascular grafts and then transport of implants to patients [169].

Bioreactors for Cornea Tissue Engineering

Corneal diseases and injuries are prevalent worldwide and can lead to vision impairment or blindness if left untreated. One of the treatment methods is to replace the damaged corneal tissue with a healthy cornea from a donor, but there is a limited resource for donor tissue [91]. There are several studies to develop a tissue engineered cornea, which could potentially decrease the need for donor tissue and result in fewer post-transplant rejection rates. Although it is not widely utilized for the cornea, bioreactors have been reported in studies to assist to repopulate decellularized corneas with cells or as a culture method after initial seeding [46]. Also, in a study, the use of different materials in corneal tissue engineering bioreactors was investigated, considering that culture configuration, autoclaving and material surface preparation are important factors affecting cell viability [113]. The use of a rotary cell culture system for repopulation has been demonstrated in the literature, thereby encouraging cells to colonize the scaffold as they cannot attach elsewhere [30]. In an another study, a more sophisticated bioreactor system, a dynamic culture system for epithelial repopulation that mimics the in vivo air–liquid interface, has been reported [171].

It is known that the protection of ex vivocorneas in an environment that mimics natural physiological conditions allows the measurement of corneal thickness and its connection with cell functionality [129]. Application of mechanical stress to the cells is a potential method to control the in vitro phenotype of cells. Research has demonstrated that placing cells in an environment that simulates in vivo stress conditions can lead to the development of functional tissue equivalents [97]. A bioreactor has developed to obtain the possibility of using ex vivo corneas for functionality testing [54]. This study aimed to evaluate the survival of cells and tissue preservation of tissue structure in porcine corneas stored in a bioreactor that regenerates an intraocular pressure equivalent transcorneal pressure gradient and regenerates the corneal environment [54].

The tissue engineering approach is not limited to the cornea but has also come a long way for ocular tissues such as the lens and retina. There is a clinical need for ocular tissue substitutes [72]. In studies involving the combination of retinal organoid production with bioreactor technology, a bioprocess using RWV bioreactors to culture pluripotent stem cells sourced retinal organoids has been reported [37].

Bioreactors for Tissue Engineered Uterine and Bladder

Uterine histoarchitecture is highly complex due to the wide range of cellular and ECM molecules that include stratification of all uterine layers [3]. The process of decellularized tissues and organs, which involves the complete elimination of cellular components by forming a scaffold while preserving the structural, mechanical and biological properties of the ECM, is one of the most important stages of tissue engineering. Removal of cellular components is important to prevent immunogenicity. Maintaining the vascular network is important for the nutrient and oxygen supply to the uterus. Recellularization of the ECM construct is the next step of tissue reconstruction. To recellularize the organ, endothelial cells are perfused by the vascular network of the scaffold, usually through a bioreactor system, and must remain viable to induce cell growth in a controlled manner [3]. Although in vitro tissue growth seems to be successful in different scaffolds, it cannot show the same mechanical effects when transferred to an in vivo environment. This can because contraction forces in vitro cannot mimic the compression forces exerted by the surrounding environment after implantation. Additional mechanical properties necessary to enhance the urological outcomes of transplantation of cell seeded scaffolds can be achieved with the use of an in vitro bioreactor [148].

Similarly, in the field of bladder tissue engineering, simulating the normal physiological functions of filling and excretion with an in vitrobioreactor can improve the additional mechanical performance, tissue organization and maturation required to improve functional outcomes after implantation [139]. In the study by Niall F. Davis and Anthony Callan, a bladder bioreactor which consists of sealed pressure chamber with a pressurized gas containers, transparent window and silicone tubing was designed to physiologically mimic bladder dynamics [32]. Another widely used bioreactor to mimic bladder physiological conditions such as pressure, the modified BOSE BioDynamic® bioreactor, has been used in different studies in the literature [26, 90, 155].

Bioreactors for Kidney Tissue Engineering

The global prevalence of chronic kidney disease is increasing, and its therapeutic options are limited to peritoneal-dialysis, hemodialysis and kidney transplantation. Kidney transplantation is the most appropriate treatment as it improves long-term survival and is cost-effective compared to long-term dialysis. However, an insufficient number of donors is a major obstacle. In order to overcome this obstacle, the concept of creating an optional functional kidney graft using patient-specific stem cells has emerged and progress has been made in the last 10 years [163]. Tissue engineering has emerged as potential solutions to address the challenges in restoring kidney function. The kidney is a highly intricate organ comprising more than 30 distinct cell types with each type meticulously organized and functionally separated to create numerous nephrons, the fundamental working units of the kidney. Therefore, various types of 3D kidney structures have been developed using appropriate scaffolding systems and cell sources to replace such complex kidney tissues and restore kidney function [73].

Both organoid-based and decellularization-based construction strategies of stem cell types, including embryonic stem cells (ESCs) and induced pluripotent stem cells (iPSCs), have been widely studied for kidney tissue engineering [183].

To develop functional bioengineered kidneys for transplantation, an efficient recellularization strategy must be established. A recent approach for kidney tissue engineering involves the use of a bioreactor system that combines a cell infusion and perfusion culture system. This system can utilize various pumps such as a syringe pump, peristaltic pump, or pulsatile pump to allow for continuous cell infusion. The bioreactor system can also promote cell viability, nutrition, proliferation, and differentiation within the scaffold. The study by Song et al. summarized the production and transplantation of a rat kidney structure into rats using decellularization/recellularization techniques with a perfusion bioreactor system [147]. Briefly, the study involved, a decellularized rat kidney recellularized using human umbilical vein endothelial cells via the renal artery and rat neonatal cells via the ureter. Negative pressure was applied over the entire kidney chamber, followed by arterial perfusion culture in a bioreactor. The researchers were able to achieve a high recellularization rate, with 70% glomeruli present in the bioengineered kidney. In another study by Peloso et al., the decellularized kidney was recellularized in a customized pulsatile perfusion bioreactor providing optimal cell culture conditions [119]. Przepiorski et al., on the other hand, developed a strategy using spinner flask bioreactors to produce kidney organoids from iPSCs that is simple, strong, cost-effective, and allows large-scale organoid production [124]. Spinner flask bioreactors have been shown to enhance nutrient and oxygen perfusion [98, 115].

Bioreactors for Lung Tissue Engineering

Lung tissue engineering is an area of interest that promises a potential option for transplantation and pulmonary research. Lung biofabrication relies on seeding cells into a cell-free organ scaffold which is then cultured in a specialized bioreactor. Cell-free lung scaffold is achieved through conventional procedures that utilize physical, enzymatic and chemical agents. Similar to other organ tissue engineering, lung progenitor cells, autologous bone marrow/adipose tissue-derived MSCs or iPSCs are used for the biofabrication of the lung. A specialized bioreactor is employed to create an environment for circulatory perfusion and mechanical ventilation with physiological parameters to support the growth and function of the lung [45].

Significant advances have been made in bioreactors for lung engineering, both at the micro and macro scale. These are closed systems with pressure-controlled perfusion and ventilation. Ex vivo lung perfusion systems are systems developed for the protection and regeneration of the lungs [116].

Small-scale bioreactors have been improved to more accurately predict the lung environment than static air–liquid interface cultures. These bioreactors, described as "lung-on-a-chip", are to work as pharmacokinetic models for studying behavior of lung cells in drug discovery and studying drug toxicities [80]. Horizontal and vertical bioreactors have been developed for the upper respiratory tract (trachea, bronchus) [104]. For whole organ tissue engineering, intact lung scaffolds have been used and bioreactors have been developed for decellularization.

Bioreactors for Skin Tissue Engineering

Tissue-engineered organotypic full-thickness skin grafts may overcome delayed wound healing problems as they offer immediate coverage of the lesion by replacing both the dermal and epidermal layers of the skin. Many tissue-engineered skin graft studies have been successful, but since it has limited commercial utility in the clinic. The major limitation of available tissue-engineered skin substitutes is known as vascularization [58]. The application of bioreactors for more efficient graft production has been suggested. In a study conducted by Helmedag et al [60], the effect of a constant-flow bioreactor system on organotypic skin grafts was investigated and its use in the production of prevascularized organotypic full-thickness skin grafts was evaluated [60]. Also, there is a need for the development of bioreactors for tissue-engineered leather to advance its production to the clinic and for its production. Bioreactors must be able to culture skin structures at the air–liquid interface, due to the maturation of the epithelial layer to produce proper barrier function [85]. For the culture of metabolically challenging tissues, continuous perfusion with the medium is known to better support metabolic activity, rather than replacing new medium once per several passing days. In addition, the risk of contamination increases in long-term culture of tissue engineered constructs that require repetitive processing. Therefore, a closed system bioreactor at an air–liquid interface was designed for the production of autologous re-established skin, which would be suitable for both clinical and experimental use [151].

Conclusion

It is necessary to understand how complex physiological pathways work in the physical context of cells, tissues, and the interaction between different culture parameters to successfully continue tissue culture and tissue engineering applications and producing in vitro 3D tissues starting from isolated cells [79, 97]. Bioreactors have been used to produce vaccines and other drugs since the 1980s and have been evaluated to use in tissue engineering, allowing the application of robust, reproducible, and controllable culture conditions and making significant improvements in the design of the reactors [138]. Bioreactors stimulate cells to grow on a scaffold and produce an

ECM by mimicking their natural niche in vitro [121, 138]. The existence of complex biochemical, metabolic and mechanical stimuli and signals between cells and the cellular environment in tissue development requires an understanding of the specific cell behavior cultured at the molecular level to improve performance in tissue culture [121].

In tissue engineering, bioreactors focus on adequately mimicking the tissue's natural system, as described in the previous sections, so that natural tissue growth can be achieved. The reason why various bioreactor designs differ from each other specifically for the target tissue is due to the variety of tissues seen in the body [14]. In tissue engineering, especially micro and small-scale bioreactor designs allow examining and understanding the behavior of tissue cells at the molecular level, and to complete process development studies with microfluidic methods [97].

The main advantages of using bioreactors in tissue culture are improved mass carrying capacity, controlled and simultaneous traceable culture conditions, providing relevant stimuli in the environment, continuous media feeding and waste removal, reducing process steps to be processed, facilitating sampling and quality controls, preventing contamination and ensuring standardization [20, 164]. Thanks to these advantages, bioreactors are preferred, and progress is made in understanding the development of tissues in product development, research and clinical research. In addition to all these, the importance of bioreactors arises because the production under standard two-dimensional, static cell culture conditions cannot provide the stimuli and 3D space needed by the cells in the tissue regeneration process. The ability of bioreactors to simultaneously control and monitor many parameters (pH, O_2, CO_2, temperature) with sensors and detectors in tissue cultures provides significant convenience to researchers. However, all parameters should be optimized to meet tissue-specific physical, biochemical and mechanical requirements, and the right scaffold structures on which cells can expand should be selected. It is clear that no single design will fit for all tissues [138]. Identifying optimization requirements and equipment design for bioreactors requires an interdisciplinary approach [8, 121]. Considering the increasing technologies and application areas, bioreactors have an important place in the treatment of diseased or injured organs and tissues, in regenerative medicine, understanding tissues and cells, tissue engineering and accordingly, in the improvement and quality of human life.

References

1. Abousleiman RI, Sikavitsas VI (2006) Bioreactors for tissues of the musculoskeletal system. Adv Exp Med Biol 585:243–259. https://doi.org/10.1007/978-0-387-34133-0_17/COVER
2. Ahmed S, Bui MPN, Abbas A (2016) Paper-based chemical and biological sensors: engineering aspects. Biosens Bioelectron 77:249–263. https://doi.org/10.1016/J.BIOS.2015.09.038
3. Almeida GHD, Iglesia RP, Araujo MS, Carreira ACO, Dos Santos EX, Calomeno CVAQ, Miglino MA (2022) Uterine tissue engineering: where we stand and the challenges ahead. Tissue Eng Part B Rev 28(4):861–890. https://doi.org/10.1089/TEN.TEB.2021.0062

4. Amrollahi P, Tayebi L (2016) Bioreactors for heart valve tissue engineering: a review. J Chem Technol Biotechnol 91(4):847–856. https://doi.org/10.1002/JCTB.4825
5. An Y, Li D (2014) Engineering skeletal muscle tissue in bioreactor systems. Chin Med J 127(23):4130–4139. https://doi.org/10.3760/CMA.J.ISSN.0366-6999.20141076
6. Ashammakhi N, Nasiri R, Barros NR de, Tebon P, Thakor J, Goudie M, Shamloo A, Martin MG, Khademhosseni A (2020) Gut-on-a-chip: current progress and future opportunities. Biomaterials 255. https://doi.org/10.1016/J.BIOMATERIALS.2020.120196
7. Barron V, Lyons E, Stenson-Cox C, McHugh PE, Pandit A (2003) Bioreactors for cardiovascular cell and tissue growth: a review. Ann Biomed Eng 31(9):1017–1030. https://doi.org/10.1114/1.1603260
8. Bayir E, Sahinler M, Celtikoglu MM, Sendemir A, Sendemir A (2020) Bioreactors in tissue engineering: mimicking the microenvironment. Biomater Organ Tissue Regenerat New Technol Future Prosp 709–752. https://doi.org/10.1016/B978-0-08-102906-0.00018-0
9. Benam KH, Villenave R, Lucchesi C, Varone A, Hubeau C, Lee HH, Alves SE, Salmon M, Ferrante TC, Weaver JC, Bahinski A, Hamilton GA, Ingber DE (2015) Small airway-on-a-chip enables analysis of human lung inflammation and drug responses in vitro. Nature Methods 13(2):151–157. https://doi.org/10.1038/nmeth.3697
10. Bernaerts K, Servaes RD, Kooyman S, Versyck KJ, Van Impe JF (2002) Optimal temperature input design for estimation of the square root model parameters: parameter accuracy and model validity restrictions. Int J Food Microbiol 73(2–3):145–157
11. Beşkardeş IG, Demirtaş TT, Durukan MD, Gümüşderelioğlu M (2015) Microwave-assisted fabrication of chitosan–hydroxyapatite superporous hydrogel composites as bone scaffolds. J Tissue Eng Regen Med 9(11):1233–1246. https://doi.org/10.1002/TERM.1677
12. Bilodeau K (2004) Conception et validation d'un bioréacteur spécifique à la régénération du tissu artériel sous contraintes mécaniques.(unpublished master's thesis.) à la Faculté des études supérieures de l'Université Laval
13. Bilodeau K, Couet F, Boccafoschi F, Mantovani D (2005) Design of a perfusion bioreactor specific to the regeneration of vascular tissues under mechanical stresses. Artif Organs 29(11):906–912. https://doi.org/10.1111/J.1525-1594.2005.00154.X
14. Blose KJ, Krawiec JT, Weinbaum JS, Vorp DA (2014) Bioreactors for tissue engineering purposes. Regenerat Med Appl Organ Transplant 177–185. https://doi.org/10.1016/B978-0-12-398523-1.00013-6
15. Boni R, Ali A, Shavandi A, Clarkson AN (2018) Current and novel polymeric biomaterials for neural tissue engineering. J Biomed Sci 25(1):1–21. https://doi.org/10.1186/S12929-018-0491-8
16. Brown TD (2000) Techniques for mechanical stimulation of cells in vitro: a review. J Biomech 33(1):3–14. https://doi.org/10.1016/S0021-9290(99)00177-3
17. Bruijns B, van Asten A, Tiggelaar R, Gardeniers H (2016) Microfluidic devices for forensic DNA analysis: a review. Biosensors 6(3). https://doi.org/10.3390/BIOS6030041
18. Burk J, Plenge A, Brehm W, Heller S, Pfeiffer B, Kasper C (2016) Induction of tenogenic differentiation mediated by extracellular tendon matrix and short-term cyclic stretching. Stem Cells Int. https://doi.org/10.1155/2016/7342379
19. Carpentier B, Layrolle P, Legallais C (2011) Bioreactors for bone tissue engineering. Int J Art Organs 34(3):259–270. https://doi.org/10.5301/IJAO.2011.6333
20. Castro N, Ribeiro S, Fernandes MM, Ribeiro C, Cardoso V, Correia V, Minguez R, Lanceros-Mendez S (2020) Physically active bioreactors for tissue engineering applications. Adv Biosyst 4(10). https://doi.org/10.1002/ADBI.202000125
21. Catapano G, Czermak P, Eibl R, Eibl D, Pörtner R (2009) Bioreactor design and scale-up, pp 173–259. https://doi.org/10.1007/978-3-540-68182-3_5
22. Cei D, Costa J, Gori G, Frediani G, Domenici C, Carpi F, Ahluwalia A (2016) A bioreactor with an electro-responsive elastomeric membrane for mimicking intestinal peristalsis. Bioinspir Biomim 12(1):016001. https://doi.org/10.1088/1748-3190/12/1/016001
23. Chen HC, Hu YC (2006) Bioreactors for tissue engineering. Biotechnol Lett 28(18):1415–1423. https://doi.org/10.1007/S10529-006-9111-X

24. Chen J, Ding J, Wu Y, Zhang S, Zheng N, Yang J, Xu J (2021) Chromium oxide nanoparticle impaired osteogenesis and cellular response to mechanical stimulus. Int J Nanomed 16:6157–6170. https://doi.org/10.2147/IJN.S317430
25. Chen J, Yuan Z, Liu Y, Zheng R, Dai Y, Tao R, Xia H, Liu H, Zhang Z, Zhang W, Liu W, Cao Y, Zhou G (2017) Improvement of in vitro three-dimensional cartilage regeneration by a novel hydrostatic pressure bioreactor. Stem Cells Transl Med 6(3):982–991. https://doi.org/10.5966/SCTM.2016-0118
26. Chen L, Wei TQ, Wang Y, Zhang J, Li H, Wang KJ (2012) Simulated bladder pressure stimulates human bladder smooth muscle cell proliferation via the PI3K/SGK1 signaling pathway. J Urol 188(2):661–667. https://doi.org/10.1016/J.JURO.2012.03.112
27. Cho S, Islas-Robles A, Nicolini AM, Monks TJ, Yoon JY (2016) In situ, dual-mode monitoring of organ-on-a-chip with smartphone-based fluorescence microscope. Biosens Bioelectron 86:697–705. https://doi.org/10.1016/J.BIOS.2016.07.015
28. Cimetta E, Figallo E, Cannizzaro C, Elvassore N, Vunjak-Novakovic G (2009) Micro-bioreactor arrays for controlling cellular environments: design principles for human embryonic stem cell applications. Methods (San Diego, Calif.) 47(2):81–89
29. Collier CA, Mendiondo C, Raghavan S (2022) Tissue engineering of the gastrointestinal tract: the historic path to translation. J Biol Eng 16(1):1–12. https://doi.org/10.1186/S13036-022-00289-6
30. Dai Y, Chen J, Li H, Li S, Chen J, Ding Y, Wu J, Wang C, Tan M (2012) Characterizing the effects of VPA, VC and RCCS on Rabbit Keratocytes onto Decellularized Bovine Cornea. PLoS ONE 7(11):e50114. https://doi.org/10.1371/JOURNAL.PONE.0050114
31. Darling EM, Athanasiou KA (2003) Articular cartilage bioreactors and bioprocesses. Tissue Eng 9(1):9–26. https://doi.org/10.1089/107632703762687492
32. Davis NF, Callanan A (2016) Development of a bladder bioreactor for tissue engineering in urology. Methods Mol Biol 1502:213–221. https://doi.org/10.1007/7651_2015_309
33. de Lucena-Thomas JP, Boonprasirt P, Luetchford K, De Bank P, Ellis M (2020) Bed expansion properties of tissue engineering particles in a fluidized bed bioreactor. Biochem Eng J 160:107632
34. Deng D, Liu W, Xu F, Yang Y, Zhou G, Zhang WJ, Cui L, Cao Y (2009) Engineering human neo-tendon tissue in vitro with human dermal fibroblasts under static mechanical strain. Biomaterials 30(35):6724–6730. https://doi.org/10.1016/J.BIOMATERIALS.2009.08.054
35. Dermenoudis S, Missirlis Y (2010) Design of a novel rotating wall bioreactor for the in vitro simulation of the mechanical environment of the endothelial function. J Biomech 43(7):1426–1431. https://doi.org/10.1016/J.JBIOMECH.2010.01.012
36. Detamore MS, Athanasiou KA (2005) Use of a rotating bioreactor toward tissue engineering the temporomandibular joint disc. Tissue Eng 11(7–8):1188–1197. https://doi.org/10.1089/TEN.2005.11.1188
37. DiStefano T, Chen HY, Panebianco C, Kaya KD, Brooks MJ, Gieser L, Morgan NY, Pohida T, Swaroop A (2018) Accelerated and improved differentiation of retinal organoids from pluripotent stem cells in rotating-wall vessel bioreactors. Stem Cell Rep 10(1):300–313. https://doi.org/10.1016/J.STEMCR.2017.11.001
38. Donato D, De Napoli IE, Catapano G (2014) Model-based optimization of scaffold geometry and operating conditions of radial flow packed-bed bioreactors for therapeutic applications. Processes 2(1):34–57. https://doi.org/10.3390/PR2010034
39. Dutt K, Harris-Hooker S, Ellerson D, Layne D, Kumar R, Hunt R (2003) Generation of 3D retina-like structures from a human retinal cell line in a NASA bioreactor. Cell Transplant 12(7):717–731. https://doi.org/10.3727/000000003108747334
40. Eibl R, Werner S, Eibl D (2009) Bag bioreactor based on wave-induced motion: characteristics and applications. Adv Biochem Eng Biotechnol 115:55–87. https://doi.org/10.1007/10_2008_15
41. El Haj AJ, Cartmell SH (2010) Bioreactors for bone tissue engineering. Proc Inst Mech Eng Part H J Eng Med 224(12):1523–1532. https://doi.org/10.1243/09544119JEIM802

42. Elder SH, Goldstein SA, Kimura JH, Soslowsky LJ, Spengler DM (2001) Chondrocyte differentiation is modulated by frequency and duration of cyclic compressive loading. Ann Biomed Eng 29(6):476–482. https://doi.org/10.1114/1.1376696
43. Elomaa L, Yang YP (2017) Additive manufacturing of vascular grafts and vascularized tissue constructs. Tissue Eng Part B Rev 23(5):436–450. https://doi.org/10.1089/TEN.TEB.2016.0348
44. Ertl P, Sticker D, Charwat V, Kasper C, Lepperdinger G (2014) Lab-on-a-chip technologies for stem cell analysis. Trends Biotechnol 32(5):245–253. https://doi.org/10.1016/J.TIBTECH.2014.03.004
45. Farré R, Otero J, Almendros I, Navajas D (2018) Bioengineered lungs: a challenge and an opportunity. Archivos de Bronconeumología (English Edition) 54(1):31–38. https://doi.org/10.1016/J.ARBR.2017.09.010
46. Fernández-Pérez J, Ahearne M (2020) Decellularization and recellularization of cornea: progress towards a donor alternative. Methods 171:86–96. https://doi.org/10.1016/J.YMETH.2019.05.009
47. Freed LE, Guilak F, Guo XE, Gray ML, Tranquillo R, Holmes JW, Radisic M, Sefton MV, Kaplan D, Vunjak-Novakovic G (2006) Advanced tools for tissue engineering: scaffolds, bioreactors, and signaling. Tissue Eng 12:3285–3305
48. Gelinsky M, Bernhardt A, Milan F (2015) Bioreactors in tissue engineering: advances in stem cell culture and three-dimensional tissue constructs. Eng Life Sci 15(7):670–677. https://doi.org/10.1002/ELSC.201400216
49. Ghosh S, Srivastava N, Jha S, Jana Kumar N (2022) Spinner Flask Bioreactor in Tissue Engineering. YMER Digital 21(06):611–626. https://doi.org/10.37896/YMER21.06/61
50. Goodhart JM, Cooper JO, Smith RA, Williams JL, Haggard WO, Bumgardner JD (2014) Design and validation of a cyclic strain bioreactor to condition spatially-selective scaffolds in dual strain regimes. Processes 2(2):345–360. https://doi.org/10.3390/PR2020345
51. Griffiths B, Noe W (1998) Scale-up of animal cell, pp 1–2
52. Groeber F, Kahlig A, Loff S, Walles H, Hansmann J (2013) A bioreactor system for interfacial culture and physiological perfusion of vascularized tissue equivalents. Biotechnol J 8(3):308–316. https://doi.org/10.1002/biot.201200160
53. Grosberg A, Alford PW, McCain ML, Parker KK (2011) Ensembles of engineered cardiac tissues for physiological and pharmacological study: heart on a chip. Lab Chip 11(24):4165–4173
54. Guindolet D, Crouzet E, He Z, Herbepin P, Perrache C, Garcin T, Gauthier AS, Forest F, Peoc'h M, Gain P, Gabison E, Thuret G (2021) Epithelial regeneration in human corneas preserved in an active storage machine. Translat Vis Sci Technol 10(2):31–31. https://doi.org/10.1167/TVST.10.2.31
55. Günal G, Zihna G, Akel H, Okan M, Karaaslan C, Aydin HM (2022) Synthesis of hybrid myocardium constructs and in vitro characterization under mechanical stimulation. Mater Today Commun 33:104477. https://doi.org/10.1016/J.MTCOMM.2022.104477
56. Halberstadt CR, Hardin R, Bezverkov K, Snyder D, Allen L, Landeen L (1994) Biotechnol Bioeng 43:740
57. Hami LS, Green C, Leshinsky N, Markham E, Miller K, Craig S (2004) GMP production and testing of Xcellerated T Cells for the treatment of patients with CLL. Cytotherapy 6(6):554–562. https://doi.org/10.1080/14653240410005348
58. Hansmann J, Groeber F, Kahlig A, Kleinhans C, Walles H (2013) Bioreactors in tissue engineering—principles, applications and commercial constraints. Biotechnol J 8(3):298–307
59. Hao S, Ha L, Cheng G, Wan Y, Xia Y, Sosnoski DM, Mastro AM, Zheng SY (2018) A spontaneous 3D bone-on-a-chip for bone metastasis study of breast cancer cells. Small (Weinheim an Der Bergstrasse, Germany) 14(12). https://doi.org/10.1002/SMLL.201702787
60. Helmedag MJ, Weinandy S, Marquardt Y, Baron JM, Pallua N, Suschek CV, Jockenhoevel S (2015) The effects of constant flow bioreactor cultivation and keratinocyte seeding densities on prevascularized organotypic skin grafts based on a fibrin scaffold. Tissue Eng Part A 21(1–2):343. https://doi.org/10.1089/TEN.TEA.2013.0640

61. Henstock JR, Rotherham M, Rose JB, El Haj AJ (2013) Cyclic hydrostatic pressure stimulates enhanced bone development in the foetal chick femur in vitro. Bone 53(2):468–477. https://doi.org/10.1016/J.BONE.2013.01.010
62. Hoerstrup SP, Sodian R, Sperling JS, Vacanti JP, Mayer JE (2000) New pulsatile bioreactor for in vitro formation of tissue engineered heart valves. Tissue Eng 6(1):75–79. https://doi.org/10.1089/107632700320919
63. Huang CC, Hagar KL, Frost LE, Sun Y, Cheung HS (2004) Effects of cyclic compressive loading on chondrogenesis of rabbit bone-marrow derived mesenchymal stem cells. Stem Cells 22(3):313–323. https://doi.org/10.1634/STEMCELLS.22-3-313
64. Huang CP, Lu J, Seon H, Lee AP, Flanagan LA, Kim HY, Putnam AJ, Jeon NL (2009) Engineering microscale cellular niches for three-dimensional multicellular co-cultures. Undefined 9(12):1740–1748. https://doi.org/10.1039/B818401A
65. Huang JH, Harris JF, Nath P, Iyer R (2016) Hollow fiber integrated microfluidic platforms for in vitro co-culture of multiple cell types. Biomed Microdev 18(5). https://doi.org/10.1007/S10544-016-0102-Y
66. Huh D, Hamilton GA, Ingber DE (2011) From 3D cell culture to organs-on-chips. Trends Cell Biol 21(12):745–754. https://doi.org/10.1016/J.TCB.2011.09.005
67. Hundt B, Best C, Schlawin N, Kaßner H, Genzel Y, Reichl U (2007) Establishment of a mink enteritis vaccine production process in stirred-tank reactor and Wave® bioreactor microcarrier culture in 1–10 L scale. Vaccine 25(20):3987–3995. https://doi.org/10.1016/J.VACCINE.2007.02.061
68. Jang KJ, Suh KY (2010) A multi-layer microfluidic device for efficient culture and analysis of renal tubular cells. Lab Chip 10(1):36–42. https://doi.org/10.1039/B907515A
69. Jasuja H, Kar S, Katti DR, Katti KS (2021) Perfusion bioreactor enabled fluid-derived shear stress conditions for novel bone metastatic prostate cancer testbed. Biofabrication 13(3). https://doi.org/10.1088/1758-5090/abd9d6
70. Jeong SI, Kwon JH, Lim JI, Cho SW, Jung Y, Sung WJ, Kim SH, Kim YH, Lee YM, Kim BS, Choi CY, Kim SJ (2005) Mechano-active tissue engineering of vascular smooth muscle using pulsatile perfusion bioreactors and elastic PLCL scaffolds. Biomaterials 26(12):1405–1411. https://doi.org/10.1016/J.BIOMATERIALS.2004.04.036
71. Jungbauer S, Gao H, Spatz JP, Kemkemer R (2008) Two characteristic regimes in frequency-dependent dynamic reorientation of fibroblasts on cyclically stretched substrates. Biophys J 95(7):3470–3478. https://doi.org/10.1529/BIOPHYSJ.107.128611
72. Karamichos D (2015) Ocular tissue engineering: current and future directions. J Func Biomater 6(1):77–80. https://doi.org/10.3390/JFB6010077
73. Kim JH, Atala A, Yoo JJ (2020) Tissue engineering of the kidney. Principles Tissue Eng 825–843. https://doi.org/10.1016/B978-0-12-818422-6.00047-2
74. Kim JJ, Ellett F, Thomas CN, Jalali F, Anderson RR, Irimia D, Raff AB (2019) A microscale, full-thickness, human skin on a chip assay simulating neutrophil responses to skin infection and antibiotic treatments. Lab Chip 19(18):3094–3103. https://doi.org/10.1039/C9LC00399A
75. Kim SS, Penkala R, Abrahimi P (2007) A perfusion bioreactor for intestinal tissue engineering. J Surg Res 142(2):327–331. https://doi.org/10.1016/J.JSS.2007.03.039
76. Kim S, Kim W, Lim S, Jeon JS (2017) Vasculature-on-a-chip for in vitro disease models. Bioengineering (Basel, Switzerland) 4(1). https://doi.org/10.3390/BIOENGINEERING4010008
77. Ko IK, Atala A, Yoo JJ (2018) Bioreactors for regenerative medicine in urology. In: Clinical regenerative medicine in urology. Singapore, Springer Singapore, pp 87–104. https://doi.org/10.1007/978-981-10-2723-9_4
78. Kopp MRG, Arosio P (2018) Microfluidic approaches for the characterization of therapeutic proteins. J Pharm Sci 107(5):1228–1236. https://doi.org/10.1016/J.XPHS.2018.01.001
79. Korossis S, Bolland F, Kearney JN, Fisher J (2005) Bioreactors in tissue engineering. Top Tissue Eng 2:1–23
80. Kumar Mahto S, Tenenbaum-Katan J, Sznitman J (2012) Respiratory physiology on a chip. Scientifica 2012:1–12. https://doi.org/10.6064/2012/364054

81. Kumar A, Starly B (2015) Large scale industrialized cell expansion: producing the critical raw material for biofabrication processes. Biofabrication 7(4). https://doi.org/10.1088/1758-5090/7/4/044103
82. Latifi N, Heris HK, Thomson SL, Taher R, Kazemirad S, Sheibani S, Li-Jessen NYK, Vali H, Mongeau L (2016) A flow perfusion bioreactor system for vocal fold tissue engineering applications. Tissue Eng Part C Methods 22(9):823–838. https://doi.org/10.1089/TEN.TEC.2016.0053
83. Lee CF, Haase C, Deguchi S, Kaunas R (2010) Cyclic stretch-induced stress fiber dynamics—dependence on strain rate, Rho-kinase and MLCK. Biochem Biophys Res Commun 401(3):344–349. https://doi.org/10.1016/J.BBRC.2010.09.046
84. Lee CY, Chang CL, Wang YN, Fu LM (2011) Microfluidic mixing: a review. Int J Mol Sci 12(5):3263. https://doi.org/10.3390/IJMS12053263
85. Lee DY, Ahn HT, Cho KH (2000) A new skin equivalent model: dermal substrate that combines de-epidermized dermis with fibroblast-populated collagen matrix. J Dermatol Sci 23(2):132–137. https://doi.org/10.1016/S0923-1811(00)00068-2
86. Lichtenberg A, Tudorache I, Cebotari S, Ringes-Lichtenberg S, Sturz G, Hoeffler K, Hurscheler C, Brandes G, Hilfiker A, Haverich A (2006) In vitro re-endothelialization of detergent decellularized heart valves under simulated physiological dynamic conditions. Biomaterials 27(23):4221–4229. https://doi.org/10.1016/J.BIOMATERIALS.2006.03.047
87. Lim D, Renteria ES, Sime DS, Ju YM, Kim JH, Criswell T, Shupe TD, Atala A, Marini FC, Gurcan MN, Soker S, Hunsberger J, Yoo JJ (2022) Bioreactor design and validation for manufacturing strategies in tissue engineering. Bio-Design Manuf 5(1):43–63
88. Liu D, Zhang H, Fontana F, Hirvonen JT, Santos HA (2017) Microfluidic-assisted fabrication of carriers for controlled drug delivery. Lab Chip 17(11):1856–1883. https://doi.org/10.1039/C7LC00242D
89. Liu L, Wu W, Tuo X, Geng W, Zhao J, Wei J, Yan X, Yang W, Li L, Chen F (2010) Novel strategy to engineer trachea cartilage graft with marrow mesenchymal stem cell macroaggregate and hydrolyzable scaffold. Artif Organs 34(5):426–433. https://doi.org/10.1111/J.1525-1594.2009.00884.X
90. Luo DY, Wazir R, Du C, Tian Y, Yue X, Wei TQ, Wang KJ (2015) Magnitude-dependent proliferation and contractility modulation of human bladder smooth muscle cells under physiological stretch. World J Urol 33(11):1881–1887. https://doi.org/10.1007/S00345-015-1509-4
91. Maghsoudlou P, Sood G, Akhondi H (2022) Cornea transplantation
92. Mahdinia E, Cekmecelioglu D, Demirci A (2019) Bioreactor scale-up, pp 213–236. https://doi.org/10.1007/978-3-030-16230-6_7
93. Mandenius CF (2016) Challenges for bioreactor design and operation. Bioreactors 1–34. https://doi.org/10.1002/9783527683369.ch1
94. Marei I, Abu Samaan T, Al-Quradaghi MA, Farah AA, Mahmud SH, Ding H, Triggle CR (2022) 3D tissue-engineered vascular drug screening platforms: promise and considerations. Front Cardiov Med 9:355. https://doi.org/10.3389/FCVM.2022.847554/BIBTEX
95. Marrero D, Pujol-Vila F, Vera D, Gabriel G, Illa X, Elizalde-Torrent A, Alvarez M, Villa R (2021) Gut-on-a-chip: mimicking and monitoring the human intestine. Biosens Bioelectron 181:113156. https://doi.org/10.1016/J.BIOS.2021.113156
96. Marsh D (2017) Engineering characterisation of a rocked bag bioreactor for improved process development and scale-up. University College London
97. Martin I, Wendt D, Heberer M (2004) The role of bioreactors in tissue engineering. Trends Biotechnol 22(2):80–86. https://doi.org/10.1016/J.TIBTECH.2003.12.001
98. Martin Y, Vermette P (2005) Bioreactors for tissue mass culture: design, characterization, and recent advances. Biomaterials 26(35):7481–7503. https://doi.org/10.1016/J.BIOMATERIALS.2005.05.057
99. Masoumi N, Howell MC, Johnson KL, Niesslein MJ, Gerber G, Engelmayr GC (2014) Design and testing of a cyclic stretch and flexure bioreactor for evaluating engineered heart valve tissues based on poly(glycerol sebacate) scaffolds. Proc Inst Mech Eng Part H J Eng Med 228(6):576–586. https://doi.org/10.1177/0954411914534837

100. Matsuura K, Wada M, Konishi K, Sato M, Iwamoto U, Sato Y, Tachibana A, Kikuchi T, Iwamiya T, Shimizu T, Yamashita JK, Yamato M, Hagiwara N, Okano T (2012) Fabrication of mouse embryonic stem cell-derived layered cardiac cell sheets using a bioreactor culture system. PLoS ONE 7(12):e52176. https://doi.org/10.1371/JOURNAL.PONE.0052176
101. Mauck RL, Soltz MA, Wang CCB, Wong DD, Chao PHG, Valhmu WB, Hung CT, Ateshian GA (2000) Functional tissue engineering of articular cartilage through dynamic loading of chondrocyte-seeded agarose gels. J Biomech Eng 122(3):252–260. https://doi.org/10.1115/1.429656
102. McLoughlin ST, Mahadik B, Fisher J (2022) Bioreactors and scale-up in bone tissue engineering. Bone Tissue Eng 225–247. https://doi.org/10.1007/978-3-030-92014-2_10
103. Melke J, Zhao F, Rietbergen B, Ito K, Hofmann S (2018) Localisation of mineralised tissue in a complex spinner flask environment correlates with predicted wall shear stress level localisation. Europ Cells Mater 36:57. https://doi.org/10.22203/ECM.V036A05
104. Miller C, George S, Niklason L (2010) Developing a tissue-engineered model of the human bronchiole. J Tissue Eng Regen Med 4(8):619–627. https://doi.org/10.1002/TERM.277
105. Mirzabe AH, Hajiahmad A, Fadavi A, Rafiee S (2022) Design of nutrient gas-phase bioreactors: a critical comprehensive review. Bioprocess Biosyst Eng 45(8):1239–1265. https://doi.org/10.1007/s00449-022-02728-6
106. Mofazzal Jahromi MA, Abdoli A, Rahmanian M, Bardania H, Bayandori M, Moosavi Basri SM, Kalbasi A, Aref AR, Karimi M, Hamblin MR (2019) Microfluidic brain-on-a-chip: perspectives for mimicking neural system disorders. Mol Neurobiol 56(12):8489–8512. https://doi.org/10.1007/S12035-019-01653-2
107. Mol A, Bouten CVC, Zünd G, Günter CI, Visjager JF, Turina MI, Baaijens FPT, Hoerstrup SP (2003) The relevance of large strains in functional tissue engineering of heart valves. Thoracic Cardiov Surg 51(2):78–83. https://doi.org/10.1055/S-2003-38993
108. Moysidou CM, Barberio C, Owens RM (2021) Advances in engineering human tissue models. Front Bioeng Biotechnol 8:1566. https://doi.org/10.3389/FBIOE.2020.620962
109. Nakazato T, Kawamura T, Uemura T, Liu L, Li J, Sasai M, Harada A, Ito E, Iseoka H, Toda K, Sawa Y, Miyagawa S (2022) Engineered three-dimensional cardiac tissues maturing in a rotating wall vessel bioreactor remodel diseased hearts in rats with myocardial infarction. Stem Cell Rep 17(5):1170–1182. https://doi.org/10.1016/J.STEMCR.2022.03.012
110. Niklason LE, Gao J, Abbott WM, Hirschi KK, Houser S, Marini R, Langer R (1999) Functional arteries grown in vitro. Science 284(5413):489–493. https://doi.org/10.1126/SCIENCE.284.5413.489
111. Obregón R, Ramón Azcón J, Ahadian S (2017) Bioreactors in tissue engineering. Tissue engineering for artificial organs: regenerative medicine, smart diagnostics and personalized medicine. In Hasan A (ed) Tissue engineering for artificial organs: regenerative medicine, smart, vol 2, pp 169–213
112. Ortega MA, Fernández-Garibay X, Castaño AG, De Chiara F, Hernández-Albors A, Balaguer-Trias J, Ramón-Azcón J (n.d.) Muscle-on-a-chip with an on-site multiplexed biosensing system for in situ monitoring of secreted IL-6 and TNF-α. Pubs.Rsc.Org. https://doi.org/10.1039/x0xx00000x
113. Orwin E, Shah A, Voorhees A, Ravi V (2007) Bioreactor design for cornea tissue engineering: material–cell interactions. Acta Biomater 3(6):1041–1049. https://doi.org/10.1016/J.ACTBIO.2007.04.008
114. Paez-Mayorga J, Hernández-Vargas G, Ruiz-Esparza GU, Iqbal HMN, Wang X, Zhang YS, Parra-Saldivar R, Khademhosseini A (2019) Bioreactors for cardiac tissue engineering. Adv Healthcare Mater 8(7):1701504. https://doi.org/10.1002/ADHM.201701504
115. Pagliuca FW, Millman JR, Gürtler M, Segel M, Van Dervort A, Ryu JH, Peterson QP, Greiner D, Melton DA (2014) Generation of functional human pancreatic β cells in vitro. Cell 9, 159(2):428–39. https://doi.org/10.1016/j.cell.2014.09.040
116. Panoskaltsis-Mortari A (2015) Bioreactor development for lung tissue engineering. Curr Transplant Rep 2(1):90. https://doi.org/10.1007/S40472-014-0048-Z

117. Pasirayi G, Auger V, Scott SM, Rahman PKSM, Islam M, O'hare L, Ali Z (2011) Microfluidic bioreactors for cell culturing: a review. Micro Nanosyst 3(2):137–160. https://doi.org/10.2174/1876402911103020137
118. Pei M, Solchaga LA, Seidel J, Zeng L, Vunjak-Novakovic G, Caplan AI, Freed LE (2002) Bioreactors mediate the effectiveness of tissue engineering scaffolds. Wiley Online Library 16(12):1691–1694. https://doi.org/10.1096/fj.02-0083fje
119. Peloso A, Ferrario J, Maiga B, Benzoni I, Bianco C, Citro A, Currao M, Malara A, Gaspari A, Balduini A, Abelli M, Piemonti L, Dionigi P, Orlando G, Maestri M (2015) Creation and implantation of acellular rat renal ECM-based scaffolds. Organogenesis 11(2):58–74. https://doi.org/10.1080/15476278.2015.1072661
120. Pinto DS, da Silva CL, Cabral JM (2019) Scalable expansion of mesenchymal stem/stromal cells in bioreactors: a focus on hydrodynamic characterization. https://doi.org/10.1016/B978-0-12-801238-3.65541-1
121. Plunkett N, O'Brien FJ (2011) Bioreactors in tissue engineering. Technol Health Care 19(1):55–69. https://doi.org/10.3233/THC-2011-0605
122. Podichetty JT, Bhaskar PR, Singarapu K, Madihally SV (2015) Multiple approaches to predicting oxygen and glucose consumptions by HepG2 cells on porous scaffolds in an axial-flow bioreactor. Biotechnol Bioeng 112(2):393–404. https://doi.org/10.1002/bit.25355
123. Pörtner R, Nagel-Heyer S, Goepfert C, Adamietz P, Meenen NM (2005) Bioreactor design for tissue engineering. J Biosci Bioeng 100(3):235–245. https://doi.org/10.1263/jbb.100.235
124. Przepiorski A, Sander V, Tran T, Hollywood JA, Sorrenson B, Shih JH, Wolvetang EJ, McMahon AP, Holm TM, Davidson AJ (2018) A simple bioreactor-based method to generate kidney organoids from pluripotent stem cells. Stem Cell Rep 11(2):470. https://doi.org/10.1016/J.STEMCR.2018.06.018
125. Rafiq QA, Coopman K, Hewitt CJ (2013) Scale-up of human mesenchymal stem cell culture: current technologies and future challenges. Curr Opin Chem Eng 2(1):8–16. https://doi.org/10.1016/j.coche.2013.01.005
126. Ratcliffe A, Niklason LE (2002) Bioreactors and bioprocessing for tissue engineering. Ann N Y Acad Sci 961:210–215. https://doi.org/10.1111/j.1749-6632.2002.tb03087.x
127. Reinwald Y, Leonard KH, Henstock JR, Whiteley JP, Osborne JM, Waters SL, Levesque P, El Haj AJ (2015) Evaluation of the growth environment of a hydrostatic force bioreactor for preconditioning of tissue-engineered constructs. Tissue Eng Part C Methods 21(1):1–14. https://doi.org/10.1089/ten.tec.2013.0476
128. Rivron N, Rouwkema J, Truckenmüller R, Karperien M, De Boer J, Van Blitterswijk CA (2009) Tissue assembly and organization: developmental mechanisms in microfabricated tissues. Elsevier 30(28):4851–4858. https://www.sciencedirect.com/science/article/pii/S0142961209006486
129. Rolev K, O'Donovan DG, Coussons P, King L, Rajan MS (2018) Feasibility study of human corneal endothelial cell transplantation using an in vitro human corneal Model. Cornea 37(6):778–784. https://doi.org/10.1097/ICO.0000000000001555
130. Ruhrberg C, Gerhardt H, Golding M, Watson R, Ioannidou S, Fujisawa H, Betsholtz C, Shima DT (2002) Spatially restricted patterning cues provided by heparin-binding VEGF-A control blood vessel branching morphogenesis. Genes Dev 16(20):2684–2698. https://doi.org/10.1101/GAD.242002
131. Sabatino MA, Santoro R, Gueven S, Jaquiery C, Wendt DJ, Martin I, Moretti M, Barbero A (2015) Cartilage graft engineering by co-culturing primary human articular chondrocytes with human bone marrow stromal cells. J Tissue Eng Regenerative Med 9(12):1394–1403. https://doi.org/10.1002/term.1661
132. Sagita ID, Whulanza Y, Dhelika R, Nurhadi I (2018) Designing electrical stimulated bioreactors for nerve tissue engineering. AIP Conf Proc 1933(1):040019. https://doi.org/10.1063/1.5023989
133. Saini S, Wick TM (2003) Concentric cylinder bioreactor for production of tissue engineered cartilage: effect of seeding density and hydrodynamic loading on construct development. Biotechnol Prog 19(2):510–521. https://doi.org/10.1021/BP0256519

134. Salazar BH, Cashion AT, Dennis RG, Birla RK (2015) Development of a cyclic strain bioreactor for mechanical enhancement and assessment of bioengineered myocardial constructs. Cardiovasc Eng Technol 6(4):533–545. https://doi.org/10.1007/S13239-015-0236-8
135. Salehi-Nik N, Amoabediny G, Pouran B, Tabesh H, Shokrgozar MA, Haghighipour N, Khatibi N, Anisi F, Mottaghy K, Zandieh-Doulabi B (2013) Engineering parameters in bioreactor's design: a critical aspect in tissue engineering. BioMed Res Int 2013(3). https://doi.org/10.1155/2013/762132
136. Schuerlein S, Schwarz T, Krziminski S, Gätzner S, Hoppensack A, Schwedhelm I, Schweinlin M, Walles H, Hansmann J (2017) A versatile modular bioreactor platform for tissue engineering. Biotechnol J 12(2):1600326. https://doi.org/10.1002/BIOT.201600326
137. Schwarz RP, Goodwin TJ, Wolf DA (1992) Cell culture for three-dimensional modeling in rotating-wall vessels: an application of simulated microgravity. J Tissue Cult Methods 14(2):51–57. https://doi.org/10.1007/BF01404744
138. Selden C, Bioengineering BF (2018) Role of bioreactor technology in tissue engineering for clinical use and therapeutic target design. Mdpi.Com. https://doi.org/10.3390/bioengineering5020032
139. Serrano-Aroca Á, Vera-Donoso CD, Moreno-Manzano V (2018) Bioengineering approaches for bladder regeneration. Int J Mol Sci 19(6):1796. https://doi.org/10.3390/IJMS19061796
140. Shachar M, Benishti N, Cohen S (2012) Effects of mechanical stimulation induced by compression and medium perfusion on cardiac tissue engineering. Biotechnol Prog 28(6):1551–1559. https://doi.org/10.1002/btpr.1633
141. Shields CW IV, Reyes CD, López GP (2015) Microfluidic cell sorting: a review of the advances in the separation of cells from debulking to rare cell isolation. Lab Chip 15(5):1230–1249. https://doi.org/10.1039/C4LC01246A
142. Shoichet MS, Tate CC, Douglas Baumann M, LaPlaca MC (2008) Strategies for regeneration and repair in the injured central nervous system. Indwelling Neur Implant Strat Contend In Vivo Environ 221–244. https://doi.org/10.1201/9781420009309.ch8
143. Shrestha J, Razavi Bazaz S, Aboulkheyr Es H, Yaghobian Azari D, Thierry B, Ebrahimi Warkiani M, Ghadiri M (2020) Lung-on-a-chip: the future of respiratory disease models and pharmacological studies. Taylor & Francis 40(2):213–230. https://doi.org/10.1080/07388551.2019.1710458
144. Sikavitsas VI, Bancroft GN, Mikos AG (2002) Formation of three-dimensional cell/polymer constructs for bone tissue engineering in a spinner flask and a rotating wall vessel bioreactor. J Biomed Mater Res 62(1):136–148. https://doi.org/10.1002/JBM.10150
145. Singh V (1999) Disposable bioreactor for cell culture using wave-induced agitation. Cytotechnology 30(1–3):149–158. https://doi.org/10.1023/A:1008025016272
146. Somerville RPT, Devillier L, Parkhurst MR, Rosenberg SA, Dudley ME (2012) Clinical scale rapid expansion of lymphocytes for adoptive cell transfer therapy in the WAVE ®bioreactor. J Trans Med 10(1). https://doi.org/10.1186/1479-5876-10-69
147. Song JJ, Guyette JP, Gilpin SE, Gonzalez G, Vacanti JP, Ott HC (2013) Regeneration and experimental orthotopic transplantation of a bioengineered kidney. Nature Med 19(5):646–651. https://doi.org/10.1038/nm.3154
148. Southgate J, Cross W, Eardley I, Thomas DFM, Trejdosiewicz LK (2005) Bladder reconstruction—from cells to materials. Proc Inst Mech Eng Part H J Eng Med 217(4):311–316. https://doi.org/10.1243/095441103322060776
149. Spier MR, Vandenberghe LPS, Medeiros ABP, Soccol CR (2011) Application of Different Types of Bioreactors in Bioprocesses, Bioreactors. Bioreactors Des Prop Appl pp 55–90
150. Stiehler M, Bünger C, Baatrup A, Lind M, Kassem M, Mygind T (2009) Effect of dynamic 3-D culture on proliferation, distribution, and osteogenic differentiation of human mesenchymal stem cells. J Biomed Mater Res Part A 89(1):96–107. https://doi.org/10.1002/JBM.A.31967
151. Sun T, Norton D, Haycock JW, Ryan AJ, MacNeil S (2006) Development of a closed bioreactor system for culture of tissue-engineered skin at an air–liquid interface. Tissue Eng 11(11–12):1824–1831. https://doi.org/10.1089/TEN.2005.11.1824

152. Sun T, Norton D, Vickers N, McArthur SL, Neil SM, Ryan AJ, Haycock JW (2008) Development of a bioreactor for evaluating novel nerve conduits. Biotechnol Bioeng 99(5):1250–1260. https://doi.org/10.1002/BIT.21669
153. Tao Y, Shih J, Sinacore M, Ryll T, Yusuf-Makagiansar H (2011) Development and implementation of a perfusion-based high cell density cell banking process. Biotechnol Prog 27(3):824–829. https://doi.org/10.1002/BTPR.599
154. Theodoridis K, Aggelidou E, et al (2020) An effective device and method for enhanced cell growth in 3D scaffolds: investigation of cell seeding and proliferation under static and dynamic conditions. Elsevier. https://doi.org/10.1016/j.msec.2020.111060
155. Tiemessen D, de Jonge P, Daamen W, Feitz W, Geutjes P, Oosterwijk E (2017) The effect of a cyclic uniaxial strain on urinary bladder cells. World J Urol 35(10):1531–1539. https://doi.org/10.1007/S00345-017-2013-9/FIGURES/6
156. Timmins NE, Palfreyman E, Marturana F, Dietmair S, Luikenga S, Lopez G, Fung YL, Minchinton R, Nielsen LK (2009) Clinical scale ex vivo manufacture of neutrophils from hematopoietic progenitor cells. Biotechnol Bioeng 104(4):832–840. https://doi.org/10.1002/BIT.22433
157. Titze I, Hitchcock R, Broadhead K, Webb K, Li W, Gray SD (2004) Design and validation of a bioreactor for engineering vocal fold tissues under combined tensile and vibrational stresses. J Biomech 37:1521–1529. https://doi.org/10.1016/j.jbiomech.2004.01.007
158. Todros S, Spadoni S, Maghin E, Piccoli M, Pavan PG (2021) A novel bioreactor for the mechanical stimulation of clinically relevant scaffolds for muscle tissue engineering purposes. Processes 9(3):474. https://doi.org/10.3390/PR9030474
159. Tondon A, Kaunas R (2014) The direction of stretch-induced cell and stress fiber orientation depends on collagen matrix stress. PLoS ONE 9(2). https://doi.org/10.1371/JOURNAL.PONE.0089592
160. Tondreau MY, Laterreur V, Gauvin R, Vallières K, Bourget JM, Lacroix D, Tremblay C, Germain L, Ruel J, Auger FA (2015) Mechanical properties of endothelialized fibroblast-derived vascular scaffolds stimulated in a bioreactor. Acta Biomater 18:176–185. https://doi.org/10.1016/J.ACTBIO.2015.02.026
161. Tran SC, Cooley AJ, Elder SH (2011) Effect of a mechanical stimulation bioreactor on tissue engineered, scaffold-free cartilage. Biotechnol Bioeng 108(6):1421–1429. https://doi.org/10.1002/BIT.23061
162. Ulbrich C, Wehland M, Pietsch J, Aleshcheva G, Wise P, Van Loon J, Magnusson N, Infanger M, Grosse J, Eilles C, Sundaresan A, Grimm D (2014) The impact of simulated and real microgravity on bone cells and mesenchymal stem cells. BioMed Res Int. https://doi.org/10.1155/2014/928507
163. Uzarski JS, Xia Y, Belmonte JCI, Wertheim JA (2014) New strategies in kidney regeneration and tissue engineering. Curr Opin Nephrol Hypertens 23(4):399–405. https://doi.org/10.1097/01.MNH.0000447019.66970.EA
164. Vrana N, Knopf-Marques H, Barthes J (Eds) (2020) Biomaterials for organ and tissue regeneration: new technologies and future prospects. Woodhead Publishing
165. Watanabe S, Inagaki S, Kinouchi I, Takai H, Masuda Y, Mizuno S (2005) Hydrostatic pressure/perfusion culture system designed and validated for engineering tissue. J Biosci Bioeng 100:105–111. https://doi.org/10.1263/jbb.100.105
166. Webster A, Dyer CE, Haswell SJ, Greenman J (2010) A microfluidic device for tissue biopsy culture and interrogation. Anal Methods 2(8):1005–1007. https://doi.org/10.1039/C0AY00293C
167. Weston MW, Yoganathan AP (2001) Biosynthetic activity in heart valve leaflets in response to in vitro flow environments. Ann Biomed Eng 29(9):752–763. https://doi.org/10.1114/1.1397794
168. Wolf DA, Kleis SJ (2016) Principles of analogue and true microgravity bioreactors to tissue engineering. Effect Spaceflight Spaceflight Anal Cult Human Microbial Cells Novel Insights Dis Mech 39–60. https://doi.org/10.1007/978-1-4939-3277-1

169. Wolf F, Rojas González DM, Steinseifer U, Obdenbusch M, Herfs W, Brecher C, Jockenhoevel S, Mela P, Schmitz-Rode T (2018) VascuTrainer: a mobile and disposable bioreactor system for the conditioning of tissue-engineered vascular grafts. Ann Biomed Eng 46(4):616–626. https://doi.org/10.1007/S10439-018-1977
170. Wong AK, Llanos P, Boroda N, Rosenberg SR, Rabbany SY (2016) A parallel-plate flow chamber for mechanical characterization of endothelial cells exposed to laminar shear stress. Cell Molecul Bioeng 9(1):127–138. https://doi.org/10.1007/S12195-015-0424-5
171. Wu Z, Zhou Q, Duan H, Wang X, Xiao J, Duan H, Li N, Li C, Wan P, Liu Y, Song Y, Zhou C, Huang Z, Wang Z (2014) Reconstruction of auto-tissue-engineered lamellar cornea by dynamic culture for transplantation: a rabbit model. PLoS ONE 9(4):e93012. https://doi.org/10.1371/JOURNAL.PONE.0093012
172. Wung N, Acott SM, Tosh D, Ellis MJ (2014) Hollow fibre membrane bioreactors for tissue engineering applications. Biotech Lett 36(12):2357–2366. https://doi.org/10.1007/s10529-014-1619-x
173. Xiang Y, Wen H, Yu Y, Li M, Fu X, Huang S (2020) Gut-on-chip: recreating human intestine in vitro. J Tissue Eng 11. https://doi.org/10.1177/2041731420965318
174. Xie Y, Lu J (2016) Bioreactors for bone tissue engineering. Biomech Biomater Orthoped 115–122. https://doi.org/10.5301/ijao.2011.6333
175. Yeatts AB, Fisher JP (2011) Bone tissue engineering bioreactors: dynamic culture and the influence of shear stress. Bone 48(2):171–181. https://doi.org/10.1016/j.bone.2010.09.138
176. Yin CH, Chen W, Hsiao CC, Kuo CY, Chen CL, Wu WT (2007) Production of mouse embryoid bodies with hepatic differentiation potential by stirred tank bioreactor. Biosci Biotechnol Biochem 71(3):728–734. https://doi.org/10.1271/bbb.60568
177. Yoon HH, Bhang SH, Shin JY, Shin J, Kim BS (2012) Enhanced cartilage formation via three-dimensional cell engineering of human adipose-derived stem cells. Tissue Eng Part A 18(19–20):1949–1956. https://doi.org/10.1089/TEN.TEA.2011.0647
178. You JB, Kang K, Tran TT, Park H, Hwang WR, Kim JM, Im SG (2015) PDMS-based turbulent microfluidic mixer. Lab Chip 15(7):1727–1735. https://doi.org/10.1039/c5lc00070j
179. Yu H, Chong SK, Hassanbhai AM, Teng Y, Balachander G, Muthukumaran P, Wen F, Teoh SH (2020) Principles of bioreactor design for tissue engineering. Principles Tissue Eng 179–203. https://doi.org/10.1016/B978-0-12-818422-6.00012-5
180. Zhang B, Montgomery M, Chamberlain MD, Ogawa S, Korolj A, Pahnke A, Wells LA, Masse S, Kim J, Reis L, Momen A, Nunes SS, Wheeler AR, Nanthakumar K, Keller G, Sefton MV, Radisic M (2016) Biodegradable scaffold with built-in vasculature for organ-on-a-chip engineering and direct surgical anastomosis. Nature Mater 15(6):669–678. https://doi.org/10.1038/nmat4570
181. Zhao J, Griffin M, Cai J, Li S, Bulter PEM, Kalaskar DM (2016) Bioreactors for tissue engineering: an update. Biochem Eng J 109:268–281. https://doi.org/10.1016/J.BEJ.2016.01.018
182. Zhao Y, Kankala RK, Wang SB, Chen AZ (2019) Multi-organs-on-chips: towards long-term biomedical investigations. Molecules 24(4):675. https://doi.org/10.3390/molecules24040675
183. Zheng CX, Sui BD, Hu CH, Qiu XY, Zhao P, Jin Y (2018) Reconstruction of structure and function in tissue engineering of solid organs: toward simulation of natural development based on decellularization. J Tissue Eng Regen Med 12(6):1432–1447. https://doi.org/10.1002/TERM.2676
184. Zhou W, Chen Y, Roh T, Lin Y, Ling S, Zhao S, Lin JD, Khalil N, Cairns DM, Manousiouthakis E, Tse M, Kaplan DL (2018) Multifunctional bioreactor system for human intestine tissues. ACS Biomater Sci Eng 4(1):231. https://doi.org/10.1021/ACSBIOMATERIALS.7B00794

Chapter 10
Scaffolds Fabrication Processes: From Classical to Advanced Techniques

Román A. Pérez and Miguel A. Mateos-Timoneda

Abstract The scaffold is one of the most important pillars of Tissue Engineering (TE). Therefore, it is very important to pay particular attention to their fabrication. Thus, in this chapter different scaffolds fabrication techniques are presented, going from the classical fabrication techniques to the most recent ones, such as 3D printing and 3D Bioprinting. Firstly, the most important parameters for the scaffolds design are given. Afterwards, the different fabrication techniques are presented, paying particular attention to the different fabrication design parameters that can be modify/control for the successful fabrication of scaffolds for TE.

Keywords Fabrication techniques · 3D printing · Scaffold materials · Tissue engineering

Introduction

The scaffold is one of three main pillars of Tissue Engineering (TE), the triad of classical TE [66] (Hutmacher et al.): cells [27], signaling [1] and, scaffold [31]. David F. Williams defined the scaffold as "the construct that is intended to support cell migration, growth, and differentiation, and guide tissue development and organization into a mature and healthy state" [67]. These properties are related, as well, to the biomaterial of choice for the fabrication of the scaffold [37]. Nevertheless, there are several requirements that both, the biomaterial and the scaffold, should be incorporated during the design step, in order to obtain the ideal scaffold/cell or scaffold/neo-tissue construct. These requirements can be divided in three main categories: Mechanical/geometric, Surface-related and, size and manufacturing [11, 23]. The scaffold should provide initial mechanical strength and stiffness because it will act as temporary

R. A. Pérez · M. A. Mateos-Timoneda (✉)
Bioengineering Institute of Technology (BIT), Universitat Internacional de Catalunya (UIC), c/Josep Trueta S/N, 08195 Sant Cugat del Vallès, Barcelona, Spain
e-mail: mamateos@uic.es

Deparment of Basic Sciences, Faculty of Medicine and Health Sciences, Universitat Internacional de Catalunya (UIC), c/Josep Trueta S/N, 08195 Sant Cugat del Vallès, Barcelona, Spain

O. Gunduz et al. (eds.), *Biomaterials and Tissue Engineering*, Stem Cell Biology and Regenerative Medicine 74, https://doi.org/10.1007/978-3-031-35832-6_10

support, i.e. provide appropriate mechanical sustenance/function of the diseased or damaged tissue. The mechanical support provided by the scaffold needs to be sustained during both in vitro and in vivo tissue growth and remodeling processes. Additionally, the scaffold's architecture must consider the geometry to enable initial cell attachment, subsequent migration, and proliferation within the matrix [22]. This is accomplished by improving the transfer of nutrients and metabolites on a larger scale [68]. In addition, the scaffold must have sufficient volume to support the growth and remodeling of organized tissue. Another crucial factor is the scaffold's surface, which must be carefully controlled to promote cell adhesion and migration of the various cell types involved in tissue regeneration [45]. The scaffold's external size and shape are also important, particularly in personalized medicine when the scaffold must be tailored to a specific patient's defect. Finally, it's essential to manufacture the scaffold under good manufacturing practice (GMP) conditions in a reproducible, quality-controlled manner while keeping costs low and production speed high [29].

Moreover, the regeneration process is highly dynamic in nature, so it is very important to take into account the changes of this properties over time and how they can impact the tissue regeneration [64].

Hollister summarized the most important properties of scaffold design in the 4F concept: Form, Function, Formation, Fixation [26]: *Form* indicates that the scaffold should have a suitable shape and size fill complex 3D defects The scaffold should support tissue *function* of the surrounding environment, especially the mechanical demand. It should also enhance tissue *formation*, i.e. control the regeneration process through the delivery of specific (either chemical of physical) cues, and by providing appropriate mass transport. Finally, *fixation* means that it should be provided in such a way that the surgeon can readily implant and attach to tissues surrounding the defect.

As previously discussed, one of the key objectives in scaffold design is to facilitate mass transfer of nutrients and metabolites and promote cell migration, ultimately leading to successful vascularization of the resulting neo-tissue [8, 48]. Porosity and pore size are critical considerations in achieving these goals [25, 36]. Pores refer to empty spaces within the scaffold, while porosity refers to the collection of pores.

Pore size is a critical factor in scaffold design and can vary significantly. Macropores, which are larger than 50 microns, have a significant impact on tissue function and the overall success of tissue regeneration [34]. Micropores, ranging from 1 to less than 50 microns, primarily influence cell function, especially cell attachment, as mammalian cells typically fall within the range of 10–20 microns in size [72]. Nanopores, ranging from 1 to 1000 nm, can impact surface texture, affecting the adsorption of ligands and, consequently, cell attachment [63].

The accessibility and interconnectivity of the pores within a scaffold are critical factors in ensuring their optimal performance [10]. Accessibility refers to the ability of cells, nutrients, and oxygen to penetrate and reach all the pores within the scaffold. Interconnectivity, on the other hand, refers to the extent to which the pores are connected and can facilitate cell migration and proliferation.

However, achieving both high accessibility and interconnectivity can be challenging, as there is often a compromise between porosity and mechanical support provided by the scaffold. Increasing porosity may offer greater free volume for cell migration, ECM deposition, and nutrient diffusion, but it may also lead to a decrease in the scaffold's mechanical properties (Hutmacher et al.) [19]. Therefore, an ideal scaffold design should balance porosity with sufficient mechanical strength to provide structural support for the growing tissue.

Thus, to address the challenges of scaffold fabrication, various techniques have been developed to create 3D scaffolds with high porosity and surface area, particularly for polymeric scaffolds [28, 46]. However, it is worth noting that these techniques can also be applied to other materials such as ceramics and composites [4].

In this chapter, we aim to provide a comprehensive overview of the most relevant methods for scaffold-based tissue engineering. We will start with the classic methods that have been utilized for several decades and progress to the more advanced techniques that are primarily based on additive manufacturing (AM) or biofabrication, which involves the printing of materials embedded with cells. While we will touch upon these latest developments, for a more detailed description, interested readers are directed to other chapters or reviews found in the literature [21, 24, 62]. By presenting this information, we hope to provide a better understanding of the range of techniques available for scaffold fabrication and their respective strengths and limitations.

Classical Scaffold Fabrication Techniques

Classical methods for scaffold fabrication rely on basic physic-chemical techniques to incorporate porosity, which is a crucial characteristic for scaffolds. These techniques have been utilized for many years in the production of 3D porous scaffolds, primarily using polymeric and composite materials (mostly polymer-ceramic). The most frequently employed classical scaffold fabrication techniques are outlined in Table 10.1.

Solvent Casting/Porogen Leaching

Solvent Casting/Porogen Leaching is one of the earliest techniques for scaffold fabrication [42]. It involves dissolving a polymer in an appropriate solvent and mixing it with template particles of a specific size and shape. The mixture is then cast into a suitable mold, and as the solvent evaporates, the polymeric matrix with the embedded template particles is left behind. The removal of the particles creates the porous structure of the scaffold [60]. This processing technique offers several benefits, including the ease of manufacturing highly porous scaffolds without specialized equipment,

Table 10.1 Classical scaffold fabrication techniques [57]

Techniques	Advantages	Disadvantages	References
Solvent casting/ particulate leaching	Highly porous scaffolds, high range of pore sizes	Poor mechanical properties, low control of pore interconnectivity	[16, 60]
Gas foaming	Control of pore size and porosity	Poor mechanical properties, low control of pore interconnectivity, non-porous surface	[12, 52, 55]
Thermal induced phase separation	Possibility to incorporate bioactive molecules (growth factors)	Selection of solvents composition	[14, 35]
Uniaxial freezing/ freeze drying	Alignment of the porosity	Small pore size and long processing time	[5, 7, 40]
(Melt)Electrospinning	Mimics the fibrous structure of ECM, down to the nanoscale	Low pore size, limited thickness of the scaffolds	[9, 15, 53]

and the ability to precisely control the size and shape of the porosity by selecting the appropriate template particles. However, there are some disadvantages to this technique, such as the possibility of retaining toxic solvents within the polymer, and the limited ability to incorporate bioactive molecules using solvents.

Gas Foaming

This technique is solvent-free and relies on the creation of gas bubbles within a polymer [52, 38]. Various polymers are pressurized with gases such as CO_2 and N_2. When the pressure is suddenly released, the solubility of the gas decreases, resulting in the formation of gas bubbles ranging from 100 to 500 μm. The main benefit of this approach is that is solvent-free technique. However, a significant drawback of this technique is its poor interconnectivity.

Thermal Induced Phase Separation

Phase separation and evaporation can also be utilized to create three-dimensional porous structures. One way to induce phase separation is by reducing the temperature of the polymeric suspension, which is known as thermally induced phase separation (TIPS) [14]. After the solvent solidifies, the polymer is compelled to fill the interstitial gaps. The frozen mixture is then subjected to lyophilization using a freeze dryer, where the ice solvent sublimates, resulting in porosity. This process is performed at low temperatures, enabling the integration of bioactive molecules such as proteins and growth factors. However, the choice of solvent and temperature are crucial factors in the manufacture of scaffolds [47].

Uniaxial Freezing/Freeze Drying

Hydrogels have been subjected to controlled freezing to generate pores and channels [7, 32]. This cooling process results in the creation of solvent ice crystals. When the solvent is removed through sublimation, an interconnected porous structure is formed (Zhang et al. 1999). When the freezing process is performed in a controlled uniaxial manner, the ice crystal structures align with the temperature gradient, resulting in the presence of interconnected channels in the scaffold structure. The temperature determines the pore size, with larger ice crystals producing larger pores [49].

(Melt)Electrospinning

The production of micron and sub-micron diameter fibrous scaffolds can be achieved through (melt)electrospinning, which is a relatively straightforward technique. Although electrospinning was developed in the 1930s, its use in tissue engineering has significantly increased since thc 1990s [13]. The high porosities, surface area-to-volume ratios, and topographical features of electrospun scaffolds promote cellular adhesion, migration, and proliferation [56, 59].

(Melt)Electrospinning involves applying high voltage to either a polymeric solution (electrospinning) or melt (meltelectrospinning). When the voltage reaches a critical point, the electrostatic repulsion generated by localized charges surpasses the surface tension of the polymer solution or melt, causing the droplet to stretch into a (Taylor) cone and ejecting a continuous jet. If the flying time of the continuous jet, from the spinneret to the collector, allows for solvent evaporation or polymer cooling, electrospun fibers with sizes typically ranging from 200 nm to 5 μm can be produced, depending on the conditions [58]. Several factors, such as applied electric voltage, polymer solution or melt flow rate, collector and spinneret architecture, among others, have a significant impact on the resulting fiber morphology and dimensions.

Electrospun scaffolds have found various applications in tissue engineering, including skin [65], tendon [69], and nerve [17] tissue engineering.

This technique is only applicable for the fabrication of fibrous scaffolds.

Advanced Scaffold Fabrication Techniques

All these techniques are based in the additive manufacturing (AM) or Rapid Prototyping. These group of techniques fabricates 3D objects through the iterative deposition of materials layers using computer-controlled equipment (CAM) [6, 41]. AM is also commonly named as 3D printing. These techniques offer a precise control over the external macro-shape and internal microstrucuture (porositiy and interconnectivity) of the 3D scaffolds when compared to the classical scaffold fabrication techniques.

AM for TE can be divided into three main approaches: (1) laser-based, (2) nozzle-based, and (3) printer-based [3, 25]. They all selectively insert material, layer-by-layer, controlled by a previously programmed deposition path. Each layer corresponds to the cross section of the model at a specific height of the designed object.

Laser-Based Systems

The laser based techniques, such selective laser sintering (SLS) and stereolithography (SLA), are based in the projection of a focused laser beam into a loosely compacted powder or photoreactive resin, for SLS and SLA, respectively [2, 54]. In SLS, the laser beam sinters the powder particles, leaving the non-irradiated areas disconnected. Successive layers of powder are deposited and scanned with the laser until the entire scaffold or object is complete. In SLA, the laser polymerizes the photoreactive resin only where the beam strikes, at the surface of the bath, resulting in the creation of the first solid plastic layer. This laser-induced polymerization process is repeated to generate subsequent layers until the desired scaffold architecture is achieved. Recently, the use of a multiphoton femtosecond laser has been explored due to its high resolution (up to the nanoscale) [50, 51].

The primary advantage of these techniques is their high resolution and reproducibility. However, the primary disadvantages include the need for post-processing (when dealing with ceramics) or the inability to incorporate bioactive molecules (for SLS), as well as the limited availability of materials with proven bioactivity and biocompatibility (for SLA).

Powder-Based Systems

These techniques rely on selectively depositing a binder onto a thin layer of powder material using inkjet printing technology [74]. Once a layer is printed, new powder is added on top and the process is repeated. The powder bed is placed on a piston that moves down to spread and print each layer until the scaffold is completed. The main advantage (and drawback) is that powder material is required, making it possible to print virtually any material that exists in powder form. However, the resolution of the printed structure is limited by the size and shape of the powder particles, and there is a risk of material becoming trapped within the scaffold. In addition, many binders are toxic organic solvents, which can be problematic.

Nozzle-Based Systems

Nozzle-based systems utilize liquefied/melted polymeric filaments that are extruded through a nozzle connected to a carriage that moves horizontally along the x–y plane. Subsequently, the platform descends one level in the z-direction to deposit the next layer, and this process is repeated until the entire scaffold is fabricated [61, 71]. The use of melt extrusion is also called fused deposition modelling (FDM) and it is one of the most affordable AM technologies. The main advantage of this techniques is that no material is trapped within the structure. However, the main limitation is the difference in porosity in the three dimensions, i.e. the resolution and porosity in the z-direction is lower than in the x and y directions, and the possible degradation of the polymer due to the high extrusion temperature.

3D Bioprinting

All the previous techniques allow to have an excellent control of the fabrication process and the architecture and shape of the obtained scaffolds. However, they do not permit the fabrication of cell-laden scaffolds. Using 3D printing as inspiration, 3D bioprinting allows the fabrication of cell-laden scaffolds with absolute control of the architecture (similar to 3D printing) and the cellular position. Thus, 3D bioprinting has been defined as the process to create functional tissue constructs with precise placement of cells and biomolecules in a 3D pattern, mimicking the microstructure and function of native tissues. Moreover, cell function and viability are preserved within the printed construct [18, 62].

One of the significant advantages of bioprinting over traditional scaffold-based tissue engineering is the ability to create complex and heterogeneous tissues with precise control over cell placement. For example, a bioprinted tissue construct can have multiple cell types, each located in specific regions of the scaffold, and differentially treated with growth factors to promote specific differentiation and tissue development.

Taking that is needed to preserve cell viability and functionality, it is important to notice that not all the 3D printing techniques are suitable for 3D bioprinting, as they involved the use of organic solvents and/or high temperatures.

Bioprinting mainly consist on three different techniques: inkjet, microextrusion and laser-assisted bioprinting [39, 44]. The first one took the inspiration for commercial paper printers, in which drops of ink are deposited in the paper using a piezoelectric or thermal actuator [70]. In the microextrusion, filaments of the cell-laden ink are deposited using pressure by a piston, screw or pneumatic [33, 43]. Finally, the laser-assisted bioprinting (LAB) techniques are based the use of an energy-absorbing layer which receives the laser and helps in the controlled evaporation of the bioink, resulting in the formation of drops [33].

A new bioprinting technique has been developed in the last years, the so-called volumetric bioprinting [20]. It is an optical-tomography-inspired printing approach, based on visible light projection into photoresponsive hydrogels, with allows the fabrication of cell-laden scaffolds with clinically relevant size in a time frame ranging from seconds to tens of seconds.

Conclusions and Future Perspectives

The fabrication and design of scaffolds is one of the most important factors for the future advancement of Tissue Engineering and its long-lasting promise to revolutionize the field of medicine. Thus, it is important to study and develop new approaches in the 3D scaffold fabrication. It has already been show that 3D printing is revolutionizing medicine, especially in the teaching and surgical planning with the development of phantoms with similar properties as native tissues. Moreover, this set of technologies has already surpass the traditional scaffold fabrication techniques, because all the different biomaterials can be virtually processed by 3D printing. A key challenge in scaffold-based TE is the vascularization throughout the entire scaffold volume, while cells are distributed in the required locations. 3B bioprinting is an advancement in this direction because it already permits the incorporation of cells into the scaffold fabrication process. However, these techniques require cellular compatibility at the stage of inclusion and printing.

References

1. Abdollahiyan P, Oroojalian F, Mokhtarzadeh, A (2021) The triad of nanotechnology, cell signalling, and scaffold implantation for the successful repair of damaged organs: an overview on soft-tissue engineering. Elsevier
2. Antonov EN, Bagratashvili VN, Whitaker MJ et al (2005) Three-dimensional bioactive and biodegradable scaffolds fabricated by surface-selective laser sintering. Adv Mater 17:327–330. https://doi.org/10.1002/ADMA.200400838
3. Billiet T, Vandenhaute M, Schelfhout J et al (2012) A review of trends and limitations in hydrogel-rapid prototyping for tissue engineering. Biomaterials 33:6020–6041. https://doi.org/10.1016/j.biomaterials.2012.04.050
4. Boccaccini A (2005) Bioactive composite materials for tissue engineering scaffolds. Taylor Fr 2:303–317. https://doi.org/10.1586/17434440.2.3.303

5. Boland E, Espy P, Bowlin G (2004) Tissue engineering scaffolds. Enc Biomater Biomed Eng
6. Bourell DL (2016) Perspectives on additive manufacturing. Annu Rev Mater Res 46:1–18. https://doi.org/10.1146/ANNUREV-MATSCI-070115-031606
7. Bozkurt A, Lassner F, O'dey D et al (2012) The role of microstructured and interconnected pore channels in a collagen-based nerve guide on axonal regeneration in peripheral nerves. Biomaterials 33:1363–1375. https://doi.org/10.1016/j.biomaterials.2011.10.069
8. Bramfeld H, Sabra G, Centis V (2010) Scaffold vascularization: a challenge for three-dimensional tissue engineering. Curr Med Chem 3944–3967
9. Brown T, Slotosch A et al (2012) Design and fabrication of tubular scaffolds via direct writing in a melt electrospinning mode. Biointerphases 7:13. https://doi.org/10.1007/s13758-011-0013-7
10. Bružauskaitė I, Bironaitė D, Bagdonas E, Bernotienė E (2016) Scaffolds and cells for tissue regeneration: different scaffold pore sizes—different cell effects. Cytotechnology 68:355–369. https://doi.org/10.1007/S10616-015-9895-4
11. Chen G, Ushida T, Tateishi T (2002) Scaffold design for tissue engineering. Macromol Biosci
12. Dehghani F, Annabi N, Bornscheuer UT, Khademhosseini A (2011) Engineering porous scaffolds using gas-based techniques. Curr Opin Biotechnol 22:661–666. https://doi.org/10.1016/j.copbio.2011.04.005
13. Doshi J, Reneker DH (1995) Electrospinning process and applications of electrospun fibers. Elsevier 35:151–160. https://doi.org/10.1109/IAS.1993.299067
14. Elbert D (2011) Liquid–liquid two-phase systems for the production of porous hydrogels and hydrogel microspheres for biomedical applications: a tutorial review. Acta Biomater 7:31–56
15. Fridrikh S, Jian H, Brenner M et al (2014) Controlling the fiber diameter during electrospinning. APS 90:4. https://doi.org/10.1103/PhysRevLett.90.144502
16. Fukano Y, Usui ML, Underwood RA et al (2010) Epidermal and dermal integration into sphere-templated porous poly(2-hydroxyethyl methacrylate) implants in mice. J Biomed Mater Res Part A 94:1172–1186. https://doi.org/10.1002/JBM.A.32798
17. Ghane N, Khalili S, Nouri Khorasani S et al (2021) Regeneration of the peripheral nerve via multifunctional electrospun scaffolds. J Biomed Mater Res Part A 109:437–452. https://doi.org/10.1002/JBM.A.37092
18. Groll J, Boland T, Burdick J et al (2016) Biofabrication: reappraising the definition of an evolving field. Biofabrication 8
19. Guarino V, Causa F, Ambrosio L (2007) Porosity and mechanical properties relationship in PCL porous scaffolds. J Appl Biomat Biomech 5:149–157
20. Guillemot F, Souquet A, Catros S, Guillotin B (2010) Laser-assisted cell printing: principle, physical parameters versus cell fate and perspectives in tissue engineering. Nanomedicine 5:507–515. https://doi.org/10.2217/NNM.10.14
21. Harley W, Li C, Toombs J et al (2021) Advances in biofabrication techniques towards functional bioprinted heterogeneous engineered tissues: a comprehensive review. Bioprinting 23
22. Haynesworth S, Reuben D, Caplan A (1998) Cell-based tissue engineering therapies: the influence of whole body physiology. Elsevier
23. Henkel J, Hutmacher DW (2013) Design and fabrication of scaffold-based tissue engineering. BioNanoMaterials 14:171–193. https://doi.org/10.1515/BNM-2013-0021/HTML
24. Hill J, Wildman R, Mata A (2022) Exploiting the fundamentals of biological organization for the advancement of biofabrication. Biotechnology 74:42–54
25. Hollister S (2005) Porous scaffold design for tissue engineering. Nature 4:518–524. https://doi.org/10.1038/nmat1421
26. Hollister SJ (2009) Scaffold design and manufacturing: from concept to clinic. Adv Mater 21:3330–3342. https://doi.org/10.1002/ADMA.200802977
27. Howard D, Buttery L et al (2008) Tissue engineering: strategies, stem cells and scaffolds. Wiley Online Libr 213:66–72. https://doi.org/10.1111/j.1469-7580.2008.00878.x
28. Hutmacher D (2000) Scaffolds in tissue engineering bone and cartilage. Biomaterials 21:2529–2543
29. Hutmacher D, Cool S (2007) Concepts of scaffold-based tissue engineering—the rationale to use solid free-form fabrication techniques. Wiley Online Libr 11:654–669. https://doi.org/10.1111/j.1582-4934.2007.00078.x

30. Hutmacher D, Sittinger M, Risbud MV (2004) Scaffold-based tissue engineering: rationale for computer-aided design and solid free-form fabrication systems. Elsevier
31. Hutmacher DW (2001) Scaffold design and fabrication technologies for engineering tissues—state of the art and future perspectives. J Biomater Sci Polym Ed 12:107–124. https://doi.org/10.1163/156856201744489
32. Jana S, Cooper A, Zhang M (2013) Chitosan scaffolds with unidirectional microtubular pores for large skeletal myotube generation. Adv Healthc Mater 2:557–561. https://doi.org/10.1002/ADHM.201200177
33. Jones N (2012) The print revolution: three-dimensional printers are opening up new worlds to research. Nature 487:22–23
34. Kasten P, Beyen I, Niemeyer P et al (2008) Porosity and pore size of β-tricalcium phosphate scaffold can influence protein production and osteogenic differentiation of human mesenchymal stem cells: an in vitro. Biomaterialia 4:1904–1915
35. lo H, Ponticiello MS, Leong KW (1995) Fabrication of controlled release biodegradable foams by phase separation. Tissue Eng 1:15–28. https://doi.org/10.1089/TEN.1995.1.15
36. Loh Q, Choong C (2013) Three-dimensional scaffolds for tissue engineering applications: role of porosity and pore size
37. Lutolf M, Hubbell J (2005) Synthetic biomaterials as instructive extracellular microenvironments for morphogenesis in tissue engineering. Nat Biotechnol 23:47–55. https://doi.org/10.1038/nbt1055
38. Maio E Di, Mensitieri G et al (2005) Structure optimization of polycaprolactone foams by using mixtures of CO_2 and N_2 as blowing agents. Polym Eng Sci 45:432–441. https://doi.org/10.1002/pen.20289
39. Malda J, Visser J, Melchels FP et al (2013) 25th anniversary article: engineering hydrogels for biofabrication. Adv Mater 25:5011–5028. https://doi.org/10.1002/ADMA.201302042
40. Maquet V, Boccaccini AR, Pravata L et al (2003) Preparation, characterization, and in vitro degradation of bioresorbable and bioactive composites based on Bioglass®-filled polylactide foams. J Biomed Mater Res Part A 66:335–346. https://doi.org/10.1002/JBM.A.10587
41. Melchels F, Domingos M, Science TK (2012) Additive manufacturing of tissues and organs. Prog Polym Sci 37:1079–1104. https://doi.org/10.1016/j.progpolymsci.2011.11.007
42. Mikos AG, Thorsen AJ, Czerwonka LA et al (1994) Preparation and characterization of poly (L-lactic acid) foams. Polymer (Guildf) 35:1068–1077
43. Mironov V, Visconti R, Kasyanov V, Forgacs G (2009) Organ printing: tissue spheroids as building blocks. Biomaterials 30:2164–2174
44. Murphy S, Atala A (2014) 3D bioprinting of tissues and organs. Biotechnology 32:773–785. https://doi.org/10.1038/nbt.2958
45. Nair A, Tang L (2017) Influence of scaffold design on host immune and stem cell responses. Elsevier
46. Nair L, Laurencin C (2007) Biodegradable polymers as biomaterials. Prog Polym Sci 32:762–798
47. Nam Y, Park TG (1999) Porous biodegradable polymeric scaffolds prepared by thermally induced phase separation. J Biomed Mater Res 47:8–17
48. Novosel E, Kleinhans C, Kluger P (2011) Vascularization is the key challenge in tissue engineering. Adv drug Deliv 63:300–311
49. O'Brien F, Harley B, Yannas I et al (2004) Influence of freezing rate on pore structure in freeze-dried collagen-GAG scaffolds. Elsevier 25:1077–1086
50. Ovsianikov A, Mironov V et al (2012) Engineering 3D cell-culture matrices: multiphoton processing technologies for biological and tissue engineering applications. Taylor Fr 9:613–633. https://doi.org/10.1586/erd.12.48
51. Ovsianikov A, Mü S, Torgersen J et al (2013) Laser photofabrication of cell-containing hydrogel constructs. ACS Publ 30:3787–3794. https://doi.org/10.1021/la402346z
52. Parks KL, Beckman EJ (1996) Generation of microcellular polyurethane foams via polymerization in carbon dioxide. II: foam formation and characterization. Polym Eng Sci 36:2417–2431. https://doi.org/10.1002/PEN.10640

53. Pham QP, Sharma U, Mikos AG (2006) Electrospinning of polymeric nanofibers for tissue engineering applications: a review. Tissue Eng 12:1197–1211. https://doi.org/10.1089/TEN.2006.12.1197
54. Profile S, Narayan J, Skoog SA et al (2014) Stereolithography in tissue engineering. Springer 25:845–856. https://doi.org/10.1007/s10856-013-5107-y
55. Quirk R, France R, Shakesheff KM, Howdle SM et al (2004) Supercritical fluid technologies and tissue engineering scaffolds. Mater Sci 8:313–321
56. Rahmati M, Mills D, Urbanska A et al (2021) Electrospinning for tissue engineering applications. Prog Mater 117
57. Ratheesh G, Venugopal JR, Chinappan A et al (2017) 3D Fabrication of polymeric scaffolds for regenerative therapy. ACS Biomater Sci Eng 3:1175–1194. https://doi.org/10.1021/ACSBIOMATERIALS.6B00370
58. Reneker DH, Yarin AL, Fong H, Koombhongse S (2000) Bending instability of electrically charged liquid jets of polymer solutions in electrospinning. J Appl Phys 87:4531–4547. https://doi.org/10.1063/1.373532
59. Robinson TM, Hutmacher DW, Dalton PD et al (2019) The next frontier in melt electrospinning: taming the jet. Wiley Online Libr 29. https://doi.org/10.1002/adfm.201904664
60. Sachlos E, Czernuszka JT (2003) Making tissue engineering scaffolds work. Review: the application of solid freeform fabrication technology to the production of tissue engineering scaffolds. Eur Cell Mater 5:29–40. https://doi.org/10.22203/eCM.v005a03
61. Salentijn GI, Oomen PE, Grajewski M, Verpoorte E (2017) Fused deposition modeling 3D printing for (bio) analytical device fabrication: procedures, materials, and applications. ACS Publ 89:7053–7061. https://doi.org/10.1021/acs.analchem.7b00828
62. Schwab A, Levato R, D'este M et al (2020) Printability and shape fidelity of bioinks in 3D bioprinting. ACS Publ 120:11028–11055. https://doi.org/10.1021/acs.chemrev.0c00084
63. Sgarminato V, Tonda-Turo C, Ciardelli G (2020) Reviewing recently developed technologies to direct cell activity through the control of pore size: From the macro- to the nanoscale. J Biomed Mater Res Part B Appl Biomater 108:1176–1185. https://doi.org/10.1002/JBM.B.34467
64. Spector M (2014) The tissue engineer's toolbox manifesto. Biomed Mater 9. https://doi.org/10.1088/1748-6041/9/1/010201
65. Sundaramurthi D, Krishnan UM, Sethuraman S (2014) Electrospun nanofibers as scaffolds for skin tissue engineering. Polym Rev 54:348–376. https://doi.org/10.1080/15583724.2014.881374
66. Vacanti C (2006) The history of tissue engineering. J Cell Mol Med 10:569–376. https://doi.org/10.1111/j.1582-4934.2006.tb00421.x
67. Williams D (2009) On the nature of biomaterials. Biomaterials 30:5897–5909. https://doi.org/10.1016/j.biomaterials.2009.07.027
68. Woodfield T, Malda J, De WJ et al (2004) Design of porous scaffolds for cartilage tissue engineering using a three-dimensional fiber-deposition technique. Elsevier
69. Xie X, Chen Y, Wang X et al (2020) Electrospinning nanofiber scaffolds for soft and hard tissue regeneration. J Mater 59:243–261
70. Xu T, Olson J, Zhao W et al (2008) Characterization of cell constructs generated with inkjet printing technology using in vivo magnetic resonance imaging. J Manuf Sci Eng 130
71. Zein I, Hutmacher D, Tan K, Teoh S (2002) Fused deposition modeling of novel scaffold architectures for tissue engineering applications. Biomaterials 23:1169–1185
72. Zhang K, Fan Y, Dunne N, Li X (2018) Effect of microporosity on scaffolds for bone tissue engineering. Biomaterials 5:115–124
73. Zhang R, Ma PX (1999) Poly (α-hydroxyl acids)/hydroxyapatite porous composites for bone tissue engineering. I. Preparation and morphology. Wiley Online Libr 44:446–455
74. Zhang Y, Tse C, Rouholamin D et al (2012) Scaffolds for tissue engineering produced by inkjet printing. Springer 2:325–335. https://doi.org/10.2478/s13531-012-0016-2

Chapter 11
Characterization of the Biological Response to Scaffolds

Luis Maria Delgado

Abstract Characterizing the biological response of biomaterials, scaffolds or medical devices is crucial to understanding and assure their functionality and safety. Commonly, biological characterization can be performed to assess the potential biological risks arising from the use of new biomaterials, surfaces, encapsulated drugs, among others. Moreover, biological characterization allows progress in Tissue Engineering and cell-biomaterial interaction understanding. The biological characterization should not only be focused on assuring the biocompatibility of the biomaterial, scaffold or medical devices. Moreover, it also has to provide a greater understanding of cell toxicity, cell-biomaterial interactions, protein-biomaterials, biomaterial resorption or degradation, and how scaffolds are infiltrated or replaced by new tissue, if applicable. The biological characterization can be preclinical and clinical, in vitro and in vivo with preclinical tools. This chapter is focused on in vitro and in vivo characterization techniques that have importance to fundamental biomaterials research and industry.

Keywords Biological characterization · Biological response · Cell-biomaterial interaction · Cell toxicity

L. M. Delgado (✉)
Bioengineering Institute of Technology (BIT), Universitat Internacional de Catalunya (UIC), Sant Cugat del Vallès, 08195 Barcelona, Spain
e-mail: lmdelgado@uic.es

Basic Sciences Department, Universitat Internacional de Catalunya (UIC), Sant Cugat del Vallès, 08195 Barcelona, Spain

O. Gunduz et al. (eds.), *Biomaterials and Tissue Engineering*, Stem Cell Biology and Regenerative Medicine 74, https://doi.org/10.1007/978-3-031-35832-6_11

Introduction to Biological Response Characterization

The characterization of biomaterials, scaffolds or medical devices is crucial to understanding and to assure the biological and functional safety of these devices. Commonly, biological characterization can be performed to assess new biomaterials, surface, properties, encapsulated ions and drugs [1],to progress in Tissue Engineering and cell-biomaterial interaction understanding [2], and to assess the potential biological risks arising from the use of medical devices [3] (Fig. 11.1).

Typically, initial characterization is focused on physical, chemical and stability assays and, subsequently, the data obtained during this initial characterization together with the intended clinical application determine the required biological characterisation.

The biological characterization should not only be focused on assuring the biocompatibility of the biomaterial, scaffold or medical devices. Moreover, it also has to provide greater understanding about cell toxicity, cell-biomaterial interactions, protein-biomaterials, biomaterial resorption or degradation, and how scaffolds

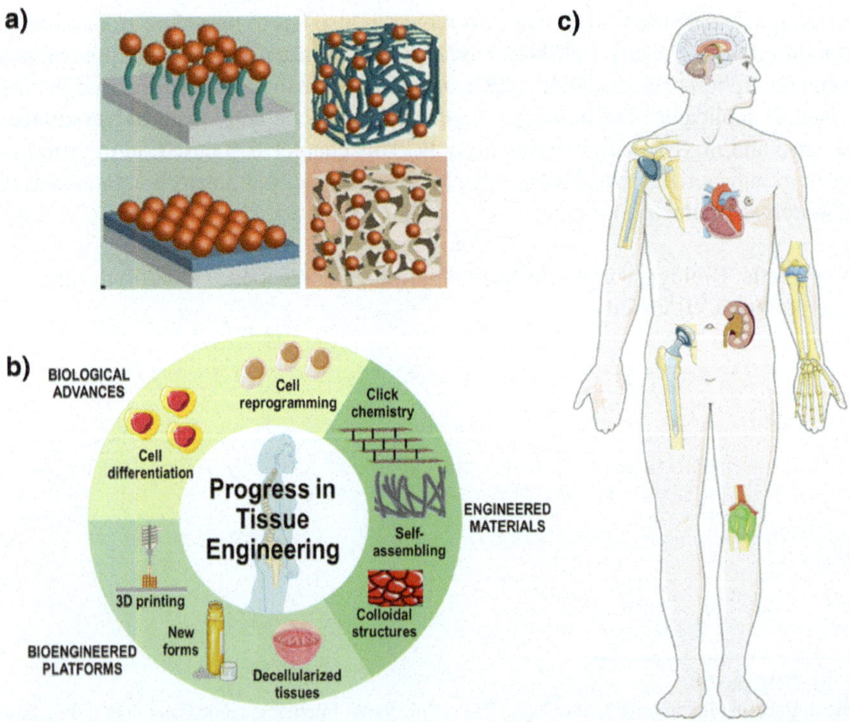

Fig. 11.1 Biological characterization can be performed: (a) to assess new biomaterials, surface, properties, encapsulated ions and drugs; (b) to progress in Tissue Engineering and cell-biomaterial interaction understanding; and (c) to assess the potential biological risks arising from the use of medical devices. Figure was prepared using images from https://smart.servier.com/

are infiltrated or replaced by new tissue, if applicable. The biological characterization can be preclinical and clinical, being in vitro and in vivo characterization the preclinical tools.

Nowadays, in vitro biological assays are a very important part of biomaterial characterization due to the refinement of old techniques, i.e. scanning electron microscopy [4] or immunohistochemistry [5], and the development of new methods that, in some cases, have been able to partially or completely replace some in vivo assays. It is worth pointing out that there has been a great pressure from society to reduce animal experimentation, driving the in vitro assay development and refinement [6]. In this chapter, we will focus on in vitro and in vivo characterization techniques that have importance to fundamental biomaterials research and industry.

Protein- and Blood-Biomaterial Interactions

Once a biomaterial or scaffold is implanted, a dynamic crosstalk between its surface and blood is one of the first the very first steps on the material-body interaction and a temporary protein matrix is formed as result, phenomenon that is known as the Vroman effect [7]. This provisional protein matrix, that depends on the chemical composition, wettability, surface charge, topography or stiffness of the biomaterial surface [8–10],plays a fundamental protagonist in the interaction between biomaterials surface and cells, which is a key factor for subsequent tissue regeneration phases [11].

Initial Protein-Biomaterial Interactions

Some research within the field of Tissue Engineering are highly interested on study protein adsorption on biomaterial surfaces to describe and to understand the interactions between biomaterials and cells, as cells and biomaterials hardly ever is directly between them and it is mediated by adsorbed proteins onto biomaterial surface that can controlled cell adhesion, proliferation and differentiation, even foreign body response [12].

Therefore, some researchers are interested on the quantity and properties of the adsorbed proteins using several techniques and points of view. For example, electrophoresis and immunoassays are suitable for the identification of protein adsorbed on surface, limiting the number of samples and proteins to be identified. Instead, mass spectrometry offers large-scale proteomics studies of proteins but it requires higher resources [13].

Regarding the quartz crystal microbalance (QCM), it is commonly accepted to assess adsorption of nucleic acids, proteins, protein-receptor pairs, and reaction between antigen and antibody onto surfaces [14]. The main advantage is that QCM allows the detection of molecules without label [15] and the study of molecular orientation and conformation can be determined using different models based on frequency resonance [16]. Moreover, QCM allows the modification of media ion strength and pH to further analyse protein adsorption/desorption and conformation [17].

Another alternatives to study protein adsorption and conformation is ellipsometry that is based on the variations of polarized light beam upon reflection from a surface [18]. The models are based on the phase difference and reflectance of the parallel and perpendicular components [19]. As quartz crystal microbalance, ellipsometry allows to study protein adsorption under pH and temperature stimuli and, furthermore, recent technical improvement using infrared light allows to assess ultrathin layers with higher sensitivity or allowing to develop in situ experiments [20].

Blood-Biomaterial Interactions

Hemocompatibility assays are used to study the effects of medical devices on blood using appropriate in vitro models. These experiments are mandatory before clinical use of any medical device and, therefore, experiment design and parameters have be conducted in accordance with [3, 21].

All in vitro models to assess blood-biomaterial interactions are based on simulating geometrical and contact conditions of the intended use and blood is analysed before and after the incubation to evaluate haemolysis, cell number, coagulation, complement system activation, platelet activation and leukocyte activation (Fig. 11.2). On the other hand, biomaterial surface is analysed in terms of blood cell attachment, protein adsorption, fibrin clot and thrombus formation [22].

Moreover, there are three different models of incubation that can mimic the conditions of the intended use of the medical device: static or dynamic (agitation or shear flow). Although static blood incubation models are very simple and requires lower

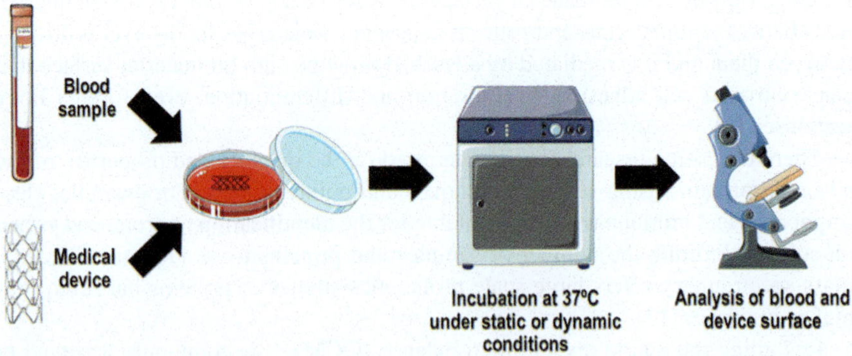

Fig. 11.2 Flow chart of hemocompatibility for medical devices. Figure was prepared using images from https://smart.servier.com/

investment, they only provide reliable data regarding thrombogenicity, while the other hemocompatibility parameters are elementary [23]. Dynamic models based on agitation only introduce a shaker or an overhead rotator, preventing cell sedimentation and, therefore, increasing data reliability. Remarkably, dynamic models that mimic blood shear flow are the most reliable and these models use different engineered systems, such as flat flow chambers, viscometer or tubular loops, to obtain shear flows in vitro [22].

In Vitro Assessment of Cell Response to Scaffolds

In vitro biological assessment of scaffolds is commonly performed by research laboratories worldwide, using immortalized or primary cells lines that are analysed in terms of cytocompatibility, cell morphology or differentiation by gene expression or protein secretion. However, these in vitro experiment have to be designed and the result interpreted according to the intended use and the cascade of biological phases that will occur upon device implantation. In vitro experimental designing is crucial as a recent multicenter study demonstrated unexpectedly poor correlation between in vitro results of biomaterials and in vivo results for bone regeneration strategies [24]. The different in vitrobiological characterization commonly used in Tissue Engineering are presented below and summarized in Table 11.1.

Cell Morphology, Adhesion and Spreading

Cell adhesion and spreading are important interaction parameters when cells are in contact with biomaterials. For cell morphology analysis, it is important to know that the cell adhesion process consists in three phases: cell-substrate contact, cell spreading and cytoskeleton reorganization [25].

Cell morphology analysis can be difficult and time-consuming depending on the desired staining and the substrate where cells are cultures. Tissue culture plastic or glass, that are flat and transparent substrates, allows simple but highly-cost benefit imaging such as phase contrast microscopy, acquiring several phase images without any staining or even time-lapse sequence images [26]. Since no toxic or fluorescent dies are involved, unlimited images can be obtained within few minutes. Scanning electron microscopy (SEM) also allows the observation of cell morphology and surface at very high magnifications of non-transparent or rough surfaces and 3D substrates. However, sample preparation is required for SEM, including cell fixation, dehydration and optional staining with osmium tetroxide, a highly toxic reagent. It is worth noting that cell fixation and dehydration could alter cell morphology, but can be minimised using environmental scanning electron microscopy with wet samples [27]. Regarding the health and safety issues related to osmium tetroxide, this step can be avoided if using a field-emission SEM [28].

On the other hand, immunofluorescence (IF) and immunohistochemistry (IHC) methods allows the cell morphology analysis of non-transparent or rough surfaces and 3D substrates. In addition, IF and IHC are more sensitive identifying morphological

Table 11.1 Summary of in vitro method for the assessment of cell response to scaffolds

Biological feature	Methods	Main description
Cell morphology, adhesion and spreading	Phase contrast microscopy	Simple method to obtain several phase images without any staining or even time-lapse images
	Scanning electron microscopy	Observation of cell morphology and surface at very high magnifications. It requires sample preparation that can alter morphology
	Immunofluorescence (IF) and Immunohistochemistry (IHC)	Combination of fluorescent staining with high sensitivity. It allows staining of adhesion proteins. It is time consuming and require epifluorescence or confocal microscopes
Cell migration	Scratching assay	Simple method based on generate a scratch in a cell monolayer, mimicking the cell migration that occurs during tissue regeneration. It is only a 2D method
	Scaffold colonisation with immunofluorescence	Cell colonisation is frequently evaluated with immunofluorescent methods: IF and IHC
	Boyden chamber	Evaluate the chemoattractant properties or cell homing capacity
	Microfluidics	Mimic cellular microenvironment, creating physiological-like models
Cell proliferation, metabolic activity and viability	Picogreen	Proliferation assays based on the quantification of dsDNA with high sensitivity
	MTT, WST-1, alamarBlue	Metabolic activity based on colorimetric changes when salts are reduced by viable cells
	LDH assay	Viability test based on the quantification of LDH enzyme by colorimetric assays that also are based on salt reductions
	Live/Dead staining	Fluorescent staining of live and dead cells also allows a qualitative and quantitative determination of cell viability
ECM secretion and remodelling	Histological staining (Sudan III, toluidine, haematoxylin–eosin, Picro Sirius Red…)	Broad staining methods depending on the target tissue
	SEM and AFM	High magnification characterisation in the scale of micro- or nano- for fibrillar structure, orientation…
	Immunohistochemistry	Immunocytochemistry analysis allows to specifically study ECM components
	PCR	Quantitative analysis of genes
	Western blot	Quantitative analysis of proteins
	Zymography	Quantitative analysis of enzyme activity

changes than simple phase contrast microscopy [29]. In Tissue Engineering, cell morphology is commonly assessed with the combination of phalloidin, a highly selective and fluorescent peptide that binds to actin filaments, with 4′,6-diamidino-2-phenylindole (DAPI), a blue fluorescent dye that binds to DNA [30]. Moreover, certain adhesion proteins (vinculin, focal adhesion kinase or integrins) can be stained with antibodies.

These cell morphologies analysis have some significant drawbacks. They are mainly qualitative and require some relatively expensive equipment such as epifluorescence microscopes for flat surfaces or confocal microscopes for rough or 3D substrates. It is worth pointing out that the adhesion proteins can assessed by Western Blot [31] and the morphology analysis can be objectivised using specific software, i.e. plugins from ImageJ or CellProfiler [32]. Furthermore, big data and machine learning are contributing to the increase of quantification tools, reducing time and errors associated with manual analysis [33].

Cell Migration

During the different tissue regeneration phases, cell homing, migration and infiltration plays a crucial role on the early stages. Therefore, Tissue Engineering together with Biological Science have developed in vitro models to study cell migration of single or grouped cells in response to chemical, biochemical and mechanical signals.

Within cell migration assays, the in vitro scratching assay is the simplest, cheapest, and well-established method for assessing cell migration in vitro. Basically, it involves to generate a scratch in a cell subconfluent monolayer, mimicking the cell migration that occurs during tissue regeneration that could be assessed by comparing the cell migration rate at each time point. The in vitro scratching assay allows to study the effects of cell–matrix and cell–cell interactions [34, 35]. This assay can be monitored using phase contrast microscopy. Furthermore, it can be combined with free plugins that automatically recognize the scratch dimensions and quantify objective parameters [36].

Cell migration or colonisation within 3D porous scaffold is also mandatory to assess in vitro. Cell seeding usually results in a preferential cell adhesion that progressively infiltrate scaffolds depending on structural, biochemical and biophysical scaffold features [37]. Due to the structural characteristics of porous scaffolds, cell colonisation is frequently evaluated with immunofluorescent methods introduced above. Moreover, scaffolds can possess chemoattractant properties or can be functionalised to gain this property. In this cases, cells can be seeded onto a porous membrane in an adjacent chamber and cells are allowed to migrate through the pores. This is known as Boyden chamber migration assay [38].

In the last decade, microfluidics has transformed cell migration assays since it allowed to mimic cellular microenvironment, creating physiological-like models. Moreover, advances in microscopy and machine learning have further revolutionised this cell migration assays [39].

Cell Proliferation, Metabolic Activity and Viability

Cell proliferation, metabolic activity and viability are commonly study to monitor the in vitrobiological response to biomaterials, scaffolds and various stimuli. The appropriate selection of a method depends on the expected outcomes, cell type, biomaterial nature and scaffold structure.

Cell proliferation can be measured through the quantification of dsDNA where advanced fluorophores become fluorescent upon binding to DNA; the resulting fluorescence intensity is proportionate to the amount of DNA of each sample. This methodology is up to 1,000 times more sensitive than DNA quantification using UV absorbance [40] and can be completely quantitative using a DNA and/or cell standard curve [41].

An indirect test to assess proliferation, viability, and cytotoxicity is measuring the metabolic activity. These assays are based on reduction reaction of tetrazolium salt to formazan, i.e. MTT and WST-1, or reduction of resazurin, i.e. alamarBlue™, producing a colorimetric change measurable using spectrophotometry [42]. Cells with high proliferation ratio have a high metabolic activity, while non-viable cells have reduced metabolic activity [43].

Finally, cell viability can be assessed through the damage of plasma membrane in a quantitative manner. Lactate dehydrogenase (LDH) is an enzyme present in the cytosol of several cell types and LDH is released to culture media upon cell. Some investigation discovered that LDH catalyses the conversion of lactate to pyruvate through the reduction of NAD+ into NADH, which in turn reduces tetrazolium salt to formazan that can be measured using spectrophotometry [44]. This a colorimetric, simple and reliable method for determining cell viability. On the other hand, a simultaneous fluorescent staining of live and dead cells also allows a qualitative and quantitative determination of cell viability. In this method, calcein-AM is stained viable cells, emitting a strong green fluorescence, while propidium iodine pass through damaged cell membrane and binds to double helix DNA, emitting red fluorescence [45]. Then, stained cells can be visualised using an epifluorescence microscope or a confocal microscope, being possible to count live and dead cells with specific software, i.e. plugins from ImageJ or CellProfiler.

ECM Secretion and Remodelling

The extracellular matrix (ECM) is the major component of the cellular microenvironment, greatly dynamic structure and under constant remodelling. ECM is composed of collagen, non-collagen proteins and proteoglycans. During the tissue regeneration phases, the ECM is under a continuous and high evolution. At early stage, a fibrin clot is formed that is rapidly infiltrated by inflammatory cells and progenitor cells that proliferate, differentiate and secrete ECM proteins to substitute the fibrin matrix and form the new tissue, mainly constituted by collagen type III with some fibronectin, elastin, and proteoglycans [46]. Then, this provisional ECM is remodelled by cells and collagen type III is gradually replaced by fibrous proteins, mainly controlled by matrix metalloproteinases (MMPs).

Regarding the characterisation of ECM secretion and remodelling, there are several options. The most traditional methods are based on histological staining such as haematoxylin–eosin, toluidine, Sudan III, oil red O or Picro Sirius Red that usually they are used to stain fibrous connective tissues, fat and cells [47, 48]. For high magnification characterisation in the scale of micro- or nano-, scanning electron microscopy (SEM) and atomic force microscopy (AFM) can provided detailed information regarding density, orientation, protein aggregation, and fibrillar structure [49, 50]. Immunocytochemistry analysis allows to specifically study ECM components such as collagen type I, III, IV, V, VI, VII, laminin, elastin, fibronectin, α-smooth muscle actin, tenomodulin, epithelial keratin, tubulin, hyaluronic acid, chondroitin sulphate, keratin sulphate, heparin sulphate, aggrecan, biglycan, decorin, endosialin, lysyl oxidase, transglutaminase-2, among others [51]. Instead, polymerase chain reaction (PCR) and western blot permit a quantitative analysis of genes, proteins and pathways of the ECM components.

In Vivo Assessment of Tissue and Host Response to Scaffolds

In vivo assessment is part of the preclinical experimentation for evaluating the biological response, host response and potential biological risks arising from the use of medical devices. This is an essential experimental study before proceeding to clinical testing.

Ethics and Regulatory

Animal experimentation helps to advance in treatments to benefit humans and animals around us. Although computer models, in vitro and ex vivo experimentation help to advance scientific knowledge related to disease and regeneration. Nonetheless, properly designed and performed animal experimentation with timely and responsible us of animals provides valuable information that serve to improve the knowledge. This idea is the base of the ethics and regulatory framework in Europe and other countries.

All scientific research strategies must minimise the number of studies based on animal experimentation and the animal stress during the experimental procedures. For this reason, the mainstream consider that research must be performed following the "3Rs": replacement, reduction and refinement. Replacement considers avoiding the use of animals and instead of using animals, chips, computer models or ex vivo models must be used. If a total replacement cannot be performed, a partial replacement can be considered by using more immature animal species, such as drosophila or zebrafish. If replacement cannot be performed, reduction is compulsory by using the minimal number of animals in each experiment. Furthermore, if animals have to be used, refinement is also mandatory, minimising the stress that suffers animals when a research procedure is performed. Subsequently, adequate housing, adequate anaesthesia and adapt analgesia after the procedure have to be used.

These principles were promulgated by Russel and Burch in 1960 [52] and they were adopted by the EU many years ago [53]. The European regulatory framework establishes the animal protection when used for teaching and scientific experimental procedures, avoiding procedure duplications, preventing animals from suffering unjustified pain or distress and minimising the used animals. The Directive 2010/

Table 11.2 Summary of animal models for clinical application

Clinical target	Animal model	References
Heart valves and vascular grafts	Rat, rabbit, sheep, dog and pig	[55, 56]
Stents	Rabbit, sheep, dog and pig	[57]
Bone scaffolds and substitutes	Mouse, rat, rabbit, dog and pig	[58, 59]
Tendon substitutes	Rat and goat	[60, 61]
Peripheral nerve regeneration	Mouse, rat and rabbit	[62, 63]

63/EU on the protection of research animals is the core of the European regulation [54] and each EU member country has its transposition, having regions with their own even more restrictive guidelines. While the 3Rs are mainly focused on the methods and technical procedures, the Directive 2010/63/EU extends the application of refinement to cover all animal housing, breeding and care, even for non-currently experimental animals.

Selection of in Vivo Tests According to Intended Usage

When an in vivo model has to be selected, three main concepts have to be considered: contact time, tissue to be in contact and available animal model for the intended application. For the contact time, it can be limited, less than 24 h; prolonged, between 24 h and less than 30 days; or permanent, more than 30 days. For the tissue to be in contact, researchers should know if the if the final device is going to be in contact with skin, mucosa or internal tissues. If the medical device is going to be an external communicated device, a transmucosal device that breaks protective barrier of skin, or an implanted device that is going to substitute or replace part of tissue. Some examples of animal models currently use for testing medical devices are summarised in Table 11.2.

Since no model can cover all parameters (anatomic defect, physiological stimuli, biomechanical requirement and functional environment) to be studied, animal experimentation often starts with smaller animal models for understanding disease or for screening therapeutic treatments. When an animal model has to be selected, the know-how of the research team and collaborator should not only drive the decision, rather the selection has to be decided according a scientifically integral question.

Planning Biological Characterization Towards Commercialization

Scaffolds as other products with medical purpose are regulated by the European Union (EU), assuring that the medical devices meet the minimal legal requirements, are safe and accomplish the intended purpose. Although each EU member state establishes its medical devices regulatory, the European Medicines Agency (EMA) is involved in the common regulatory process. The European legal framework is regulated by the Regulation (EU) 2017/745 for the clinical investigation and sale of medical devices for human use [64].

Each medical device manufacturer has to CE mark its medical devices. For this purpose, each manufacturer has to pass a conformity assessment consisting of an audit of its quality system by a notified body. In addition, a review of the device technical documentation (safety and success with the intended performance) is required, depending on the type of medical device. In USA, manufacturers have to submit a 510(k) premarket notification for each medical device to U.S. Food and Drug Administration (FDA) that is the legislative authority to regulate medical devices [65].

The main objective of the current regulatory framework is the protection of humans from potential risks arising from the use of medical devices. This regulatory framework is constituted by several standards and guidelines, being ISO 10993 one of the most important and used standards [66]. The standards and guidelines define the biological evaluation of medical devices within a risk management process. In general, the evaluation of the proposed medical devices is determined by the nature and contact duration with human tissues. In addition, the justification of the material choice with respect to its biocompatibility, functionality and biological response is part of the technical file and the evaluation. With regards to the biological response, in vitro, in vivo and ex vivo results are crucial evidence to anticipate the medical device behavior when used in humans. Therefore, planning the biological characterization before the commercialization is crucial and well-designed and planned tests can help during the regulatory affairs stages. In Europe, the ISO 10993 standard provides some guidelines to plan the in vitro and in vivo assessment.

All the premarket experimentation is a small round of the whole adventure before to be able to launch an approved medical device to the market. Even more, the assessment of the medical devices never stops, the medical device manufacturer has to maintain the regulatory affairs tasks associated with the medical devices after the product. This is known as the postmarket surveillance, ensuring that the device continues being safe and any adverse event or deviation is reported to the affected national medical devices agencies.

References

1. Williams DF (2008) On the mechanisms of biocompatibility. Biomaterials 29:2941–2953. https://doi.org/10.1016/j.biomaterials.2008.04.023
2. Khademhosseini A, Langer R (2016) A decade of progress in tissue engineering. Nat Protoc 11:1775–1781. https://doi.org/10.1038/nprot.2016.123
3. ISO (2018) ISO 10993–1:2018—Biological evaluation of medical devices—part 1: evaluation and testing within a risk management process
4. Courtland R (2018) The microscope revolution that's sweeping through materials science. Nature 563:462–464
5. Dixon AR, Bathany C, Tsuei M et al (2015) Recent developments in multiplexing techniques for immunohistochemistry. Expert Rev Mol Diagn 15:1171–1186
6. The 3Rs | NC3Rs

7. Vroman L, Adams A, Fischer G, Munoz P (1980) Interaction of high molecular weight kininogen, factor XII, and fibrinogen in plasma at interfaces. Blood 55:156–159. https://doi.org/10.1182/blood.v55.1.156.156
8. Hallab NJ, Bundy KJ, O'Connor K et al (2001) Evaluation of metallic and polymeric biomaterial surface energy and surface roughness characteristics for directed cell adhesion. Tissue Eng 7:55–70. https://doi.org/10.1089/107632700300003297
9. MacDonald DE, Deo N, Markovic B et al (2002) Adsorption and dissolution behavior of human plasma fibronectin on thermally and chemically modified titanium dioxide particles. Biomaterials 23:1269–1279. https://doi.org/10.1016/S0142-9612(01)00317-9
10. Xu LC, Siedlecki CA (2007) Effects of surface wettability and contact time on protein adhesion to biomaterial surfaces. Biomaterials 28:3273–3283. https://doi.org/10.1016/j.biomaterials.2007.03.032
11. Bosch E, Diez L, Giordano B et al (2019) Biological roles and delivery strategies for ions to promote osteogenic induction. Front Cell Dev Biol 8(614545):1–19
12. Swartzlander MD, Barnes CA, Blakney AK et al (2015) Linking the foreign body response and protein adsorption to PEG-based hydrogels using proteomics. Biomaterials 41:26–36. https://doi.org/10.1016/j.biomaterials.2014.11.026
13. Aebersold R, Mann M (2003) Mass spectrometry-based proteomics. Nature 422:198–207. https://doi.org/10.1038/nature01511
14. Janshoff A, Steinem C (2005) Label-free detection of protein-ligand interactions by the quartz crystal microbalance. Methods Mol Biol 305:47–64. https://doi.org/10.1385/1-59259-912-5:047
15. Su J, Esmaeilzadeh H, Zhang F et al (2018) An ultrasensitive micropillar-based quartz crystal microbalance device for real-time measurement of protein immobilization and protein-protein interaction. Biosens Bioelectron 99:325–331. https://doi.org/10.1016/j.bios.2017.07.074
16. Hampitak P, Melendrez D, Iliut M et al (2020) Protein interactions and conformations on graphene-based materials mapped using quartz-crystal microbalance with dissipation monitoring (QCM-D). Carbon N Y 165:317–327. https://doi.org/10.1016/j.carbon.2020.04.093
17. Delcroix MF, Demoustier-Champagne S, Dupont-Gillain CC (2014) Quartz crystal microbalance study of ionic strength and pH-dependent polymer conformation and protein adsorption/desorption on PAA, PEO, and mixed PEO/PAA brushes. Langmuir 30:268–277. https://doi.org/10.1021/la403891k
18. Mora MF, Wehmeyer JL, Synowicki R, Garcia CD (2009) Investigating protein adsorption via spectroscopic ellipsometry. Biological Interactions on Materials Surfaces. Springer, US, pp 19–41
19. Elwing H (1998) Protein absorption and ellipsometry in biomaterial research. Biomaterials 19:397–406. https://doi.org/10.1016/S0142-9612(97)00112-9
20. Kroning A, Furchner A, Aulich D et al (2015) In situ infrared ellipsometry for protein adsorption studies on ultrathin smart polymer brushes in aqueous environment. ACS Appl Mater Interfaces 7:12430–12439. https://doi.org/10.1021/am5075997
21. ISO (2017) ISO 10993–4:2017—Biological evaluation of medical devices—part 4: selection of tests for interactions with blood
22. Weber M, Steinle H, Golombek S et al (2018) Blood-contacting biomaterials: in vitro evaluation of the hemocompatibility. Front Bioeng Biotechnol 6:99. https://doi.org/10.3389/FBIOE.2018.00099/BIBTEX
23. Haycox CL, Ratner BD (1993) In vitro platelet interactions in whole human blood exposed to biomaterial surfaces: insights on blood compatibility. J Biomed Mater Res 27:1181–1193. https://doi.org/10.1002/JBM.820270909
24. Hulsart-Billström G, Dawson JI, Hofmann S et al (2016) A surprisingly poor correlation between in vitro and in vivo testing of biomaterials for bone regeneration: results of a multicentre analysis. Eur Cell Mater 31:312–322. https://doi.org/10.22203/ECM.V031A20
25. Khalili AA, Ahmad MR (2015) A review of cell adhesion studies for biomedical and biological applications. Int J Mol Sci 2015 16:18149–18184. https://doi.org/10.3390/IJMS160818149

26. Kong Z, Zhu X, Zhang S et al (2012) Phase contrast microscopy of living cells within the whole lens: spatial correlations and morphological dynamics. Mol Vis 18:2165–2173. https://doi.org/10.1007/BF00606669
27. Thiberge S, Nechushtan A, Sprinzak D et al (2004) Scanning electron microscopy of cells and tissues under fully hydrated conditions. Proc Natl Acad Sci 101:3346–3351. https://doi.org/10.1073/pnas.0400088101
28. Goldberg MW (2019) Field emission scanning electron microscopy in cell biology featuring the plant cell wall and nuclear envelope. In: Biological field emission scanning electron microscopy. Wiley, Chichester, UK, pp 343–362
29. Mori H, Cardiff RD (2016) Methods of immunohistochemistry and immunofluorescence: converting invisible to visible, pp 1–12
30. Pintado-Sierra M, Delgado L, Aranaz I et al (2014) Surface hierarchical porosity in poly (ε-caprolactone) membranes with potential applications in tissue engineering prepared by foaming in supercritical carbon dioxide. J Supercrit Fluids 95:273–284. https://doi.org/10.1016/j.supflu.2014.09.019
31. Guillem-Marti J, Delgado L, Godoy-Gallardo M et al (2013) Fibroblast adhesion and activation onto micro-machined titanium surfaces. Clin Oral Implants Res 24:770–780. https://doi.org/10.1111/j.1600-0501.2012.02451.x
32. Carpenter AE, Jones TR, Lamprecht MR et al (2006) Cell profiler: image analysis software for identifying and quantifying cell phenotypes. Genome Biol 7:R100. https://doi.org/10.1186/gb-2006-7-10-r100
33. Theriault DH, Walker ML, Wong JY, Betke M (2012) Cell morphology classification and clutter mitigation in phase-contrast microscopy images using machine learning. Mach Vis Appl 23:659–673. https://doi.org/10.1007/s00138-011-0345-9
34. Bobadilla AVP, Arévalo J, Sarró E et al (2019) In vitro cell migration quantification method for scratch assays. J R Soc Interface 16:20180709. https://doi.org/10.1098/rsif.2018.0709
35. Liang C-C, Park AY, Guan J-L (2007) In vitro scratch assay: a convenient and inexpensive method for analysis of cell migration in vitro. Nat Protoc 2:329–333. https://doi.org/10.1038/nprot.2007.30
36. Suarez-Arnedo A, Figueroa FT, Clavijo C et al (2020) An image J plugin for the high throughput image analysis of in vitro scratch wound healing assays. PLoS ONE 15:e0232565. https://doi.org/10.1371/journal.pone.0232565
37. Intranuovo F, Howard D, White LJ et al (2011) Uniform cell colonization of porous 3-D scaffolds achieved using radial control of surface chemistry. Acta Biomater 7:3336–3344. https://doi.org/10.1016/j.actbio.2011.05.020
38. Chen H-C (2005) Boyden chamber assay. Methods Mol Biol 294:15–22. https://doi.org/10.1385/1-59259-860-9:015
39. Ren J, Wang N, Guo P et al (2022) Recent advances in microfluidics-based cell migration research. Lab Chip 22:3361–3376. https://doi.org/10.1039/D2LC00397J
40. Ahn S (1996) PicoGreen quantitation of DNA: effective evaluation of samples pre- or post-PCR. Nucleic Acids Res 24:2623–2625. https://doi.org/10.1093/nar/24.13.2623
41. Delgado LM, Fuller K, Zeugolis DI (2018) Influence of cross-linking method and disinfection/sterilization treatment on the structural, biophysical, biochemical, and biological properties of collagen-based devices. ACS Biomater Sci Eng 4:2739–2747. https://doi.org/10.1021/acsbiomaterials.8b00052
42. Perrot S, Dutertre-Catella H, Martin C et al (2003) Resazurin metabolism assay is a new sensitive alternative test in isolated pig cornea. Toxicol Sci 72:122–129. https://doi.org/10.1093/toxsci/kfg014
43. Díez-Tercero L, Delgado LM, Bosch-Rué E, Perez RA (2021) Evaluation of the immunomodulatory effects of cobalt, copper and magnesium ions in a pro inflammatory environment. Sci Rep 11:1–13. https://doi.org/10.1038/s41598-021-91070-0
44. Smith SM, Wunder MB, Norris DA, Shellman YG (2011) A simple protocol for using a LDH-based cytotoxicity assay to assess the effects of death and growth inhibition at the same time. PLoS ONE 6:e26908. https://doi.org/10.1371/journal.pone.0026908

45. Kaneshiro ES, Wyder MA, Wu Y-P, Cushion MT (1993) Reliability of calcein acetoxy methyl ester and ethidium homodimer or propidium iodide for viability assessment of microbes. J Microbiol Methods 17:1–16. https://doi.org/10.1016/S0167-7012(93)80010-4
46. Delgado LM, Bayon Y, Pandit A, Zeugolis DI (2015) To cross-link or not to cross-link? Cross-linking associated foreign body response of collagen-based devices. Tissue Eng Part B Rev 21:298–313. https://doi.org/10.1089/ten.TEB.2014.0290
47. Fischer AH, Jacobson KA, Rose J, Zeller R (2008) Hematoxylin and eosin staining of tissue and cell sections. CSH Protoc 2008:pdb.prot4986. https://doi.org/10.1101/pdb.prot4986
48. Lattouf R, Younes R, Lutomski D et al (2014) Picrosirius red staining. J Histochem Cytochem 62:751–758. https://doi.org/10.1369/0022155414545787
49. Jorba I, Uriarte JJ, Campillo N et al (2017) Probing micromechanical properties of the extracellular matrix of soft tissues by atomic force microscopy. J Cell Physiol 232:19–26. https://doi.org/10.1002/jcp.25420
50. Leonard AK, Loughran EA, Klymenko Y et al (2018) Methods for the visualization and analysis of extracellular matrix protein structure and degradation. Methods Cell Biol 143:79–95. https://doi.org/10.1016/bs.mcb.2017.08.005
51. Satyam A, Kumar P, Fan X et al (2014) Macromolecular crowding meets tissue engineering by self-assembly: a paradigm shift in regenerative medicine. Adv Mater 26:3024–3034. https://doi.org/10.1002/adma.201304428
52. Russell WMS, Burch RL (1960) The principles of humane experimental technique. Med J Aust 1:500–500. https://doi.org/10.5694/J.1326-5377.1960.TB73127.X
53. European Commission (n.d.) Animals used for scientific purposes—environment. Retrieved from https://ec.europa.eu/environment/chemicals/lab_animals/3r/alternative_en.htm. Accessed on 25 Feb 2023
54. European Parliament (2010) Directive 2010/63/EU—On the protection of animals used for scientific purposes. Off J Eur Union, pp 33–79
55. Kheradvar A, Zareian R, Kawauchi S et al (2017) Animal models for heart valve research and development. Drug Discov Today Dis Models 24:55–62. https://doi.org/10.1016/j.ddmod.2018.04.001
56. Swartz DD, Andreadis ST (2013) Animal models for vascular tissue-engineering. Curr Opin Biotechnol 24:916–925. https://doi.org/10.1016/j.copbio.2013.05.005
57. Iqbal J, Chamberlain J, Francis SE, Gunn J (2016) Role of animal models in coronary stenting. Ann Biomed Eng 44:453–465. https://doi.org/10.1007/s10439-015-1414-4
58. Hatt LP, Thompson K, Helms JA et al (2022) Clinically relevant preclinical animal models for testing novel cranio-maxillofacial bone 3D-printed biomaterials. Clin Transl Med 12(2):1–33. https://doi.org/10.1002/ctm2.690
59. McGovern JA, Griffin M, Hutmacher DW (2018) Animal models for bone tissue engineering and modelling disease. Dis Model Mech 11(4):1–14. https://doi.org/10.1242/dmm.033084
60. Hast MW, Zuskov A, Soslowsky LJ (2014) The role of animal models in tendon research. Bone Joint Res 3:193–202. https://doi.org/10.1302/2046-3758.36.2000281
61. Warden SJ (2007) Animal models for the study of tendinopathy. Br J Sports Med 41:232–240. https://doi.org/10.1136/bjsm.2006.032342
62. Geuna S, Raimondo S, Fregnan F et al (2016) In vitro models for peripheral nerve regeneration. Eur J Neurosci 43:287–296. https://doi.org/10.1111/ejn.13054
63. Vela F, Martínez-Chacón G, Ballestín A et al (2020) Animal models used to study direct peripheral nerve repair: a systematic review. Neural Regen Res 15:491. https://doi.org/10.4103/1673-5374.266068
64. Regulation (EU) 2017/ 745 of the European Parliament and of the Council—of 5 April 2017—on medical devices, amending Directive 2001/83/EC, Regulation (EC) No 178/2002 and Regulation (EC) No 1223/2009 and repealing Council Directives 90/385/EEC and 93/42/EEC
65. Yano H, Abe SI (2005) Regulations of assisted reproduction devices in the United States, premarket notification [510 (k)]. J Mamm Ova Res 22(4):255–259
66. ISO (2009) ISO 10993-1:2009—Biological evaluation of medical devices—part 1: evaluation and testing within a risk management process

Chapter 12
Additive Manufacturing of Biomaterials

F. Otaola, C. de Lartigue, V. Fitzpatrick, D. Luart, M. Leturia, E. Guenin, and C. Egles

Abstract The use of Additive Manufacturing (AM), also called 3D printing, has increased in recent years. These processes are applied in many different fields such as aerospace, motor vehicles, electronics, and medical fields among others. In particular, additive manufacturing has been used for the fabrication of biomaterials to create products for biomedical applications such as prototypes, implants, scaffolds for tissue engineering, models, or drug-delivery systems. Its versatility allows the personalization of the object to the specific needs of each patient based on anatomical data. Furthermore, AM enables the production of highly complex objects that can not be realized with traditional techniques such as subtractive manufacturing. Additive manufacturing is gaining in popularity thanks to its adaptability in terms of fabrication materials, such as polymers, metals or ceramics, depending on the needs of the application, as well as the fast transition from a numerical model to the final object. In the present review, we summarize additive manufacturing techniques used to create biomaterials focusing on their advantages and drawbacks and the reasons why they can be preferred to traditional ones. Some biomedical applications are presented for each technique.

Keywords Additive manufacturing · Biomaterials · 3D printing techniques · 3D scanning

F. Otaola and C. de Lartigue these authors worked equally.

F. Otaola (✉) · D. Luart · M. Leturia · E. Guenin
Alliance Sorbonne Université, Université de Technologie de Compiègne (UTC), TIMR EA 4297 UTC/ESCOM, CS 60319, Compiègne Cedex 60203, France
e-mail: franco.otaola@utc.fr

C. de Lartigue (✉) · C. Egles
Normandie Université, UNIROUEN, INSA Rouen, CNRS, PBS (UMR 6270), 55 Rue Saint-Germain, Évreux 27000, France
e-mail: claire.de-lartigue@univ-rouen.fr

V. Fitzpatrick
Alliance Sorbonne Université, Université de Technologie de Compiègne, CNRS, UMR 7338 Biomécanique Et Bioingénierie (BMBI), Centre de Recherche Royallieu, CS 60319, Compiègne Cedex 60203, France

O. Gunduz et al. (eds.), *Biomaterials and Tissue Engineering*, Stem Cell Biology and Regenerative Medicine 74, https://doi.org/10.1007/978-3-031-35832-6_12

Introduction

Additive manufacturing (AM) is the process of creating an object through incremental addition of material. This method, very different from conventional manufacturing approaches like machining, which are usually subtractive. There are increasing applications of additive manufacturing in multiple industrial fields, including biomedical engineering. Indeed, obtaining a functional medical device or implant requires the combination of manufacturing processing and biomaterials engineering, and the considerable interest in personalized medicine has fueled research and development into customizable approaches like those allowed by additive manufacturing.

AM is composed of a very large set of technologies that allow the fabrication of objects made of many different materials, such as metal, ceramics, or polymers (natural or synthetic). This wide range of choices allows engineers to choose the material that best fits the needs for the fabrication of the part and its desired properties.

Moreover, AM allows us to design parts at multiple scales of organization, from the nano to the macroscale. For AM approaches to tissue engineering, for example, the macroscale gives the general shape of the part, allowing a custom fit to the patient's anatomy the microscale can be designed to mimic the architecture of the replaced tissue; and the surface of the implant can be controlled at the nanoscale for cell mechanotransduction [1].

AM allows fast prototyping, meaning a fast transition between the numerical model and the fabricated part. This characteristic, combined with its geometrical freedom allowed by AM, have allowed the fabrication of patient-specific parts. As such, AM has truly changed the paradigm, from the application being adapted to the generic part, to the part adapted to the desired application.

Of note, there are some limitations to the materials that can be used for AM in the medical field. Indeed, to be implanted, a biomaterial must meet certain criteria such as biocompatibility, bioactivity, biodegradability, immunocompatibility and mechanical properties in accordance with the tissue to be replaced [2, 3]. AM of biomaterials can be used in various medical fields. From orthopedics to the vascular system. AM approaches are particularly valuable for parts with specific and complex geometries that cannot be fabricated with conventional manufacturing techniques.

This chapter is devoted to an overview of all additive manufacturing techniques that are used in the field of biomaterials. After a preliminary description of what is additive manufacturing for biomaterials, we will present the advantages and drawbacks as well as the main applications of the most common AM methods employed for biomedical engineering. We will also develop the reason why these processes are preferred over more traditional ones.

What is Additive Manufacturing for Biomaterials?

Biomaterials

A biomaterial is a material that can be used to create a device in order to replace a function or a part of the body. This system has to perform its role in a safe, economic and physiologically acceptable manner [4]. These considerations are governed and enforced by regulatory agencies like the FDA, that lay down specific conditions biomaterials must meet before reaching the market.

There are many techniques to design and manufacture biomaterials that are used to create devices. In this chapter, we will develop the additive manufacturing of biomaterials.

Additive Manufacturing

There are three main types of manufacturing techniques: formative, subtractive, and additive manufacturing (Fig. 12.1) [5].

Formative Manufacturing (FM) consists in molding the material to its final shape and is commonly done by heat and pressure. FM includes different techniques such as injection, molding, casting, stamping, vacuum forming, and forging. FM is mostly used to fabricate parts with a simple geometry which are usually made of polymers or metals.

Subtractive Manufacturing (SM) uses cutting tools to remove material from a block, called blank, to achieve the final shape. SM includes various techniques such as Computer Numerical Control (CNC), turning (lathe), and drilling. This process is able to produce parts made of non-brittle materials, such as metals and polymers.

Additive Manufacturing (AM), which is frequently referred to as 3D printing, is a relatively new fabrication technique [6]. In this approach, the material is

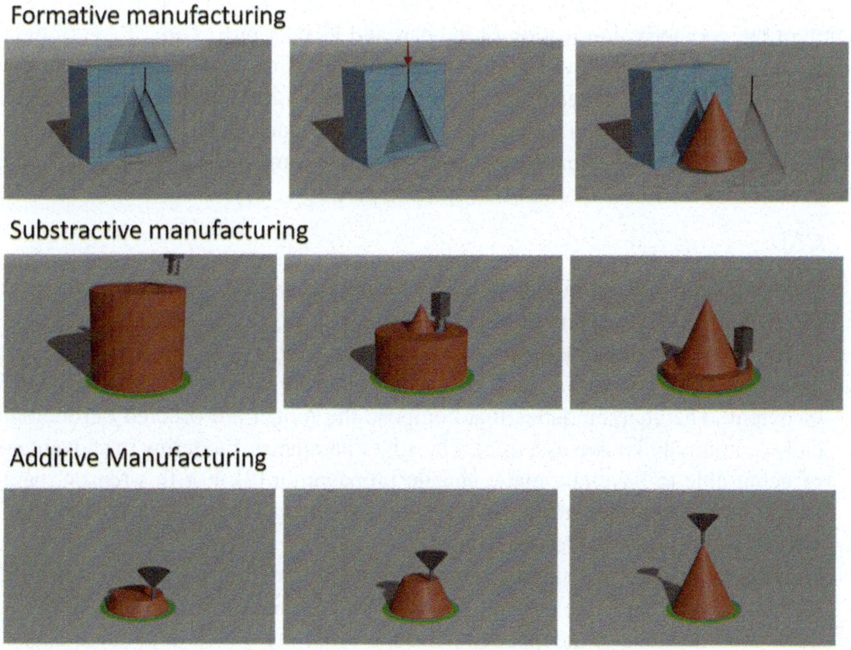

Fig. 12.1 Schematic of the three main types of manufacturing techniques

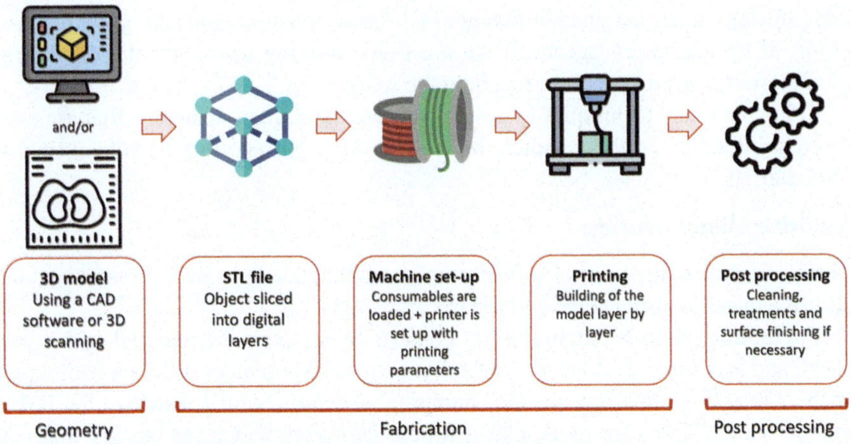

Fig. 12.2 Manufacturing process of a biomaterial by additive manufacturing

added, generally in a layer-by-layer manner. This approach presents several advantages, including allowing a higher geometrical freedom compared to the previous techniques.

Workflow of Additive Manufacturing

The workflow of additive manufacturing can be divided into three main steps: generation of the geometry, fabrication of the part and its post-processing. Each of these steps can be furthermore divided in several sub-steps (Fig. 12.2).

The first step is the **generation of the geometry**. Indeed, it is necessary to have a numerical approximation of the geometry of the part. There are two different approaches for the generation of the model, either by Computer Aided Design (CAD) software or by the 3D scanning of an existing part.

- CAD

  When generating a 3D model with CAD software, two different methods can be used: parametric modeling or direct modeling.

  In parametric modeling, the geometry is described by several features, such as extrusion or revolution, and by fixing dimensional constraints, like length or height. The characteristics that compose the model are ordered chronologically (commonly known as a tree). This gives parametric modeling the advantage of being able to have automatic change propagation [7]. For this reason, parametric modeling is generally used in the engineering fields where several slight modifications to the model are needed.

  Direct modeling is generally compared with modeling with clay, where the user pushes and pulls the geometry to model the desired final form. Each one of these modifications is independent, and therefore there is no automatic change propagation (unlike parametric modeling). This gives the advantage of allowing fast model generation and the possibility to generate more complex geometries, with the detriment that geometry is not parameterized.

Parametric and direct modeling is the more common and classic modeling tools. Nevertheless there are new emerging modeling approaches [7], like: tool path generator [8], generative design which uses a CAD model that is optimized [9], or CAD combined with artificial intelligence [10].

- 3D scanning

It is also possible to 3D scan a part. In this case, the model of the object is generated by positioning millions of points that will digitally recreate the surface of the object. The points are used for the generation of triangles that will approximate the different surfaces of the object. The combination of these triangles will create a numerical model, which is commonly known as a mesh. The main advantage of this approach is that it allows us to obtain a model rapidly and easily. Nevertheless, in contrast with CAD modeling, the geometry is an approximation of the real/desired geometry. Furthermore, depending on the original object's material, color, transparency, or surface finish, scanning the object can be difficult or costly.

It should be noted that CAD and 3D scanning are commonly combined. For example, 3D scanning helps to generate the model of the arm of a patient and CAD software is used to generate a custom cast around it. Another example is the reconstruction of an object, such as the fabrication of a bone prosthesis. The original bone is 3D scanned to create a base model and by using CAD software (generally by direct modeling), the bone is reconstructed to its original form.

Fabrication is the second step of the process. First, the model generated in the previous step has to be sliced, generating the steps for the 3D printer. The slicing software takes the mesh of the geometry as an input (.stl or .obj file) and creates a set of instructions for the 3D printer of what to do at each layer. If necessary, supports are generated to help the fabrication of the object. Even though limited, there is some emerging software that directly generates the output file of the slicing step, without the need of any CAD model [8].

After the generation of the sliced model, the printer is loaded and set up. The fabrication is then carried out by the 3D printer. In general, no user intervention is needed during this step.

Post-processing of the part is the last step of the workflow. Depending on the AM technique and the desired properties of the part, different post-processing approaches can be implemented. There are three main categories of post-processing operations: the removal of supports, treatments to achieve the final mechanical properties; and surface finish (polishing, smoothing, coating).

Support removal: as explained before, it could be necessary to print support structures for the object. This sacrificial support can be made from the same material as the part (breakaway type), or a material that is soluble to facilitate support removal.

Treatments for mechanics: some AM techniques require heat treatment of the object so it can reach its final mechanical properties.

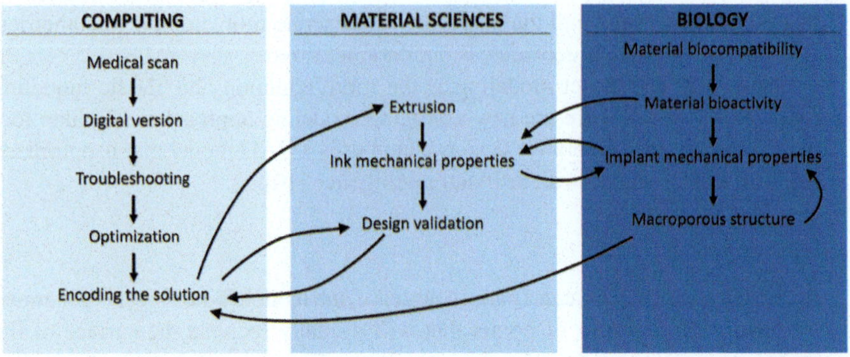

Fig. 12.3 Communication in the multidisciplinary field of additive manufacturing for biomaterials

Surface finishing: surface preparation (sanding, gap filling, blasting) is sometimes necessary. Surface finishing is not compulsory, but can be carried out for both functional or aesthetic purposes. It can be composed of a machining step (polishing) and/or a coating can be added (metal or epoxy coating, painting, lacquering) [5].

Debris or undesired material (e.g., unpolymerized resin, unsintered powder) can have very significant effects on general biocompatibility, as would the presence of microorganisms like bacteria. As such, in the case of implantable materials (e.g., medical devices, prostheses), post-processing steps ensuring the cleanliness and sterility of the part will frequently be required before use.

The workflow depicted in Fig. 12.2 shows the steps required to generate a 3D printed part. This requires the intervention of:

- In silico work (designing the model, slicing it with adequate parameters, generating commands for the 3D printer)
- Material selection (for processability, mechanics, biocompatibility)
- Fabrication and post-fabrication operations.

As such additive manufacturing for biomaterials is a multidisciplinary field where computing, material sciences and biology are in permanent and dynamic communication (Fig. 12.3) to achieve the best result to answer clinical needs.

Different Types of Additive Manufacturing

Additive manufacturing for biomaterials can be divided into two categories: acellular and cellular. The second category, also called bioprinting, involves the incorporation of living cells into the bioinks used for printing. This section is developed in more detail in an upcoming chapter, this chapter will focus exclusively on acellular printing.

AM is composed of several fabrication techniques, each with its own constraints, capabilities and materials. As presented in Table 12.1, AM can be divided into four main families: extrusion, liquid, powder, and sheet-based techniques. As AM is a growing field, with new and more complex technologies developed every year, the previously mentioned families are a mere attempt to classify the different technologies. Nevertheless, this rough classification can fall short. A brief description of each family can be given as follows.

- **Extrusion-based**: the fabrication material is extruded, either by heating a solid material (*e.g.*, thermoplastic polymers), or as a highly concentrated suspension of particles in an ink (*i.e.*, slurry state).
- **Liquid-based**: the fabrication material is in a liquid state, and it is solidified through different processes. These processes can be photopolymerization, evaporation of the liquid (for suspensions), cooling down, or electrodeposition.
- **Powder-based**: the fabrication material is in a powder state. This family can be divided into two different groups. In the first group, the object is fabricated by selectively sintering/fusing or binding the particles inside a bed of powder. In the second group, the powder is deposited on demand only over the printing location.
- **Sheet-based**: these techniques use sheets for the fabrication of the object. There are two different groups: in the first group, the final part is composed of a stack of these sheets, and in the second group, the sheets are only used as a support/binder for particles of other materials and they are then burned away.

1. **Extrusion based**

   a. **Solid material**

   Fused Filament Fabrication (FFF), commonly called Fused Deposition Modeling (FDM), is a technique where a filament of thermoplastic polymer (PLA, PCL, PEEK, etc.) is melted and added layer-by-layer on the build platform (Fig. 12.4). The material then cools down and solidifies [11].

   The way the parts are printed implies that they have an anisotropy in their mechanical properties, specifically in the printing direction due to low adhesion between the layers. Moreover, when printing large parts or fine details, the cooling of the sections takes place at different speeds, which can lead to the deformation of the part. It is therefore necessary to take these parameters into account when designing the parts. Additionally, the surface of an FFF printed part is rough due to the process itself. To improve layer adhesion, the melted material is pressed onto the previous layer, generating an ellipsoid cross-section of the deposed filament. This causes a visible layer distinction. Combined with the use of printing supports (depending on the geometry of the part), the surface could require a post-production step to obtain a smooth surface. To reduce post-processing steps, a dissolvable support can be used. The parts can be fabricated partially (or completely) hollow to save time and material. Naturally, this will affect the strength of the printed object. This technique is the most common among AM technologies, probably due to the ease of operation and the low cost of the machines and materials [5].

Table 12.1 Main additive manufacturing techniques

Family	Sub family	Group	Technology	S/M material[1]	Material	Examples of bio applications
Extrusion based	Solid		Fused filament fabrication (FFF)	M	TP^2 polymers, ceramics* and metals*	Orthopedic implants or prosthesis [13, 55]bioreactors [14], fixtures [15]
			Wire laser additive manufacturing (WLAM)	M	Metals	Implants and prototype [20]
	Slurry		Direct ink writing (DIW)	M*	Pastes and hydrogels	Tissue engineering [22], drug delivery [32]
Liquid based	VAT polymerization	UV-wavelength	Stereolithography (SLA)	S	TS^3 polymers, ceramics*, metals* and hydrogels*	Dentistry [56], tissue engineering [36, 37]
			Digital light processing (DLP)	S		
		Visible-wavelength	Daylight polymer printing (DPP)	S		
	Jetting techniques		Drop on demand (DOD)	S	Wax, ceramics	Tissue engineering with cells, see following chapter
			PolyJet	M	TS^3 polymers	Anatomical models for surgical planning [42]
			NanoParticle jetting (NPJ)	S	Ceramics and metals	
			Liquid printing metal	S	Metals	

(continued)

Table 12.1 (continued)

Family	Sub family	Group	Technology	S/M material[1]	Material	Examples of bio applications
	Others		Electrochemical additive manufacturing (ECAM)	M	Metals	Antibacterial surface for dental and orthopedic implants [57]
			Rapid liquid printing (RLP)	S	Polymers	None (in study) [58]
Powder based	Inside a bed of powder	Melting/sintering techniques	Selective laser sintering (SLS)	S	Ceramics and metals	Orthopedics, bone implants and tissue engineering, prosthetics [43]
			Selective electron beam melting (SEBM)	S	Metals	
			Selective laser melting (SLM)	S	Metals	
		Binder jetting	Multi jet fusion (MJF)	S	TP[2] polymers	Pharmacological and drug delivery [44]
			Binder jetting (BJ)	S	Metals, ceramics, sand and composites	Anatomical models for surgical planning [42]
		Others	Selective powder deposition (SPD)	M	Metals	
	Direct deposition		Cold spray additive manufacturing (CSAM)	S	Metals	Orthopedics in study [46]
			Laser powder deposition (LPD)	S	Metals	Orthopedics in study

(continued)

Table 12.1 (continued)

Family	Sub family	Group	Technology	S/M material[1]	Material	Examples of bio applications
Sheet based	Direct fabrication		Laminated object manufacturing (LOM)	S	TP^2 polymers and paper	
			Ultrasonic additive manufacturing (UAM)	M	Metals	
	Fabrication of support for final part		Roll powder sintering (RPS)	S	Ceramics, metals and TP^2 polymers, hydrogels	Potentially many [51]
			Composite based additive manufacturing (CBAM)	S	TP^2 polymers combined with carbon or glass	Surgical staples [54]

* Less common cases, S/M material[1] = Single/Multi material, TP^2 = Thermoplastic, TS^3 = Thermosetting

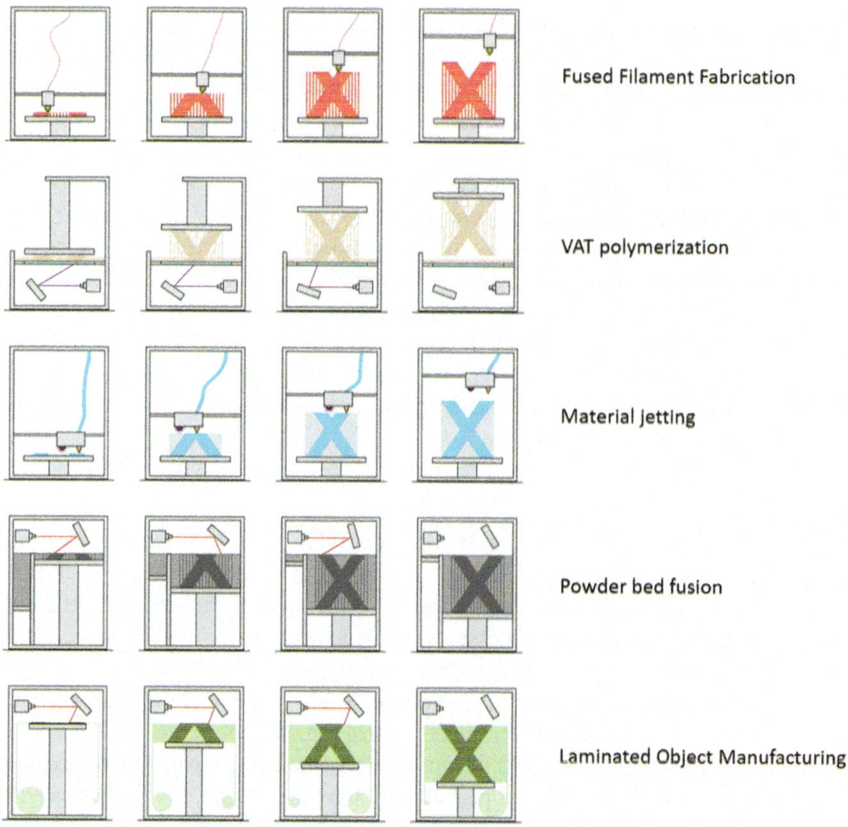

Fig. 12.4 Steps to create a part with different additive manufacturing techniques

In the field of biomedical engineering, FFF is frequently used for the prototyping or manufacturing of surgical tools [12] but also customized implants or prostheses [13], parts of bioreactors [14], or fixtures that are adapted to biological samples (e.g., for mechanical testing). Similarly, recent papers have reported the design and manufacturing of 3D-printed stereotaxic fixtures for surgeries on small animals (e.g., infant mice), to improve precision of the surgical operation [15]. Because of its low cost and ease of use, FFF can be used to troubleshoot issues or determine tolerances before the final part is produced.

In addition to these applications, tissue engineering approaches have also involved the use of FFF. Most notably, the manufacturing of bone implants using PEEK has been reported [16], but also other biocompatible polymers like PLA [17] or PCL [18], to manufacture implantable scaffolds that are well-suited for orthopedic applications.

Wire-Feed Additive Manufacturing can be divided into two main groups: **Wire Arc Additive Manufacturing** and **Laser Additive Manufacturing** (WAAM/WLAM). They are a similar technology to FFF, where fabrication materials are metal wires that are melted using a laser or an electric arc as heat source [19].

As this technology is quite expensive, it is generally used for prototyping or repairing fabricated parts instead of the fabrication of new ones. Furthermore, this approach generates parts with a rough surface finish. For this reason, they are generally combined with a subtractive head which, once several layers have been printed, removes a thin layer on the surface to give the desired surface finish.

Applications of these additive manufacturing approaches are generally limited to metal objects and scaffolds, in particular Ti-6Al-4 V-based implants and prototypes [20, 21].

b. **Slurry**

Robocasting or **Direct Ink Writing** (DIW) is the last of techniques inside the extrusion-based family. They consist of the deposition of a paste (also called ink) generally using a syringe. Depending on the ink, the solidification can be done by several processes: cooling down, photopolymerization, drying or the material has enough mechanical strength to support itself.

As this technology can use any type of ink, it is one of the most versatile technologies among AM, from the fabrication of hydrogels [22], to the fabrication of bone scaffolds [23]. Furthermore, the inks can also be mixed with fillers to obtain specific properties [24]. On the contrary, the main disadvantage of this technology is its low resolution, as the inks have a high viscosity which limits the size of the syringe nozzle, and by extension the resolution.

This approach has been widely used in the field of tissue engineering, using syringe extruders (pneumatically or motor-driven), or screw extruders. Compared to most additive manufacturing approaches, the mild conditions of deposition and solidification frequently allow the 3D printing of cell-laden structures [25–27]. This approach can also be used for the controlled deposition of polymer-based scaffolds for tissue engineering [28, 29] and/or drug delivery applications [30–32]. Of note, recent work as even applied these approaches to in situ 3D printing, where the material is deposited directly inside the body, opening up exciting new possibilities for the future of this technology [33].

2. **Liquid based**

a. **Vat polymerization**

Vat polymerization is a technology where a liquid thermoset photo-polymer resin is polymerized inside a tank with a specific light source (Fig. 12.4). In general, the light source used is a UV light. The most common technologies of VAT polymerization are **stereolithography** (SLA) and **digital light**

processing (DLP). The difference between these two technologies is the light source: stereolithography uses a laser, while digital light processing uses a projector.

In SLA, for each layer the laser follows the cross-section of the part, similar to the paths in extrusion-based technologies. In DLP, the entire cross-section of the part is photopolymerized simultaneously. Since the entire layer is printed at the same time, the printing time for one or N parts is the same with DLP. In contrast, in SLA the printing time will be extended according to the number of objects to be produced. Nevertheless, SLA has some advantages over DLP technology. In general, the laser spot is smaller than the projector resolution, giving a more detailed object. Furthermore, as the light source is concentrated in a smaller spot, more energy is provided to the liquid resin. For this reason, SLA can use resins that require more power. In both (SLA and DLP), depending on the light position of the source in regards to the resin vat, the technologies can be classified in bottom up or top-down. In the bottom-up approach, the most common one, the light source is placed under a transparent section of the vat tank. In this approach, after each layer is finished, the freshly polymerized layer is attached not only to the previous layer (or the build plate if it is the first layer), but also to the transparent section of the vat. The build plate is raised, to peel off the part, generating high stresses on the part, which can cause warping of the part. In the top-down approach, the light source is placed over the vat. This makes the design of 3D printer more complex, but eliminates the peeling step after each layer, as the new layer will be only attached to the previous one (or the build plate). This reduces the stress experienced by the parts, and enables a better quality of the part. As the resin is contained inside the vat, these technologies are monomaterial, which means that the support will be constructed from the same material as the part, and they will therefore be of the breakaway type. Vat polymerization technologies present a great surface quality with similar results to other classic technologies such as formative manufacturing. However, it is important to note that photopolymers tend to be brittle, which implies a low mechanical resistance and a short lifetime due to mechanical fatigue and sensitivity to sunlight. Lastly, it is possible to use resins charged with different particles such as metal, ceramics, or even amorphous silica (i.e., glass). Once the part is fabricated, a post processing step is done where the parts are sintered to burn away the polymeric matrix and the final part is composed only by the filler material. Of note, this sintering step is usually accompanied with a shrinking of the object, which may not be uniform in all directions, and needs to be taken into account when designing the part.

Daylight Polymer Printing (DPP), also referred to liquid crystal display (LCD) technology, is a method that, unlike SLA and DLP, uses a light source in the visible spectrum (400—800 nm) to polymerize resin. A LCD screen is placed under the tank and can block the light from the diodes, resulting in the polymerization of only the points not masked. Using daylight renders this technique cheaper compared to the previous methods.

Vat polymerization approaches to additive manufacturing have had a profoundly transformative effect on several biomedical fields. Most notably, in the mid-2000s nearly 100% of hearing aids went from being manufactured using conventional methods to being 3D printed (https://cepr.org/), and new approaches to improve these devices are still being implemented [34]. Likewise, clear dental aligners like those manufactured by Invisalign were only made possible by the development of new biocompatible resins and SLA/DLP polymerization approaches, combined with in silico models for orthodontic applications [30, 35]. Naturally, the high resolution of vat polymerization techniques, combined with the development of biocompatible resins, has led to increased interest in tissue engineering, most notably for cartilage [36, 37]. There has even been recent work using cell-laden photocurable resins for the manufacturing of cartilage scaffolds [38].

Other less common technologies include **Masked Stereolithography** (MSLA) where the light source is an array of LEDs and the light is filtered by a screen that lets the light go through in the desired section (*i.e.*, cross section of the object for that layer). This technology has the same advantages as DLP, but with a cheaper price.

b. **Jetting techniques**

Material Jetting (MJ) is a technology where photosensitive resins are deposited in a drop on demand approach to create the cross section of the object layer by layer. As the droplets of the resin are deposited a light source photopolymerized it (Fig. 12.4).

This technology shows the best surface finish of the different AM technologies. Furthermore, different resins can be used in the same part. This allows us to manufacture objects made of several materials, with different mechanical properties and visual appearance. This also enables the use of soluble supports, completely removing the use of breakaway supports which can affect the surface finish of the part. The main disadvantage of this technology is the low mechanical properties of the fabricated parts, especially regarding mechanical fatigue and sensitivity to sunlight, as in the case of the vat polymerization resins.

Other AM techniques based on liquid polymerization can be used to produce objects with various materials such as polymers or metals (see Table 12.1).

Despite its promising features for the building of multimaterial constructs, the technical constraints of material jetting mean that it is not commonly used for implants and biomanufacturing [39]. As technologies move forward and the needs in the field evolve, it is possible that the positive features of material jetting will reignite an interest in the use of this technology in the biomedical field. This technology is nevertheless used in the medical field to print anatomical models for surgical planning or medical models [40–42].

3. **Powder-based**

   a. **Inside a bed of powder**

In this group, different approaches are possible, either by direct consolidation of the powder by an energy source, or by using a binder. The most common technologies for direct consolidation of the powder are **Selective Laser Sintering** (SLS) and **Selective Laser Melting** (SLM), where a laser is used to selectively sinter (or melt) a layer of particles to create the cross-section of the object. **Selective Electron Beam Melting** (SEBM) is a similar technique where an electron beam is used to melt the powder at each layer. Once finished, the layer is covered with a new layer of fresh powder, and the process is repeated (Fig. 12.4).

In this family, the particles can be polymers, metals, or ceramics, each of them requiring higher power than the previous to sinter/melt the particles. The heat inside the powder bed can accumulate and generate temperature gradients that can become a limiting factor of the process. This accumulation of heat inside the bed degrades the unused powder around the fabricated object. In the case of polymers, the unused powder serves as the support for the construction of the parts, which is advantageous for manufacturing by avoiding the need for support removal This is not the case for the ceramics and metals, where as the temperature gradients get higher, the presence of supports to anchor the part to the build plate becomes necessary. Because of the use of powder, the surface finishing will be matte and grainy. Surface finish depends on powder size. The smaller the powder, the smoother the surface, but smaller powder sizes make the process harder to master. With this technique a high level of accuracy can be reached. One of the most important limitations of the technique is time, as printing is a long step, but cooling of the cake of powder is also very time consuming. Depending on the material, different percentages of non-used power is recyclable. As such, to reduce the costs it is highly recommended to maximize the number of parts and minimize the volume occupied by them. Like other printing methods, shrinking and warping can occur due to different temperatures in the powder bed. These phenomena can be reduced with the use of a heating tank [5].

Additive manufacturing approaches using metal powders, in particular titanium alloys, are promising in the field of orthopedics and bone implants, due to the ability to generate high strength macroporous structures that are well suited to bone tissue engineering and prosthetics [43].

On the other side, there is the binder jetting family where, instead of sintering/melting the particles to form the object, the particles are bound together during the fabrication by a binder material. It is necessary to highlight that the fabricated parts will have the mechanical properties of the binder material and not the ones from the material of the particles. For this reason, to obtain the final parts with the desired mechanical properties a post process of sintering is necessary to burn away the binder matrix to obtain a final part in the same material as the particles. In the case of metals, a second post-processing step can be carried out in which a metal is infiltrated into the part to achieve higher mechanical properties and a lower porosity. This approach is used by **Multi Jet Fusion** (MJF) and **Binder Jetting** (BJ).

In the medical field, additive manufacturing approaches like binder jetting have mostly found pharmacological and drug delivery applications. More specifically, binder jetting 3D printed drug products have been approved by the FDA since 2015, and their interest for personalized medicine has been steadily increasing in recent years [44]. This approach to drug formulation is compatible with many FDA-approved excipients, further increasing the potential to bring patient-tailored and challenging drugs to market [45]. Binder jetting is also used in the medical field to print anatomical models for surgical planning or medical models [40–42].

b. **Direct deposition**

The second group of the powder-based family is direct deposition, where the powder is transported and deposited in the desired spot. These technologies are limited to metals. This group can be divided into two different subgroups. The first is when the powder is projected to the fabrication spot and at the same time an energy source melts the powder to form a solid layer of material. This includes **Laser Powder Deposition** (LPD) and **Direct Energy Deposition** (DED). Another technology in this second subgroup is **Cold Spray Additive Manufacturing** (CSAM), where metal powder particles are projected at high velocities to a build plate (and then to the previous layer of deposed material) and are fused instantaneously with the rest of the object.

While medical applications of CSAM [46], DED [47] or LPD [48] are hinted at and explored in the literature, these approaches are still far from being commonplace, and only time will tell whether these strategies are truly well suited to the field of biomedical engineering.

4. **Sheet-based**

a. **Direct fabrication**

The last group of additive manufacturing techniques is sheet-based AM. One of the most common is **Laminated Object Manufacturing** (LOM). In this process, material sheets or rolls are cut, with a cutter or a laser, and each sheet is laminated with the previous one. These two steps, cutting and lamination, can be inverted, they are called “cut-then-bond” or “bond-then-cut” [49].

This speedy process allows rapid prototypes at low cost. LOM enables the production of large parts but with a lack of microstructure control and limited design. However, this method is wasteful, similar to subtractive fabrication, and implies a waste removal process that requires human intervention and can deteriorate the part. To reduce the piece damaging it, one significantly method could be to cut and realize a fine cross hatching on the waste but this would increase the whole process time [50].

Another technology in this family is the **Ultrasonic Additive Manufacturing** (UAM). In contrast with LOM technology, UAM uses metal sheets that are welded together by an ultrasonic roller.

Unlike other technologies capable of fabricating objects with metal, UAM shows several advantages, such as the production of fully dense parts. Furthermore, the complete process does not require heat sources, which allows the placement of internal sensors inside the object during the fabrication without damaging them. Another benefit is the possibility to alternate the metals during the printing process achieving a multi material metal part. UAM has the main disadvantage of being a wasteful AM technology. To obtain the final cross section at each layer, the part needs to be machined to remove excess material.

To this date, no biomaterial application can be found, for both LOM and UAM, in the literature.

b. **Fabrication of support for final part**

Roll Porous Scaffold (RPS) is a process where a support ribbon is perforated with a laser while it is rolled. The scaffold is progressively filled with a bioink [51] or a powder. When a powder is used the object is sintered. The ribbon and the support are then removed. The RPS is a fast technique that allows microscale work [52].

The many possibilities of hydrogel bioinks that can be used with RPS would allow the technique to produce biomaterials for tissue engineering for different tissues such as skin, bone or muscle. However, studies still need to be conducted on these topics [51].

In **Composite Based Additive Manufacturing** (CBAM) a binder solution is printed on a matrix sheet composed of randomly oriented fibers. A thermoplastic powder is then attached to the sheet thanks to the previously applied binder. Once the excess of powder is removed, the sheets are stacked together and heated in an oven for their consolidation. A sandblast step is finally realized to remove the excess of material.

This method is known for being a fast and cheap technique. No support is needed as the part is supported by the sheets itself. However, because of the difficulty of removing unwanted material, parts can only present simple shapes and none of the unused material can be recycled [53]. Due to layer stacking the mechanical properties are anisotropic. Finally, unlike other techniques, CBAM is not subject to shrinkage or warping.

In the medical field, this technique has been explored to produce surgical staples. However, they presented less implantable characteristics than other 3D printing methods [54].

Biomaterials

Most AM methods can be used to produce biomaterials. As reported in Table 12.1, for each family of AM techniques and even within each family, the range of fabrication materials and the fabrication constraints can be quite different. Even though the range of available fabrication materials has considerably increased in recent years and is expected to increase even more, the printing of complex materials is still challenging. The main AM fabrication materials can be divided into four families: polymers, metals, ceramics, and hydrogels. The understanding of the chemistry, mechanical properties and biocompatibility of these materials is crucial for a better use of AM. A description of each family of materials is given in the following paragraph, with their corresponding advantages and disadvantages, as well as the different approaches which can be followed for their application in biology.

The first and most common family of materials corresponds to polymers. They offer some advantages, notably a wide range of mechanical properties and characteristics, such as flexibility and impact resistance. Also, the AM techniques related to these materials have generally lower costs and are easier to use. Different AM technologies allow the use of thermoplastic or thermosetting polymers. The most used AM technique for the fabrication with thermoplastic parts is the FFF. This technology has been used in biological applications, mostly for bone scaffolds, either from pure polymer based materials, such as biocompatible PEEK [55], or by composite materials (polymers with fillers) such as hydroxyapatite composites [59, 60]. Another application of AM thermoplastic polymers is biocompatible Shape Memory Polymers (SMP) for the fabrication of stents [61]. The thermosetting polymers on the other hand are mostly used with vat polymerization technologies. The most common thermosetting polymers used in this technique are not biocompatible, nevertheless due to its high precision and inexpensive cost, in recent years biocompatible resins have been developed [62]. Another advantage in AM of thermosetting polymers versus thermoplastics is that vat-polymerization techniques do not require high temperatures. This opens up the possibility of loading biocompatible resins with cells, for tissue engineering applications for example [63].

The second family of materials is metals. For biomedical applications, the only metals and alloys that can be used are the biocompatible ones. AM of metals generally relies on powder-based techniques, where a bed of powder is selectively consolidated (either by a binder or by energy) to generate the desired form. In comparison with the AM of polymers, the AM of metals is orders of magnitude more expensive. Nevertheless, the mechanical properties of metals open the possibility to other applications. The metals used in AM are mainly for the fabrication of permanent implants and precision chirurgical tools. In other cases, their biodegradability and absorption inside the body opens up the possibility for biodegradable implants and scaffolds [64].

The third family of materials are ceramics, due to their high mechanical stability and their biocompatibility. This family is present in different biomedical applications, mostly in the fabrication of scaffolds, bone grafts and implants. Lastly, AM of ceramics can be used for the coating of metallic implants for improved biocompatibility and cellular growth [65]. Due to their stiffness, the direct fabrication of ceramic parts can be challenging. The most common approach today is fabricating the object using a polymeric matrix that is highly loaded with ceramic particles. Once fabricated, the part is placed in an oven, where the polymeric matrix is burned away and the ceramic particles are sintered, resulting in a ceramic part. Ceramic AM has had a big impact on the dental industry, where patient-specific prostheses are fast generated by vat polymerization.

The AM of hydrogels has been developed specifically for biological applications, the main reasons are their biocompatibility and their fast biodegradability. Moreover, hydrogels can present a shape memory [66, 67]. However, the low mechanical properties of the hydrogels make them a rather complex material to use in AM.

When to Choose Additive Manufacturing

Process

As shown in Fig. 12.1, there are three main types of manufacturing techniques: Formative, Subtractive and Additive manufacturing [5].

Formative manufacturing is mostly used to fabricate parts with a simple geometry, commonly made of polymers or metals, and in high production volume as it presents the lowest production cost. The main limitations of this technique are the high initial investment, as molds and dies can be expensive (fixed cost), and the fact that the geometrical freedom of the part is rather limited (many design constraints).

The main limitations of **subtractive manufacturing** are the higher time of production, the design limitations in size or geometry, and the higher cost, since a lot of raw material is lost (high production cost). This technique is the most precise among the different manufacturing techniques.

The three main strengths of additive manufacturing are the almost complete geometrical freedom, the possibility to fabricate in almost any existing material and the capability of rapid transition from a model to a real object (also known as rapid prototyping) (see Table 12.2). AM is composed of several techniques and each of them has its own capabilities and limitations, from design constraints and resolution to fabrication materials.

An important aspect of AM is that thanks to the geometrical freedom and the possibility for rapid prototyping, it enables patient-personalized objects such as implants or casts. For example, AM can use data from 3D medical imaging methods, such as X-ray computed tomography (CT) and magnetic resonance imaging, directly for the

Table 12.2 Strengths and weaknesses of traditional and additive manufacturing techniques

Process	Traditional manufacturing	Additive manufacturing
Geometry	Limitations	No limitations, flexible and complex parts, infill options
N° of processes needed to get a raw part	One or more	One
Stocks needed	Yes	No
Profitability	Based on large batches	Independent of number of units
Prototypage speed	Slow	Fast
Production speed	Fast	Slow
Weight	Fixed	Can easily be reduced
Goal	Mass production	Mass customization

fabrication of the objects. This enables the possibility of lower operation and hospitalization times, reducing costs while improving the performance of the implants [55].

Economy

Production costs are a major concern when selecting a manufacturing technique for industrial applications. In the case of additive manufacturing, one of the limiting factors is that there is almost no reduction of costs linked to the number of parts produced. As described in Fig. 5a, when comparing additive to subtractive and formative manufacturing, the two latter techniques see a steep reduction of costs correlated to the increase in the number of units produced [5]). As a result, additive manufacturing is mostly used for fast prototyping and production of complex-shaped structures as finished products (Fig. 5b).

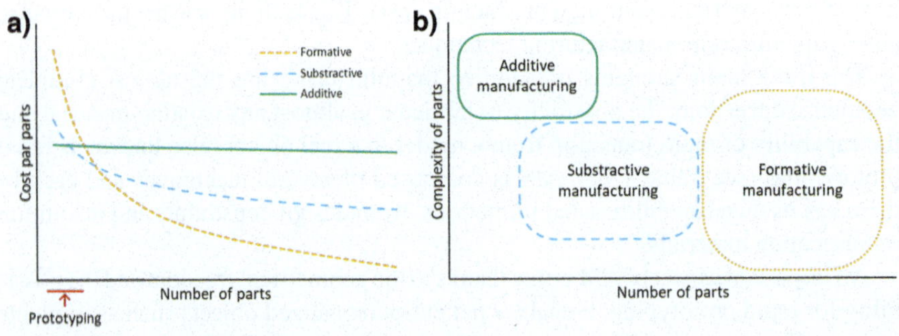

Fig. 12.5 **a** Evolution of the production costs of a part according to the number of parts produced. **b** Complexity of the parts according to the production capacity

When considering costs, one must also consider the material costs as it often represents a large proportion of the final cost of a product [68]. In the case of AM, this can be an important element of choice, especially considering the important loss if material is linked to subtractive manufacturing.

The choice of the right printing machine and its costs is the third factor to be taken into account. Using additive manufacturing for biomaterial design and tissue engineering offers simple solutions for the creation of complex (in shape, in composition, in heterogeneity of the organization) scaffolds. The recent ability to create matrices with organized cells already in the structure has opened many possibilities for researchers, but it has also created a large spang in the prices of the machines. Today, the market is offering many commercial bioprinting platforms, with prices ranging from $5,000 to over $500,000. The right choice for the material needed for specific applications will therefore also determine the economic impact on the production costs of the biomaterial [69].

Finally, the human cost represents a large portion of the final price of a biomaterial created by AM. The multiplicity of potential parameters linked to the production of the part (often more than 250 possible choices) creates the need for highly trained users of the technology, keeping production costs high [58].

Logistic

Nowadays, AM is an ubiquitous technique that can be easily developed at local production sites close to the consumer. Theoretically, this offers many logistical advantages compared to traditional manufacturing methods. For medical devices, however, the production is bound by specific regulations created by the International Standard Organization, such as the ISO 13485. The purpose of ISO 13485 is to ensure the consistent design, development, production, installation and delivery (and even disposal) of safe medical devices for their intended purpose. This necessary framework by the regulatory agencies is an issue for the rapid deployment of multiple production sites for biomaterials by AM [70]. The possibility that each hospital could have a local production site for patient-specific 3D printed biomaterials directly at the patient bedside is compromised by the aptitude to secure and adapt the place to the previously mentioned regulations.

Conclusion

As outlined in this chapter, additive manufacturing covers a very large range of techniques and materials. This versatility is of course well adapted to the creative design of new biomaterials and offers a great adaptation to any applicative question. Moreover, this field is constantly evolving with new progress and advances appearing at a fast rate. As an example, some printers on the market in 2023 allow to simultaneously use six independent printheads as different as a heated or cooled heads, cell electrowriting printhead, melt electrowriting printhead, or UV curing toolhead. This technique will therefore help the design of heterogeneous biomaterials in terms of structure (hard or soft scaffolds) of materials (metal or polymeric material together) or of cell distribution (acellular or cellularized areas in the same biomaterial).

In the specific case of medical devices, the additive manufacturing workflow offers the flexibility and the accessibility of allowing a personalization of implants and its ancillaries for each patient. Moving from mass production to personalized and patient-specific biomaterial design is a major trend in medical devices development. This approach not only facilitates surgical implantation, but also reduces stress for the patient as well as post-implantation adaptation time.

Additive manufacturing is also evolving towards better resolutions, to realize smaller parts while maintaining the possibility of complex shapes. This could increase the quality of the surface of the created pieces, and therefore reduce the number of post-production treatments needed to have new material with better biocompatibility.

All these possibilities come back to the researcher's choice of the right technology of additive manufacturing, consistent with the structure of the biomaterial to be developed or the application to be targeted. This choice has to be carried out in accordance with the material selected, the appropriate mechanical properties, surface finishing, interaction with cells or surrounding tissues, and the needs of the patient and the medical practitioner in charge of the implantation.

References

1. Chia HN, Wu BM (2015) Recent advances in 3D printing of biomaterials. J Biol Eng 9(1):4. https://doi.org/10.1186/s13036-015-0001-4
2. ASTM International (ASTM) (2016) Standard practice for selecting generic biological test methods for materials and devices (ASTM Standard F748–16). https://doi.org/10.1520/F0748-16. Retrieved from www.astm.org
3. ISO/TC 194 (2020) Évaluation biologique des dispositifs médicaux (ISO 10993:2020). Retrieved from https://www.iso.org/fr/ics/11.100.20/x/
4. Park J, Lakes RS (2007) Biomaterials: an introduction. Springer Science & Business Media
5. Redwood B, Schöffer F, Garret B (2017) The 3D printing handbook: technologies, design and applications. In: 3D Hubs. Retrieved from https://lib.hpu.edu.vn/handle/123456789/31395
6. Kodama H (1998) Automatic method for fabricating a three-dimensional plastic model with photo-hardening polymer. Rev Sci Instrum 52(11):1770. https://doi.org/10.1063/1.1136492
7. Zou Q (2022) Parametric/direct CAD integration. https://doi.org/10.48550/arXiv.2203.02252
8. Gleadall A (2021) FullControl GCode designer: open-source software for unconstrained design in additive manufacturing. Addit Manuf 46:102109. https://doi.org/10.1016/j.addma.2021.102109
9. Rajput S, Burde H, Singh US, Kajaria H, Bhagchandani RK (2021) Optimization of prosthetic leg using generative design and compliant mechanism. Mater Today Proc 46:8708–8715. https://doi.org/10.1016/j.matpr.2021.04.026
10. Hunde BR, Woldeyohannes AD (2022) Future prospects of computer-aided design (CAD)—a review from the perspective of artificial intelligence (AI), extended reality, and 3D printing. Results Eng 14:100478. https://doi.org/10.1016/j.rineng.2022.100478
11. Chakraborty D, Reddy BA, Choudhury AR (2008) Extruder path generation for curved layer fused deposition modeling. Comput-Aided Des 40(2):235–243. https://doi.org/10.1016/j.cad.2007.10.014

12. Rankin TM, Giovinco NA, Cucher DJ, Watts G, Hurwitz B, Armstrong DG (2014) Three-dimensional printing surgical instruments: are we there yet? J Surg Res 189(2):193–197. https://doi.org/10.1016/j.jss.2014.02.020
13. van der Stelt M, Stenveld F, Bitter T, Maal TJJ, Janssen D (2022) Design evaluation of FFF-printed transtibial prosthetic sockets using follow-up and finite element analysis. Prosthesis 4(4):589–599. https://doi.org/10.3390/prosthesis4040048
14. Gensler M et al (2020) 3D printing of bioreactors in tissue engineering: a generalised approach. PLOS ONE 15(11):e0242615. https://doi.org/10.1371/journal.pone.0242615
15. Steffens S et al (2022) 3D-printed design of a stereotaxic adaptor for the precision targeting of brain structures in infant mice. Eur J Neurosci 55(3):725–732. https://doi.org/10.1111/ejn.15588
16. Baek I, Kwon O, Lim C-M, Park KY, Bae C-J (2022) 3D PEEK objects fabricated by fused filament fabrication (FFF). Materials 15(898):1–9. https://doi.org/10.3390/ma15030898
17. Schlégl ÁT, Told R, Kardos K, Szőke A, Ujfalusi Z, Maróti P (2022) Evaluation and comparison of traditional plaster and fiberglass casts with 3D-printed PLA and PLA–$CaCO_3$ composite splints for bone-fracture management. Polymers 14(3571):1–17. https://doi.org/10.3390/polym14173571
18. Rezania N, Asadi-Eydivand M, Abolfathi N, Bonakdar S, Mehrjoo M, Solati-Hashjin M (2022) Three-dimensional printing of polycaprolactone/hydroxyapatite bone tissue engineering scaffolds mechanical properties and biological behavior. J Mater Sci Mater Med 33(3):31. https://doi.org/10.1007/s10856-022-06653-8
19. Xia C et al (2020) A review on wire arc additive manufacturing: monitoring, control and a framework of automated system. J Manuf Syst 57:31–45. https://doi.org/10.1016/j.jmsy.2020.08.008
20. Pixner F et al (2020) Wire-based additive manufacturing of Ti-6Al-4V using electron beam technique. Materials 13(3310):1–23. https://doi.org/10.3390/ma13153310
21. Murr LE et al (2009) Microstructures and mechanical properties of electron beam-rapid manufactured Ti–6Al–4V biomedical prototypes compared to wrought Ti–6Al–4V. Mater Charact 60(2):96–105. https://doi.org/10.1016/j.matchar.2008.07.006
22. Mondal D, Willett TL (2020) Mechanical properties of nanocomposite biomaterials improved by extrusion during direct ink writing. J Mech Behav Biomed Mater 104:103653. https://doi.org/10.1016/j.jmbbm.2020.103653
23. Monfared MH et al (2022) A deep insight into the preparation of ceramic bone scaffolds utilizing robocasting technique. Ceram Int 48(5):5939–5954. https://doi.org/10.1016/j.ceramint.2021.11.268
24. Lewis JA (2006) Direct ink writing of 3D functional materials. Adv Funct Mater 16(17):2193–2204. https://doi.org/10.1002/adfm.200600434
25. Agarwal S, Saha S, Balla VK, Pal A, Barui A, Bodhak S (2020) Current developments in 3D bioprinting for tissue and organ regeneration—a review. Front Mech Eng 6(589171):1–22. https://doi.org/10.3389/fmech.2020.589171
26. Persaud A, Maus A, Strait L, Zhu D (2022) 3D bioprinting with live cells. Eng Regen 3(3):292–309. https://doi.org/10.1016/j.engreg.2022.07.002
27. Dey M, Ozbolat T (2020) 3D bioprinting of cells, tissues and organs. Sci Rep 10:14023. https://doi.org/10.1038/s41598-020-70086-y
28. Zhang B, Cristescu R, Chrisey DB, Narayan RJ (2020) Solvent-based extrusion 3D printing for the fabrication of tissue engineering scaffolds. Int J Bioprinting 6(1):211. https://doi.org/10.18063/ijb.v6i1.211
29. Zhu W, Ma X, Gou M, Mei D, Zhang K, Chen S (2016) 3D printing of functional biomaterials for tissue engineering. Curr Opin Biotechnol 40:103–112. https://doi.org/10.1016/j.copbio.2016.03.014
30. Mohammed AA, Algahtani MS, Ahmad MZ, Ahmad J, Kotta S (2021) 3D printing in medicine: technology overview and drug delivery applications. Ann 3D Print Med 4:100037. https://doi.org/10.1016/j.stlm.2021.100037

31. Mahmood MA (2021) 3D printing in drug delivery and biomedical applications: a state-of-the-art review. Compounds 1(3):94–115. https://doi.org/10.3390/compounds1030009
32. Fitzpatrick V et al (2021) Functionalized 3D-printed silk-hydroxyapatite scaffolds for enhanced bone regeneration with innervation and vascularization. Biomaterials 276:120995. https://doi.org/10.1016/j.biomaterials.2021.120995
33. Zhu Z, Park HS, McAlpine MC (2020) 3D printed deformable sensors. Sci Adv 6(25):eaba5575. https://doi.org/10.1126/sciadv.aba5575
34. Vivero-Lopez M et al (2021) Anti-biofilm multi drug-loaded 3D printed hearing aids. Mater Sci Eng C 119:111606. https://doi.org/10.1016/j.msec.2020.111606
35. Barone S, Neri P, Paoli A, Razionale AV, Tamburrino F (2019) Development of a DLP 3D printer for orthodontic applications. Procedia Manuf 38:1017–1025. https://doi.org/10.1016/j.promfg.2020.01.187
36. Kim SH, Kim DY, Lim TH, Park CH (2020) Silk fibroin bioinks for digital light processing (DLP) 3D bioprinting. In: Chun HJ, Reis RL, Motta A, Khang G (eds) Bioinspired biomaterials: advances in tissue engineering and regenerative medicine. Singapore: Springer, pp 53–66. https://doi.org/10.1007/978-981-15-3258-0_4
37. Kim SH et al (2018) Precisely printable and biocompatible silk fibroin bioink for digital light processing 3D printing. Nat Commun 9(1620):1–14. https://doi.org/10.1038/s41467-018-03759-y
38. Hong H et al (2020) Digital light processing 3D printed silk fibroin hydrogel for cartilage tissue engineering. Biomaterials 232:119679. https://doi.org/10.1016/j.biomaterials.2019.119679
39. Salmi M (2021) Additive manufacturing processes in medical applications. Materials 14(191):1–16. https://doi.org/10.3390/ma14010191
40. Mitsouras D et al (2015) Medical 3D printing for the radiologist. Radiographics 35(7):1965–1988. https://doi.org/10.1148/rg.2015140320
41. Tappa K, Jammalamadaka U (2018) Novel biomaterials used in medical 3D printing techniques. J Funct Biomater 9(17):1–16. https://doi.org/10.3390/jfb9010017
42. Touri M, Kabirian F, Saadati M, Ramakrishna S, Mozafari M (2019) Additive manufacturing of biomaterials—the evolution of rapid prototyping. Adv Eng Mater 21(2):1800511. https://doi.org/10.1002/adem.201800511
43. Guo AXY et al (2022) Biomedical applications of the powder-based 3D printed titanium alloys: a review. J Mater Sci Technol 125:252–264. https://doi.org/10.1016/j.jmst.2021.11.084
44. Rahman Z, Charoo NA, Kuttolamadom M, Asadi A, Khan M (2020) Chapter 46—printing of personalized medication using binder jetting 3D printer. In: Faintuch J, Faintuch S (eds) Precision medicine for investigators, practitioners and providers. Academic Press, pp 473–481. https://doi.org/10.1016/B978-0-12-819178-1.00046-0
45. Kozakiewicz-Latała M et al (2022) Binder jetting 3D printing of challenging medicines: from low dose tablets to hydrophobic molecules. Eur J Pharm Biopharm 170:144–159. https://doi.org/10.1016/j.ejpb.2021.11.001
46. Ashokkumar M, Thirumalaikumarasamy D, Sonar T, Deepak S, Vignesh P, Anbarasu M (2022) An overview of cold spray coating in additive manufacturing, component repairing and other engineering applications. J Mech Behav Mater 31(1):514–534. https://doi.org/10.1515/jmbm-2022-0056
47. Freire B, Babcinschi M, Ferreira L, Señaris B, Vidal F, Neto P (2020) Direct energy deposition: a complete workflow for the additive manufacturing of complex shape parts. Procedia Manuf 51:671–677. https://doi.org/10.1016/j.promfg.2020.10.094
48. Costello AC, Koduri SK, Sears JW (2003) Optimization of laser powder deposition for 316L stainless steel. Int Congr Appl Lasers Electro-Opt p 701. https://doi.org/10.2351/1.5060072
49. Dermeik B, Travitzky N (2020) Laminated object manufacturing of ceramic-based materials. Adv Eng Mater 22(9):2000256. https://doi.org/10.1002/adem.202000256
50. Chiu YY, Liao YS, Hou CC (2003) Automatic fabrication for bridged laminated object manufacturing (LOM) process. J Mater Process Technol 140(1):179–184. https://doi.org/10.1016/S0924-0136(03)00710-6

51. Shulunov V (2019) A novel roll porous scaffold 3D bioprinting technology. Bioprinting 13:e00042. https://doi.org/10.1016/j.bprint.2019.e00042
52. Shulunov V (2014) A high performance, high precision, low cost rapid prototyping and manufacturing technology. Int J Autom Smart Technol 4:179–184. https://doi.org/10.5875/ausmt.v4i3.718
53. Goh GD, Yap YL, Agarwala S, Yeong WY (2019) Recent progress in additive manufacturing of fiber reinforced polymer composite. Adv Mater Technol 4(1):1800271. https://doi.org/10.1002/admt.201800271
54. Al Takhayneh O (2021) Rapid prototyping of surgical staples. Univ Wollongong Thesis Collect 2017. Retrieved from https://ro.uow.edu.au/theses1/1422
55. Zhao Y, Zhao K, Li Y, Chen F (2020) Mechanical characterization of biocompatible PEEK by FDM. J Manuf Process 56:28–42. https://doi.org/10.1016/j.jmapro.2020.04.063
56. Milovanović A et al (2021) The effect of time on mechanical properties of biocompatible photopolymer resins used for fabrication of clear dental aligners. J Mech Behav Biomed Mater 119:104494. https://doi.org/10.1016/j.jmbbm.2021.104494
57. Ge X et al (2020) Cicada-inspired fluoridated hydroxyapatite nanostructured surfaces synthesized by electrochemical additive manufacturing. Mater Des 193:108790. https://doi.org/10.1016/j.matdes.2020.108790
58. Li C, Zhou X, Liu X (2021) Integrating the bio-digital aesthetic value and architecture potential of RLP-concrete 3D printing technology. IOP Conf Ser Earth Environ Sci 768(1):012156. https://doi.org/10.1088/1755-1315/768/1/012156
59. Pu'ad NASM, Haq RHA, Noh HM, Abdullah HZ, Idris MI, Lee TC (2020) Review on the fabrication of fused deposition modelling (FDM) composite filament for biomedical applications. Mater Today Proc 29:228–232. https://doi.org/10.1016/j.matpr.2020.05.535
60. Sharma R, Singh R, Penna R, Fraternali F (2018) Investigations for mechanical properties of HAP, PVC and PP based 3D porous structures obtained through biocompatible FDM filaments. Compos Part B Eng 132:237–243. https://doi.org/10.1016/j.compositesb.2017.08.021
61. Yeazel TR, Becker ML (2020) Advancing toward 3D printing of bioresorbable shape memory polymer stents. Biomacromolecules 21(10):3957–3965. https://doi.org/10.1021/acs.biomac.0c01082
62. González G et al (2020) Materials testing for the development of biocompatible devices through vat-polymerization 3D printing. Nanomaterials 10(1788):1–13. https://doi.org/10.3390/nano10091788
63. Murphy CA, Lim KS, Woodfield TBF (2022) Next evolution in organ-scale biofabrication: bioresin design for rapid high-resolution vat polymerization. Adv Mater 34(20):2107759. https://doi.org/10.1002/adma.202107759
64. Velásquez-García LF, Kornbluth Y (2021) Biomedical applications of metal 3D printing. Annu Rev Biomed Eng 23(1):307–338. https://doi.org/10.1146/annurev-bioeng-082020-032402
65. Ly M, Spinelli S, Hays S, Zhu D (2022) 3D printing of ceramic biomaterials. Eng Regen 3(1):41–52. https://doi.org/10.1016/j.engreg.2022.01.006
66. Chen J, Huang J, Hu Y (2020) 3D printing of biocompatible shape-memory double network hydrogels. ACS Appl Mater Interfaces 13(11):12726–12734. https://doi.org/10.1021/acsami.0c17622
67. Shiblee MNI, Ahmed K, Khosla A, Kawakami M, Furukawa H (2018) 3D printing of shape memory hydrogels with tunable mechanical properties. Soft Matter 14(38):7809–7817. https://doi.org/10.1039/C8SM01156G
68. Thomas DS, Gilbert SW (2014) Costs and cost effectiveness of additive manufacturing. In: National Institute of Standards and Technology, NIST SP 1176. https://doi.org/10.6028/NIST.SP.1176
69. Tong A et al (2021) « Review of low-cost 3D bioprinters: state of the market and observed future trends. SLAS Technol Transl Life Sci Innov 26(4):333–366. https://doi.org/10.1177/24726303211020297
70. Verboeket V, Khajavi SH, Krikke H, Salmi M, Hoolmström J (2021) Additive manufacturing for localized medical parts production: a case study. IEEE Access 9:25818–25834. https://doi.org/10.1109/ACCESS.2021.3056058

Chapter 13
Bioprinting

Musa Ayran, Berrak Bulut, and Songul Ulag

Abstract This chapter provides an extensive overview of bioprinting technology, its operating principles, different types of bioprinting, challenges faced by the field, and specific applications that hold potential for future advancements. The chapter begins with a definition of bioprinting and the printing process, including an introduction to the different types of bioprinters available. It then highlights the importance of selecting the most appropriate technique and bioink for different application situations, including an assessment of various bioprinting processes and bioinks, such as inkjet, extrusion, and laser-assisted bioprinting. It further examines the challenges facing the field of bioprinting, including the need for more efficient bioprinting processes and enhanced bioinks. It delves into the various types of bioinks used for bioprinting, including hydrogel-based bioinks, protein-based bioinks, polysaccharides, decellularized extracellular matrix (dECM)-based bioinks, and synthetic polymer-based bioinks, highlighting their specific properties suitable for different tissue types and applications. The advancements made in 3D bioprinting using bioinks for producing specific tissues, such as cornea, skin, bone, and cartilage, are also explored. Finally, the chapter provides a comprehensive overview of specific applications with potential for significant future advancements and notable progress. These areas of bioprinting research and development are discussed in relation to their potential impact on various industries, including regenerative medicine, organ transplantation, and tissue engineering. In summary, this chapter presents a detailed overview of bioprinting technology, its various components, challenges, and potential applications. It provides a valuable resource for researchers, academics, and practitioners in the field of bioprinting, as well as those interested in the areas of regenerative medicine, organ transplantation, and tissue engineering.

Keywords Bioprinting · Bone · Cartilage · Cornea · Skin · Tissue engineering

M. Ayran · B. Bulut · S. Ulag (✉)
Center for Nanotechnology and Biomaterials Application and Research (NBUAM), Marmara University, 34722 Istanbul, Turkey
e-mail: ulagitu1773@gmail.com

M. Ayran · S. Ulag
Department of Metallurgical and Materials Engineering, Faculty of Technology, Marmara University, 34722 Istanbul, Turkey

O. Gunduz et al. (eds.), *Biomaterials and Tissue Engineering*, Stem Cell Biology and Regenerative Medicine 74, https://doi.org/10.1007/978-3-031-35832-6_13

Introduction

Tissue and organ damage, such as cancer and heart attacks, affect vast numbers of people. Although organ and tissue transplantation is the preferred method for treating as a therapy option for some of these ailments, it is severely limited because to the lack of donors. The biomedical industry attempts to resolve this problem by creating tissue replacements that outperform existing therapy modalities and offer injured sides a long-term fix. A cutting-edge technique in the biomedical industry called layer-by-layer deposition of bioprinting, three-dimensional (3D) bioprinting enables the production of biological tissue constructs. Considering how complicated human organs, which showed the medically undiscovered method of organ growth, The expectation that 3D bioprinting can effectively mitigate the inadequacy of organ supply for transplantation is excessively optimistic. However, in a certain amount of time, 3D bioprinting will undoubtedly play a more significant part in creating in vitro organ models. With regards to 3D bioprinting, Scientists may now alter living cells, their biological and biochemical surroundings, and other components of complex biological structures thanks to the invention of bioink materials.

Conventional approaches utilized for drug testing and studying biological mechanisms, such as two-dimensional (2D) cell culture or animal trials, have several drawbacks. The 2D cell culture is far simpler than the microenvironment in vivo, and 2D models may provide the contradictory outcome. These outcomes increase the urgency of the need for more precise in vitro models, which 3D bioprinting is adept at. 3D bioprinting is currently the most effective technique for producing living, 3D cell-laden objects in vitro, which allows for the fine structural construction of diverse cells. Owing to its ability to completely manage the shape and content of the bioprinted structures, 3D bioprinting has been highly embraced as a replacement for conventional manufacturing methods [40]. In addition, when contrasted with traditional techniques for scaffold production such as solvent casting particle leaching, gas foaming, and molding [67], 3D bioprinting offers several advantages. By employing suitable design, it can fabricate a porous tissue structure with high precision, thereby providing the optimal microenvironment for biological constituents, which is crucial for successful tissue and organ regeneration. A key advantage of 3D bioprinting is its capacity to directly incorporate patient-specific data into the biofabrication process to produce accurate designs.

This chapter begins with the history and innovations of bioprinting. Following the first section, it will introduce the various types of bioprinting methods and identify the bioinks for bioprinting, which represent bioink evaluation standards, and several typical bioprinting methods in order to select the most appropriate method and bioink for various application scenarios. We also compare several bioprinting methodologies, where the choice of printing techniques for diverse applications will be covered. Subsequently, we will delve into the progress on specific tissues such as the cornea, skin, bone, and cartilage that were produced using bioinks in 3D bioprinting applications. Lastly, we will discuss specific applications that show the most potential for future work of a substantial magnitude and the tremendous advancement in their respective fields.

The History and Innovations of Bioprinting

Charles Hull applied for a patent on the stereolithography method in 1984. They were to be the progenitors of a brand-new production technique that would transform the industry. Bioprinting was first demonstrated using cytoscribing technology [44], a technique for placing cells in micropositions and creating 2D synthetic tissues. The study utilized an HP inkjet printer to achieve cytoscribing at a low-level placement of cells, along with a graphics plotter for precise high-level positioning. With the initial effort to grow cartilage tissue resembling an ear on a mouse's dorsa in 1996 [75]. In addition, the first bladder produced in three dimensions using patient cells was implanted in 1999, sparking a scientific revolution since it guaranteed no graft rejection. Researchers also found that stem cells had the capacity to develop into specialized cells that may later be used to cure a wide range of illnesses.

The first bioprinter was created by Thomas Boland in 2000 when he adapted an inkjet printer to bioprinted cells onto a Petri dish, marking a significant development in this area [88]. The first extrusion-based bioprinting technique was described and later made available for purchase under the name 3D-Bioplotter [50]. Wilson and Boland created the initial inkjet bioprinter by adapting a typical HP inkjet printer [85]. It was invented in 2004 to create 3D tissue using just cells and no scaffold. By utilizing bioprinting, Norotte et al. produced vascular tissue without a scaffold [60].

In the years that followed, there was a lot of progress and study being done in the field of science. In 2012, structures like liver and baptismal tissue were produced using bioprinting [14]. In 2014, tissues with blood vessels, and in 2016, the heart valve were all being developed [39]. Noor and colleagues created a perfusable scale-down heart [61]. These developments of bioprinting from the past to the present are exhibited in the Fig. 13.1.

Strkingly, the characterization or progress of innovative approaches for 3D bioprinting requires greater attention from the literature. This is mostly due to a shift in focus away from engineering and toward fields like biology, medicine, or material science. Additionally, there is a severe shortage of research, particularly for the majority of the complicated technology. This is clear from the literature, where the most frequently researched techniques are laser-based techniques and stereolithography. Extrusion 3D bioprinting initially emerged in the 2000s, but it took off after 2015 when commercial bioprinters entered the market [73].

A second extremely novel field of applications is the magnetic levitation method, which was developed in 2020 [69]. The initial magnetic-based bioprinter investigations, however, revealed a drawback: the bioinks must be able to endure the force of Earth's gravity. Regarding this subject, space agencies are also looking into the possibility of enhancing 3D printing of delicate human tissues like blood arteries and muscles utilizing microgravity.

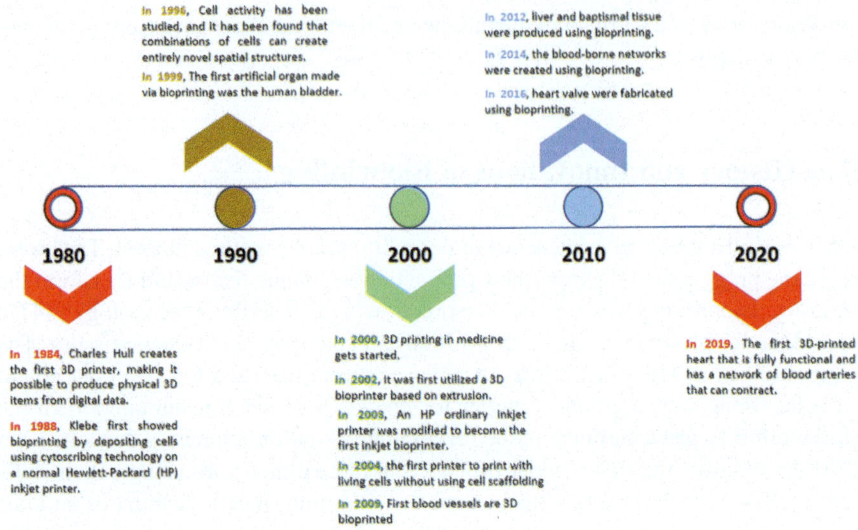

Fig. 13.1 Brief history of bioprinting

Introduce the Different Types of Bioprinting

Extrusion-based, droplet-based, and laser-based bioprinting are the three main methods used in 3D bioprinting. Figure 13.2 shows examples of three different types of bioprinting: extrusion-based bioprinting, droplet-based bioprinting, and laser-based bioprinting. Extrusion-based bioprinting uses continuous filaments made of bioinks to create structures; droplet-based bioprinting produces discrete droplets to stack into structures; and laser-based bioprinting optically captures and then directs cells onto a substrate. Cells and biocompatible materials are utilized as the "ink" for 3D tissue engineering in 3D bioprinting, a form of additive manufacturing. Bio-inks are printed in layers to form scaffolds of the necessary size and shape while cells are kept within for functional integration, maturation, and tissue synthesis. These bioprinting methods involve printing biomaterial scaffolds that are seeded with cells in vitro or in vivo after printing, or printing biomaterials with cells for encapsulated cell constructions.

Extrusion-Based Bioprinting

An extrusion-based bioprinting system typically comprises a dispensing head, positioning control unit, and temperature control unit. The printing stage serves as a support structure for the scaffold being constructed, whereas the dispensing head houses a dispenser that can move along three axes. A host computer controls three

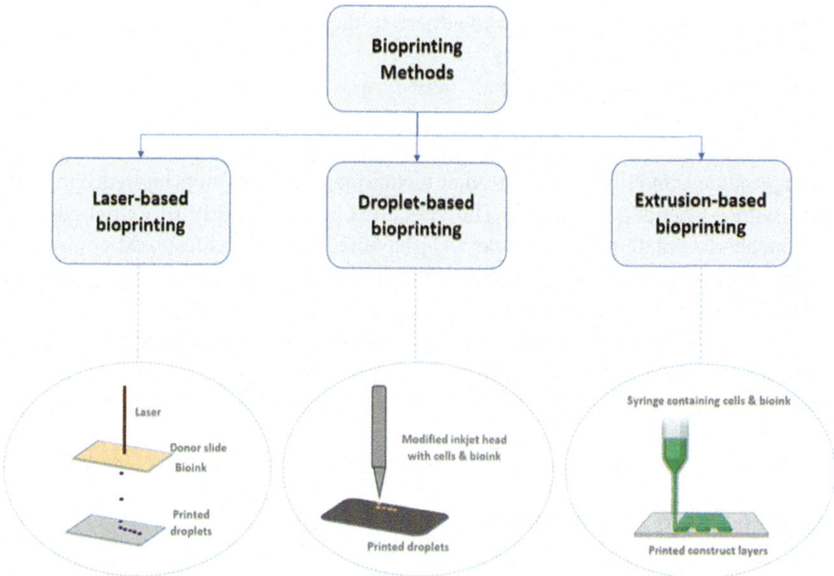

Fig. 13.2 The classification of bioprinting

controllers responsible for regulating temperature, positioning, and dispensing. In order to fabricate a multi-layered scaffold during the bioprinting process, the bioink is loaded into a syringe and then mechanically extruded through a needle onto the printing stage.

Hydrogels utilized in extrusion-based bioprinting are considered non-Newtonian fluids, where the shear rate significantly impacts viscosity [38]. Extrusion-based methods are particularly fitted for hydrogels with shear thinning and thixotropic behavior. The random polymer chains in shear-thinning hydrogels are organized in a way that makes them extrudable by the shear pressures. The essential parameters that influence the physical and biological qualities of the printed construct are the extrusion temperature, nozzle type, applied pressure, and the biomaterial's rheological properties, which are relevant to the viscosity and shear thinning of hydrogels. The system must apply tremendous pressure to the hydrogels during the printing process. The shear force acting on the cells could be so great as to result in cell death depending on the viscosity of the printed hydrogel. Additionally, this printing technique is now limited to a few materials, reducing the range of possible applications. It has been able to create 3D-printed tissues that closely resemble the extracellular matrix (ECM) in terms of separate functional layers and the capacity to apply flow to the inner channel through careful material selection and engineering.

Biomaterials must be able to undergo room-temperature gelation or must be able to conduct chemical cross-linking either before or during the material extrusion [11]. Additionally, the bioink needs to gel quickly in order to maintain its structure without spreading. In order for the bioink to adhere to the substrate and maintain its shape, a suitable substrate must be used.

In comparison to ink-jet and laser-based bioprinting methods, extrusion bioprinting presents several advantages. It allows for the dispensing of a diverse range of biomaterials and cells, including native and synthetic hydrogel polymers, cell aggregates, and dECM, while other techniques can only bioprint hydrogel polymers with suspended cells [70]. The versatility and capability of extrusion-based approaches to create pore networks [41] have led to their widespread commercial adoption. Furthermore, extrusion-based bioprinting is the most effective strategy for fabricating large-scale constructs with structural integrity. Nonetheless, the printing substantial constructs may have negatively impact the cell survival due to prolonged exposure to dehydration and nutrient deprivation throughout the printing process [66].

Droplet-Based Bioprinting

Droplet-based bioprinting is derived from conventional paper printing, with Klebe's work showing for the first time that cells could be printed [44]. In order to create constructions by manipulating fluid qualities like surface tension and viscosity, droplet-based bioprinting uses bioink that is stored in a cartridge to create droplets under the influence of gravity, ambient pressure, and fluid mechanics [21]. Its basic operating concept relies on the application of changing potentials to generate the pressure required to force the discharge of the bioink as droplets from the nozzle.

Droplet-based bioprinting, among other methods, has emerged as a viable method of bioprinting because of its ease of use, high spatial resolution, high throughput, non-contact printing, and capacity to create concentration gradients of bioactive components [21]. In contrast to extrusion-based bioprinting, which uses continuous filaments as its basic unit, droplet-based bioprinting contain multiple, independent droplets as its fundamental unit, which generally results in higher resolution than extrusion-based bioprinting. Additionaly, the resolution of droplet-based bioprinting is higher than that of extrusion-based bioprinting and it can incorporate more biologics than extrusion-based bioprinting. Moreover, the technology enables bioprinting by depositing regulated amounts of bioink at predetermined sites, allowing for spatially heterocellular constructions with well-defined cell placement. On the other hand, the main disadvantage is that droplet-based bioprinting arises due to the clogging of orifice which barely maintain an consistent flow throughout the bioprinting procedure. Such properties restrict the number of biomaterials that may be used for droplet-based bioprinting [67]. Furthermore, in droplet-based bioprinting, the range of available materials that can be used as bioink is limited. However, if the bioink used in droplet-based bioprinting is non-fibrous and has low viscosity, the tube system and nozzle enable smooth substance flow without encountering blockages.

Droplet-based bioprinting is widely used in tissue engineering, regenerative medicine, transplantation, clinical, medicine, high-throughput monitoring, and medical research due to its ease of use and capacity to precisely control biologics, such as cells, growth factors, genes, meds, biopolymers, etc. Furthermore, bioprinting technology has attracted the attention of researchers in a broad range of studies, such as pharmaceutics and regenerative medicine, as evidenced by recent studies [28, 67]. For example, Xu and colleagues exhibited that bioprinted stem cells preserved their functional and differentiating properties, both in vitro and in vivo [89].

Laser-Based Bioprinting

Laser-based bioprinting employs mirrors to focus a laser pulse onto a layer of bioink located above the substrate. The heat generated by the laser creates a high-pressure bubble, which dislodges some of the ink and prompts cell-laden droplets to be deposited onto the substrate. To create the final construct, this process is repeated numerous times in a layer-by-layer method [24, 46].

The main benefit of laser-based bioprinting is its capacity to print well-defined tissue constructions, despite its high precision and resolution, the complex setup of this bioprinting method has restricted its usage in the bioprinting field compared to other more easily accessible bioprinting approaches. With such incredible accuracy, it is possible to recreate the heterogeneity and diversity of tissues utilizing tissue-engineered cell arrays that integrate cells with pertinent biological elements. Another significant advantage of laser-based bioprinting, similar to droplet-based bioprinting, is its non-contact physical nature. Also, laser-based bioprinting can precisely print various tissue constituents, particularly cells, and replicate their spatial organization. However, due to the cytotoxic nature of the photoinitiators, structurally simple and weak constructions may be constructed in general. The combination of numerous materials in the bioprinted construct is one of the primary constraints of laser-based bioprinting. Most importantly, laser-based bioprinting lacks a broader range of bioink materials. In addition, due to the high shear stress and droplet formation process, only a small number of cells can be enclosed in each droplet. Laser-based bioprinting, on the other hand, enables mass bioprinting of cells since it does not suffer from shear stress [22, 63, 74].

Laser-based bioprinting naturally avoids issues like nozzle clogging, non-reproducibility due to solution viscosity, cross-infection, or incurring substrate damage because it is a non-contacting and nozzle-free technique. Unlike droplet-based bioprinting, it has superior accuracy and creates smaller droplets; also, its bioink has a higher cell concentration, shortening the biological maturation period. Other research has demonstrated that laser-based bioprinting is more appropriate for in situ and in vivo bioprinting and advantageous for the development of multi-layer cell structures [6, 46].

Identifying the Bioinks for Bioprinting

The Enhancement of 3D bioprinting ink materials with the necessary mechanical and biological properties has relied on the use of two main groups of material precursors. In the field of bioprinting, the prevailing type of bioink utilized is known as support material-based bioink. This approach involves the encapsulation of cells within hydrogels or other extracellular matrices, which are subsequently printed into complex 3D structures. Hydrogels that include cells promote cell growth and proliferation and aid in developing new tissue. The second kind of bioink mimics embryonic development by bioprinting cells without the need for a scaffold in a manner similar to that of embryonic development.

The selection of an optimal biofabrication technique for diverse biomaterial inks is widely acknowledged as a crucial and challenging procedure to attain elevated levels of manufacturing process efficiency. The objective of the application is the primary factor in determining the best biofabrication technique for a particular bioink. Additionally, the ink's chemical, biological, and mechanical characteristics must be determined since they will affect the benefits and drawbacks of each biofabrication method in different ways. In order to promote cell growth and retain shape fidelity after printing, the bioink needs to have the biomechanical qualities that make extrusion simple and require little shear stress [35].

Hydrogel-Based Bioinks

A group of polymeric materials that are crosslinked are known as hydrogels. They have the capacity of absorbing and being hold enormous amounts of water. In tissue engineering, hydrogels are divided into four categories: Protein-based, polysaccharides, synthetic polymer-based and dECM-based bioinks as indicated in the Fig. 13.3. Some types of hydrogels can replicate the natural tissue environment because they have some key native ECM component characteristics [81]. The hydrogels are composed of extremely hydrated three-dimensional environments that resemble the ECM. Moreover, they exhibit permeability to a range of substances, including nutrients, oxygen, and other molecules. that are soluble in water as well as cellular movement and dynamic communication network with porous connectivity due to the peculiar nature of their architecture [53]. Although both natural and manufactured hydrogels have major drawbacks, these ECM-mimicking properties enable the encapsulation of cells within a mechanically sturdy, highly hydrated 3D milieu.

In general, the formulation of hydrogel ink needs a particular thick polymeric solution that may quickly generate well-connected after printing by the thermal, physical and enzymatic cross-linking of polymers, as shown in Fig. 13.4. Hydrogels have the ability to be created using a various types of crosslinking techniques. Ionic crosslinkers like can be used to start the physical crosslinking process. Hydrogel with crosslinkers like photoinitiators can be used to build the scaffold layer by layer if you

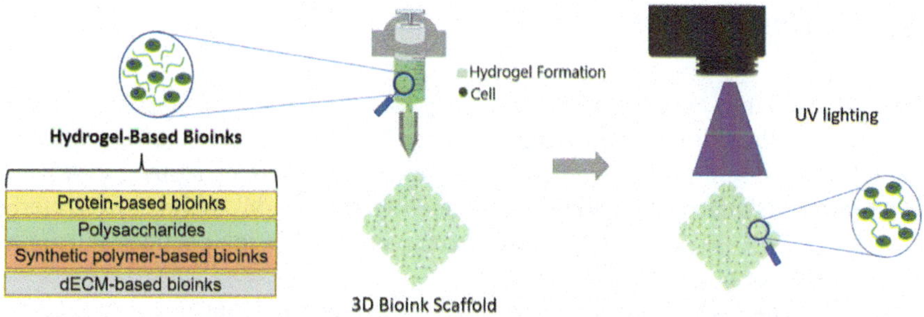

Fig. 13.3 Bioink types and their crosslinked strue after exposing UV-light

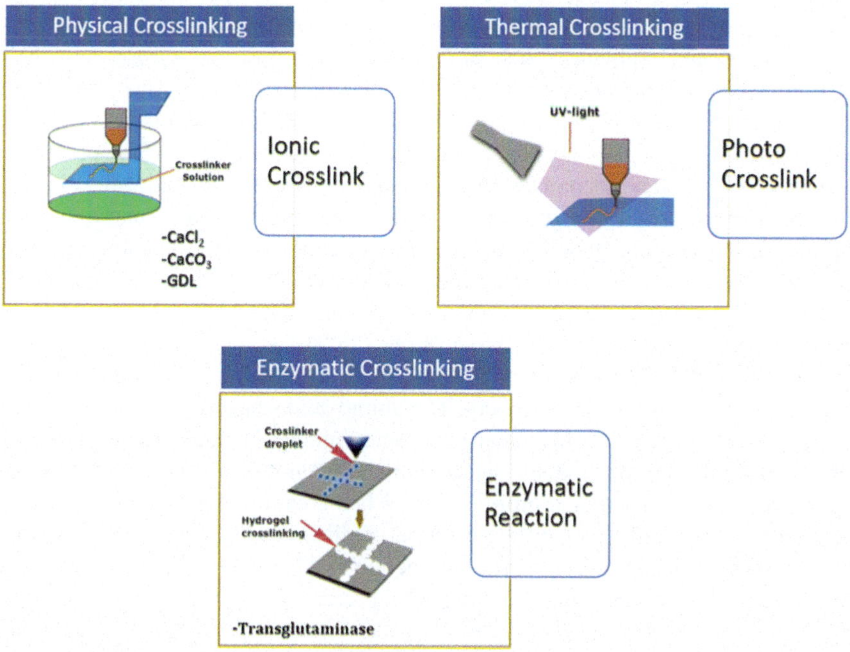

Fig. 13.4 The crosslinking mechanism of bioinks

employ a photoinitiator (LAP or IRGACURE) to create crosslinked hydrogel with high mechanical qualities. In order to aid in the formation of covalent links in protein-based polymers, we can also use another crosslinker for enzymes. The delicate enzymatic activities within cells serve to mitigate excessive cell death. Transglutaminase is an illustrative example of such enzymes [64].

Protein-Based Bioinks

Protein-based inks are particularly appealing for producing scaffolds that duplicate the protein composition of genuine organs and tissues because they are plentiful in biological systems. These protein-based compositions are being used in place of other materials in an effort to reduce immunological reactions, promote biomaterial host integration, and eventually dissolve completely, releasing peptides and amino acids that the cells can use for nutrition.

Owing to their widespread availability, cost-effectiveness, and adjustable physicochemical, mechanical, and biological traits, as well as their exceptional biocompatibility and biodegradability, protein-based biomaterials such as fibrin, collagen, and keratin have been extensively employed across various biomedical domains. More recently, they have also been integrated with 3D bioprinting systems [82].

Various tissue analogues, such as the heart, meniscus, and cornea, have been 3D printed using collagen inks. On the other hand, it is rarely utilized because of its poor mechanical qualities and quick biodegradation rate. Therefore, it has to be combined with other polymers to enhance its mechanical qualities while preserving the material's overall biocompatibility and capacity for cell growth [58].

The architecture of silk protein is characterized by a lengthy central segment surrounded by terminal domains, and the Central segment comprises short, repetitive hydrophobic regions that are intercalated with hydrophilic domains. Silk-based inks are typically generated using recombinant DNA methods with spider or silkworm templates, or sourced from silkworm cocoons [26, 80].

Keratin, a fibrous protein found in wool, hair, and feathers, is frequently utilized for tissue scaffolds and drug delivery. The amino acid sequences of glutamic acid-aspartic acid-serine (EDS), leucine-aspartic acid-valine (LDV), and arginine-glycine-aspartic acid (RGD) are three cell-binding amino acid motifs found in keratin that help to increase cell adhesion and proliferation. When creating inks for 3D bioprinting, keratin was oxidized to lessen the creation of disulfate linkages. The medication halofuginone, which prevents tissue contraction and is used to treat cutaneous burn wounds, has been loaded onto keratin scaffolds created via 3D printing [58].

Polysaccharides

The accessibility of hydrogels made of polysaccharides and their suitable biophysical and biomimetic characteristics for the progression of innovative bioinks make them desirable bioprinting materials. The most prominent polysaccharides from diverse origins, including seaweed (carrageenan and alginate), higher plants (starch and cellulose), Bacterial in nature (xanthan gum and dextran), and Animal origin, primarily from crustaceans, are used to demonstrate The versatility and adaptability of polysaccharides for the formation of Hydrogel-based bioinks (chitin and chitosan).

The enhancement of bioinks for diverse biomedical implementation has generated tremendous interest because cellulose is a polysaccharide that is abundant, inexpensive, renewable, and has great stiffness and biocompatibility [9]. The beneficial properties like large area, excellent mechanical stability, tuneable surface properties, good biocompatibility, cellular identification, and biodegradability of cellulose make it possible to bioprint numerous complex biologically organized with various printing solution and structure fidelity [13].

A common natural source of a wide variety of polysaccharides in marine is algae, sometimes referred to as seaweed. Consider alginic acid (and notably its salt form, sodium alginate), carrageenan, and agarose as examples of polysaccharides made from seaweed. Up to 70% of the dry weight of seaweeds is made up of polysaccharides, which are mostly found on the cell walls of these organisms.

One of Seaweed-derived polysaccharide is alginate that are mostly used as hydrogels bioinks in the recent years. Using alginate to create more intricate Living scaffolds in 3D that permit the production of both soft tissue[77] and "hard" (bone and cartilage biomimetic tissue structures [52, 90], designing disease models has essentially been a key aspect of the largest attempts to create innovative bioinks formulated with alginate.

Polysaccharides possess unique characteristics such as biodegradability, low toxicity, flexible chemistry, and numerous crosslinking methods, making them an excellent class of polymeric source materials to create innovative bioinks based on hydrogels. However, it is evident that Polysaccharides exhibit significant potential in the advancement of hydrogel-based biomaterials for 3D bioprinting, and the exploration of this area is making noteworthy progress due to the regular publication of novel methods and biomaterials. In fact, there are currently a number of commercial bioinks based on polysaccharides [29]. However, there are concerns about the repeatability and mass production of polysaccharides. While some, such as cellulose and chitin, are naturally sourced from abundant and renewable materials, their properties are significantly influenced by their origin, and the extraction processes may introduce various variations [79].

Synthetic Polymer-Based Bioinks

Some of the earliest biomaterials utilized as scaffolds in tissue engineering are synthetic hydrogels. The resources of biomaterials from their natural origin are scarce; however, the synthesized polymers are more easily accessible. Using synthetic polymers as future ink materials for 3D printing has enormous promise for creating sophisticated and highly customizable structures for hard and soft tissues, tissue scaffolds, and medical equipments [59, 93].

By choosing the right polymer, controlling the architecture, and taking use of post-polymerization functionalization options, synthetic polymers also provide the opportunity to fine-tune rheological and functional behaviors, such as the inks' ability to print and mechanical integrity. Although synthetic polymer-based biomaterials are currently utilized, their degradation rate is slow and the process of creating these biological constructs through 3D bioprinting does not achieve an exact replication of the natural tissue regeneration process. Therefore, the advancement of bioprinting technology for medical applications depends on the development of innovative biomaterials. In particular, great consideration must be given to the choice of synthetic polymer in order to retain biocompatibility and biodegradation difficulties. Synthetic polymers-based such as polyethylene glycol (PEG), polycaprolactone (PCL), polyvinylpyrrolidone (PVP), poly(lactic-co-glycolic) acid (PLGA) and poly(l-lactic) acid (PLA) which they are frequently utilized in 3D bioprinting, are also synthetic polymers that the FDA has authorized for use in healthcare technology [48].

Compared to conventional manufacturing methods, synthetic polymers as ink ingredients can advance 3D printing technology by enabling the production of well-specified biofabricated constructions with enhanced size, resolution, integrity, and feature control. Moreover, by utilizing synthetic polymers as a bioink and through careful formulation, functionalization, and cross-linking chemistry, various innovative bioink materials can confer the desired biological, chemical, and mechanical properties necessary for successful fabrication of functional organs via 3D bioprinting. As such, synthetic polymers hold potential as a critical component for advancing the field of in-human organ fabrication.

dECM-Based Bioinks

Hydrogel-based bioinks have been created using a wide range of hydrogels, but dECM-based bioinks have gained popularity as a means of creating hydrogel-based bioinks that are tailored to certain tissues. Decellularization is primarily used to preserve the content and structure of the original ECM while depleting host cells from tissues and organs [70]. They are created by separating the cells from the tissues and organs while retaining the natural ECM elements and the dECM-based bioinks include a special blend of structural and functional native tissue ECM elements [1].

dECM-based bioinks, which have a similar component and content to native tissues and organs, have recently been thought of as the ideal biomaterial to mimic the complexity of the natural microenvironment. dECM-based bioinks also are a favorable choice in the application of 3D bioprinting of tissue constructions because of their complex composition that is tissue-specific and offers great biochemical functioning and biocompatibility. Particularly, dECM-based bioinks offer a tremendous advantage for application due to the numerous collagenous proteins' physical cross-linking through intermolecular interaction as temperature-sensitive hydrogels [43].

Despite these encouraging developments and their widespread use, the dECM-based bioinks are now constrained by a lack of mechanical stability and printability, which could jeopardize their ability to be used in practical uses. They also frequently need labor-intensive postprocessing, such as removing the supporting polymer, which poses a risk of the designed construct collapsing. Additionally, the challenge of replicating the biomechanical characteristics of stiffer tissues at higher scales remains under investigation. Therefore, numerous physical solutions have been tried to improve the stability and accuracy of the form throughout the use of the bioprinting technique to create useful large-volume tissue and organ constructs at stiffer tissues [10, 36, 47].

The use of dECM-based bioinks presents a novel technique for constructing 3D Bio-inspired designs. Unlike native or synthetic polymers, dECM retains the unique composition and structure of the extracellular matrix that is specific to each tissue. Tissues are decellularized employing decellularization in order to create tissue-specific dECM-based bioinks [1]. Recent developments in dECM-based bioink design have increased their printability. Combining cross-linkable hydrogels [76, 92] and dECM-based bioinks is a method that is frequently used. In addition, methacrylation [3, 42] or the inclusion of a photocross-linker [33] can be used to effectively cross-link dECM-based bioinks. These alterations enhance the mechanical characteristics of dECM-based bioinks, allowing for the production of stable 3D structures with biomechanical characteristics that closely resemble those of native tissues. In future perspectives, it is crucial to develop optimal decellularization techniques for tissues to prevent compromising the extracellular matrix (ECM) characteristics in dECM-based bioink design. When using dECM-based bioinks to fabricate 3D bioprinted structures, it is vital to prevent ECM degradation to achieve the most authentic tissue replication [1].

Bioprinted Tissues

Bioprinting technology has the possibility to analyze biomaterials, as it is a computer-aided technology that can be written simultaneously with a predetermined arrangement of stacking layers. For more than a decade, the several different tissue types, including connective, epithelial, muscle, and nervous tissues has been produced via cell-loaded bioprinting technology. Bioprinting has significantly surpassed the primary drawbacks of conventional scaffold approaches due to its substantial advantage in pattern creation and exact placement of numerous cell types. This powerful technology is promising for the future of the fabricating tissue and organ. Extensive research has been conducted on bioprinting for animal transplants, which involves in vivo implantation of bioprinted tissues and organs in various related areas. The present section of the chapter aims to comprehensively examine the plausible applications of bioprinting technology, as illustrated in Fig. 13.5.

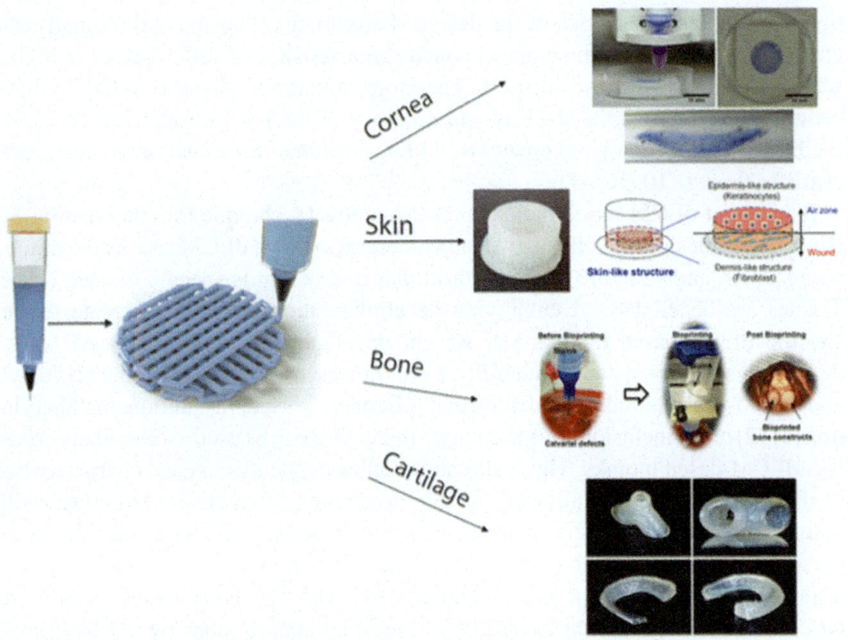

Fig. 13.5 Bioprinted tissue applications

Cornea

The cornea, the eye's outermost layer, serves as a barrier to shield the eye from outside influences. [8, 20]. Between 200 and 250 lamellae are heterogeneously distributed across the corneal stroma, running parallel in the middle and posterior stroma and intertwined in the anterior stroma [4, 17, 27]. Deformation of the perfect spherical anterior surface of the cornea or thinning and weakening of the cornea first causes blurring of vision and progresses to blindness [72, 84].

Due to the nearly perfect round shape and lamellar structure of the cornea, production was not possible before the 3D printing techniques. In a recent research about the 3D bioprinting of artificial cornea studied by Isaacson and his colleagues [31], the 3D bioprinting procedure was carried out using a template created from a patient's unique digital representation of their cornea. A patient-specific digital model of the cornea was developed using a Scheimpflug camera with a Placido disc and discretized with the Finite Element Method (FEM), as depicted in Fig. 13.6. Two bioinks, namely sodium alginate and methacrylated type I collagen, were utilized in the study. The rationale for combining collagen and alginate is to take advantage of collagen's tensile strength and alginate's biomechanical properties concurrently.

An important point mentioned in the research is that 3D bioprinting technique for the cornea requires the usage of biomaterials with appropriate rheological properties for high resolution of bioink. In this direction, gelatin was included to the bioink in

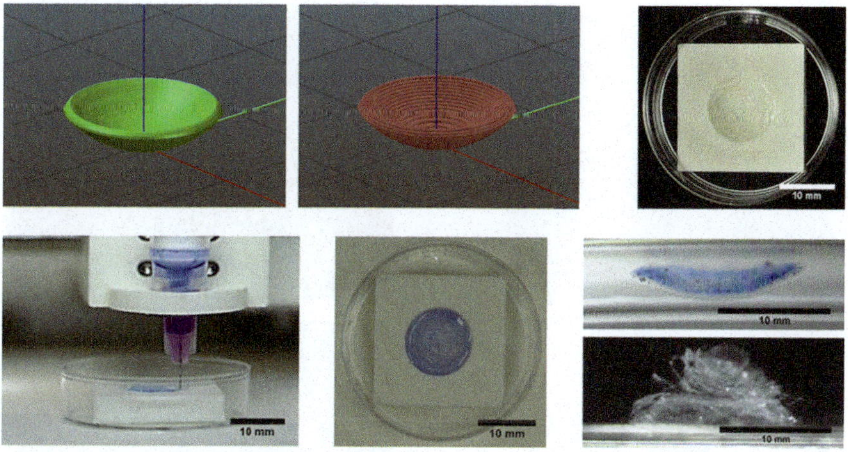

Fig. 13.6 Design and fabrication of corneal structure with bioprinters

order to adjust the viscosity. In addition, the use of gelatin accelerated the migration of bioinks to the corneal center during 3D bioprinting. This research serves as solid evidence that employing low-viscosity bioinks and 3D bioprinting technology may generate artificial human corneas quickly and effectively. The outcomes demonstrated that composite collagen and alginate bioinks had a high starting vitality of corneal keratocytes in protective shell and visible spreading. The initial cell survival was likely also influenced by the usage of bioprintable gelatin and the avoidance of dehydration because the produced tissue was so thin. The cells that were encapsulated exhibited a strong cellular activity even after 7 days following their application through 3D bioprinting as seen in Fig. 13.7 was stated as a benefit in the preference of composite collagen and alginate bioinks. This study proved that with the bioprinting parameters as speed, diameter of the needle, and the rheological properties of the bioink have an effect on the printing accuracy and mechanical stability.

Skin

Treatments for skin burns and wounds that cause significant skin lesions are minimal, and autologous split-thickness skin grafts, which are also a common form of therapy [25, 49]. Skin tissue fabrication and tissue replacement have used a variety of tissue engineering techniques, including autologous split-thickness skin grafts, allografts, cell-free dermal replacements, and commercial goods with cellularized graft-like properties [16, 37]. However, these skin substitutes have drawbacks such as having little therapeutic effect and post-administration immune rejection responses [86]. Therefore, there is an urgent need for skin substitutes that indicate a more complicated system that modifies the skin's whole functionality and combines the skin's

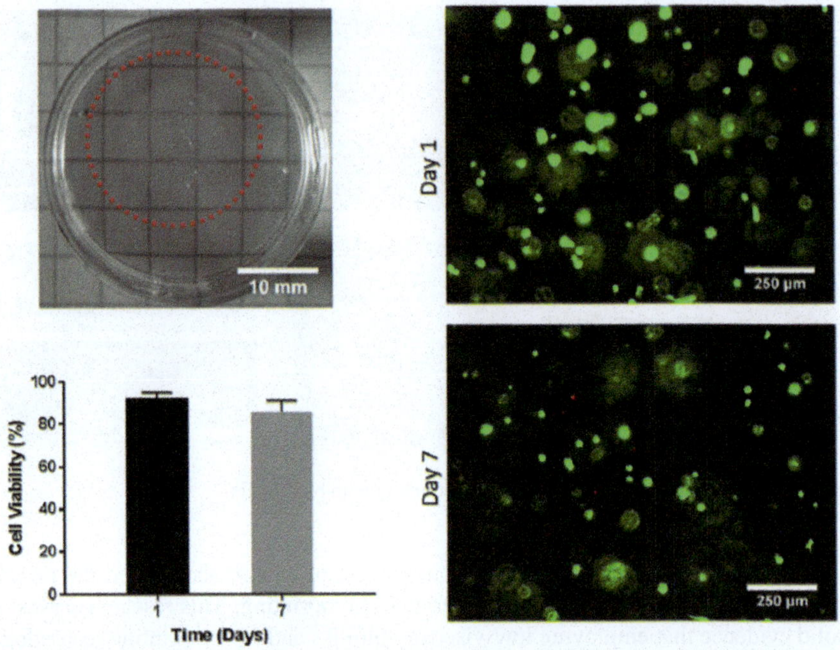

Fig. 13.7 Cell viability was assessed at days 1 and 7 through the acquisition of fluorescence microscopy images utilizing a live/dead stain. These images were representative of the population of cells under examination

structure with various cellular phenotypes. Recently, bioprinting technology for skin tissue adopted that has a remarkable impact on wound regeneration, encouraging skin healing and collagen restructuring in the injured area [2, 65, 70]. The flow characteristics of bioinks, particularly their fluidity and print performance, are crucial features in the field of 3D bioprinting. The use of dECM has gained widespread recognition as a biomimitic materials that accurately emulate the cellular microenvironment. This is attributed to its capacity to offer a conducive milieu for cellular growth and proliferation, thus rendering it a compelling candidate for tissue engineering endeavors [91]. Using dECM produced from porcine, keratinocytes, and human fibroblasts as skin-forming cells in a skin defect model, scientists created 3D-printed skin replacements. This research examined the curative advantages of bioprinted skin replacements in a setting that resembled the course of real wound healing [34].

Three-dimensional bioprinters were used to create the chimney structures. In the chimney wound model, a structure known as a chimney was created to stop mice's wounds from shrinking too quickly by taking the human tissue repair into account. The traditional chimney wound model is made from a microtube, a non-biocompatible substance. In this research, a noncytotoxic, biocompatible Polylactic Acid (PLA) component was used to strengthen the chimney construction as seen in Fig. 13.8.

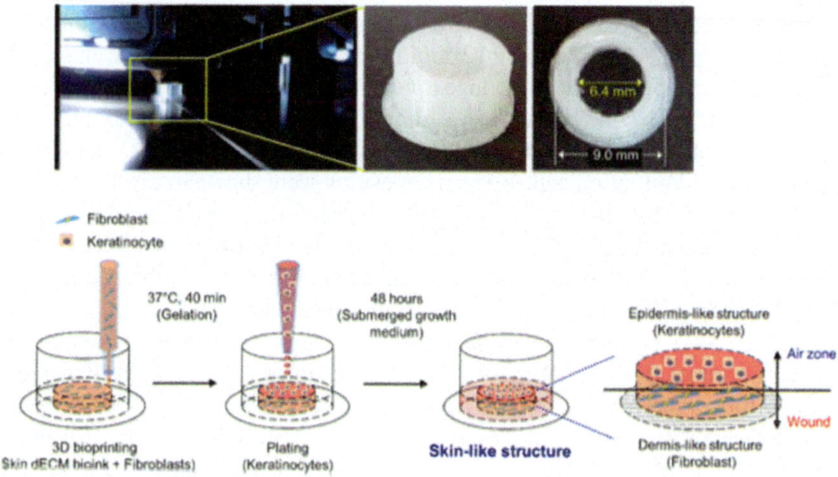

Fig. 13.8 3D printing technology enabled the creation of bioengineered skin substitutes and chimney structures

Decellularization process was used to make skin dECM bioink in sponge form from porcine skin. The human epidermal keratinocytes were coated on gelatinous bioink and immersed in keratinocyte growth medium for 48 h. It was tested in a rheological analyzer to analyze the properties of the prepared bioink, considering variables like viscoelasticity and viscosity. According to the viscoelasticity findings, the bioink thermally gelled at 37 °C and kept its structure following gelation. The bioink has demonstrated printability based on the results of its viscosity analysis. To investigate the potential influence of 3D printing on cell proliferation, a study was conducted using a skin bioink blended with fibroblasts. The objective of this study was to determine whether printing the bioink affects the rate of cell proliferation.

The skin texture is homogeneous thanks to 3D printing technology. Using a biopsy punch, an excisional wound was produced in the dorsal area of sedated Balb/c nude mice. The generated skin tissue was then transplanted into the mouse's dorsal region. Ten mice were compared in the skin-dECM, skin-dECM + fibroblasts, and skin-dECM + fibroblasts + keratinocyte experimental groups to examine the therapeutic effectiveness of skin replacements. By taking samples from the center and periphery on days 3, 6, and 12 to examine histological changes and collagen deposition in the wound, the experiment was carried out for a total of 12 days. The immune reaction and tissue repair based on cell transplant that was conducted at the margins of the wound on days 6 and 12 were also examined.

In addition, immunological reactions and re-epithelialization were evaluated by CK14, F4/80, and Ki67 staining. The analysis of affected site and collagen deposition were determined by H&E and MT staining (injured edge or center). All dECM transplanted groups exhibited quicker cell infiltration at the injured site on day 3 compared to normal control group.

This is used as proof that the dECM causes cell infiltration and accelerates the inflammatory phase. On days 6 and 12, infiltration of these cells was seen, and in mice that had the transplanted cells, collagen precipitation was found in the middle and at the edges of the wound tissues. It was shown that the skin-dECM + keratinocytes group + fibroblasts had the greatest percentage of collagen regions. Histological analysis of cell infiltration and collagen deposition were shown in Fig. 13.9.

Through an exhaustive evaluation of the research data, two crucial features of dECM derived from porcine were identified. Firstly, the artificial skin comprising keratinocytes and fibroblasts produces a substantial amount of fibronectin in the dermal-epidermal basement membrane region, which aids in re-epithelialization. Secondly, it is worth noting that the dECM derived from porcine sources encompasses a total of 24 residual growth factors, including keratocyte growth factor, basic fibroblast growth factor, and fibroblast growth factor. These growth factors possess the potential to promote tissue regeneration. The dECM from porcine demonstrates inherent healing potential even in the absence of surrounding structures. The outcomes of the experimental research reveal that the 3D-printed skin, which mimics the structure of the skin layer, displays prompt re-epithelialization and remarkable tissue restoration, as demonstrated by histological and immunohistochemical analyses on animals. Hence, utilizing cells from 3D-printed skin tissue could emerge as

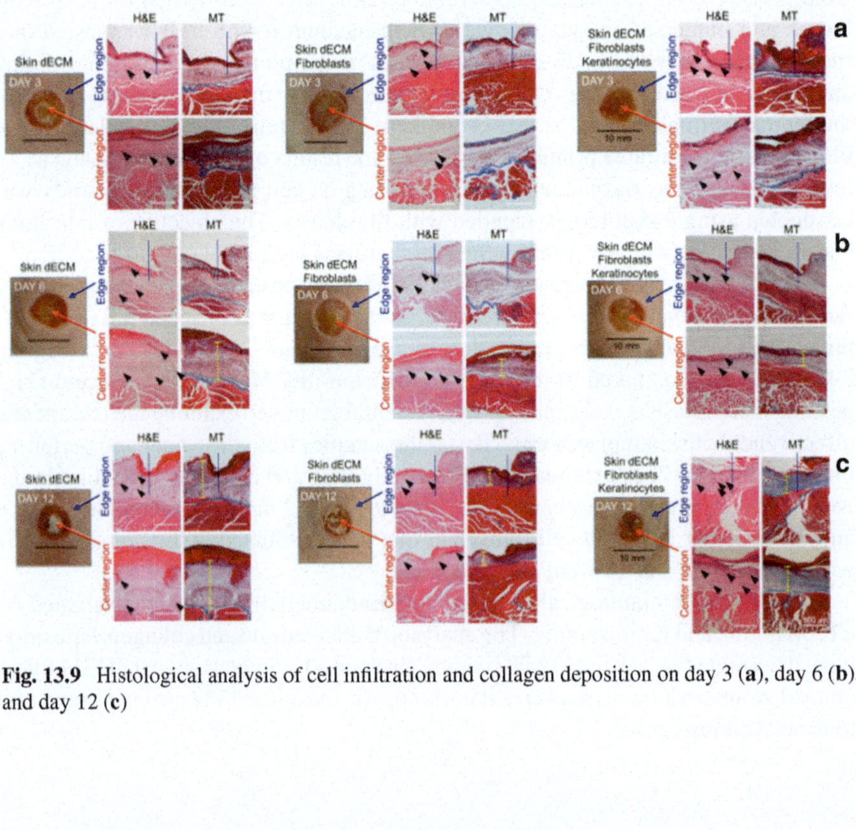

Fig. 13.9 Histological analysis of cell infiltration and collagen deposition on day 3 (**a**), day 6 (**b**), and day 12 (**c**)

a novel approach for skin grafts aimed at treating burns and wounds, particularly in areas where the availability of current resources is restricted.

Bone

Numerous people across the world are affected each year by cranial-maxillofacial (CMF) anomalies, including genetic mutations, chromosomal abnormalities, and battle wounds [78]. Patients with craniofacial disorders usually undergo more than one surgical operation for the reconstruction of the crania, and as a result of these operations, they often face poor aesthetic outcomes and functional losses [5, 51]. Intraoperative bioprinting (IOB) is a promising technology that enables the precise delivery of regenerative structures to damaged areas using real-time digital data processing. By capturing and transmitting defect information with high accuracy, IOB allows for the immediate delivery of anatomically accurate bone structures to the affected region for craniofacial (CMF) repair. This advanced technique has the potential to revolutionize CMF surgery by facilitating faster and more accurate treatment delivery [87]. In a recent research, Moncal et al. [57] investigated IOB technology in situ to repair critical sized rat calvarial defects.

Plasmid-DNAs (pDNAs) expressing bone morphogenetic protein-2 (BMP-2) and platelet-derived growth factor (PDGF) were chosen as the in situ delivery reservoir for pDNAs in a gene-activated matrix (an osteogenic bioink filled with pDNAs). BMP-2 has been shown to promote bone regeneration both in vitro and in vivo. PDGF possesses angiogenic properties. Enhanced vascular endothelial growth factor (VEGF) expression boosts cell migration (chemotaxis), proliferation (mitogenesis), and osteoblast cell growth. Due to their viability, safety, and promise for clinical translation, genealogical growth factors have been employed in applications of gene therapy without viral vectors for promoting bone regeneration, but their relatively poor transfection efficiency has limited their usage as an ideal gene transfection carrier. It was determined that more vascularized bone tissue was formed with the application of PDGF first and then BMP-2, rather than the simultaneous application of PDGF and BMP-2 two growth factors. Through the construction of our gene-activated matrix (bioink), this research utilized the method of first introducing PDGF and then BMP-2 to examine its capability in bone regeneration in vivo. This research aimed to investigate the regulated co-delivery of plasmid DNAs (pDNAs) in situ using IOB technology, for the purpose of repairing critical-sized rat calvarial injuries. This study is the first of its kind to explore this novel approach to pDNA delivery, which has the potential to revolutionize regenerative medicine. The osteogenic bioink was loaded with pBMP-2 and pPDGF-B encapsulated in chitosan nanoparticles (CS-NPs) and then directly bioprinted into critical sized calvarial injury in a rat model. In vivo and in vitro evaluations of the regulated co-delivery of pDNAs from bioprinted bone structures were conducted, and functional characterization of intraoperatively bioprinted rat calvarial injuries was used to assess potential therapeutic effectiveness for cranial injuries. Using a custom-built Multi-Arm BioPrinter (MABP), bone formations were 3D bioprinted. An air distributor was used to load the bioink before mechanically extruding it. Figure 13.10 present a schematic overview and the application steps of the process. Male Fischer white 4-week-old rats were used to harvest rat bone marrow mesenchymal stem cells (rBMSCs) for cell culture.

This findings from this research show that controlled co-delivery of pPDGF-B and pBMP-2 from intraoperatively bioprinted constructs resulted in the highest amount of formation of a new mineralized bone tissue and bone coverage in comparison to the other groups, allowing for better maintenance of calvarial bone defects at 6 weeks. According to research, PDGF stimulates early phases of cell recruitment during bone regeneration, while BMP-2 then activities and promote osteoblast proliferation and mineralization. Thus, the significance of order in the distribution of growth factors is highlighted. There may be an antagonistic impact between PDGF

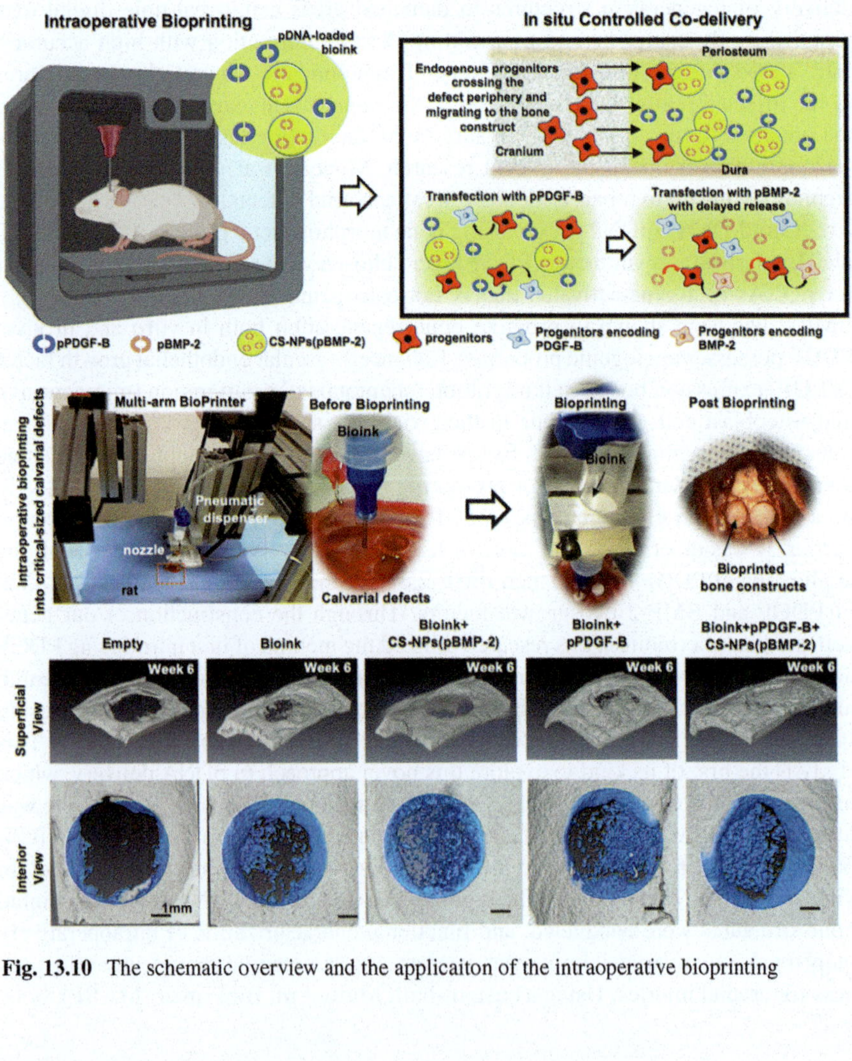

Fig. 13.10 The schematic overview and the applicaiton of the intraoperative bioprinting

and BMP-2 signaling since co-administration of growth factors resulted in much less tubule development than with controlled growth factor. The findings demonstrate that regulated co-administration of pPDGF-B and pBMP-2 is advantageous for the cellular migration, proliferation, and mineralization progression of healing, particularly when addressing bone repair. The recommended technique could be advantageous at promoting quick therapy and rapid bone tissue restoration.

Cartilage

Articular cartilage injury is a common clinical orthopedic disease [30]. The biggest problem of cartilage healing is the absence of cells that will provide healing [12]. Chondrocytes maintain the intermediate in their lacunae. When the cartilage is injured, some matrix flow towards the defect, but this cannot fill the defect. Chondrocytes do not have the ability to leave their lacunae and migrate to the defect. Stem cells within the bone marrow cannot reach the defect as long as the integrity of the subchondral bone is preserved [32, 56, 62]. While various tissues are healing, there is a vascular entry, and there is a migration of cells that will form the tissue and matrix to the healing area. However, there is no vascular tissue in cartilage healing. Therefore, cells must be obtained from another source [19, 54].

There are several conventional methods for repairing articular cartilage, including osteochondral transplantation, osteotomy, autologous matrix-induced chondrogenesis (AMIC) technology, autologous chondrocyte implantation (ACI), microfracture technology [71]. The developed ACI method has technical disadvantages. In this method, it is necessary to use a sensitive cell suspension to seal the periosteum in a watertight manner and, therefore, to suture the cartilage [7, 55]. In standard monolayer cell culture, chondrocytes are phenotypically forced to differentiate back. Injection of cell suspension causes inhomogeneous distribution and cell loss by leaking from the periosteal graft, and also the surgery time is long [83]. In order to eliminate these disadvantages, AMIC was defined in 1999. Since 2002, the second generation of ACI, the AMIC method, which uses three-dimensional biodegradable matrices, has entered clinical use [68]. Nevertheless, hyaline cartilage regeneration, which has a structure like natural cartilage, is still a significant issue that has to be tackled. AMIC technology is currently unable to completely heal cartilage abnormalities. To solve these problems, Zhou et al. [94] aimed to develop AMIC technology in articular cartilage repair using active biofilm produced with a 3D bioprinter.

At the beginning of the research, Alginate/Gelatin/Hyaluronic acid hybrid solution was prepared. Alginate can fast gel and has a good biocompatibility and biodegradability. Figure 13.11 presents the schematic illustration of the preparing the biofilm for repair of cartilage defects. In addition to its numerous benefits, the use of hydrogels in tissue engineering has some limitations. One of these limitations is the inability to induce cartilage formation due to the lack of appropriate cell adhesion sites. Additionally, hydrogels may not have the necessary components to support cartilage formation. However, hydrogels derived from algae and gel are often used within the scope of cartilage tissue regeneration owning to their strong adhesion to cells. Furthermore, hyaluronic acid, a naturally occurring substance found abundantly in cartilage tissue, is widely utilized in medical settings for the treatment of osteoarthritis.

Male Sprague Dawley rats that were 2 weeks old were used to extract bone mesenchymal stem cells (BMCSs). BMCSs were grown to passage three. Chondrogenic progenitor cells (CPCs) were obtained using the BMCSs technique, and they were grown till passage two. After the cells to be added to the composition were prepared, fibronectin and then cells were added to the Alginate/Gelatin/Hyaluronic acid composites to obtain a cell-loaded bio ink. A pneumatic 3D bioprinter was used to create biofilm scaffolds using a direct extrusion method. On days 1, 7, and 14, cell viability was evaluated. The images of cartilage injury model during surgery and grossmorphology of regenerated cartilage tissues at 6 and 12 weeks after surgery. Using Transwell culture plates, an indirect contact co-culture system was used to investigate the impact of the biofilm Alginate/Gelatin/Hyaluronic acid + fibronectin + CPCs on the differentiation of BMSCs released by microfracture.

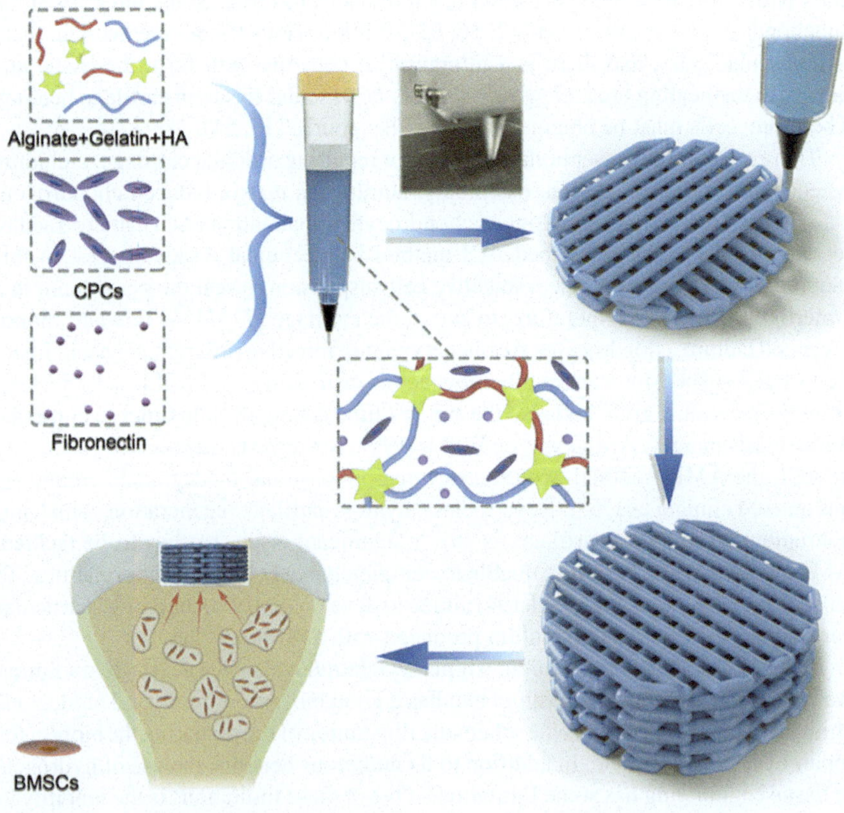

Fig. 13.11 The schematic illustration of the preparing the biofilm for repair of cartilage defects

This research's biofilm preparation featured a stinct network of structure, uniform pores, and consistent filament spacing. In 3D printing processes, the viscosity of the material is the primary factor that impacts printing quality. It is crucial to ensure that the material has a low enough viscosity to flow smoothly through the nozzle during extrusion, making it suitable for 3D printing. However, the viscosity should also be high enough to support the layer-by-layer structure required for building a robust and stable structure. To this end, recent research has demonstrated that using low-viscosity hydrogels can facilitate high-resolution fabrication by enabling better model-assisted bioink interaction.

This research's findings indicated that Alg/Gel/HA composite hydrogel outperformed AMIC technology in terms of maintaining cartilage. The active biofilm's uniform porosity and strong structure allowed for optimal nutrient exchange and cell-to-cell communication, which successfully promoted growth and proliferation and provided unique mechanical support prior to the formation of new cartilage. By supplying BMSCs released from subchondral bone with a stable microenvironment and enough mechanical support and architectural integrity, biofilm was able to cover a defect. This research showed that the rat cartilage injury model reproduces a laminar structure comparable to normal hyaline cartilage. With this Alginate/Gelatin/Hyaluronic acid + CPCs + fibronectin biofilm, The interaction between the CPC-loaded biofilm and BMSCs secreted from the bone marrow throughout the healing phase vastly improved the AMIC result. The results of this research can be utilized to create AMIC technology to repair full-thickness cartilage lesions and to serve as a theoretical foundation for future advancements in clinical practice.

Conclusion and Future Perspectives

This chapter provided a thorough comparative analysis of biomaterials for printing utilized in various 3D tissue bioprinting, including protein, dECM, hydrogels, polysaccharides and synthetic bioinks, as well as their application in specific tissues such as cornea, skin, bone and cartilage. The technology of 3D bioprinting, which combines biomaterials and cells with various methodologies, has advanced significantly in recent years. Nevertheless, several issues still need to be resolved in further studies about bioprinting methods, cellular sources and biomaterials choices.

Future research into the advancements in bioink materials must find ways to improve mechanical qualities to support bioprinted structures, provide biological substances to enhance cellular connections and establish an environment conducive to the implantation of 3D bioprinting. Bioinks are a crucial component in bioprinting as they play a significant role in their development and utilization. When developing bioinks, it is important to consider both the biological aspects that support cell survival, differentiation, and proliferation, and the biomechanical properties that enable the printed gel to disperse and eventually form a durable and enduring hard tissue. Despite the difficulties that 3D bioprinting currently faces, without a doubt, the foundation of the application of bioprinting is the creation and stable synthesis of innovative bioinks that seems to be on track to develop tissue engineering and enable organ regeneration, which will eventually reduce the demand for organs and save lives.

References

1. Abaci A, Guvendiren M (2020) Designing decellularized extracellular matrix-based bioinks for 3D bioprinting. Adv Healthcare Mater 9(24):2000734
2. Ahn G, Min KH, Kim C, Lee JS, Kang D, Won JY, Cho DW, Kim JY, Jin S, Yun WS (2017) Precise stacking of decellularized extracellular matrix based 3D cell-laden constructs by a 3D cell printing system equipped with heating modules. Sci Rep 7:8624
3. Ali M, Pr AK, Yoo JJ, Zahran F, Atala A, Lee SJ (2019) A photo-crosslinkable kidney ECM-derived bioink accelerates renal tissue formation. Adv Healthcare Mater 8(7):1800992
4. Ambati BK, Nozaki M, Singh N (2006) Corneal avascularity is due to soluble VEGF receptor-1. Nature 443(7114):993–997
5. Baer PRB, Wenke JC, Thomas SJ, Hale CRG (2012) Investigation of severe craniomaxillofacial battle injuries sustained by U.S. Service members: a case series. Craniomaxillofac Trauma Reconstr 5:243–252
6. Barron JA, Spargo BJ, Ringeisen BR (2004) Biological laser printing of three dimensional cellular structures. Appl Phys A-Mat Sci Process 79(4–6):1027–1030
7. Becher C, Malahias MA, Ali MM, Maffulli N, Thermann H (2019) Arthroscopic microfracture vs. arthroscopic autologous matrix-induced chondrogenesis for the treatment of articular cartilage defects of the talus. Knee Surgery, Sports Traum, Arthro 27(9):2731–2736
8. Bektas CK, Hasirci V (2020) Cell loaded 3D bioprinted GelMA hydrogels for corneal stroma engineering. Biomat Sci 8:438
9. Chinga-Carrasco G (2018) Potential and limitations of nanocelluloses as components in biocomposite inks for three-dimensional bioprinting and for biomedical devices. Biomacromol 19(3):701–711
10. Choi YJ, Jun YJ, Kim DY, Yi HG, Chae SH, Kang J, Lee J, Gao G, Kong JS, Jang J, Chung WK, Rhie JW, Cho DW (2019) 3D cell printing of functional skeletal muscle constructs using skeletal muscle-derived bioink. Biomaterials 206:160–169
11. Colosi C, Shin SR, Manoharan V et al (2016) Microfluidic bioprinting of heterogeneous 3D tissue constructs using low-viscosity bioink. Adv Mater 28:677–684
12. Coutts RD, Healey RM, Ostrander R, Sah RL, Goomer R, Amiel D (2001) Matrices for cartilage repair. Clin Orthop Relat Res 9:271
13. Curvello R, Raghuwanshi VS, Garnier G (2019) Engineering nanocellulose hydrogels for biomedical applications. Adv Coll Interface Sci 267:47–61
14. Dababneh AB, Ozbolat IT (2014) Bioprinting technology: a current state-of-the-art review. J Manuf Sci Eng 136(6)
15. Duan B (2017) State-of-the-art review of 3D bioprinting for cardiovascular tissue engineering. Ann Biomed Eng 45(1):195–209
16. Dzobo K, Motaung KSC, Adesida A (2019) Recent trends in decellularized extracellular matrix bioinks for 3D printing: an updated review. Int J Mol Sci 20(18):4628
17. Eghrari AO, Riazuddin SA, Gottsch JD (2015) Overview of the cornea: structure, function, and development. Prog Mol Biol Transl Sci 134:7–23
18. Foty RA, Pfleger CM, Forgacs G, Steinberg MS (1996) Surface tensions of embryonic tissues predict their mutual envelopment behavior. Development 122(5):1611–1620
19. Galois L, Freyria AM, Grossin L, Hubert P, Mainard D, Herbage D (2004) Cartilage repair: surgical techniques and tissue engineering using polysaccharide and collagen based biomaterials. Biorheology 41(3–4):433–443

20. Golchet G, Carr J, Harris MG (2000) Why don't we have enough cornea donors? A literature review and survey. Optometry 71(5):318–328
21. Gudapati H, Dey M, Ozbolat I (2016) A comprehensive review on droplet-based bioprinting: past, present and future. Biomaterials 102(2016):20–42
22. Guillemot F, Souquet A, Catros S, Guillotin B (2010) Laser-assisted cell printing: principle, physical parameters versus cell fate and perspectives in tissue engineering. Nanomedicine 5(3):507–515
23. Guillemot F, Souquet A, Catros S, Guillotin B, Lopez J, Faucon M et al (2010) High-throughput laser printing of cells and biomaterials for tissue engineering. Acta Biomater 6(7):2494–2500
24. Guillotin B et al (2013) Laser-assisted bioprinting for tissue engineering. In: Forgacs G, Sun W (eds) Biofabrication micro- and nano-fabrication, printing, patterning and assemblies. William Andrew Publishing, Boston, pp 95–118
25. He P, Zhao J, Zhang J, Li B, Gou Z, Gou M, Li X (2018) Bioprinting of skin constructs for wound healing. Burns and Trauma 6:5
26. Heidebrecht A, Eisoldt L, Diehl J, Schmidt A, Geffers M, Lang G, Scheibel T (2015) Biomimetic fibers made of recombinant spidroins with the same toughness as natural spider silk. Adv Mater 27(13):2189–2194
27. Hogan MJ, Alvarado JA, Weddell JE (1971) The cornea. histology of the human eye. Saunders, Philadelphia 55:111
28. Horvath L et al (2015) Engineering an in vitro air-blood barrier by 3D bioprinting. Sci Rep 5:7974
29. Hölzl K, Lin S, Tytgat L, Van Vlierberghe S, Gu L, Ovsianikov A (2016) Bioink Properties before, during and after 3D Bioprinting. Biofabrication 8:032002
30. Huey DJ, Hu JC, Athanasiou KA (2012) Unlike bone, cartilage regeneration remains elusive. Science 338(6109):917–921
31. Isaacson A, Swioklo S, Connon CJ (2018) 3D bioprinting of a corneal stroma equivalent. Exp Eye Res 173:188–193
32. Iwasa J, Engebretsen L, Shima Y, Ochi M (2009) Clinical application of scaffolds for cartilage tissue engineering. Knee Surg Sports Traumatol Arthrosc 17(6):561–577
33. Jang J, Kim TG, Kim BS, Kim SW, Kwon SM, Cho DW (2016) Tailoring mechanical properties of decellularized extracellular matrix bioink by vitamin B2-induced photo-crosslinking. Acta Biomater 33:88–95
34. Jang KS, Park SJ, Choi JJ, Kim HN, Shim KM, Kim MJ, Jang IH, Jin SW, Kang SS, Kim SE, Moon SH (2021) Therapeutic efficacy of artificial skin produced by 3D bioprinting. Materials 14(17):5177
35. Jessop ZM, Al-Sabah A, Gardiner MD, Combellack E, Hawkins K, Whitaker IS (2017) 3D bioprinting for reconstructive surgery: principles, applications and challenges. J Plastic Reconstruct Aesth Surg 70:1155–1170
36. Jiang T, Munguia-Lopez JG, Flores-Torres S, Kort-Mascort J, Kinsella JM (2019) 3D printing of biological matters. Appl Phys Rev 6(1):011310
37. Jorgensen AM, Varkey M, Gorkun A, Clouse C, Xu L, Chou Z, Murphy SV, Molnar J, Lee SJ, Yoo JJ (2020) Bioprinted skin recapitulates normal collagen remodeling in full-thickness wounds. Tissue Eng Part A 26(9–10):516–526
38. Jungst T, Smolan W, Schacht K, Scheibel T, Groll J (2016) Strategies and molecular design criteria for 3D printable hydrogels. Chem Rev 116:1496–1539
39. Karzyński K, Kosowska K, Ambrożkiewicz F, Berman A, Cichoń J, Klak M, Wszoła M et al (2018) Use of 3D bioprinting in biomedical engineering for clinical application. Med Stud/Studia Medyczne 34(1):93–97
40. Khalil S, Nam J, Sun W (2005) Multi-nozzle deposition for construction of 3D biopolymer tissue scaffolds. Rapid Proto J 11(1):9–17
41. Khoda AKM, Ozbolat IT, Koc B (2013) Designing heterogeneous porous tissue scaffolds for additive manufacturing processes. Comput Aided Des 45:1507–1523
42. Kim W, Lee H, Lee J, Atala A, Yoo JJ, Lee SJ, Kim GH (2020) Efficient myotube formation in 3D bioprinted tissue construct by biochemical and topographical cues. Biomaterials 230:119632

43. Kim H, Kang B, Cui X, Lee SH, Lee K, Cho DW, Jang J et al (2021) Light-activated decellularized extracellular matrix-based bioinks for volumetric tissue analogs at the centimeter scale. Adv Func Mater 31(32):2011252
44. Klebe RJ (1988) Cytoscribing: a method for micropositioning cells and the construction of two- and three-dimensional synthetic tissues. Exp Cell Res 179(2):362–373
45. Koch L, Kuhn S, Sorg H, Gruene M, Schlie S, Gaebel R, Polchow B, Reimers K, Stoelting S, Ma N et al (2009) Laser printing of skin cells and human stem cells. Tissue Eng Part C Methods 16(5):847–854
46. Koch L, Gruene M, Unger C, Chichkov B (2013) Laser-assisted cell printing. Curr Pharm Biotechnol 14(1):91–97
47. Kolesky DB, Homan KA, Skylar-Scott MA, Lewis JA (2016) Three-dimensional bioprinting of thick vascularized tissues. Proc Natl Acad Sci USA 113(12):3179–3184
48. Kumar S (2021) Synthetic polymer-derived single-network inks/bioinks for extrusion-based 3D printing towards bioapplications. Materials Advances 2(21):6928–6941
49. Kuna VK, Padma AM, Hakansson J, Nygren J, Sjöback R, Petronis S, Sumitran-Holgersson S (2017) Significantly accelerated wound healing of full-thickness skin using a novel composite gel of porcine acellular dermal matrix and human peripheral blood cells. Cell Transplant 26(2):293–307
50. Landers R, Hubner U, Schmelzeisen R, Mulhaupta R (2002) Rapid prototyping of scaffolds derived from thermoreversible hydrogels and tailored for applications in tissue engineering. Biomaterials 23(23):4437–4447
51. Li Z, Yang B, Weng XS, Tse G, Chan MTV, Wu WKK (2018) Emerging roles of microRNAs in osteonecrosis of the femoral head. Cell Prolif 51(1):e12405
52. Luo Y, Li Y, Qin X, Wa Q (2018) 3D Printing of concentrated alginate/gelatin scaffolds with homogeneous nano apatite coating for bone tissue engineering. Mater Des 146:12–19
53. Mahinroosta M, Jomeh Farsangi Z, Allahverdi A, Shakoori Z (2018) Hydrogels as intelligent materials: a brief review of synthesis, properties and applications. Today Chem 8:42–55
54. Mastbergen SC, Saris DB, Lafeber FP (2013) Functional articular cartilage repair: here, near, or is the best approach not yet clear? Nat Rev Rheumatol 9(5):277–290
55. Makris EA, Gomoll AH, Malizos KN, Hu JC, Athanasiou KA (2015) Repair and tissue engineering techniques for articular cartilage. Nat Rev Rheumatol 11(1):21–34
56. Marcacci M, Kon E, Zaffagnini S, Filardo G, Delcogliano M, Neri MP (2007) Arthroscopic second generation autologous chondrocyte implantation. Knee Surg Sports Traumatol Arthrosc 15:610–619
57. Moncal KK, Aydın RST, Godzik KP, Acri TM, Heo DN, Rizk E, Wee H, Lewis GS, Salem AK, Ozbolat IT (2022) Controlled Co-delivery of pPDGF-B and pBMP-2 from intraoperatively bioprinted bone constructs improves the repair of calvarial defects in rats. Biomaterials 281:121333
58. Mu X, Agostinacchio F, Xiang N, Pei Y, Khan Y, Guo C, Kaplan DL et al (2021) Recent advances in 3D printing with protein-based inks. Prog Polym Sci 115:101375
59. Narupai B, Nelson A (2020) 100th anniversary of macromolecular science viewpoint: Macromolecular materials for additive manufacturing. ACS Macro Lett 9(5):627–638
60. Norotte C, Marga FS, Niklason LE, Forgacs G (2009) Scaffold-free vascular tissue engineering using bioprinting. Biomaterials 30(30):5910–5917
61. Noor N, Shapira A, Edri R, Gal I, Wertheim L, Dvir T (2019) 3D printing of personalized thick and perfusable cardiac patches and hearts. Adv Sci 6(11):1900344
62. Ochi M, Uchio Y, Kawasaki K, Wakitani S, Iwasa J (2002) Transplantation of cartilage-like tissue made by tissue engineering in the treatment of cartilage defects of the knee. J Bone Joint Surg 84:571–578
63. Odde DJ, Renn MJ (2000) Laser-guided direct writing of living cells. Biotechnol Bioeng 67(3):312–318
64. Ooi HW, Mota C, Ten Cate AT, Calore A, Moroni L, Baker MB (2018) Thiol–ene alginate hydrogels as versatile bioinks for bioprinting. Biomacromol 19(8):3390–3400

65. Ouyang L, Yao R, Zhao Y, Sun W (2016) Effect of bioink properties on printability and cell viability for 3D bioplotting of embryonic stem cells. Biofabrication 8:035020
66. Ozbolat IT, Hospodiuk M (2016) Current advances and future perspectives in extrusion-based bioprinting. Biomaterials 76(2016):321–343
67. Ozbolat IT (2016) 3D Bioprinting: fundamentals, principles and applications. Academic Press
68. Panni AS, Regno CD, Mazzitelli G, D'Apolito R, Corona K, Vasso M (2018) Good clinical results with autologous matrix-induced chondrogenesis (Amic) technique in large knee chondral defects. Knee Surg Sports Traumatol Arthrosc 26(4):1130–1136
69. Parfenov VA, Khesuani YD, Petrov SV, et al (2020). Magnetic levitational bioassembly of 3D tissue construct in space. Sci Adv 6(29)
70. Pati F, Jang J, Ha DH, Won KS, Rhie JW, Shim HH, Kim DH, Cho DW (2014) Printing three-dimensional tissue analogues with decellularized extracellular matrix bioink. Nat Commun 5:1–11
71. Richter DL, Schenck RC, Wascher DC, Treme G (2016) Knee articular cartilage repair and restoration techniques: a review of the literature. Sports Health 8(2):153–160
72. Ruberti JW, Zieske JD (2008) Prelude to corneal tissue engineering- gaining control of collagen organization. Prog Retin Eye Res 27(5):549–577
73. Santoni S, Gugliandolo SG, Sponchioni M, Moscatelli D, Colosimo BM. (2021). 3D bioprinting: current status and trends—a guide to the literature and industrial practice. Bio-Design Manufact, 1–29
74. Serra P, Fernandez-Pradas JM, Colina M, Duocastella M, Dominguez J, Morenza JL (2006) Laser-induced forward transfer: a direct-writing technique for biosensors preparation. J Laser Micro/Nanoeng 1(3):236–242
75. Sharma P, Kumar P, Sharma R (2018) Chimeric antigen receptors; their current and futuristic scope. Curr Signal Transduct Ther 13(2):153–158
76. Skardal A, Devarasetty M, Kang HW, Mead I, Bishop C, Shupe T, Lee SJ, Jackson J, Yoo J, Soker S, Atala A (2015) A hydrogel bioink toolkit for mimicking native tissue biochemical and mechanical properties in bioprinted tissue constructs. Acta Biomater 25:24–34
77. Somasekharan LT, Raju R, Kumar S, Geevarghese R, Nair RP, Kasoju N, Bhatt A (2021) Biofabrication of skin tissue constructs using alginate, gelatin and diethylaminoethyl cellulose bioink. Int J Biol Macromol 189:398–409
78. Szpalski C, Barr J, Wetterau M, Saadeh PB, Warren SW (2010) Cranial bone defects: current and future strategies. Neurosurg Foc 29(6)
79. Teixeira MC, Lameirinhas NS, Carvalho JP, Silvestre AJ, Vilela C, Freire CS (2022) A guide to polysaccharide-based hydrogel bioinks for 3D bioprinting applications. Int J Mol Sci 23(12):6564
80. Teulé F, Cooper AR, Furin WA, Bittencourt D, Rech EL, Brooks A, Lewis RV (2009) A protocol for the production of recombinant spider silk-like proteins for artificial fiber spinning. Nat Protoc 4(3):341–355
81. Tibbitt M, Anseth K (2009) Hydrogels as extracellular matrix mimics for 3D cell culture. Biotechnol Bioeng 103(4):655–663
82. Veiga A, Silva IV, Duarte MM, Oliveira AL (2021) Current trends on protein driven bioinks for 3d printing. Pharmaceutics 13(9):1444
83. Weigelt L, Hartmann R, Pfirrmann C, Espinosa N, Wirth SH (2019) Autologous matrix induced Chondrogenesis for Osteochondral lesions of the talus: a clinical and radiological 2- to 8-year follow-up study. Am J Sports Med 47(7):1679–1686
84. Whitcher JP, Srinivasan M, Upadhyay MP (2001) Corneal blindness: a global perspective. Bull World Health Organ 79(3):214–221
85. Wilson WC Jr, Boland T (2003) Cell and organ printing 1: protein and cell printers. Anat Rec A Discov Mol Cell Evol Biol 272(2):491–496
86. Won JY, Lee MH, Kim MJ, Min KH, Ahn G, Han JS, Jin S, Yun WS, Shim JH (2019) A potential dermal substitute using decellularized dermis extracellular matrix derived bio-ink. Artif Cells, Nanomed Biotech 47:644–649

87. Wu Y, Ravnic DJ, Ozbolat IT (2020) Intraoperative bioprinting: repairing tissues and organs in a surgical setting. Trends Biotechnol 38(6):594–605
88. Wu Y, Fuh J, Ozbolat IT (2022) 3D bioprinting in tissue and organ regeneration. Academic Press
89. Xu T et al (2013) Hybrid printing of mechanically and biologically improved constructs for cartilage tissue engineering applications. Biofabrication 5(1):015001
90. Yang X, Lu Z, Wu H, Li W, Zheng L, Zhao J (2018) Collagen-alginate as bioink for three-dimensional (3D) cell printing based cartilage tissue engineering. Mater Sci Eng, C 83:195–201
91. Yi S, Ding F, Gong L, Gu X (2017) Extracellular matrix scaffolds for tissue engineering and regenerative medicine. Curr Stem Cell Res Ther 12(3):233–246
92. Yu C, Ma X, Zhu W, Wang P, Miller KL, Stupin J, KorolevaMaharajh A, Hairabedian A, Chen S (2019) Inkjet-printed scaffold incorporating ultrafine decellularized matrix for tissue engineering with mesenchymal stem cells. Biomaterials 194:1–14
93. Zhou LY, Fu J, He Y (2020) A review of 3D printing technologies for soft polymer materials. Adv Func Mater 30(28):2000187
94. Zhou Y, Qin R, Chen T, Zhang K, Gui J (2021) 3D bioprinting modified autologous matrix-induced chondrogenesis (AMIC) technique for repair of cartilage defects. Mater Des 203:109621

Index

O. Gunduz et al. (eds.), *Biomaterials and Tissue Engineering*, Stem Cell Biology and Regenerative Medicine 74, https://doi.org/10.1007/978-3-031-35832-6

The manufacturer's authorised representative in the EU is Springer Nature Customer Service Centre GmbH, Europaplatz 3, 69115 Heidelberg, Germany. If you have any concerns regarding our products, please contact ProductSafety@springernature.com

Printed and bound by CPI Group (UK) Ltd, Croydon, CR0 4YY
28/09/2026
02224085-0013